Methods in Enzymology

Volume XLVII
ENZYME STRUCTURE
Part E

METHODS IN ENZYMOLOGY

EDITORS-IN-CHIEF

Sidney P. Colowick   Nathan O. Kaplan

Methods in Enzymology

Volume XLVII

Enzyme Structure

Part E

EDITED BY

C. H. W. Hirs

DIVISION OF BIOLOGICAL SCIENCES
INDIANA UNIVERSITY
BLOOMINGTON, INDIANA

Serge N. Timasheff

GRADUATE DEPARTMENT OF BIOCHEMISTRY
BRANDEIS UNIVERSITY
WALTHAM, MASSACHUSETTS

1977

ACADEMIC PRESS New York San Francisco London
A Subsidiary of Harcourt Brace Jovanovich, Publishers

ACADEMIC PRESS, INC.
111 Fifth Avenue, New York, New York 10003

United Kingdom Edition published by
ACADEMIC PRESS, INC. (LONDON) LTD.
24/28 Oval Road, London NW1

Library of Congress Cataloging in Publication Data

Main entry under title:

Enzyme structure.

    (Methods in enzymology, v. 11, 25-26, 47)
    Part B-   edited by C. H. W. Hirs and S. N. Tim-
asheff.
    Includes bibliographical references.
    1.   Enzymes—Analysis.   I.   Hirs, Christophe Henri
Werner,   Date   ed.
QP601.M49   vol. 11, etc.   547′.758   79-26910
ISBN 0-12-181947-7

Table of Contents

Section I. Amino Acid Analysis

Section II. End Group Methods

Section III. Chain Separation

Section IV. Cleavage of Disulfide Bonds

Section V. Selective Cleavage by Chemical Methods

Section IX. Chemical Modification

Section X. Methods of Peptide Synthesis

Contributors to Volume XLVII

Article numbers are in parentheses following the names of contributors.
Affiliations listed are current.

ETTORE APPELLA (37), *National Institute of Allergy and Infectious Diseases and the National Cancer Institute, National Institutes of Health, Bethesda, Maryland*

GARY BALIAN (14), *Departments of Biochemistry and Medicine, University of Washington, Seattle, Washington*

DIANA C. BARTELT (19), *Department of Biological Sciences, Hunter College of the City University of New York, New York, New York*

JAMES R. BENSON (2), *Durrum Chemical Corporation, Palo Alto, California*

R. BERTRAND (16), *Centre de Recherches de Biochimie Macromoleculaire, Montpellier-Cedex, France*

RAYMOND L. BLAKLEY (28), *Biochemistry Department, University of Iowa, Iowa City, Iowa*

PAUL BORNSTEIN (14), *Departments of Biochemistry and Medicine, University of Washington, Seattle, Washington*

JOHN BRIDGEN (33, 38), *Laboratory of Molecular Biology, Medical Research Council, Cambridge, England*

THOMAS W. BRUICE (40), *Department of Biochemistry and Biophysics, University of California, San Francisco, California*

CHRISTOPHER C. Q. CHIN (23), *Department of Biochemistry, University of Minnesota, St. Paul, Minnesota*

A. DARBRE (35), *Department of Biochemistry, University of London, King's College, London, England*

MARILYNN S. DOSCHER (48), *Department of Biochemistry, Wayne State University School of Medicine, Detroit, Michigan*

GABRIEL R. DRAPEAU (21), *Department of Microbiology, University of Montreal, Montreal, Quebec, Canada*

ANGELO FONTANA (42, 43, 44), *Institute of Organic Chemistry, University of Padua, Padua, Italy*

ROGER W. GIESE (3), *College of Pharmacy and Allied Health Professions, Northeastern University, Boston, Massachusetts*

ROBERT W. GRACY (22), *Departments of Chemistry and Biochemistry, North Texas State University, Denton, Texas*

LEWIS J. GREENE (19), *Department of Biochemistry, Faculty of Medicine of Ribeirão Preto, University of São Paolo, São Paolo, Brazil*

P. E. HARE (1), *Geophysical Laboratory, Carnegie Institution, Washington, D.C.*

FRED C. HARTMAN (46), *Biology Division, Oak Ridge National Laboratory, Oak Ridge, Tennessee*

RIKIMARU HAYASHI (8), *The Research Institute for Food Science, Kyoto University, Kyoto, Japan*

ROBERT L. HEINRIKSON (20), *Department of Biochemistry, The University of Chicago, Chicago, Illinois*

ALAN C. HERMAN (25), *Department of Surgery, Duke University Medical Center, Durham, North Carolina*

C. H. W. HIRS (9), *Division of Biological Sciences, Indiana University, Bloomington, Indiana*

JOHN K. INMAN (37), *National Institute of Allergy and Infectious Diseases and the National Cancer Institute, National Institutes of Health, Bethesda, Maryland*

PANAYOTIS G. KATSOYANNIS (47), *Department of Biochemistry, Mount Sinai School of Medicine, City University of New York, New York, New York*

GEORGE L. KENYON (40), *Department of Pharmaceutical Chemistry, University of California, San Francisco, California*

HENRY C. KRUTZSCH (39), *Section on*

Physiological Chemistry, National Heart, Lung, and Blood Institute, National Institutes of Health, Bethesda, Maryland

C. Y. LAI (26, 36), *Roche Institute of Molecular Biology, Nutley, New Jersey*

MICHAEL LANDON (15), *Department of Biochemistry, University Hospital and Medical School, Nottingham, England*

RICHARD A. LAURSEN (30), *Department of Chemistry, Boston University, Boston, Massachusetts*

FREDERICK A. LIBERATORE (4), *Department of Biochemistry, The Ohio State University, Columbus, Ohio*

WERNER MACHLEIDT (24, 29), *Institut für Physiologische Chemie, Physikalische Biochemie und Zellbiologie der Universität München, Munich, Germany*

GARY E. MEANS (45), *Department of Biochemistry, The Ohio State University, Columbus, Ohio*

EDITH WILSON MILES (41), *Laboratory of Biochemical Pharmacology, National Institute of Arthritis, Metabolism, and Digestive Diseases, National Institutes of Health, Bethesda, Maryland*

WILLIAM M. MITCHELL (18), *Department of Pathology, Vanderbilt University, Nashville, Tennessee*

I. LUCILE NORTON (46), *Biology Division, Oak Ridge National Laboratory, Oak Ridge, Tennessee*

R. E. OFFORD (6), *Laboratory of Molecular Biophysics, Department of Zoology, Oxford, England*

JOACHIM OTTO (24), *Institut für Physiologische Chemie, Physikalische Biochemie und Zellbiologie der Universität München, Munich, Germany*

J-F. PECHÈRE (16), *Centre de Recherches de Biochimie Macromoleculaire, Montpellier-Cedex, France*

JOHN J. PISANO (5, 39), *Section on Physiological Chemistry, National Heart, Lung, and Blood Institute, National Institutes of Health, Bethesda, Maryland*

DENNIS A. POWERS (32), *Department of Biology, The Johns Hopkins University, Baltimore, Maryland*

ALDO PREVIERO (31), *Unite de Recherches, INSERM U 147, Montpellier, France*

JAMES F. RIORDAN (3), *Biophysics Research Laboratory, Department of Biological Chemistry, Harvard Medical School, and Division of Medical Biology, Peter Bent Brigham Hospital, Boston, Massachusetts*

GARFIELD P. ROYER (4), *Department of Biochemistry, The Ohio State University, Columbus, Ohio*

JOSEF RUDINGER* (10, 11), *Department of Biochemistry, Faculty of Science, University of Geneva, Geneva, Switzerland*

URS TH. RÜEGG (10, 11, 12), *Department of Biochemistry, Faculty of Science, University of Geneva, Geneva, Switzerland*

WALTER E. SAVIGE (42, 43, 44), *Division of Protein Chemistry, Commonwealth Scientific and Industrial Research Organization, Parkville, Victoria, Australia*

GERALD P. SCHWARTZ (47), *Department of Biochemistry, Mount Sinai School of Medicine, City University of New York, New York, New York*

WARREN E. SCHWARTZ (4), *Department of Biochemistry, The Ohio State University, Columbus, Ohio*

JACK SILVER (27), *Department of Cellular and Developmental Immunology, Scripps Clinic and Research Foundation, La Jolla, California*

EMIL L. SMITH (17), *Department of Biological Chemistry, School of Medicine, The Center for the Health Sciences, University of California, Los Angeles, California*

GEORGE R. STARK (13), *Department of Biochemistry, Stanford University, School of Medicine, Stanford University Medical Center, Stanford, California*

* Deceased.

GEORGE E. TARR (34), *Department of Biochemistry and Molecular Biology, Northwestern University, Evanston, Illinois*

HARALD TSCHESCHE (7), *Lehrstuhl für Organische Chemie und Biochemie, Technische Universität, Munich, Germany*

THOMAS C. VANAMAN (25), *Department of Microbiology and Immunology, Duke University Medical Center, Durham, North Carolina*

ELMAR WACHTER (24, 29), *Institut für Physiologische Chemie, Physikalische Biochemie und Zellbiologie der Universität München, Munich, Germany*

FINN WOLD (23), *Department of Biochemistry, University of Minnesota, St. Paul, Minnesota*

CARL L. ZIMMERMAN (5), *Section on Physiological Chemistry, Laboratory of Chemistry, National Heart, Lung, and Blood Institute, National Institutes of Health, Bethesda, Maryland*

Preface

"Enzyme Structure," the eleventh volume in this series, was published ten years ago. Three supplements, Parts B, C, and D, appeared between 1972 and 1974 and served to update and expand the methodology covered in the original volume. This book, Part E, represents the first of additional supplements and relates directly to the topics previously included in Part B, i.e., those methods primarily chemical in nature. The supplements to follow will relate to the physical-chemical methods previously dealt with in Parts C and D.

This volume retains the organization adopted in Parts A and B, with emphasis on those areas in which substantial development has occurred. In particular, amino acid sequence determination has been given coverage representative of the many advances in this field. The importance of synthetic methods in the investigation of enzyme structure is recognized by the inclusion for the first time in these volumes of extensive articles on peptide synthesis.

We are grateful to the authors for their effective cooperation and for their many suggestions for improving the volume. As before, the staff of Academic Press has provided inestimable help in the assembly of this volume. We thank them for their many courtesies.

C. H. W. HIRS
SERGE N. TIMASHEFF

METHODS IN ENZYMOLOGY

EDITED BY

Sidney P. Colowick and Nathan O. Kaplan

VANDERBILT UNIVERSITY
SCHOOL OF MEDICINE
NASHVILLE, TENNESSEE

DEPARTMENT OF CHEMISTRY
UNIVERSITY OF CALIFORNIA
AT SAN DIEGO
LA JOLLA, CALIFORNIA

I. Preparation and Assay of Enzymes
II. Preparation and Assay of Enzymes
III. Preparation and Assay of Substrates
IV. Special Techniques for the Enzymologist
V. Preparation and Assay of Enzymes
VI. Preparation and Assay of Enzymes (*Continued*)
Preparation and Assay of Substrates
Special Techniques
VII. Cumulative Subject Index

METHODS IN ENZYMOLOGY

EDITORS-IN-CHIEF

Sidney P. Colowick   Nathan O. Kaplan

VOLUME VIII. Complex Carbohydrates
Edited by ELIZABETH F. NEUFELD AND VICTOR GINSBURG

VOLUME IX. Carbohydrate Metabolism
Edited by WILLIS A. WOOD

VOLUME X. Oxidation and Phosphorylation
Edited by RONALD W. ESTABROOK AND MAYNARD E. PULLMAN

VOLUME XI. Enzyme Structure
Edited by C. H. W. HIRS

VOLUME XII. Nucleic Acids (Parts A and B)
Edited by LAWRENCE GROSSMAN AND KIVIE MOLDAVE

VOLUME XIII. Citric Acid Cycle
Edited by J. M. LOWENSTEIN

VOLUME XIV. Lipids
Edited by J. M. LOWENSTEIN

VOLUME XV. Steroids and Terpenoids
Edited by RAYMOND B. CLAYTON

VOLUME XVI. Fast Reactions
Edited by KENNETH KUSTIN

VOLUME XVII. Metabolism of Amino Acids and Amines (Parts A and B)
Edited by HERBERT TABOR AND CELIA WHITE TABOR

VOLUME XVIII. Vitamins and Coenzymes (Parts A, B, and C)
Edited by DONALD B. MCCORMICK AND LEMUEL D. WRIGHT

VOLUME XIX. Proteolytic Enzymes
Edited by GERTRUDE E. PERLMANN AND LASZLO LORAND

VOLUME XX. Nucleic Acids and Protein Synthesis (Part C)
Edited by KIVIE MOLDAVE AND LAWRENCE GROSSMAN

Section I
Amino Acid Analysis

[1] Subnanomole-Range Amino Acid Analysis

By P. E. HARE

Background. The detection and quantitative analysis of amino acids in subnanomole quantities can be accomplished by any of several methods. Thin-layer chromatography (TLC) has been used to detect PTH-derivatives at subnanomole levels.[1] Combined with autoradiography, TLC techniques have been described for subpicomole levels of tritiated dansyl amino acids.[2] Gas chromatography (GLC) of amino acid derivatives has been used to analyze amino acids in the picomole range.[3] High-performance liquid chromatography (HPLC) of amino acid derivatives,[4] although developed primarily for monitoring amino acid sequencing, may offer a sensitive method for amino acid analysis from protein and peptide hydrolysates. When amino acid derivatives must be synthesized prior to analysis it is necessary to ensure quantitative and reproducible derivative yields as well as derivative stability. Ideally, the less sample preparation necessary, the less the risk of sample degradation and contamination. Ion-exchange chromatography provides a system capable of subnanomole sensitivity in which there is a minimum of sample preparation, since the free amino acids rather than derivatives are separated.

Ion-exchange chromatography, as developed by Moore and Stein[5] and Hamilton[6] and subsequently automated by Spackman, Stein, and Moore,[7] has long been the method of choice and standard of comparison for other methods of amino acid analysis. Improvements in cation-exchange resins, the use of narrow-bore columns, more sensitive flow cells, and electronic amplification have made it possible to use ion-exchange column chromatography for amino acid analysis in the subnanomole range. This section will detail the instrumentation and techniques needed for reliable and quantitative amino acid analysis at as low as picomole levels. It must be recognized that there is a significant difference between the analysis of

---

[1] M. R. Summers, G. W. Smythers, and S. Oroszlan, *Anal. Biochem.* **53**, 624 (1973).
[2] S. R. Burzynski, *Anal. Biochem.* **65**, 93 (1975).
[3] R. W. Zumwalt, K. Kuo, and C. W. Gehrke, *J. Chromatogr.* **57**, 193 (1971).
[4] C. L. Zimmerman, E. Appella, and J. J. Pisano, *Anal. Biochem.* **75**, 77 (1976).
[5] S. Moore and W. H. Stein, *J. Biol. Chem.* **192**, 663 (1951); S. Moore and W. H. Stein, *J. Biol. Chem.* **211**, 893 (1954).
[6] P. B. Hamilton, *Anal. Chem.* **30**, 914 (1958); P. B. Hamilton, *Anal. Chem.* **35**, 2055 (1963).
[7] D. H. Spackman, W. H. Stein, and S. Moore, *Anal. Chem.* **30**, 1190 (1958).

a standard amino acid mixture in the subnanomole range and the analysis of the amino acids prepared from subnanomole quantities of a peptide or protein. The sample preparation steps including hydrolysis are critical and may add or destroy some amino acids, so that the amino acid analysis of the final mixture may not accurately represent the composition of the starting peptide or protein. Factors that are of minor consequence in the preparation of milligram quantities of protein material may be of extreme importance in the hydrolysis of microgram amounts of this material.

Instrumentation. Virtually all instrumentation for ion-exchange chromatography of amino acids uses the same basic approach that was used initially by Spackman, Stein, and Moore.[7] Figure 1 shows two of several

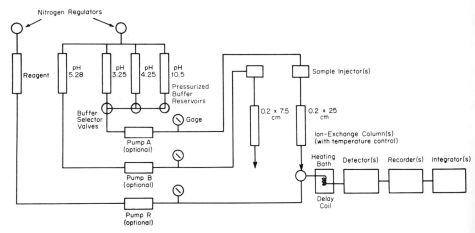

Fɪɢ. 1. Diagram of a high-sensitivity amino acid analyzer (2-column system). For low pressure (up to 500 psi) operation, reservoirs are of Teflon tubing (⅛ inch × 100 feet has capacity of about 60 ml) or cross-linked polystyrene cylinders. Buffer selector valves are on-off pressure-operated slider valves (Chromatronix) automated by a timer and solenoids. A timer determines buffer sequence times. Teflon tubing (22-gauge) connects reservoirs to sample injector. Teflon tubing (32-gauge) connects sample injector to column, and column to detector. The sample injector is made of two 3-way manual slide valves for single injection or an automated 20-port valve (Chromatronix) for multiple injection. The column is 2-mm bore stainless steel (type 316) with Swagelok reducing fittings on both ends. The lower end is fitted with a stainless steel filter disk (0.5 μm retention). Resin is Durrum DC4A (8 μm diameter).

For high-pressure operation (up to 5000 psi) reservoirs (at 20 psi) are fed into displacement pumps (Milton Roy minimpump), and stainless steel tubing connects pump and sample injector (Valco, 7000 psi rating) and connects injector to column.

For both high- and low-pressure systems post-column tubing is Teflon (32-gauge), including time-delay coil for heating bath.

Detector for ninhydrin is shown in Fig. 3.

Detector for fluorescence system is Aminco Fluoromonitor.

possible configurations for high-sensitivity amino acid analysis. Buffer and reagent (usually ninhydrin) are stored in pressurized reservoirs and metered directly by pressure or optionally by pumps through the sample injector and column. The reagent is metered by pressure (or pump) to the mixer, where it is mixed with the column effluent to produce a detectable reaction product. The absorbance or fluorescence of the reaction product is measured in the detector and recorded on a strip chart device.

In the original Spackman, Stein, and Moore instrument the ion-exchange column bed was 9 mm in diameter by 150 cm long. The flow-cell was 2.3 mm long, and a sample load of 1 μmol of aspartic acid gave an absorbance of approximately 1. The lower limit of detection was 10–20 nmol. Using a system similar to that in Fig. 1 gives an absorbance of approximately 1 with only 1 nmol of aspartic acid (Fig. 2) and a lower limit of detection of somewhat more than 1 pmol. This increase in sensitivity of over 10^3 requires a number of modifications, particularly in the configuration of the column and the detector. The volume of the

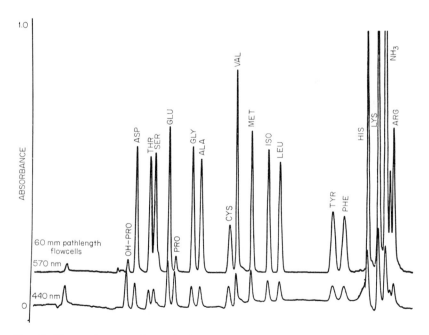

Fig. 2. Chromatogram of standard amino acid mixture (500 pmol each). Total run time is 90 min. Nitrogen pressure is 500 psi (no pumps). Ninhydrin was used as the detection reagent for single-column run. Additional buffer at pH 2.90 was included to resolve hydroxyproline from aspartic acid. Flow-cell pathlength was 60 mm (Fig. 3). Column: 25 cm × 2 mm; reagent: ninhydrin; buffers: (1) 0.2 N Na⁺ at pH 2.90, (2) pH 3.25, (3) pH 4.25, (4) pH 10.5; temperature, 55°. Iso = isoleucine.

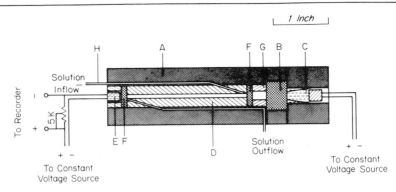

Fig. 3. Diagram of ninhydrin system flow-cell for pathlengths up to 60 mm. A, Delrin housing block; B, interference filter; C, lens end-type lamp (CM 20-3); D, Teflon rod with 0.5-mm diameter flow channel and diagonal channels (0.5 mm) for pull fit of 32-gauge Teflon tubing for entrance and exit of fluid stream; E, photo-conductor cell (Clairex 905L); F, quartz windows pressure-sealed by threaded removable delrin inserts G in housing block A. Power supply: 1.8 V dc.

column bed (0.2 × 25 cm) is approximately 0.75 ml, less than $\frac{1}{100}$ of the volume of the original design. The maximum path length of the flow-cell used is 60 mm, an increase of approximately a factor of 25 over the original design (Fig. 3). To prevent diffusion and loss of resolution, Teflon tubing lines from the column to the detector should be 0.3 mm or less. The delay coil can be shortened to improve resolution, and the color yield can be maintained by using a heating bath (silicon oil) temperature of 115° or even higher. It is important to control the temperature with a suitable device, such as a proportional temperature controller.[8]

The cation-exchange resin used in the 0.2 × 25-cm column is Durrum DC-4a and is graded to 8 ± 0.5 μm.[9] Hamilton[10] showed the importance of grading the resin used in the column, and his development of a hydraulic system (shown in reference 7) for the fractionation of the resin into homogeneous-sized fractions has been one of the most important contributions to the development of automatic ion-exchange amino acid chromatography. The development of pumps that will deliver reproducible flow rates of a few milliliters per hour at back pressures up to 5000 psi or even higher make possible the potential development of analytical systems that will rapidly separate (<15 min) a standard mixture of amino acids at the level of 1 pmol.

Commercial Instrumentation. Commercial amino acid analyzers that

[8] The following device has worked well in our laboratory: Proportional Temperature Controller Model 70-115 from RFL Industries, Inc., Boonton, NJ 07005.

[9] Durrum Chemical, 1228 Titan Way, Sunnyvale, CA 94086.

[10] P. B. Hamilton, *Anal. Chem.* **30**, 914 (1958).

make use of narrow-bore columns are currently available.[11] Sensitivity to amino acid quantities of approximately 50–100 pmol is often achieved but reproducibility at these levels is marginal ($<10\%$). Some commercial instruments use fluorescent detection, but most use colorimetric detection because of problems in the fluorescence detection of proline and hydroxyproline. Path lengths of most of the colorimetric flow-cell designs vary from 5 mm to 25 mm. Many instruments use colorimetric detectors that have integral scale expansion and linear conversion so that signal output varies linearly with amino acid concentration, making automatic data reduction easier and less costly. In general, fluorescent detection is linear with concentration over a wide range of amino acid concentrations.

Use of HPLC Equipment. The proliferation of instrumentation for high-performance liquid chromatography is an indication of the widespread and growing interest in liquid chromatography. Most of these instruments at present rely on the detection of the separated compounds by absorbance, fluorescence, or refractive index. Few have the capability of adding a reagent to the column effluent, although that would be a relatively simple addition and would increase the usefulness to classes of compounds like amino acids that require the addition of a reagent for detection. Some of the newer amino acid analyzers are capable of operation in the 5000 psi range, and if equipped with suitable columns and detectors could perform virtually any high-performance liquid chromatographic separation. Conversely, most HPLC systems could be modified to do high-sensitivity amino acid chromatography by adding a reagent reservoir (pressurized or pump) and mixer together with a suitable column and detector. HPLC systems are being used for amino acid derivatives,[12] and it seems not far into the future that HPLC instrumentation will evolve into sophisticated instruments similar to some present amino acid analyzers with automatic multiple-sample injection, data reduction, *and* the ability to add one or more reagents to the column effluent!

Updating Standard Amino Acid Analyzers. The conversion of a standard amino acid analyzer into a high-sensitivity instrument capable of the detection of subnanomole amounts of amino acids requires several substantial modifications.[11] The paper by Liao *et al.*[13] describes the conversion of a standard amino acid analyzer to a narrow-bore column system, listing the necessary components, all of which are commercially available.

[11] P. E. Hare, *in* "Protein Sequence Determination," 2nd ed. (S. B. Needleman, ed.), p. 211. Springer-Verlag Berlin and New York, 1975.
[12] J. J. Pisano, *in* "Protein Sequence Determination," 2nd ed. (S. B. Needleman, ed.), p. 291. Springer-Verlag, Berlin and New York, 1975.
[13] T.-H. Liao, G. W. Robinson, and J. Salnikow, *Anal. Chem.* **45**, 2286 (1973).

Liao *et al.* also recommended that the ninhydrin reagent be diluted with water [2 parts of standard dimethyl sulfoxide (DMSO) reagent to 1 part of water] and mixed in a 1:1 ratio with the column effluent. We have found that the use of the more dilute ninhydrin reagent results in less baseline "noise" and consequently higher sensitivity. Since the diluted reagent is less viscous, it mixes more efficiently with the column effluent with less tendency to form incipient precipitates that contribute to baseline noise.

The fluorescence system described by Benson and Hare[14] uses a relatively dilute aqueous solution of *o*-phthalaldehyde (OPA) to produce a fluorescent reaction product with amino acids. It appears to give a more stable and quieter baseline than does ninhydrin or fluorescamine, and consequently can be used to increase the sensitivity of a standard-bore column (9 mm) amino acid analyzer. The only necessary modification would be the addition of a fluorometric detector. The ninhydrin pump would be used to pump the OPA reagent. This arrangement would use significantly more buffers and reagents than the narrow-bore column and would not be potentially as sensitive, although it should be capable of detecting subnanomole quantities of amino acids. Baseline artifacts due to buffer impurities and buffer changes would be more difficult to control in a standard-bore column system as compared to a narrow-bore system because all the sensitivity increase would be from the detector alone whereas in a narrow-bore column a large increase in sensitivity results from the smaller bore and volume of the column itself.

Felix and Terkelsen[15] have described a system using a narrow-bore column and a fluorescent detection system capable of detecting all the amino acids including proline and hydroxyproline. *N*-Chlorosuccinimide is used to convert secondary amines to primary amines, which then react with fluorescamine. This system requires three reagent reservoirs and pumps in addition to the three reagents and pump for the column effluent.

Proline and hydroxyproline do not react directly with *o*-phthalaldehyde but must first be partially oxidized by hypochlorite[16] or be detected by some other suitable reagent.[17] This increases the complexity of the sys-

---

[14] J. R. Benson and P. E. Hare, *Proc. Natl. Acad. Sci. U.S.A.* **72**, 619 (1975).

[15] A. M. Felix and G. Terkelsen, *Anal. Biochem.* **56**, 610 (1973).

[16] M. Roth and A. Hampai, *J. Chromatogr.* **83**, 353 (1973). Unpublished results in our laboratory show that hypochlorite can be used to react with proline and hydroxyproline to form primary amines that will react with OPA.

[17] N. B. D. chloride reacts with secondary amines to form highly fluorescent derivatives and in combination with OPA reagent may provide a suitable system for fluorescent detection of all the protein amino acids. M. Roth, *J. Clin. Chem. Clin. Biochem.* **14**, 361 (1976).

tem by requiring an additional reservoir and reagent pump. It has been our experience that, if all the amino acids (including proline and hydroxyproline) in a hydrolysate are to be detected, ninhydrin provides the simplest system and can be used for the routine detection of subnanomole quantities of amino acids. If results for proline and hydroxyproline are not necessary, then o-phthalaldehyde provides an excellent high-sensitivity system with only a single reagent reservoir needed. At 1 pmol levels of most amino acids, tht signal to noise ratio is approximately 5 to 1, significantly better than either ninhydrin or fluorescamine ($S/N \cong 1$).

Single-Column vs Two-Column Systems. It might seem that, since a two-column system requires two sample aliquots whereas the single-column system requires only one, the single-column system would be the method of choice for high-sensitivity amino acid analysis. Actually a single-column system may not be the proper choice, since contaminants that interfere with the analysis of the basic amino acids in the single-column system do not interfere in the two-column system. It is obvious that if an improvement by more than a factor of 2 can be achieved in the two-column system, then the two-column system is the method of choice. Actually the improvement far exceeds a factor of 2, since the buffer contaminants that come off the column interfere seriously at levels of 250–500 pmol and make quantitation difficult for the basic amino acids at these levels and below. In Figs. 4 and 5 can be seen the improvement in the resolution of the basic amino acids in a separate 7.5-cm column run at pH 5.28 as compared to the single-column run. In the two-column system only the buffer change between the first and second buffer is significant, but in the single-column system the contamination interference in the basic region is so bad that it is not possible to achieve accurate results below 250 pmol, depending on the degree of baseline shift. If buffer quality is poor the results may not be accurate even for much higher levels.

In a two-column system the acidic and neutral amino acids are analyzed in the longer column, and after phenylalanine is eluted the column is washed with base and regenerated for the next run. The basic amino acids, hydroxylysine through arginine, are eluted on a 7.5-cm column at pH 5.28. Since only one buffer is used, the buffer contaminants are an integral part of the baseline and do not interfere with any of the amino acids, including ammonia. For analysis of amino acids near picomole concentrations it has been our experience that a separate column for the basic amino acids provides a much more accurate amino acid pattern of the sample. When it becomes possible to achieve "cleaner" buffers and eliminate the interference in the basic region, a single-column system will

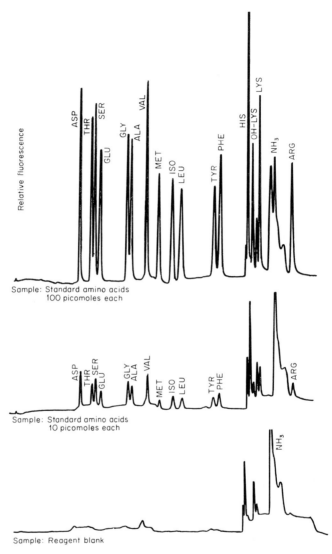

Sample: Standard amino acids
100 picomoles each

Sample: Standard amino acids
10 picomoles each

Sample: Reagent blank

Fig. 4. Chromatograms of *o*-phthalaldehyde system for 100-pmol and 10-pmol levels of a standard amino acid mixture. Lower tracing is a reagent blank and shows relatively high fluorescent impurities in basic region that interfere with single-column analysis of basic amino acids. Column: 25 cm × 2 mm; reagent: *o*-phthalaldehyde; buffers: (1) pH 3.25, 0.2 *N* Na⁺, (2) pH 4.25, 0.2 *N* Na⁺, (3) pH 10.5, 0.2 *N* Na⁺; detector: Aminco Fluoromonitor; temperature: 60°. Iso = isoleucine.

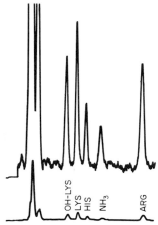

Fig. 5. Short-column chromatogram of basic amino acids (5 pmol each) showing the relatively high signal-to-noise ratio compared to the single-column system. Upper curve is 20× recorder expansion of lower curve. Length of run is 20 min. Detection limit is less than 1 pmol of most amino acids. Fluorescence detector is Aminco Fluoromonitor (American Instrument Company, Silver Spring, Maryland). Column: 7.5 cm × 2 mm; reagent: o-phthalaldehyde; buffer: pH 5.28, 0.35 N Na$^+$; temperature: 60°; sample: standard amino acids, 5 pmol each.

obviously be desirable and useful to the analysis of picomole levels of amino acids.

Preparing the Column. A Teflon extension tube (10 feet × ⅛ inch) is attached to the column and filled with buffer. A slurry of resin in buffer (2 parts buffer and 1 part resin) is drawn by syringe into the column and far enough into the extension tube to ensure enough resin to pack the column. The bottom of the column is fitted with a stainless steel filter (0.5 μm), and the slurry is forced into the column (pump or pressurized buffer). After the column is packed, it is "washed" with pH 10.5 borate buffer and regenerated with pH 3.25 buffer before a sample is injected. A small length (3–8 cm) of the Teflon tubing packed with resin is convenient for adding resin to the column after the first few runs. It is attached to the top of the column and pressurized with buffer to fill any dead space that may develop at the top of the column. Generally, after the first few runs on a newly packed column the resin is tightly packed and does not pack further. Addition of samples of high salt content (>1 M Na$^+$) will shrink the resin bed and form a dead space above the resin. This will result in a loss of resolution and will require repacking or the addition of resin to fill the void space.

Sample Injection. To ensure maximum resolution, the sample should be injected at pH 2 (±0.2) in a volume no more than 100 μl for a column

2 mm $\times$ 25 cm. Generally the smaller the sample volume injected the better. Although the sample can be injected directly through a septum onto the resin bed by a high-pressure syringe (Hamilton HP series), most sample injectors use a sample loop into which a sample is loaded at ambient pressures. A bypass valve allows buffer to flow through the column without interruption during the loading operation. To inject the sample the valve is operated to pressurize the sample loop and force the sample into this column. Manual sample injector valves are available that operate up to 7000 psi.[18] Automatic multiple sample valves are also available.[19]

Buffer Changes. The simplest system to provide for a buffer sequence consists of a 100-ft length of 22-gauge Teflon tubing into which appropriate volumes of pH 3.25, 4.25, 10.5, etc., are placed. This buffer sequence reservoir follows the sample injector and can even be part of the sample "loop." The sample is injected onto the column and is followed by the appropriate volumes of the various buffers. The column is finally regenerated and ready for the next analysis. No electronic timers or buffer change valves are necessary.

For multiple-sample analysis it is desirable to use on-off valves for each separate buffer reservoir and an automatic timer to control buffer times and injection. The use of a 20-port injection valve and solenoid-operated slider valves for each buffer all controlled by a master programmer timer provides our laboratory with a versatile system capable of analyzing up to 20 hydrolysate samples per day.

Discussion. There is no simple way to measure sensitivity and compare the sensitivities of different systems. The sensitivity limit of any system is the background noise produced by the system when no sample is added. The signal-to-noise ratio when a sample is run provides some measure of the ultimate sensitivity of the system. The background noise is a combination of several factors including electrical noise, incomplete mixing of reagent and effluent, precipitation and gas bubbles in the flow lines, as well as contaminants in the buffer. Many of these factors are not constant from day to day and some tend to increase with the age of the reagent.

Proper mixing of the reagent with the column effluent is of critical importance for high-sensitivity work. The ninhydrin reagent, even the diluted formula of Liao *et al.*,[13] is difficult to mix completely with the

---

[18] Valco Instruments Co., P. O. Box 19032, Houston, TX 77024. For 500 psi or lower operation, several kinds of valves are available.

[19] A 20-port automatic sample system that has been used in our laboratory successfully for several years is made by Chromatronix. Altex and Valco also make multiple-sample injectors.

aqueous buffers in the column effluent because of viscosity differences. Small regions of incompletely mixed buffer and reagent create "noise" in the absorbance monitor and contribute significantly to baseline fluctuations. Undamped pumps, temperature changes, particulate matter all contribute to fluctuations in the absorbance monitor.

In our experience, the fluorescamine reagent (acetone solution) also exhibits mixing difficulties and appears to have about the same signal-to-noise ratio as ninhydrin for similar-size sample loads. The limit of sensitivity of both ninhydrin and fluorescamine appears to be about 1 or 2 pmol.

The o-phthalaldehyde reagent, since it is an aqueous solution, mixes more efficiently and produces a significantly smoother baseline than does either ninhydrin or fluorescamine. At 5 pmol (Fig. 4) it can be seen that the signal-to-noise ratio is of the order of 10–20 for the basic amino acids. At this sensitivity setting, the basic region on a single-column system is entirely off-scale and it is not possible to use a single-column analysis for the basic amino acids. The two-column system is superior for levels less than about 200 pmol.

Preparation of Reagents and Samples for High-Sensitivity Analysis. The quality of the water used in the preparation of buffers and reagents can limit the potential sensitivity of an amino acid analyzer by the introduction of interfering baseline artifacts. Unfortunately, the quality of the water used for buffer preparation is not readily determined by the usual standards for water quality, such as specific conductivity. Most of the contaminants appear to be nonionized materials or weakly ionized materials (i.e., ammonia) that may be absorbed into the water during storage. Ubiquitous airborne contaminants make the storage of high-quality water very difficult. The trend in commercial pure-water systems is to purify enough for current use rather than to produce large quantities of pure water for storage, since during storage recontamination, at least to some extent, will occur.

Simple distillation (even triple distillation) or deionization does not remove all interferring materials. Unless an integral recycling mode is included with the deionization system, it is possible for microorganisms to grow on the resin surfaces and seriously contaminate the final product. The best way to ascertain the quality of the water is to use it in the preparation of the buffers with known quality hydrochloric acid and sodium citrate and observe the baseline artifacts during a blank run on the amino acid analyzer. For the analysis of subnanomole quantities of amino acids it is obviously desirable to reduce baseline artifacts to a minimum.

A procedure that has been developed in our laboratory and produces water of excellent quality for eluent buffers involves the treatment of distilled water with alkaline sodium hypochlorite (NaOCl). The alkaline hypochlorite decomposes ammonia and amino acids. After refluxing, the solution is made acidic with sulfuric acid, and stannous chloride is added to reduce any remaining hypochlorite. Distillation of the mixture produces a quality of water that has provided a minimum of baseline artifacts when used in buffers for ultramicro amino acid analysis. The following procedure is used with a 5-liter distilling flask.

1. Fill flask with 2500 ml H_2O.
2. Add 0.5 g of NaOH (pellets).
3. Add 6.5 ml of NaOCl (4–6%).
4. Reflux overnight.
5. Add 4 ml of concentrated H_2SO_4 (CAUTION!).
6. Add 2 g of $SnCl_2 \cdot 2H_2O$.
7. Distill.
8. Check water for free chlorine by adding 1 g of KI to 10 ml of distilled water. If a yellowish color appears, add more $SnCl_2 \cdot 2H_2O$. If there is no yellow color, distill the remainder into protected (citric acid trap) flask. Collect 2 liters.

Obviously, if an adequately pure water supply is already available, it will not be necessary to use this procedure to obtain good-quality water for high-sensitivity analysis. Simple precautions should be followed in preparing buffers.

The present limit of sensitivity in amino acid analysis by ion-exchange chromatography is the baseline "noise" due in large part to buffer contaminants. These contaminants appear to be trace amounts of amino acids, ammonia, and related compounds, the source of which may be the water used in the buffer preparation, the hydrochloric acid, the buffer salts and other added constituents, or even the environment in which the buffer preparation is carried out (a smoke-filled room will contribute significant amounts of contaminants to buffers prepared therein!)

Premixed buffers and buffer concentrates are available commercially[20] and may provide adequate buffers for high-sensitivity analysis; however, it is important to specify that the buffers will be used for high-sensitivity analysis and must be of suitable quality.

To ensure good quality buffers prepared in your own laboratory it is necessary to check each component for amino acid contamination levels. First, check the HCl quality by evaporating 1 ml to dryness,

---

[20] Beckman Instruments, Inc. 1117 California Avenue, Palo Alto, CA 94304; Durrum Chemical, 1228 Titan Way, Sunnyvale, CA 94086; Hamilton Company, P.O. Box 10030, Reno NV 89510; Pierce Chemical Company, Box 117, Rockford, IL 61105.

dissolving the residue in 100 μl of pH 2 solution (1 ml of 6 N HCl in 100 ml of H_2O) and injecting 50 μl onto the ion-exchange column. Comparison of the resulting chromatogram with a blank (i.e., 50 μl of pH 2 solution) indicates the quantity of free amino acids in the HCl. Repeating the test on 1 ml of HCl after hydrolysis will indicate the presence of any combined amino acid materials in the HCl or on the glassware.

After the quality of HCl has been determined to be adequate (less than picomole levels per milliliter of HCl), the quality of the buffer salts can be ascertained. Sodium citrate ($Na_3C_6H_5O_7 \cdot 2H_2O$) is usually the buffer salt used in buffers for amino acid analysis. A 0.2 N sodium concentration contains 19.6 g of sodium citrate in a liter of solution. A solution of 20 mg of sodium citrate in 1 ml of H_2O approximates 0.2 N Na^+ concentration. Add approximately 40 μl of "clean" 6 N HCl to get a final pH 2 solution. Of this pH 2 sodium citrate solution, 100 μl is injected onto the ion-exchange column and the resulting chromatogram compared with the water blank and the HCl chromatograms. The concentration of the HCl in the pH 2 sodium citrate solution is only 4 μl HCl/100 μl, or less than $\frac{1}{100}$ of the HCl test (500 μl equivalent). Results of the tests are helpful in indicating the source of possible contaminants. If thiodiglycol (TG) is to be used in the buffers then an aliquot of pH 2 buffer with TG added should be analyzed.

Hydrochloric acid. Amino acid contamination in several commercial brands of reagent grade hydrochloric acid has been reported[21] and can be the source of serious buffer contamination problems. Fortunately, distillation of 6 N HCl removes most of this unwanted contamination.[22] Triple-distilled HCl is usually adequate for buffer preparation. Storage of "clean" HCl is a serious problem because of ammonia and airborne particulate contamination. Storage of 10–15-ml quantities of clean HCl in sealed glass vials is preferable to storage of larger quantities. A simple test to determine the suitability of the HCl for hydrolysis of peptides and proteins is to heat 1 ml of it at 150° for 25 min. Evaporate to dryness on a rotary evaporation or in a stream of nitrogen, dissolve the residue in pH 2 solution (1 ml of 6 N HCl in 100 ml of H_2O) and analyze for amino acids. This test should be done routinely for HCl used in hydrolysis of samples. Appreciable concentrations of amino acids in the HCl blank will contribute to buffer contamination problems and limit the sensitivity of the system by increasing the number and amount of interfering baseline artifacts.

Another approach in the preparation of "clean" HCl involves the

---

[21] P. B. Hamilton and T. T. Myoda, *Clin. Chem.* **20,** 687 (1974).
[22] Y. Wolman and S. L. Miller, *Nature (London)* **234,** 548 (1971).

use of HCl gas (Matheson, electronic grade[23]) and a good grade of water. Small quantities of concentrated HCl can be made when needed by this method and the problems of storage are eliminated. The filtered HCl gas is slowly passed through an H_2SO_4 trap and into water (cooled by an ice bath) until the water is saturated. The concentration is approximately 12 N at room temperature.

After determining that the buffer components are satisfactory the buffers are prepared by dissolving 19.6 g of sodium citrate in a liter of "clean" water. HCl (6 N or gas) is added while monitoring the pH. At pH 4.25 (or lower if desired) 250 ml of the buffer are added to the pH 4.25 reservoir while the remainder of the buffer is titrated with HCl to pH 3.25 (or other desired pH) and placed into the pH 3.25 reservoir and capped.

The pH 10.5 buffer is made by adding 10 g of sodium chloride, 1 g of EDTA, and 2 g of boric acid to a liter of water. Add NaOH pellets to pH 10.5.

For two-column operation the pH 5.28 buffer is made by adding 34.3 g of sodium citrate to 1 liter of water and adding HCl to pH 5.28. For faster long-column wash and regeneration the pH 10.5 borate buffer may be replaced by 0.2 N NaOH solution containing 1 g per liter of EDTA.

For diluting and dissolving the samples to be analyzed it is necessary to use a solution of pH near 2.00. We use a dilute HCl solution containing 1 ml of 6 N HCl in 100 ml of H_2O.

Some procedures use a high ionic strength citrate buffer for the final buffer to elute the basic amino acids from a single-column system. The high salt content of these buffers causes the resin to shrink and increases the flow rate in a constant-pressure system or decreases the back pressure in a constant-volume pump system. A change in flow rate or back pressure usually introduces baseline artifacts, and for that reason we prefer to use the pH 10.5 borate buffer (0.2 N Na^+) or the two-column system.

Preparation of Ninhydrin Reagent. The ninhydrin reagent is conveniently made as follows:

1. Mix 3 parts of DMSO with 1 part of 4 M lithium acetate buffer.

2. Stir for 10 min with magnetic mixer while bubbling nitrogen (or argon) through the mixture to displace traces of oxygen.

3. Add 20 g of ninhydrin for each liter while maintaining the stirring and nitrogen bubbling.

[23] Matheson Gas Products, P.O. Box 85, East Rutherford, NJ 07073

4. Add 0.625 g of hydrindantin for each liter of solution. Solution turns dark, and it is difficult to tell when the hydrindantin is dissolved. It is best done by observing the bottom of the flask from time to time. Finally dilute 1 liter of reagent with 500 ml of oxygen-free water to make 1.5 liters of dilute reagent.

5. Transfer to reservoir under oxygen-free conditions.

This reagent is the diluted DMSO reagent of Liao et al.[13] Ready-to-use ninhydrin reagent is available from several sources[20] and may be diluted with oxygen-free water (2 parts of reagent to 1 part of water).

Fluorescamine. Fluorescamine (Fluram)[24] is unstable in aqueous solution and is best dissolved in acetone (150–200 mg per liter of solution). If kept in a glass reservoir under nitrogen it is stable. A borate buffer is used with fluorescamine to buffer the column effluent to pH 9.5 or so. It is made by dissolving 30 g of boric acid in a liter of water and adding potassium hydroxide (pellets or solution) to a final pH of 10. If sodium hydroxide is used there is more chance of a precipitate forming during the addition of the acetone reagent.

o-Phthalaldehyde. Since *o*-phthalaldehyde is stable in aqueous solution it can be combined with the buffer necessary for pH adjustment.[14] *o*-Phthalaldehyde[25] is dissolved in ethanol (1 g in 10 ml) to which 2-mercaptoethanol (0.5 ml) has been added. After solution of the *o*-phthalaldehyde, the mixture is added to 1 liter of boric acid–potassium hydroxide buffer of pH 10.5 (30 g of boric acid dissolved in 1 liter and titrated to pH 10.5 with potassium hydroxide). BR1J-35[26] is added to the reagent to enhance the fluorescence of some of the amino acids, such as lysine and hydroxylysine.

Sample Preparation. Subnanomole analysis of standard amino acid mixtures is relatively straightforward. However, the analysis of sub-nanomole levels of amino acids from submicrogram quantities of peptides and proteins is much more difficult. The hydrolysis procedure is the crucial step and results in the partial or complete destruction of some amino acid residues. Incomplete release during hydrolysis of amino acid residues forming the more stable peptide bonds may yield levels that are too low for amino acids like valine and isoleucine. If a sufficient quantity of sample peptide or protein is available it is desirable to hydrolyze sample aliquots for two or more different time periods: a relatively short period for determination of labile amino acids, such as serine and threonine, and a relatively long period for the

[24] Roche Diagnostics, Nutley, NJ 07110.
[25] Sigma Chemical Company, P.O. Box 14508, St. Louis, MO 63178.
[26] Pierce Chemical Company, Box 117, Rockford, IL 61105.

analysis of amino acids forming resistant peptide bands, like valine and isoleucine.

Hydrolysis. A convenient procedure for hydrolysis is to use small screw-cap vials (2 ml, 35 × 12 mm) fitted with a Teflon liner.[26] The sample may be weighed directly into the vial or, if in solution, it can be evaporated to dryness directly in the vial; 6 N HCl is added (100–500 μl). The vial is flushed with a stream of clean nitrogen and capped immediately. Hydrolysis is carried out in a heated aluminum block for 20 min at 155°. Ideally a series of samples is heated for progressively longer time periods and enables a correction to be made for the destruction of labile amino acids as well as the slower release of amino acids forming resistant peptide bonds. Results using the procedure outlined above are essentially identical to results using more traditional procedures: e.g., 22 hr at 110° and evacuating and sealing the hydrolysis tube. After hydrolysis the excess HCl can be evaporated on a rotary evaporator or in a stream of nitrogen. An alternative procedure that avoids drying the sample uses 0.6 N NaOH to neutralize the HCl (1 part of 6 N HCl + 9 parts of 0.6 N NaOH).

Contamination Problems. The most serious problem in the analysis of subnanomole quantities of amino acids is that of amino acid contamination in the HCl. It is essential that control samples of the 6 N HCl be hydrolyzed and analyzed for amino acids. Batches of 6 N HCl that have been stored in containers that are frequently opened may acquire a significant amount of amino acid material from airborne sources. It is best to store the HCl in the freezer in sealed ampules.

Another source of contamination may be the glassware used in the storage of HCl and in the hydrolysis step. Heating the glassware to 500° for a few hours is a satisfactory method for cleaning glassware. Heating vials or ampules carefully in a flame also results in adequately clean glass.

Conclusion. With the equipment and techniques outlined in this article it is possible to carry out routine amino acid analysis on samples of peptides and proteins, which on hydrolysis yield subnanomole levels of amino acids. By carrying reagent blanks and low-level standard amino acid mixtures through the hydrolysis procedures, it is possible to correct for contamination effects and for losses during hydrolysis. Hydrolysis at 155° of sample aliquots for periods of time from 20 min to 2 hr yield data nearly identical to the more commonly used, but time-consuming, 24- to 72-hr sequence. Even with less than 1 μg of protein, it is possible to run a series of three or more hydrolysis periods and get relatively accurate corrected compositional data on the amino acid composition of the material.

[2] Improved Ion-Exchange Resins

By James R. Benson

Ion-exchange chromatography is important to biochemists because of its versatility in separating biomolecules. Amino acids, peptides, proteins, carbohydrates, polyamines, and nucleotides have been successfully separated on various ion-exchange resins now commercially available. In most applications, prior derivatization of sample molecules as required for gas–liquid chromatography is unnecessary; thus, purified material can be collected without loss of biological activity. Every modern biochemical laboratory routinely employs ion-exchange resins. However, early attempts to separate biomolecules by ion exchange were fraught with difficulty because only block-polymerized, granular, polymeric material was available. These irregular particles were difficult to pack in columns, frequently yielded irreproducible results due to batch variations, required low eluent flow rates to avoid high eluent pump pressures, and resulted in variable elution patterns depending upon how a column was packed. It was not until the introduction of spherical ion-exchange resins during the 1960's that these problems began to be surmounted.

Most ion-exchange resins in use today are synthesized from divinylbenzene cross-linked polystyrene; either emulsion or suspension polymerization techniques are used to produce perfect spheres. Appropriate functional groups are then attached to the polymeric matrix to impart ion-exchange properties. These "microporous" resins are distinguished from macroporous resins and pellicular resins that contain ion-exchange groups only on surfaces of rigid spheres. By using the entire sphere volume to achieve a separation, ion-exchange capacity is dramatically increased over that of pellicular resins. In addition, nonionic interactions between solute molecules and the copolymer matrix play a more important role in the separation, significantly affecting elution profiles. For example, hydrophobic interactions of amino acid side chains with the polystyrene matrix will sufficiently retard certain amino acids so that elution does not follow the order expected solely on the basis of ion-exchange properties. This very useful property allows separation of solutes with identical pI values that would otherwise coelute.

Microporous ion-exchange resins are generally employed for biochemical applications. The importance of particle size, size distribution, and cross-linkage are discussed below.

Resin Properties

Particle-Size Distribution

When the particle-size distribution of a spherical resin is sufficiently narrow, column-packing technique is greatly simplified and elution patterns of separated biomolecules generally become independent of packing techniques. Conversely, when resins containing a wide range of particle diameters are used, packing techniques become extremely important because nonuniform distribution of particle sizes throughout the resin bed affect solute separation and elution patterns. For example, if eluents of changing molarity are utilized for a separation as required in most amino acid analysis protocols and in gradient elution systems for separating peptides, expansion and contraction of resin particles due to these molarity changes result in resin particle migration and redistribution within the bed. Elution profiles can change significantly if particle redistribution is pronounced. Resins containing many "fine" particles of very small diameter resulting from incomplete classification can contribute not only to irreproducible elution profiles as discussed above but also can be a cause of eluent pump pressures. Small resin particles tend to migrate downward within the resin bed; accumulation of these particles on the bed support can result in increased resistance to eluent flow and subsequent elevated pump pressures. Modern ion-exchange resins are sized to close tolerances so that these problems are not encountered. For most applications, resins sized to ±10% of the mean diameter are satisfactory. A resin with a narrow size distribution also permits accelerated eluent linear flow velocity, U_0, defined as

$$U_0 = F/A \qquad (1)$$

where F is the eluent flow rate in cm^3 min^{-1} and A is the cross-sectional area of the resin bed in cm^2. (It is convenient to express eluent flow rate in terms of U_0 because it is independent of column geometry.) The absence of "fine" particles assures that eluent pump pressures will remain constant from one analysis to the next and allow maximum chromatograph performance. The advantage of using an accelerated linear flow velocity is that the analysis time can be proportionately reduced. For example, the pioneering efforts of Moore and Stein made use of granular cation exchange resins at an eluent linear flow velocity of 0.105 cm min^{-1} and required in excess of 168 hr to complete a single amino acid analysis.[1] In contrast, modern amino acid analyzers com-

[1] S. Moore and W. H. Stein, *J. Biol. Chem.* **192**, 663 (1951).

monly utilize linear flow velocities of 2.5 cm min^{-1} with resultant analysis times of 1 or 2 hr.

A wide particle size distribution also adversely affects resolution. Solute molecules, carried through the resin bed by eluents, migrate into, then out of the resin matrix. These molecules spend more time in larger spheres than in smaller ones resulting in broadening of solute bands when resin particles are not of uniform size.

Particle Size

Ion-exchange resins have been improved in recent years through advances in polymer chemistry technology, by narrower size classification and by reducing the mean particle diameter of resin spheres. The importance of small particles in amino acid analysis was first noted in a theoretical treatment of liquid chromatography by Hamilton *et al.*[2] Their results were interpreted and modified by Benson, who discussed the relationship between particle size and analysis time.[3] He showed that the time for an analysis, t_2, can be expressed as

$$t_2 = t_1(Z_2\phi_2 U_{01})/(Z_1\phi_1 U_{02}) \qquad (2)$$

where t is the analysis time, Z is the resin bed length in cm, U_0 is the linear flow velocity, and ϕ is the "packing density" defined as m/V_c, where m is the mass of the resin in grams in the column and V_c is the volume of the empty column in cm^3. Subscripts 1 and 2 indicate initial and revised conditions, respectively. According to Eq. (2), analysis time can be reduced by simply shortening resin bed length or increasing eluent linear flow velocity; however, such changes result in loss of resolution between any two amino acids.[2] Resolution, R_{ab}, is given by

$$R_{ab} \propto (1/d_p)(Z/2U_0)^{1/2} \qquad (3)$$

where d_p is the resin mean particle diameter. Equation (3) shows that resolution losses can be compensated by reducing the mean particle diameter d_p. This required new diameter, d_{p2}, is given by

$$d_{p2}^2 = [(Z_2 U_{01}\phi_2)/(Z_1 U_{02}\phi_1)]d_{p1}^2 \qquad (4)$$

where d_{p1} is the mean particle diameter of resin used in the initial analysis. The above equations summarize the reasons why the mean diameter of ion-exchange resins introduced in recent years has been progressively

[2] P. B. Hamilton, D. Bogue, and R. A. Anderson, *Anal. Chem.* **32**, 1782 (1960).
[3] J. R. Benson, *in* "Applications of the Newer Techniques of Analysis" I. L. Simmons and G. W. Ewing, eds.), p. 438. Plenum, New York, 1973.

smaller. Resolution can be greatly improved and analysis times can be dramatically reduced by using small-sphere-diameter resin.

An ancillary advantage to the use of small resin spheres is increased sensitivity. With accelerated linear flow velocities and shortened resin bed lengths, solute molecules spend less time in resin beds and band broadening through diffusion is reduced. Although effluent solute peak areas remain constant, peak heights increase, resulting in more sensitive detection.

Cross-Linking

This term generally refers to the percentage of divinylbenzene present at the time of copolymerization with styrene and may not necessarily reflect actual extent of cross-linking of linear polymer. Thus, an "8% cross-linked resin" refers to copolymer containing 92 parts styrene, 8 parts divinylbenzene. Molecular structure of resultant resin is usually not investigated; the product is simply evaluated in terms of its performance in a liquid chromatograph.

Hewett and Forge studied cross-linking as it relates to amino acid analysis at accelerated eluent linear flow velocities.[4] Their results show that eluent flow resistance decreases dramatically in a nonlinear fashion as cross-linking increases (Fig. 1). Presumably, this is because of the increased rigidity of the resin spheres that results in less distortion or compression at the eluent linear flow velocities employed (greater than 6 cm min^{-1}). However, peak broadening increases as cross-linking is increased. Figure 1 illustrates this broadening effect with aspartic acid; other amino acids behave similarly. Cysteic acid, also illustrated, does not show a broadening effect. It does not interact with the resin but is eluted at the column void volume. Peak-broadening effects are less pronounced with resins containing 8–10% cross-linking; if cross-linking exceeds 10%, broadening effects render resins unacceptable for amino acid analysis.

Hewett and Forge attempted to reduce the time for an amino acid analysis by increasing eluent linear flow velocity; in all their experiments, resin bed length remained constant at 48 cm. Another approach to reducing analysis time is to reduce resin bed length, thus precluding the necessity to increase eluent linear flow velocity, and allowing lower cross-linked resins to be successfully used. These results along with those of Hewett and Forge are illustrated in the next section.

[4] G. Hewett and C. Forge, *Fed. Proc., Fed. Am. Soc. Exp. Biol.* **35**, 1382 (1976).

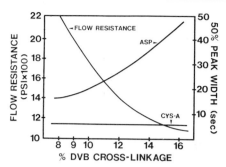

Fig. 1. Effects of resin cross-linking on eluent flow resistance and peak width. Cation-exchange resins with several degrees of cross-linking and sized to 10 ± 0.5 μm were packed in 0.172×48-cm beds in stainless steel columns. A 0.20 N sodium citrate buffer solution, pH 3.25, was supplied to each bed at a linear flow velocity of 7.17 cm min^{-1} with columns maintained at $50°$. Flow resistance (eluent pump pressure) was recorded for each resin. A sample consisting of cysteic acid and aspartic acid, 10 nmol each, was applied to each resin bed, and peak width at 50% of peak height was measured. Peak width of cysteic acid remains constant because it does not interact with cation-exchange resins and is eluted at column void volume. Analyses were accomplished on a Durrum D-500 Amino Acid Analyzer. DVB, divinylbenzene. (Data provided by Gary Hewett.)

Applications of Ion-Exchange Resins[5]

Amino Acid Analysis

High-Pressure Applications. Hewett and Forge approached the problem of reducing the time of an amino acid analysis[4] by utilizing a high-pressure amino acid analyzer, the Durrum Model D-500 (Durrum Instrument Corp., Sunnyvale, CA). This instrument is capable of routine operation at eluent pressures up to 200 atm (3000 lb in^{-2}) and has been described elsewhere.[6] This high eluent pressure capability permits greatly accelerated eluent linear flow velocities; hence, the potential for reducing analysis time. A typical analysis of protein amino acids on this instrument is shown in Fig. 2. A complete analysis on a 0.172×48-cm bed of DC-4A cation-exchange resin (Durrum Chemical Corp.) is obtained in 90 min at an eluent linear flow velocity of 7.17 cm min^{-1}. This resin is nominally 8% cross-linked and is sized to 9.0 ± 0.5 μm. Eluent

[5] In this and following sections only Durrum resins are discussed. In most cases, the Durrum resins are unique; comparable resins are not available from other manufacturers. In other cases, equivalent resins are available. Readers are referred to other manufacturers' technical data for more specific information.

[6] J. R. Benson, *Am. Lab.* (October) 51 (1972).

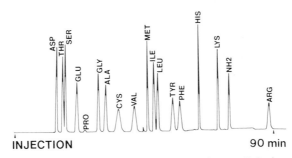

Fig. 2. High-pressure amino acid analysis using 8% cross-linked resin. A mixture containing 10 nmol of each amino acid was applied to a 0.172 × 48-cm bed of Durrum DC-4A cation-exchange resin. This is an 8% cross-linked resin sized to 9 ± 0.5 μm. An eluent linear flow velocity of 7.17 cm min⁻¹ produced a flow resistance of ca. 170 atm. Analyzed on a Durrum D-500 with photometric detection at 570 nm using ninhydrin. (Data provided by Gary Hewett.)

pump pressure is typically 170 atm (2500 lb in⁻²) at the flow rate specified.

Analysis time could be further reduced by increasing eluent linear flow velocity; however, pressure limitations of the instrument would be exceeded before substantial improvement could be made. The search for a resin that exhibits minimum flow resistance but yields acceptable resolution resulted in a compromise (Fig. 1). A special 10% cross-linked resin (Durrum DC-X10-10) sized 10 ± 0.3 μm, was developed. Results of an amino acid analysis using DC-X10-10 in a 0.172 × 48-cm bed are shown in Fig. 3. An extremely high linear flow velocity of 13 cm min⁻¹ produced an eluent pump pressure of 180 atm (2700 lb in⁻²). This is only slightly higher than the pressure from the 8% cross-linked material, yet resulted in an analysis time of only 30 min. Four discrete buffer solutions were employed for this analysis; their formulations are given in Table I. According to Eqs. (2), (3), and (4), it should be possible to achieve analyses in less than 30 min by further reducing resin mean particle diameter and by reducing the resin bed length. These possibilities are discussed in detail elsewhere.[3,7]

Intermediate Pressure Applications. Most amino acid analyzers in current use employ reciprocating piston pumps rated to about 70 atm (1000 lb in⁻²). At flow rates possible below that limit, resin bead distortion does not occur and 8% cross-linked resins can be used without difficulty.

The 9.0-μm diameter, 8% cross-linked, DC-4A resin has been used

[7] J. R. Benson, *in* "Instrumentation in Amino Acid Sequencing" (R. Perham, ed.), p. 1. Academic Press, New York, 1975.

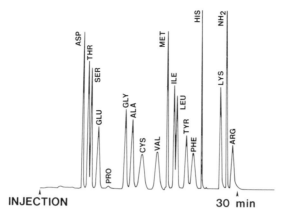

Fig. 3. High-pressure amino acid analysis using 10% cross-linked resin. A mixture containing 10 nmol of each amino acid was applied to a 0.172 × 48-cm bed of Durrum DC-X10-10 cation-exchange resin. This resin is nominally 10% cross-linked and is sized to 10 ± 0.3 µm. Eluent linear flow velocity of 13.0 cm min⁻¹ produced a flow resistance of only ca. 180 atm due to the higher resin cross-linking (cf. Fig. 2). Note also that peak heights are greater than those in Fig. 2. The amino acids spend less time in the resin bed, resulting in diminished peak broadening (see text for discussion); analysis accomplished on a Durrum D-500 Amino Acid Analyzer with photometric detection at 570 nm using ninhydrin. (Data provided by Gary Hewett.)

successfully in conventional instruments, without exceeding eluent pump pressure ratings. Two-hour amino acid analyses have been accomplished on a 0.3 × 30-cm bed; using an eluent linear flow velocity of 2.4 cm min⁻¹, eluent pump pressures of less than 35 atm (500 lb in⁻²) were produced.[7]

A new, smaller-diameter resin was developed in an effort to further reduce analysis time on conventional instruments. A 0.32 × 15-cm bed of Durrum DC-5A was used for this purpose. This is a nominally 8% cross-linked resin sized to 6.0 ± 0.5 µm. At an eluent linear flow velocity of 2.4 cm min⁻¹ an eluent pump pressure of 40 atm (600 lb in⁻²) was produced; this amino acid analysis, completed in 60 min, is illustrated in Fig. 4. Four sequential buffer solutions were employed; their formula-

TABLE I
ELUENT FORMULATION USED FOR 30-MIN AMINO ACID ANALYSIS

Eluent	pH	Na⁺ molarity	Anion	Additive
A	3.15	0.20	Citrate	5% ethanol
B	3.25	0.20	Citrate	None
C	4.25	0.20	Citrate	None
D	9.45	0.20	Borate	None

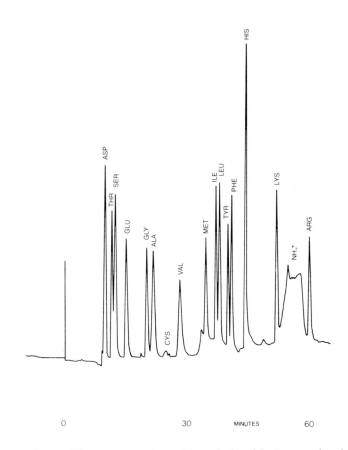

Fɪɢ. 4. Fluorescent amino acid analysis with 6-μm resin. A sample containing 100 pmol of each amino acid was applied to a 0.32 × 15-cm bed of Durrum DC-5A cation-exchange resin. This is an 8% cross-linked resin classified to 6.0 ± 0.5 μm. Eluent linear flow velocity of 2.4 cm min⁻¹ produced a flow resistance of ca. 40 atm (600 lb in⁻²). At these relatively low-flow velocities, resin cross-linkage is not important; the fast analysis time is due primarily to the short resin bed length. Column temperature was maintained at 45°, then changed to 65° after 22 min. Four buffer solutions were sequentially supplied to the column as follows: A, 25 min; B, 8 min; C, 4 min; D, 28 min. Buffer solution formulations are given in Table II. The microbore chromatograph was constructed entirely from components of Durrum's Amino Acid Analyzer Kit (see text footnote 9). Detection was accomplished with a filter fluorometer (American Instrument Company) using a Corning 7-60 primary filter and a Wratten 2E secondary filter. The recorder (Honeywell) sensitivity was set to 10 mV, full-scale deflection. Fluorogenic reagent was a 4% solution of Fluoropa Premix (Durrum) in 0.40 M potassium borate buffer solution, pH 10.4.

TABLE II
ELUENT FORMULATION USED FOR 60-MIN AMINO ACID ANALYSIS[a]

Eluent	pH	Na^+ molarity	Buffer or salt	Grams liter^{-1}
A	3.25	0.20	Sodium citrate $\cdot 2H_2O$	19.6
B	4.25	0.20	Sodium citrate $\cdot 2H_2O$	19.6
C	5.25	0.20	Sodium citrate $\cdot 2H_2O$	19.6
D	10.0	0.20	Boric acid	3.0
			NaOH	2.0
			NaCl	8.76

[a] All buffer solutions contain phenol, 1.0 g liter^{-1}.

tion is given in Table II. This method, utilizing a sequence of eluents of constant cation molarity, was first described by Hare.[8] The instrument used for automatically analyzing amino acids was a modified Durrum Amino Acid/Peptide Analyzer Kit.[9] Amino acid fluorophors were formed by reaction of column effluents with a 4% solution of "Fluoropa Premix" (Durrum) in a potassium borate buffer solution. Fluoropa Premix contains a highly purified, fluorogenic grade of o-phthalaldehyde in a specially stabilized concentrate that upon dilution with a suitable buffer solution, produces a final reagent similar to that described by Roth[10] and by Benson and Hare[11] for the fluorometric detection of primary amines.

Only 100 pmol of each amino acid were contained in the sample analyzed (Fig. 4). Exceptionally high sensitivity was achieved by using microbore chromatography techniques coupled with high-sensitivity fluorescent detection.[7] Sensitivity is also enhanced because 6-μm diameter resin beads allow resin bed length to be reduced to 15 cm while still providing adequate separation of amino acids.

Peptide Analysis

Automatic amino acid sequencing of intact protein molecules is not yet a reality so that primary structure must be determined by sequencing peptide fragments obtained by chemical or enzymic cleavage of purified proteins. Consequently, separation and purification of these fragments is a problem common to all protein laboratories. At this

[8] P. E. Hare, *Space Life Sci.* 3, 354 (1972).

[9] J. Reiland and J. R. Benson, *Durrum Resin Report No. 7*, Durrum Chemical Corp., Palo Alto, CA, 1976.

[10] M. Roth, *Anal. Chem.* 43, 880 (1971).

[11] J. R. Benson and P. E. Hare, *Proc. Natl. Acad. Sci. U.S.A.* 72, 619 (1975).

TABLE III
ANION-EXCHANGE RESINS HAVING NARROW PARTICLE-SIZE DISTRIBUTIONS[a]

Resin designation	Cross-linking (% divinylbenzene)	Particle size (μm)
DA-X12-11	12	11 ± 1
DA-X12-8	12	8 ± 1
DA-X8-11	8	11 ± 1
DA-X8-8	8	8 ± 1
DA-X4-11	4	11 ± 1
DA-X2-11	2	11 ± 1

[a] All resins are available from Durrum Chemical Corporation.

writing, the most powerful method of achieving preparative, high-resolution separation of peptide mixtures is ion-exchange chromatography. In this regard, recent improvements in ion-exchange resins augur well for protein chemists.

Because a multiplicity of possible interactions exists between the resin matrix and hydrophilic and hydrophobic moities, an environment with enormous potential for separating complex mixtures is created. Ion-exchange resins are also advantageous because separation methods can be easily scaled up to accommodate larger samples. A separation for a given peptide mixture can be developed using microbore chromatography techniques requiring only a small amount of sample material. Once the protocol is defined, large-diameter columns can be employed for preparative separations.

Recently, interest in separating and detecting peptide fragments from microgram quantities of enzymically digested proteins has emerged. If structural characterization of proteins available only in small amounts is to be realized, higher sensitivity detection methods must be developed. Use of small diameter resin particles in microbore chromatographs, coupled with the highly sensitive fluorescence assay method previously described, is one way of achieving this goal. One such procedure, using 9-μm diameter resin in a 0.20×25-cm bed, has been reported.[12] A discussion of micro methods for separating peptides using ninhydrin detection is found elsewhere in this volume [24, 25].

Traditionally, peptide separations have been performed on two types of ion-exchange resins: cation exchangers for separation of neutral and basic peptides and anion exchangers for separation of acidic and neutral peptides.

Anion-Exchange Resins. Several new anion-exchange resins have

[12] J. R. Benson, *Anal. Biochem.* **71,** 459 (1976).

TABLE IV
Eluent Formulation for Peptide Separations[a]

Eluent	pH	Buffer or salt	Grams liter^{-1}	Titrate with
A	4.25	Sodium citrate ·2H$_2$O	19.6	HCl
B	5.25	Sodium citrate ·2H$_2$O	19.6	HCl
C	6.25	Sodium citrate ·2H$_2$O	19.6	HCl
D	7.20	Sodium hydroxide	8.0	Phosphoric acid
E	8.20	Sodium hydroxide	8.0	Phosphoric acid
F	10.0	Boric acid	3.0	Not required
		NaOH	2.0	
		NaCl	8.76	

[a] All buffer solutions contain phenol, 1.0 g liter.

been shown to be particularly useful for separating mixtures of peptides. An 8% cross-linked resin sized to 11 ± 1 μm (Durrum DA-X8-11) has been used by Herman and Vanaman to separate peptides from a tryptic digest of bovine brain modulator protein (this volume [25]); they present a chromatogram showing exceptional resolution. A resin with identical composition but sized to 8 ± 1 μm (Durrum DA-X8-8) has not yet been used for peptide separations, but because of the smaller-diameter particles, should yield even greater resolution. Closely classified anion-exchange resins with several degrees of cross-linking are also available. A study of cross-linking effects on peptide separations will be published elsewhere.[13] Table III lists commercially available anion-exchange resins with exceptionally narrow particle-size distribution.

Cation-Exchange Resins. Cation-exchange resins nominally 8% cross-linked, used successfully in amino acid analysis, have also proved to be useful in peptide separations. Microbore procedures using photometric detection with ninhydrin detection have been described,[14] as has a fluorometric detection method using o-phthalaldehyde.[12] In pursuit of still greater sensitivity and shorter analysis times, a cation-exchange resin with a mean particle diameter of only 6.0 ± 0.5 μm was used for separation of peptide mixtures. This resin, Durrum DC-5A, is the same as that used in the amino acid analysis illustrated in Fig. 4.

Tryptic peptides from S-aminoethylated human globin were applied to a 0.32×15-cm bed of DC-5A; sample preparation is described elsewhere.[12] The chromatograph was identical to that used for the amino acid analysis shown in Fig. 4; eluent linear flow velocity was 2.4 cm min^{-1}. Six buffer solutions with constant sodium molarity were sequentially applied to the resin bed; their composition is indicated in Table IV. Fluorometric detection with 4% Fluoropa Premix in 0.40 M po-

[13] R. A. Bradshaw and J. R. Benson, in preparation.
[14] A. C. Herman and T. C. Vanaman, *Anal. Biochem.* **63**, 550 (1975).

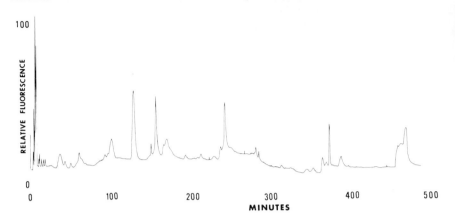

FIG. 5. Fluorescent peptide map of trypsin digested S-aminoethyl (SAE) globin. A sample containing tryptic peptides from 20 μg of SAE human globin was applied to a 0.32 × 15-cm bed of Durrum DC-5A resin, the same as that used for the analysis depicted in Fig. 4. Eluent linear flow velocity of 2.4 cm min⁻¹ produced a flow resistance of ca. 35 atm with column temperature maintained at 57°. The separation depicted was achieved by sequentially supplying six buffer solutions to the resin bed as follows: A, 135 min; B, 90 min; C, 90 min; D, 45 min; E, 90 min; F, 45 min. The formulations for these solutions are given in Table IV. Chromatograph and fluorescent detection method are identical to that indicated in Fig. 4.

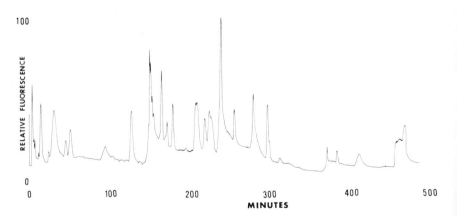

FIG. 6. Fluorescent peptide map of trypsin-digested bovine serum albumin. A sample containing tryptic peptides from 50 μg of bovine serum albumin was analyzed in an identical manner to that described for the S-aminoethyl globin illustrated in Fig. 5. Small mean-particle-diameter resin used in a short resin bed permits rapid analyses. High sensitivity is achieved by using microbore chromatography techniques coupled with fluorescent detection methods.

tassium borate solution, pH 10.4, was used, with results as shown in Fig. 5. A similar analysis of tryptic peptides from bovine serum albumin is shown in Fig. 6. Only picomole quantities of protein were required for these analyses. The short, 15-cm resin bed provided sufficient resolution, yet allowed relatively rapid separation. A summary of commercially available cation-exchange resins of narrow particle-size distribution is presented in Table V. A variety of mean particle diameters is available for analytical and preparative separations.

TABLE V
CATION-EXCHANGE RESINS FOR PEPTIDE SEPARATIONS[a]

Resin designation	Particle size (μm)	Application
DC-1A	14 ± 2	Macro (preparative)
DC-6A	11 ± 1	Macro or micro
DC-4A	9 ± 0.5	Micro
DC-5A	6 ± 0.5	Micro

[a] All resins are nominally 8% cross-linked. Available from Durrum Chemical Corporation.

[3] Internal Standards for Amino Acid Analysis

By JAMES F. RIORDAN and ROGER W. GIESE

The general purpose of internal standards is to monitor physical and chemical losses and variations during analysis of amino acids by ion-exchange and gas–liquid chromatography. The standard obviously must not interfere with the analysis (e.g., by coeluting with any of the amino acids in the sample), but it can monitor the analysis only to the degree that its characteristics match those of the sample. Thus, an amino acid standard is more similar to a mixture of free amino acids than to these same amino acids combined in a protein. This gives rise to certain shortcomings in the sample monitoring ability of an internal standard. Chemical steps tend to be especially susceptible to such limitations; no internal standard can behave chemically the same as all of the amino acids from a protein during acid hydrolysis, because the rates of release and stabilities of the amino acids vary. Thus, except for monitoring amino acids of equal stability and completely released by the hydrolytic conditions employed, an internal standard in the hydrolysis mixture serves primarily to expose and correct for mechanical losses.

Other properties of the internal standard requiring special attention are stability and means of detection. These are more critical in gas–liquid chromatography of amino acids, where greater use has been made of precolumn derivatization, high column temperatures, and non-amino acid internal standards.

Column parameters such as length, temperature, and pH can be quite important in regard to the selection and use of an internal standard in ion-exchange analysis of amino acids. Almost without exception, internal standards appropriate for the long column (150 cm), used for analysis of neutral and acidic amino acids, will not be of value on the short columns (5 to 50 cm) used for basic amino acids. Further, an internal standard that is effective on the 50-cm basic column for physiological analyses may be ineffective on the 5- to 15-cm basic columns employed for protein hydrolyzates. The transition to single-column analyzers now taking place will bring more discrepancies of this type. At the same time, some internal standards will be found to be much more subject to overlap problems with minor changes in pH and temperature than others, or the standards may be resolved only when a slightly longer version of a given column is used. Thus, for each internal standard the particular column and conditions reported must be noted. Unfortunately, this information is not always available for many of the internal standards that have been proposed. Moreover, there are conflicting claims in the literature regarding the stability and resolution of some of the internal standards. The potential user therefore should be cautious, and not assume that the mere listing, commercial or otherwise, of a substance as an internal standard for amino acid analysis assures its success in all cases.

Internal Standards for Ion-Exchange Amino Acid Analysis

Shortly after the automatic determination of amino acids was described by Spackman et al.[1] and analyzers became available commercially it was recognized that internal standards would be useful for improving the accuracy of the technique. Siegel and Roach[2] first pointed out the advantage of an internal standard in compensating for day-to-day variability in the color yield of the ninhydrin reaction. They defined criteria for identifying an ideal standard and suggested β-2-thienyl-DL-alanine for this purpose. Later, Walsh and Brown[3] introduced norleucine as a standard for monitoring mechanical losses that might occur on

[1] D. H. Spackman, W. H. Stein, and S. Moore, Anal. Chem. 30, 1190 (1958).
[2] F. L. Siegel and M. K. Roach, Anal. Chem. 33, 1628 (1961).
[3] K. A. Walsh and J. R. Brown, Biochim. Biophys. Acta 58, 596 (1962).

preparing protein hydrolyzates. Improvements in the design and operation of amino acid analyzers as well as other factors have led to recommendations of additional standards, many of which are now used routinely.

Principle

An accurately known amount of the internal standard can be added to a protein solution prior to acid hydrolysis. After hydrolysis the solution is taken to dryness, dissolved in a measured volume of citrate buffer, and an aliquot is introduced into the analyzer. The recovery of the internal standard gives the general recovery of the amino acids of the protein solution. For example, an 80% recovery of internal standard indicates that the original sample contained 100/80 times the observed yield (direct or extrapolated) of amino acids. One special application of this technique is to determine the concentration of a pure protein solution. The minimal dry weight of the protein (non-amino acid components are excluded) can be calculated by summing the weights of the individual amino acid residues (minus one water molecule for each peptide bond).[3]

Internal standards can also be added to the mixture of amino acids after hydrolysis. When this approach is employed with two-column analyzers, a different standard may be required for each column. The recovery of each standard allows compensation for any differences in sample application, column yield, and detection efficiency (especially due to differences in ninhydrin stability and pumping rates) between the two columns.

Criteria for Internal Standards

An internal standard for use with protein hydrolysates and ninhydrin detection should be an α-amino acid that does not occur naturally in protein hydrolysates. Its color yield with ninhydrin should be similar to that of most amino acids, should be evaluated on the instrument under the operating conditions scheduled for its use,[1] and should be linear with concentration. It should elute from the analyzer quantitatively and in a unique position not overlapping any of the amino acids. However, its elution should not be so delayed as to unduly extend the time of analysis. Ideally its elution position should be insensitive to minor changes in the pH or temperature of the eluting buffer. If the standard is to be added to the protein solution prior to hydrolysis, it should be stable for at least 72 hr in 6 N HCl at 110° *in vacuo* or under whatever conditions are employed for this purpose. Standards to be used in analyses of physiological fluids must not overlap any of the ninhydrin-reactive substances in such samples.

Most of the internal standards that have been suggested to date have been intended for two-column analyzers. Changing to single-column procedures should eliminate the use of internal standard pairs except to

SUBSTANCES TESTED AS INTERNAL STANDARDS FOR AMINO ACID ANALYSIS[a]

Compound	References[b]
A. *Ion-exchange chromatography*	
α-Aminobutyric acid	3
ε-Aminocaproic acid	4
L-2-Amino-4-guanidinobutyric acid	5
L-2-Amino-3-guanidinopropionic acid	3
3-Aminotyrosine	3, 6
Citrulline	3, 7
Cysteic acid	8
L-2,4-Diaminobutyric acid	9
Diaminopimelic acid	3
Glucosamine	10
Homoarginine	11
Homocitrulline	12
Hydroxylysine	3
Isovaline	3
Nitrotyrosine	13
Norleucine	3
Norvaline	3
S-β-(4-Pyridylethyl)-L-cysteine	14
S-β-(4-Pyridylethyl)-DL-penicillamine	15
Taurine	3, 10
β-(2-Thienyl)-DL-alanine	2
β-(2-Thienyl)-DL-serine	16
B. *Gas–liquid chromatography*	
Acetyl mercaptoethanol	17, 18
n-Butyl stearate	19–21
Decanoic acid	22
Dimethyldodecanedioate	18, 23
Fluorene	22, 24
Homoserine	18, 25
Hydroxyproline	18, 25
p-Methylphenylalanine	18, 25
Norleucine	17, 20, 26–28
Ornithine	20
Phenanthrene	22, 24
Tranexamic acid	21

[a] Not all of these compounds were successful. See text for details.
[b] Numbers refer to text footnotes.

monitor intrarun variations or other special circumstances. In addition, standards that elute after phenylalanine from a typical long column become even more useful if they precede the basic amino acids on a single column.

Although not yet widely employed, reagents other than ninhydrin (like fluorescamine and o-phthalaldehyde) have been introduced in the past few years to detect amino acids in ion-exchange column effluents. Use of the internal standards described in this report in analyzers equipped with such detection methods has yet to be reported. Hence their suitability for this purpose remains to be established.

Characteristics of Specific Standards

More than 20 different amino acids or amino acid analogs have been tested as internal standards 4–28 (see the tabe). The first compound

---

[4] L. S. Bates, *Anal. Biochem.* **41**, 158 (1971).

[5] T. Gerritsen, M. L. Rehberg, and H. A. Waisman, *Anal. Biochem.* **11**, 460 (1965).

[6] M. Sokolovsky, J. F. Riordan, and B. L. Vallee, *Biochem. Biophys. Res. Commun.* **27**, 20 (1967).

[7] M. C. Corfield and A. Robson, *Biochem. J.* **84**, 146 (1962).

[8] A. Wainer and J. S. King, Jr., *J. Chromatogr.* **20**, 143 (1965).

[9] C. Dennison, *J. Chromatogr.* **63**, 409 (1971).

[10] J. E. Purdie and D. B. Smith, *Can. J. Biochem.* **43**, 49 (1965).

[11] H. T. Keutmann and J. T. Potts, Jr., *Anal. Biochem.* **29**, 175 (1969).

[12] T. Gerritsen, M. L. Rehberg, and H. A. Waisman, *Anal. Biochem.* **11**, 460 (1965).

[13] R. W. Giese and J. F. Riordan, *Anal. Biochem.* **64**, 588 (1975).

[14] J. F. Cavins and M. Friedman, *Anal. Biochem.* **35**, 489 (1970).

[15] M. Friedman, A. T. Noma, and M. S. Masri, *Anal. Biochem.* **51**, 280 (1973).

[16] Pierce Chemical Co., General Catalog 1976–1977, p. 51.

[17] J. R. Coulter and C. S. Hann, Gas chromatography of amino acids, in "New Techniques in Amino Acid, Peptide, and Protein Analysis," (A. Niederwieser and G. Pataki, eds.), p. 75. Ann Arbor Sci. Publ., Ann Arbor, Michigan, 1971.

[18] A. Niederwieser, Chromatography of amino acids and oligopeptides, in "Chromatography," 3rd ed. (E. Heftmann, ed.), p. 393. Van Nostrand-Reinhold, Princeton, New Jersey, 1975.

[19] C. W. Gehrke and D. L. Stalling, *Sep. Sci.* **2**, 101 (1967).

[20] C. W. Gehrke, D. Roach, R. W. Zumwalt, D. L. Stalling, and L. L. Wall, "Quantitative Gas–Liquid Chromatography of Amino Acids in Proteins and Biological Substances," Analytical Biochemistry Laboratories, Columbia, Missouri, 1968.

[21] C. W. Gehrke, K. Kuo, and R. W. Zumwalt, *J. Chromatog.* **57**, 209 (1971).

[22] C. W. Gehrke and K. Leimer, *J. Chromatogr.* **57**, 219 (1971).

[23] M. Gee, *Anal. Chem.* **39**, 1677 (1967).

[24] C. W. Gehrke, H. Nakamoto, and R. W. Zumwalt, *J. Chromatogr.* **45**, 24 (1969).

[25] R. Pocklington, *Anal. Biochem.* **45**, 403 (1972).

[26] P. Husek and K. Macek, *J. Chromatogr.* **113**, 139 (1975).

[27] A. Darbre and A. Islam, *Biochem. J.* **106**, 923 (1968).

[28] J. Jonsson, J. Eyem, and J. Sjoquist, *Anal. Biochem.* **51**, 204 (1973).

to be proposed for this purpose was β-2-thienyl-DL-alanine.[2] Its major advantage was that it could be used on both the 50- and 150-cm columns of the original Spinco Model 120 amino acid analyzer. It elutes between leucine and tyrosine on the 150-cm column and before the combined tyrosine–phenylalanine peak on the 50-cm column. Studies on the acid hydrolysis stability of β-2-thienyl-DL-alanine have not been carried out, and its use seems to have been restricted to monitoring the daily reliability of the ninhydrin reagent or for correcting differences between long and short column runs.[29] For these purposes it is added to the protein sample after acid hydrolysis, where its recovery is better than 98%. It has been incorporated along with L-2-amino-3-guanidinopropionate, another internal standard, in the 0.2 N sodium citrate buffer, pH 2.2, used to take up samples after hydrolysis and drying.[30] Although blood plasma and protein hydrolysates contain no interfering substances, a coeluting substance occurs in urine and liver extracts. Also, 3,4-dihydroxyphenylalanine (DOPA) overlaps with β-thienylalanine.[2]

The standard most frequently used to correct for sample losses occurring during or after hydrolysis is norleucine.[3] It has been dispensed as a solution containing 10 μmol of norleucine per milliliter of 0.1 N HCl.[29] Its color yield with ninhydrin is the same[3] or 1.01 times[31] that of leucine on the Spinco Model 120 analyzer. It is completely stable during 72 hr in constant-boiling HCl at 110°[18] or during 144 hr in 6 N HCl at 105°,[3] and it has been added prior to performic acid oxidation.[31] It elutes after leucine and before tyrosine, generally without overlap in older analyzers but tending to coelute somewhat with leucine when the newer high pressure methods are employed.[13] It has been considered to have the disadvantage of late elution, thus delaying the quantitation of rapidly emerging substances, particularly in analyses of physiological fluids.[8]

Recently, 3-nitrotyrosine has been proposed as an internal standard for amino acid analysis.[13] Like norleucine, it is stable to acid hydrolysis in the presence of added protein for at least 96 hr at 110°. Its ninhydrin color yield is 1.01 relative to 1.02 for valine on the Spinco 120C analyzer, and it obeys Beer's law by manual analysis. It elutes just after, but well resolved from, phenylalanine and it displays visible color. This latter characteristic allows quantitation of stock solutions by absorbance measurements and also visible monitoring of amino acid mixtures in pro-

[29] J. P. Bargetzi, K. S. V. Sampath Kumar, D. J. Cox, K. A. Walsh, and H. Neurath, *Biochemistry* 2, 1468 (1963).
[30] J. W. Prahl and H. Neurath, *Biochemistry* 5, 2131 (1966).
[31] M. J. Crumpton and J. M. Wilkinson, *Biochem. J.* 88, 228 (1963).

cedures preceding acid hydrolysis, for example, gel filtration. Commercial samples of nitrotyrosine contain impurities, but these can be removed by chromatography on Bio-Gel P-6. A typical purification starts with 150 mg of 3-nitrotyrosine dissolved in 2 ml of 1 N NaOH. The solution is brought rapidly to pH 2 by addition of 5 N HCl, any precipitate is removed by filtration (Whatman No. 1 paper), and the clear filtrate is applied to a 1.5 $\times$ 25-cm column of Bio-Gel P-6 equilibrated at room temperature with 0.01 N HCl. Most of the yellow nitrotyrosine elutes far ahead of the dark-colored impurities. It can be crystallized as yellow needles by adjusting the pH to 5 and reducing the solution volume with a slow stream of air. High-purity product has a maximum absorption at 356 nm (ϵ2910) in 0.01 N HCl and at 428 nm (ϵ4390) in 0.015 N NaOH.[13] It has been noted that while the elution position of 3-nitrotyrosine does not interfere with that of any amino acid found in protein hydrolysates, certain protected residues frequently employed in peptide synthesis (benzylaspartate, benzylcysteine) do elute in the same region, and adequate precautions should be taken when such residues might be present.[13] Such caution also pertains to other internal standards. Indiscriminate use of any internal standard risks masking of uncommon amino acids and amino acid derivatives. Generally speaking, the internal standards have not been thoroughly evaluated from this point of view.

Walsh and Brown[3] suggested L-2-amino-3-guanidinopropionate as the counterpart to β-2-thienyl-DL-alanine in monitoring differences between long and short-column analyses. This former substance has been shown to be rather unstable to acid hydrolysis[31] and hence should not be added prior to this step. One report said that it should not be used at all, since its recovery averaged 81.4 and 87.6% on a Spinco 120 C analyzer for additions before and after hydrolysis, respectively, whether it was analyzed in the presence or absence of cereal material.[4] It elutes between ammonia and arginine both on the short column of two-column analyzers and on single-column analyzers. Certain mixed-bed resins that are used for preparing deionized water have been observed to introduce a ninhydrin reactive substance that coelutes with L-amino-3-guanido-propionate.[32] This material can be removed by filtering the water through activated charcoal.

Another standard suggested for use on a column separating basic amino acids is S-β-(4-pyridylethyl)-L-cysteine.[14] It elutes after ammonia on the short column of the Spinco Model 120 analyzer, but it may not be completely resolved from arginine. On the Durrum Model

---

[32] L. Cueni, unpublished observations.

500 it elutes well ahead of arginine, but significantly overlaps the ammonia peak. The elution position is critically dependent on the resin batch and pH of the eluting buffer. Pyridylethylcysteine has a ninhydrin color value on the Spinco Model 120 of 1.02 relative to 1.00 for leucine that follows Beer's law by a manual ninhydrin method. It is stable even after 120 hr in constant-boiling HCl and therefore can be added to the protein sample prior to acid hydrolysis. If used together with norleucine it provides simultaneous monitoring of precolumn handling errors and intercolumn yield and detection differences. It is unstable to performic acid.[33]

Friedman and co-workers[15] later prepared another vinylpyridine derivative, S-β-(4-pyridylethyl)-DL-penicillamine, which they found to be much less sensitive to small changes in pH of the citrate buffer and hence a better standard than the corresponding cysteine derivative. It is stable to acid hydrolysis, its ninhydrin color is linear with concentration, and, of importance, it elutes after arginine on both Beckman (Spinco) and Durrum analyzers, thus avoiding any overlap with ammonia. However, it does coelute with the artifact introduced by a mixed-bed ion exchange resin (see above). Since pyridylethylpenicillamine is aromatic it can, under certain circumstances, be assayed in the hydrolysate by ultraviolet spectroscopy ($\epsilon 259 = 2400$).[34] It is not particularly well suited for use with physiological fluids, since it emerges about 1 hr later than arginine and hence delays analysis.

These same workers also examined the potential of several other substituted cysteines and penicillamines for use as internal standards on basic columns. The 2-pyridylethyl, p-nitrobenzyl, and p-nitrophenethyl derivatives generally failed to elute in unique positions. However, some of these compounds might be applicable to physiological analysis of basic amino acids on a 50-cm column, where the resolution is quite different from that on a normal short column.

Several additional substances have been examined as basic column internal standards, and some of these may have an advantage in specific instances. Keutmann and Potts[11] used homoarginine, which they found to be stable to acid hydrolysis (100% recovery after 24 or 48 hr; 96% recovery after 72 or 96 hr) and to elute after arginine on a 5-cm basic column. However, Bates[4] reported that the recovery of homoarginine from hydrolyzed samples (24 hr of heating) varied from 91.6 to 96.5% and considered this, together with the extra 20 min required for elution, sufficient to eliminate homoarginine as a practical basic column stan-

---

[33] K. J. Stevenson, *Anal. Biochem.* **56**, 450 (1973).
[34] Pierce Chemical Co., Pierce Previews, April, 1973.

dard. A more acceptable choice may be ϵ-aminocaproic acid.[4] It elutes just ahead of lysine with slight overlap, and it is stable to acid hydrolysis. Recoveries ranged from 97 to 105% with a mean of 99.7%. A sample of 0.25 μmol on the column was found to be most practical. Another possibility is L-2,4-diaminobutyric acid, which elutes between lysine and histidine and hence necessitates using a slightly longer short column in order to avoid overlap on both sides. Its early elution was exploited by running two short columns at the same time, one on wash whenever the other was on monitor, thus doubling the sample throughput.[9]

Gerritsen et al.[5] have used homocitrulline and 2-amino-4-guanidino-butyric acid as internal standards for analysis of amino acids in serum. The solution used contained 1 μmol of homocitrulline and 1 μmol of 2-amino-4-guanidinobuytric acid per milliliter of citrate buffer, pH 2.2 or pH 1.5. Homocitrulline elutes just before the buffer change on the long column and the aminoguanidinobutyric acid appears between tryptophan and arginine on the 50-cm column. The elution positions of these compounds were very sensitive to slight changes in pH, and the compounds were not examined for acid hydrolysis stability. Both were recovered in essentially quantitative yield and were useful in evaluating methods for determining free amino acids in serum.

Hydroxylysine and 3-aminotyrosine both elute ahead of lysine on the 15-cm short column of the Spinco Model 120 analyzer but were considered less than ideal candidates as internal standards since tryptophan breakdown products or tyrosine chlorination products elute in the same region.[3]

Another long-column standard is cysteic acid, which can be applied to the column before the sample.[8,35] After 15 min of column operation, the column is reopened and the sample to be analyzed is introduced. The standard then emerges about 5 min ahead of aspartic acid, devoid of overlap. The color yield of cysteic acid is the same as that of leucine.[36]

Taurine has been used on occasion,[3,10] but its concentration appears to increase slightly on acid hydrolysis, a less than desirable characteristic. Other substances that have been examined, but found lacking on the Spinco Model 120 analyzer, include norvaline and isovaline which coelute with methionine, α-aminobutyric acid which overlaps cystine, diaminopimelic acid which overlaps alloisoleucine, and citrulline which coelutes with proline.[3] Citrulline has been used as a standard in the

[35] A. Wainer and J. S. King, Jr., Automat. Anal. Chem. Technicon Symp. 1965, p. 698 (1966).
[36] P. B. Hamilton and R. A. Anderson, Anal. Chem. 31, 1504 (1959).

automated analysis of amino acids using polarographic quantitation of their copper complexes.[7] Elution was carried out in a chloroacetate–acetate buffer system and under these conditions citrulline appeared between glutamic acid and proline. Glucosamine eluted between tyrosine and lysine on a 20-cm column but together with phenylalanine on the 150-cm column.[10] Finally, β-(2-thienyl)-DL-serine has been suggested as a standard,[16] but its suitability has yet to be established.

Internal Standards for Gas–Liquid Chromatography

The gas–liquid chromatography of amino acids has long offered promise of becoming a useful method of analysis.[27] The method, which recently has been reviewed,[26] requires small sample sizes and is fast and accurate. It can also be aided by internal standards and the same amino acids can be used as in ion-exchange chromatography except for most of those bearing a polar guanidino or carbamino group.[18,20,26] The particular subject of internal standards in amino acid analysis by GLC has been reviewed briefly,[18] the need for determining and using individual response factors has been cited,[26] and some of the substances that have been tested or used explicitly as internal standards in this regard are summarized in the table.

[4] Complete Hydrolysis of Polypeptides with Insolubilized Enzymes

By GARFIELD P. ROYER, WARREN E. SCHWARTZ, and FREDERICK A. LIBERATORE

Polypeptides and proteins contain amino acids and amino acid derivatives that are lost during treatment with strong acid or base prior to amino acid analysis. Typically, hydrolysis is performed with 6 N HCl at 110° in a sealed tube for 24 hr. This treatment results in the loss of tryptophan, asparagine, glutamine, glycosides, carboxylate esters, sulfate esters, and phosphate esters. Hydrolysis of proteins with *soluble* enzymes under mild conditions has long been recognized as an approach to this problem.[1-3] The use of immobilized enzymes[4] for peptide hydrolysis has many advantages over the use of soluble enzymes: the enzymes may be reused, no autolysis occurs, backgrounds are lower, the digest is readily removable from the enzyme, high enzyme concentrations may be

[1] R. L. Hill and W. R. Schmidt, *J. Biol. Chem.* **237**, 389 (1962).

[2] R. L. Hill, *Adv. Protein Chem.* **23**, 63 (1965).

[3] H. T. Keutmann, J. A. Parsons, J. T. Potts, Jr., and R. J. Schlueter, *J. Biol. Chem.* **245**, 1491 (1970).

[4] G. P. Royer and J. P. Andrews, *J. Biol. Chem.* **248**, 1807 (1973).

used, and finally, a process based on bound enzymes is readily automatable.

What combination of bound enzymes is necessary for complete hydrolysis? The answer depends, of course, on the nature of the polypeptide substrate. We showed that aminopeptidase M (APM) bound to porous glass was the only enzyme necessary for total digestion of the aminoethylated A chain of bovine insulin.[4] The B chain of insulin is not totally hydrolyzed by bound APM because its C terminus is Thr-Pro-Lys-Ala. This observation is consistent with the known specificity of soluble APM, i.e., X-Pro bonds are hydrolyzed slowly or not at all. We have subsequently shown that the C terminus of the B chain of insulin is readily digested by immobilized carboxypeptidase Y (CPY). We feel that the best general approach for an unknown polypeptide entails initial fragmentation by endopeptidases followed by treatment with bound CPY, APM, Pronase, and prolidase. This approach is illustrated with the dodecapeptide Leu-Gly-Pro-Trp-Val-Arg-Gly-Glu-Ala-Pro-Ile-Lys.[5]

The solubility of the peptide is an important consideration for the selection of the enzymes. A polypeptide that is soluble at pH 8.0 could be digested with a mixture of bound Pronase, bound prolidase, and bound APM. If the peptide is soluble in the pH range of 5–6, a mixture of bound papain and bound CPY would be used. Subsequently, treatment with bound prolidase and APM would be carried out at pH 8.0. Nearly all polypeptides are soluble in the pH range of 1–2, where bound pepsin may be used. An apolar polypeptide, insoluble in the pH range of 1–8, could be solubilized with urea or dioxane solutions. As an example we have digested N-blocked, insoluble peptides with bound CPY in 20% dioxane.

Procedures

Preparation of Bound Enzymes

Support Materials. The most convenient and versatile supports for this study are dextran-coated porous glass (Pierce Chemical Company) and CL-Sepharose 4-B (Pharmacia).[6] The CL-Sepharose may be stirred with a magnetic stirring bar. The glass is very friable and is best used in a column or rotated slowly in a tube.

---

[5] This "model peptide" was the generous gift of Drs. Hugh D. Niall and G. W. Tregear. The peptide will soon be available commercially from Pierce Chem. Co.
[6] Cross-linked agarose. This material has good dimensional stability and may be used in columns at reasonable flow rates.

OXIDATION OF SUPPORTS.[7] Both supports are oxidized with 0.01 M NaIO$_4$ (10 ml/g glass or 10 ml/g settled gel) for 2 hr at room temperature. When dextran-coated porous glass is used, a vacuum should be applied to release trapped air at the start of the reaction. This may be done with a filter bell or sidearm flask connected to a water aspirator.

PREPARATION OF HEXAMETHYLENEDIAMINE-CL-SEPHAROSE.[7] A solution of 0.1 M hexamethylenediamine (HMD) in 0.1 M borate is adjusted to pH 9.0. This solution is mixed with the oxidized gel (2 ml/ml settled gel) with stirring at room temperature. At $t = 20$ and 40 min, NaBH$_4$ (~0.25 mg/ml gel) is added. The total amount of NaBH$_4$ may be weighed and the individual portions approximated. Alternatively, a stock solution of NaBH$_4$ in isopropanol may be prepared and precise portions pipetted into the reaction mixture. Twenty minutes after the second addition of NaBH$_4$, the support is washed with the following: 0.1 M acetic acid, 1% w/v NaHCO$_3$, and distilled water (50 ml/ml settled gel in each case). The washing is done slowly on a Büchner funnel.

CPY IMMOBILIZATION[8,9] HMD-CL-Sepharose was equilibrated with 1 mM pyridine-Cl buffer, pH 4.75. To 50 ml of gel suspension (30 ml settled gel), 5 mg of CPY was added. A total amount of 1-cyclohexyl-3-(2-morpholinoethyl)-carbodiimide-metho-p-toluenesulfonate of 20 mg/mg protein was added in increments over a period of 1 hr at 4°. At the end of an additional 4-hr period the reaction was terminated by the washing procedure used earlier.

IMMOBILIZATION OF PEPSIN. Alkylamine glass (300 mg, Pierce) is degassed and washed with 1 liter of coupling buffer (0.13 M pyridine-HCl, pH 4.5). Pepsin[10] solution (3.5 ml, 2 mg/ml of coupling buffer) is combined with 14 mg of 1-cyclohexyl-3-(2-morpholinoethyl) carbodiimide-metho-p-toluenesulfonate. The pH is readjusted to pH 4.5. Damp alkylamine glass is added and the suspension is gently rotated at room temperature with a constant-torque stirrer motor. The suspension is rotated for an additional hour at 0–4°. The glass-bound enzyme is collected by filtration and washed with 0.5 l of 1 mM HCl, 0.5 l of 0.5 M KCl and 0.5 l of 1 mM HCl.

Peptide Hydrolysis

The dodecapeptide Leu-Gly-Pro-Trp-Val-Arg-Gly-Glu-Ala-Pro-Ile-Lys (0.9 mg) is dissolved in 0.9 ml of 0.1 M N-ethylmorpholine-ace-

[7] G. P. Royer, F. A. Liberatore, and G. M. Green, *Biochem. Biophys. Res. Commun.* **64**, 478 (1975). Conditions for coupling other enzymes appear in Table, p. 44.

[8] F. A. Liberatore, J. E. McIsaac, Jr., and G. P. Royer, *FEBS Lett.* **68**, 45 (1976).

[9] Isolated by the method of R. W. Kuhn, K. A. Walsh, and H. Neurath, *Biochemistry* **13**, 3871 (1974).

[10] Worthington Biochemical Corporation.

tate (NEM-Ac), pH 8.0. Damp trypsin-Sepharose (250 mg) and chymo-
trypsin-Sepharose (250 mg) are added. The digestion is done for 2 hr at
37°. Agitation by magnetic stirrer or rotation is satisfactory. The digest
suspension is then centrifuged at 2000 rpm for 10 min. The Sepharose
pellet is extracted twice with 0.5-ml portions of 0.1 M NEM-Ac, pH 6.0.
The three supernatants are combined with 500 mg of damp CPY-
Sepharose, and the pH is adjusted to 6.0. Digestion is carried out at
50° for 10 hr. The digest is centrifuged and the pellet is washed with
0.1 M NEM-Ac, pH 8.0, as before. To these combined supernatants the
following bound enzymes are added: APM (600 mg, damp), Pronase (500
mg, damp), and prolidase (1 g, damp). The digestion is done at 37° for
12 hr. The digest is treated with additional buffer as described above.
The combined supernatants are lyophilized in two tubes; the residue of
one tube is dissolved in 0.6 ml of citrate buffer, pH 2.2, and 0.5 ml is
applied to an amino acid analyzer column. The remaining sample is
hydrolyzed with 6 N HCl (Pierce) in an evacuated sealed tube at 110°
for 24 hr.

Discussion. PROLINE RELEASE. CPY readily releases proline except
when it is adjacent to glycine. We have shown that our preparation
of bound CPY rapidly releases the proline from the C terminus of some
synthetic peptides and the B chain of insulin. The results from the example described follow.

residues/mole:
Leu -Gly -Pro -Trp-Val -Arg -Gly -Glu -Ala -Pro-Ile-Lys
1.04-0.34-0.32-0.28-1.00-1.00-1.00-0.96-1.00-1.00-1.03-1.08

Clearly some of the tripeptide Gly-Pro-Trp remains at the end
of 24 hr. The mean recovery versus an HCl digest is 84%. Upon longer
digestion with bound CPY, only the dipeptide Gly-Pro remains. Although this resistant dipeptide is fully resolved on the amino acid
analyzer, it may be further digested with bound prolidase.

APPROACH. There are two possible approaches to the determination
of amino acid composition using hydrolysis by immobilized enzymes.
One involves the use of an automatic amino acid analyzer set up for
physiological fluids. Asparagine, glutamine, and serine may be determined directly. Alternatively, the analyzer designed for acid hydrolysates may be employed. Asparagine and glutamine coelute with serine.
It is therefore necessary to run a companion hydrolysis with acid. The
increases in the values of aspartic and glutamic acids over the enzymic
hydrolysate represent the asparagine and glutamine contents. This
second method is more attractive in that cyclization of glutamine is not
a problem and valuable structural information may be obtained. As an
example, evidence for the existence of peptide cross-links or D-amino

¹¹ C. C. Q. Chin and F. Wold, *Anal. Biochem.* **61**, 379 (1974).

PROTEASES BOUND TO OXIDIZED CL-SEPHAROSE AND DEXTRAN-COATED GLASS

Protease	Coupling buffer	Amount of enzyme presented to support	NaBH₄ added	Wash solutions (at least 50 ml/ml gel in each case)
Trypsin[a]	0.1 M borate, pH 9.0, 1 mM in benzamidine	20 mg/10 ml buffer/20 ml Sepharose	2 mg at $t = 20$ and $t = 40$ min	(1) 2.0 M KCl (2) 1% w/v NaHCO₃ (3) Dist. H₂O
α-Chymotrypsin[a]	0.1 M borate, pH 8.75, 1 mM in indole	20 mg/10 ml buffer/20 ml Sepharose	2 mg at $t = 20$ and $t = 40$ min	(1) 2.0 M KCl (2) 1% w/v NaHCO₃ (3) Dist. H₂O
Papain[a]	0.01 M borate, pH 8.5	5 mg/2.5 ml buffer/5 ml Sepharose	0.5 mg at $t = 20$ and $t = 40$ min	(1) 0.01 M borate, pH 8.5, 1 M in NaCl, 1.5 mM in cysteine, 0.25 mM in EDTA (2) Dist. H₂O
Thermolysin[b]	0.01 M borate, pH 8.5	5 mg/2.5 ml buffer/5 ml Sepharose	0.5 mg at $t = 20$ and and $t = 40$ min	(1) 1 mM borate, pH 8.5, 0.1 M in CaCl₂ and 3 × 10⁻⁵ M ZnCl₂ (2) Dist. H₂O
Prolidase[c]	0.01 M borate, pH 8.5, 1 mM in MnCl₂	5 mg/2 ml buffer/5 g Sepharose	0.5 mg at $t = 20$ and $t = 40$ min	(1) 1 mM borate, pH 8.5 (2) 1 M KCl (3) Dist. H₂O
Pronase[b]	0.1 M borate, pH 9.0, 1 mM in CaCl₂	20 mg/10 ml buffer/300 mg oxidized dextran-coated glass[c]	1 mg at $t = 20$ and $t = 40$ min	(1) 250 ml H₂O (2) 250 ml 1 M KCl (3) 250 ml H₂O
Aminopeptidase M[d]	0.1 M borate, pH 8.5, 1 mM in MnCl₂	5 mg/5 ml buffer/300 mg oxidized dextran-coated glass	1 mg at $t = 50$ and at $t = 75$ min	(1) 250 ml H₂O (2) 250 ml 1 M KCl (3) 250 ml H₂O

[a] Worthington Biochemical Corporation.
[b] Calbiochem.
[c] Miles.
[d] Available from Boehringer-Mannheim or Sigma. The Sigma product is sold under the name Leucine Aminopepetidase-Type IV.
The ammonium sulfate suspensions. The ammonium sulfate is removed by dialysis against coupling buffer.

acids may be seen. Chin and Wold[11] showed that the Cys-Cys-Ala-Ser-Val-Cys sequence of insulin was not hydrolyzed by a mixture of bound enzymes. Digestion before and after derivatization of cystine could conceivably aid in placement of disulfide bridges.

APPLICABILITY. The total hydrolysis of large, highly cross-linked proteins using mixtures of bound proteases as described here is not generally feasible. One problem is the inaccessibility of the peptide bonds at or near the cystine residues. There is a second important limitation. Cross-linked, apolar cores are not uncommon in proteins. Such structures may arise during the course of the digestion and precipitate. Enzymic digestion would cease at this point. We hope to be able to reductively cleave disulfide bridges under mild conditions and carry out the digestion of proteins with a high degree of apolar character in urea or water–organic solvent mixtures. It is our opinion that the most immediate application of this method is in peptide synthesis. Side reactions, racemization, and acid-sensitive groups can be detected. The use in total hydrolysis of peptides prior to sequencing is feasible. Detection of "affinity labeled" amino acids is also possible.

[5] High-Performance Liquid Chromatography of Amino Acid Derivatives

By CARL L. ZIMMERMAN and JOHN J. PISANO

In the last decade, liquid chromatographic analyses have been performed with the speed and resolving power characteristic of gas chromatography. Improved column packings, packing techniques, column designs and high-pressure, pulseless pumps are largely responsible for the superior performance. High-performance liquid chromatography (HPLC) features low dead volume, narrow-bore columns, and small, uniform packings (e.g., 5–6 μm). High-performance separations have been achieved by each of the four major separation mechanisms: adsorption, molecular exclusion, ion-exchange, and partition.

Many amino acid derivatives may be easily separated by partition chromatography and determined with high sensitivity using UV or fluorescence detectors. Silica[1-4] and reversed phase[5-9] columns have been

[1] G. Frank and W. Strubert, *Chromatographia* 6, 522 (1973).
[2] A. P. Graffeo, A. Haag, and B. L. Karger, *Anal. Lett.* 6, 505 (1973).
[3] P. Fankhauser, P. Fries, P. Stahala, and M. Brenner, *Helv. Chim. Acta* 57, 271 (1974).
[4] E. W. Mathews, P. G. H. Byfield, and I. MacIntyre, *J. Chromatogr.* 110, 369 (1975).
[5] C. L. Zimmerman, J. J. Pisano, and E. Appella, *Biochem. Biophys. Res. Commun.* 55, 1220 (1973).

especially useful for the analysis of phenylthiohydantoins (PTH) and of 2,4-dinitrophenyl (DNP) and 1-dimethylaminonaphthalene-5-sulfonyl (dansyl) amino acids.

Materials and Method

Several manufacturers offer suitable instruments.[10] A gradient maker is recommended because it shortens the analysis. We have used a DuPont Model 830 Liquid Chromatograph equipped with a DuPont 838 solvent programmer and 254-nm UV detector (DuPont Co., Instrument Products Division, Wilmington, DE 19898). Samples (1–10 μl) were injected through a septum with a Hamilton HP syringe without depressurizing the pump. Sample loop injection valves are an obvious alternative means for introducing the sample.[11]

Zorbax®-ODS columns, 0.46 × 25-cm (DuPont) are operated at 62° with an inlet pressure of 375 psi and an initial flow rate of 1.0 ml/min. Zorbax consists of totally porous silica microspheres (5–6 μm); ODS is the abbreviation for octadecylsilane, which is covalently bonded to the Zorbax.

Acetonitrile, spectral grade "distilled in glass" was purchased from Burdick and Jackson Laboratories (Muskegon, MI). Buffers were prepared with reagent-grade sodium acetate, adjusted to ±0.02 pH unit, and filtered through a 0.2-μm Millipore filter. PTH and dansyl standards were purchased from Pierce Chemical Co. (Rockford, Illinois) and DNP amino acids from the Sigma Chemical Co. (St. Louis, Missouri). Standards were usually dissolved in reagent grade methanol, 1 mg/ml. PTH samples from Edman degradations were evaporated to dryness with a nitrogen stream and dissolved in methanol.

Chromatograms

All 20 PTH's are separated in less than 20 min with the Zorbax-ODS column (Fig. 1). Normally, a gradient analysis can be completed in less than 15 min, since the slowest eluting derivative, PTH-Arg, is

[6] C. L. Zimmerman, E. Appella, and J. J. Pisano, *Anal. Biochem.* **75,** 77 (1976).

[7] C. L. Zimmerman, E. Appella, and J. J. Pisano, *Anal. Biochem.* **77,** 569 (1977).

[8] E. Bayer, E. Grom, B. Kaltenegger, and R. Uhmann, *Anal. Chem.* **48,** 1106 (1976).

[9] H. Yoshida, C. L. Zimmerman, and J. J. Pisano, *in* "Peptides: Chemistry, Structure and Biology, Proceedings of the Fourth American Peptide Symposium" (R. Walter and J. Meienhofer, eds.). Ann Arbor Sci. Publ. Ann Arbor, Michigan, 1975.

[10] H. M. McNair, *J. Chromatogr. Sci.* **14,** 477 (1976).

[11] A description of HPLC equipment may be found in L. R. Snyder and J. J. Kirkland, "Modern Liquid Chromatography," pp. 503–518. Wiley, New York, 1974.

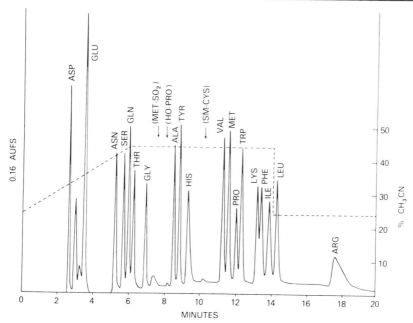

Fig. 1. Gradient elution of phenylthiohydantoins from a 25 × 0.46-cm Zorbax®-octadecylsilane column. Solvent A: 0.01 N sodium acetate, pH 4.5; B: acetonitrile. Pressure, 375 psi; initial flow rate, 1.0 ml/min; temperature 62°; sample size, 1–2 nmol of each PTH. S-Carboxymethylcysteine coelutes with Asp.

water-soluble and is not extracted with the other derivatives at the conversion step of the Edman procedure. Most of the organic-soluble derivatives may be separated in even less time and without a gradient (Fig. 2). In an analysis completed in less than 10 min, only PTH-Ser and Gln are not separated. However, they can be separated under different conditions (Fig. 2A).

Samples from the automated degradation (Beckman Sequencer) of sperm whale myoglobin are free of interfering peaks (Fig. 3) and are easily quantitated. As little as 5 pmol of derivative will give a peak which is 6 times higher than the noise level.[6]

DNP and dansyl amino acids also may be determined with the Zorbax-ODS column (Fig. 4). Although little effort has been directed to this problem, it is apparent that HPLC is a worthy alternative to the TLC and electrophoretic methods used for the analysis of these derivatives.

Comment

Silica columns also have been used for the analysis of PTH's[1-4] and dansyl[8] amino acids. Unfortunately, retention times are difficult to re-

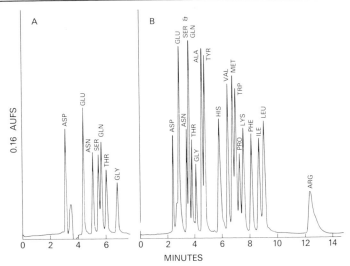

FIG. 2. Isocratic elution of phenylthiohydantoins. Panel A: Separation of early eluting derivatives with CH₃CN:0.01 N sodium acetate, pH 4.5, 24:76 (v/v). Other conditions same as Fig. 1. Panel B: CH₃CN:0.01 N sodium acetate, pH 4.5, 42:58 (v/v), other conditions same as in Fig. 1. PTH-His and -Arg may be analyzed separately using CH₃CN:0.01 N sodium acetate, pH 7.5, 50:50 (v/v).

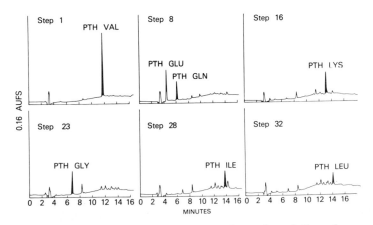

FIG. 3. Selected steps from the automated Edman degradation of sperm whale myoglobin. After the conversion step, the derivative was extracted into ethyl acetate, blown dry with a nitrogen stream, and brought up to standard volume in methanol. From 2 to 5% (2–5 μl) was injected. Step 8 shows the deamidation of PTH-Gln. Steps 28 and 32 show the unequivocal identification of PTH-Leu and -Ile.

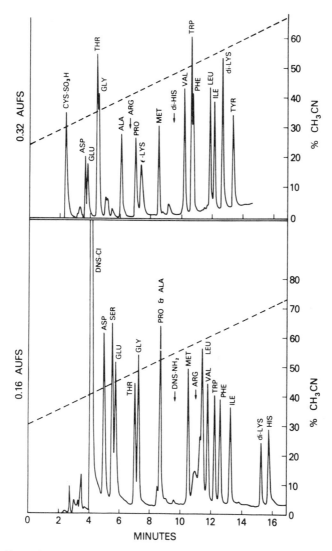

FIG. 4. *Top:* Gradient elution of 2,4-dinitrophenyl (DNP) amino acids. Solvent A: 1% acetic acid; B: acetonitrile. Sample size, ∼1 µg of each derivative. Other conditions same as in Fig. 1. *Bottom:* Gradient elution of dansyl amino acids. Conditions same as those used for DNP amino acids.

produce when gradients are used in which the amount of protic solvent is increased (to elute PTH-Asp and -Glu). Equilibration with less polar starting solvent is difficult. It is preferable to use single solvents and separately analyze the polar and nonpolar PTH's.[1] Elution of PTH-Arg and -His from silica columns has not been reported.

Reversed-phase columns tested from several manufacturers give comparable results. However, the sharpest peaks and best resolution of the PTH's have been obtained with Zorbax-ODS. Previously, using a 0.21 × 25-cm column, we were not able to separate PTH-Gly from Gln and Met from Val; also, PTH-Phe,-Ile, and -Leu were barely resolved.[6] With the 0.46 × 25-cm column, however, all 20 PTH's are separated in half the time (Fig. 1). Resolution is so superior with the wider-diameter column that, when developed without a gradient, only PTH-Ser and -Gly do not separate (Fig. 2B). They may be separated with a different solvent should the need occur (Fig. 2A). Normally, PTH-Gln may be distinguished from Ser because it is partially deaminated and is accompanied by PTH-Glu.

Large-diameter columns (>0.4 cm) previously have been shown to give unexpectedly superior resolution.[12-14] It has been suggested that, when large diameter columns are operated with high flows, solute never contacts the walls through lateral diffusion. Without wall contact, zone spreading is greatly reduced.[12] The superior results with the present wider columns also may be attributed to the 5–6 μm Zorbax used, as it is much easier to prepare stable, homogeneous columns with the material than with the <4 μm particles used in previous columns. Columns have been in regular use for 3 months with no detectable loss in efficiency. Fouled columns may be cleansed by washing overnight with 50:50 (v/v) methanol:chloroform.

Changes in temperature and flow rate can markedly influence column performance. The separation shown in Fig. 2 was significantly decreased when the flow rate was changed ±0.2 ml/min, or the temperature decreased to 50°C. Recent Zorbax-ODS columns have given higher back pressures. The inlet pressure must be 500–1000 psi to obtain a flow rate of 0.8 ml/min. Resolution with the later columns is better for PTH-Phe and -Lys, but Ala and Tyr do not separate as well. It is apparent that each investigator will have to "fine tune" his column by optimizing flow rate first, then column temperature. The gradient shown appears to be optimal for this separation on all columns tested.

The dansyl amino acids have also been well resolved on silica and reversed-phase columns containing a bonded C-8 rather than a C-18 hydrocarbon.[8] Dansyl derivatives have been widely used for end-group analysis of peptides because they are detected with high sensitivity.

[12] J. J. DeStefano and H. C. Beachell, *J. Chromatogr. Sci.* **8**, 434 (1970).
[13] H. C. Beachell and J. J. DeStefano, *J. Chromatogr. Sci.* **10**, 481 (1972).
[14] J. P. Wolf, III, *Anal. Chem.* **45**, 1248 (1973).

However, given the ease and high sensitivity of DNP analysis by HPLC, it would seem that this derivative could regain some of its early popularity for the end-group analysis of proteins, since it is more reactive with sterically hindered end groups.

Speed and resolution notwithstanding, HPLC is an attractive method for the analysis of amino derivatives because they are easily quantitated at the picomole level with a UV detector. Dansyl amino acids may be detected at femtomole levels with a fluorescence detector.[8] Laboratories dealing with radioactive derivatives may be especially attracted to the ease of sample collection.

[6] The Use of Logarithmic Plots of Electrophoretic Mobilities of Peptides

By R. E. OFFORD

Measurements of electrophoretic mobility on paper have been used for a long time to obtain qualitative estimates of the net charge of the migrating molecule (e.g., among many others, the measurements on peptides by Sanger *et al.*[1] and by Ambler[2]). The use of mobilities to give quantitative estimates of charge only became widespread after Offord[3] produced a logarithmic plot of peptide mobility at pH 6.5 versus molecular weight (Fig. 1). A survey of the literature indicates that Fig. 1 has been used to assign side-chain amide groups, or to correct assignments made by other methods, in several hundred sequence determinations. Some method of assigning amides it, of course, essential, since the distinction between asparagine and aspartic acid, and between glutamine and glutamic acid, is lost in the acid-hydrolysis step of both amino acid analysis and end-group determination by the dansyl method. In addition the use of logarithmic plots has been extended to other applications besides amide determination, such as in the discovery of new amino acids, in the location of prosthetic groups, the characterization of active-site sequences, and the characterization of synthetic products. Similar plots have been employed at pH values other than 6.5. It is the purpose of this article to set out the practical and theoretical considerations that have to be borne in mind when making and using mobility measurements, to examine the reliability of the method as a means of amide determination in the light of the data published during the first 10

[1] F. Sanger, E. O. P. Thompson, and R. Kitai, *Biochem. J.* **59**, 509 (1955).
[2] R. P. Ambler, *Biochem. J.* **89**, 349 (1963).
[3] R. E. Offord, *Nature (London)* **211**, 591 (1966).

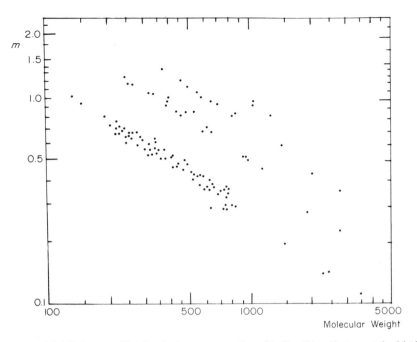

Fig. 1. Mobilities at pH 6.5 relative to aspartic acid. Peptides that contain histidine and cysteic acid have been excluded. The sign of m has been ignored in constructing the plot. The buffer system was pyridine/acetic acid/water (25:1:225 by volume), and the paper was cooled by toluene/pyridine, 23:2, v/v. Reproduced by permission of the publishers from R. E. Offord, *Nature (London)* **211**, 591 (1966). Subsequent experience has shown that the plot is applicable to measurements made in tanks cooled by White Spirit/pyridine mixtures.

years' use of the method, and to discuss other applications of logarithmic plots of mobility.

Theoretical Issues

The 1966 paper[3] pointed out that it had been generally expected that the mobility of peptides on paper would be proportional to $(mass)^{-1}$, or (following a treatment based in Stokes' law) on $(mass)^{-1/3}$.

A number of workers in the Medical Research Council Laboratory of Molecular Biology at Cambridge (in particular, R. P. Ambler) had collected sufficient mobility data while preparing peptides that had been subsequently sequenced to show that neither of the above relationships were valid. It did appear, however, that mobility, mass and charge were likely to be governed by some relationship that was consistent enough to make it possible to use mobility measurements in a more quantitative way. My own preference was toward a model for electrophoretic mobility in which the retarding force on the molecule and its accompanying ionic atmosphere was the result of shear across a very thin element of the retarding medium, rather than throughout an infinite volume as in the treatment by Stokes' law. Such a model leads to an expression of the form

$$m = k\epsilon M^{-2/3} \tag{1}$$

where m is the mobility relative to that of a standard substance, ϵ is the net charge, M is the mass, and k is a constant. Equation (1) gives rise to a logarithmic plot of slope $-\frac{2}{3}$. In practice,[3] the points in a logarithmic plot of the relative mobilities of peptides lie on a series of straight lines (Fig. 1), one for each integral value of the net charge, and their slopes are almost exactly $-\frac{2}{3}$. [Bailey and Ramshaw[4] have confirmed that this is so by extracting the data from the 1966 paper[3] and subjecting it to a least-squares analysis: they give the slope as -0.693 ± 0.012.]

Although the coincidence between the actual slopes and that predicted from the more favored model was intriguing, Fig. 1 was not represented as having established the validity of that model, if for no other reason than because several important factors had been neglected in the derivations. The algebraic treatment was given in the hope that the numerical coincidence between prediction and observation would produce a rigorous treatment in more expert hands.

Practical Considerations

It is fairly easy to obtain a measurement of electrophoretic mobility that is adequate for use on the plot in Fig. 1, even though several experimental variables in addition to the factors considered above will affect the position to which the molecule will have migrated by the end of the run. These variables are often automatically compensated for by including a standard substance in the run, but it is best to be aware of them, especially as their magnitude may be considerably affected by the design of the electrophoresis apparatus. The principal additional variables are the magnitude and direction of any bulk flow of buffer that there might be through the paper; the effect of the temperature of the buffer; and the perturbation of the electric field liable to be caused by excessively high concentrations of charged molecules at any point on the paper and by nearby conducting objects.

[4] C. J. Bailey and J. A. M. Ramshaw, *Biochem. J.* **135**, 889 (1973).

Bulk Flow of Solvent

The bulk flow has a component that is uniform in magnitude and direction along the paper, superimposed on one that varies in magnitude and direction along the paper. The uniform component is the resultant of electroendosmosis and (if the apparatus is of the vertical type due to Michl[5]) siphoning from the upper to the lower electrode compartment. The variable component is the inflow from the troughs at both ends which replaces water lost from the paper as it warms up during the run. Both charged and neutral substances are displaced from their true positions by these effects.

Effects of Inhomogeneities in Temperature

Although the temperature will be fairly uniform along the paper, the viscosity of the buffer (and thus the retarding force) varies so sharply with temperature that even minor local variations can perturb the mobility of a substance relative to any standard rather different from it in mobility.

Changes in the net charge as a result of the temperature dependence of the free energies of ionization do not appear to occur to a significant extent, presumably because of the pK values for all the groups present in the peptides plotted in Fig. 1 are distant from 6.5.

Electrostatic Perturbations

Unlike R_f values, mobilities are not appreciably affected by the presence in the sample of large quantities of soluble, neutral substances, such as urea. Charged materials, on the other hand, can present serious problems if they are present in too great an amount. The effect of a concentration of charged molecules is to perturb the electrostatic field and in general to reduce mobility. This perturbation begins to be noticeable if there is much more than, say, 0.1 μeq of a charged species per centimeter of origin, particularly if a lightweight paper, such as Whatman No. 1, is used.

The problem presents itself in three main ways: if salts are present in the samples, if there is a concentration of insoluble peptide at the origin, and if the peptides themselves are at too high a surface concentration on the paper.

Very heavy salt concentrations give a characteristic comet shape to

---

[5] H. Michl, *Monatsh. Chem.* **82**, 489 (1951).

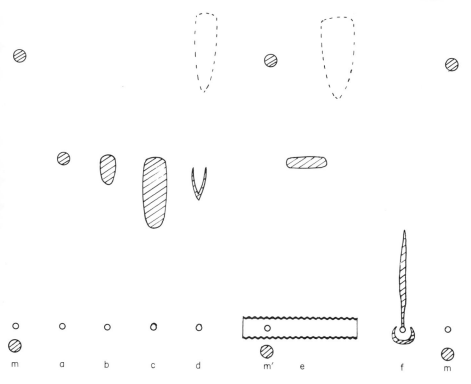

FIG. 2. Factors that affect mobility measurements. (a, b, c) The effect of progressive overloading. Note that the position of the leading edge is scarcely altered. (d) The effect of contamination by salts. (e) Salty samples: the effect of rerunning, as described in the text. (f) Effect of applying a mixture that contained a fast-running substance, an insoluble charged substance, and a neutral substance. (m) A mixture of a substance of zero mobility and the substance of standard mobility. (m′) A marker mixture applied between the first and second runs in (e). The dotted outlines represent negatively staining material.

the peptide bands (Fig. 2d) and often reveals itself in addition by the presence of fast-moving streaks that stain less strongly with ninhydrin than does the background. Even the use of internal standards of mobility is not totally effective as a means of correction for the presence of salt because the degree of perturbation in a given sample is not a simple function of the distance of migration. It may be necessary, if contamination is heavy, to run the sample for about 5 min, dry the paper, cut out the strip along which the peptides and salts will have run, turn it through 90° and sew it, by the standard procedures, to a new sheet of paper. A second run of normal length should yield a true value for the mobility, since the bulk of the salts will no longer lie directly on a line between the peptide and the electrode compartments (Fig. 2e). Contamination by salts that happen to buffer at some inappropriate pH will produce an additional perturbation, but it is usually overcome by the same means.

Insoluble peptides, or those that bind to paper and remain at the origin, have their most noticeable effect on the values for the mobilities of the more slow-moving compounds. If they produce a severe perturbation, the fact is usually revealed by distortions in the shape of the spots near the origin (Fig. 2f). True values for mobilities can usually be obtained in the manner described above for samples contaminated with salts.

The best way to deal with the effect of overloading (which is revealed by large spots, having in severe cases a ragged trailing edge) is to repeat the run with smaller quantities of material. It is not always possible to follow this advice; for instance, if it is necessary to measure the mobilities from the records of preparative electrophoreses, as had to be done for Fig. 1. It appears to be true that the position of the *leading* edge of the spot is least dependent on the degree of loading (Fig. 2a,b,c) and *m* values are customarily measured to this point.

Metal objects that are grounded can disturb the electrostatic field, and this fact should be taken into account when positioning metal cooling coils since the water constitutes a conducting path to earth. This does not apply to the flat-plate type of apparatus, because the effect there is uniform.

Correction for Unavoidable Perturbations

Not all of the above perturbations are negligible, and not all can be avoided. Their effects on the reliability of the measured mobilities can be very much mitigated be relating measurements to the mobilities of standard substances run at the same time. For the very greatest accuracy the standard substance should be mixed with the unknown on the origin before the run begins. Mobilities at pH 6.5 are calculated relative to that of aspartic acid, one of the fastest negatively charged substances. It seems to have become accepted practice to give a negative sign to mobilities of negatively charged molecules, i.e., the mobility of aspartic acid = —1.0. Mobilities at pH 1.9 are either expressed relative to that of serine or, as Bailey and Ramshaw[4] prefer, N^α-dansylarginine.

It is first necessary to correct for the displacement of the position of zero mobility by the resultant flow of solvent mentioned above. This can be done by relating all measurements to the center of a spot of an electrically neutral substance. The yellow compound N^ε-dinitrophenyl-lysine is suitable for pH 6.5. Some authors prefer a dansyl compound, e.g., N^α-dansylarginine, because these can be made visible in trace quantities by UV light. They are thus particularly suitable for inclusion in a run as an internal standard. Urea is neutral at both pH 6.5 and 1.9, and a 2-mg spot can be located either by an increase in the transparency of the paper when wet or, after drying, by the fact that it produces a white spot against the pink background that results when the paper is stained with the cadmium–ninhydrin reagent of Heilmann et al.[6] Bailey and Ramshaw[4] have shown that dansyl sulfonic acid is uncharged at pH 1.9 and can be used as a neutral marker at that pH.

Even if the zero point is established, it is a little less easy to correct the mobility of migrating substances for the inflow from the ends and the other sources of nonuniform behavior mentioned above. If the mobilities of the unknown and of the standard substances are very different, neither should be allowed to approach closer than 5 cm to either electrode trough. If this is not convenient, it is best to use a secondary standard which is closer in mobility to the unknown. Secondary standards for negatively charged compounds include dansyl sulfonic acid ($m = -0.65^7$), the blue dye Xylene Cyanol FF ($m = -0.42$), and those for positively charged compounds include lysine (m = approximately 1.0), dansyl or dinitrophenylagmatine, or Methyl Green.

Precision and Accuracy

In general, even if only the simplest of the above precautions are observed, the spread of repeated measurements of mobility is of the order of $\pm 5\%$.[8] This is usually a sufficient degree of precision, but some authors have obtained more precise figures for crucial peptides. For instance, di Donato and D'Alessio[9] were drawn to a conclusion of some general biochemical importance (see disulfide bridges below) on the basis of a comparison between two measured mobilities on Fig. 1. They repeated each of the two measurements three times and obtained m values of 0.474 ± 0.004 and 0.422 ± 0.011. The determination of m values as described above is sufficiently precise to form the basis of a method for typing the V_H subgroups of human immunoglobulin.[10]

[6] J. Heilmann, J. Barrolier, and E. Watzke, Hoppe-Seyler's Z. Physiol. Chem. 309, 219 (1957).

[7] J. R. Brown and B. S. Hartley, Biochem. J. 101, 214 (1966).

[8] R. P. Ambler and L. H. Brown, Biochem. J. 104, 784 (1967).

[9] A. di Donato and G. D'Alessio, Biochem. Biophys. Res. Commun. 55, 919 (1973).

[10] A. Moulin, D. Eskinazi, C. DePreval, and M. Fougereau, Immunochemistry 12, 883 (1975).

Applicability to Thin-Layer Electrophoresis

Plots have been constructed for peptides run at pH 6.5 on thin layers of cellulose powder. Bates *et al.*[11] find Fig. 1 directly applicable to thin-layer electrophoresis, while Vandekerckhove and Van Montagu[12] observed that the slopes of the lines on their plot were the same as those in Fig. 1, but the lines were displaced downward by about 10–15%. The explanation for the discrepancy between these two findings no doubt rests on some fine point of experimental detail. Others [e.g.,13,14] have used the plot on thin layers.

Peptides That Depart from Ideal Behavior

Figure 1 excludes peptides that contain histidine, because of the nearness of the pK of the imidazoyl side chain to 6.5. The pK will be affected by the sequence of the peptide and, by combining Eq. (1) with the Henderson–Hasselbach equation, an estimate of the value in any particular instance can be deduced from the extent to which the peptide lies off the lines in Fig. 1. Romero-Herrera and Lehmann[15] have used this approach to determine the pK of the histidine side chain in some peptides derived from human myoglobin. (As some of the buffering power of the system derives from the pyridine in the coolant, they found it essential to determine the true pH on the paper during the run by measuring m for free histidine, for which the pK is known). They found pK values that ranged from 5.65 (histidine NH_2-terminal) through 6.1 (histidine in an internal position in the peptide) to 6.50 (histidine COOH-terminal). The trend of these values with the position of histidine in the sequence is in accordance with what would be expected on the basis of tables of electrometric titrations of synthetic peptides.[16] These tables also suggest that it should be quite easy to find peptides in which the pK values lay a little outside the above range.

While Fig. 1 was being constructed, a number of peptides were found that lay nearer to the anode than expected. In other words, the mobility of negatively charged peptides in this group was enhanced, that of positively charged peptides was reduced, and nominally neutral peptides had m values of up to approximately -0.1. The effect became

[11] D. L. Bates, R. N. Perham, and J. R. Coggins, *Anal. Biochem.* **68**, 175 (1975).

[12] J. Vanderkerckhove and M. Van Montagu, *Eur. J. Biochem.* **44**, 279 (1974).

[13] E. Bause and G. Legler, *Hoppe-Seyler's Z. Physiol. Chem.* **355**, 438 (1974).

[14] E. Schiltz and J. Reinboldt, *Eur. J. Biochem.* **56**, 467 (1975).

[15] A. E. Romero-Herrera and H. Lehmann, *Proc. R. Soc. London, Ser.* **186**, 249 (1974).

[16] J. P. Greenstein and M. Winitz, "The Chemistry of the Amino Acids," p. 490. Wiley, New York, 1961.

increasingly pronounced if the pH was allowed to rise above 6.5. These phenomena are presumably due to a lowering of the pK of the α-amino group. The effect was found to be so severe with peptides that contained cysteic acid that it was necessary to segregate them onto one of two further plots, depending on whether or not the cysteic acid was NH$_2$-terminal (Figs. 3 and 4).

Figure 1 contains all peptides other than those with histidine and cysteic acid. MetSO$_2$ in the NH$_2$-terminal position might be expected to have a considerable effect on the pK, but only occurred in one peptide, which happens to contain cysteic acid. It is plotted in Fig. 4. Some of the peptides plotted in Fig. 1 behave as if they too have lowered pK values of the α-amino group, but to a less marked extent than those plotted separately. These peptides are those that have at their NH$_2$-termini basic amino acids (especially arginine), asparagine, threonine, and serine. A useful indication that one is dealing with a peptide with a

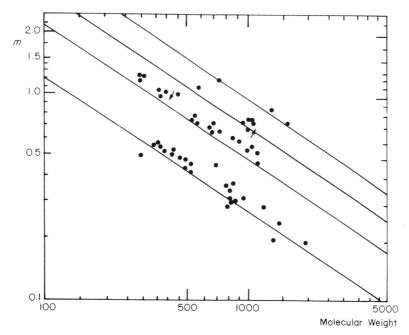

Fig. 3. Mobilities at pH 6.5 of peptides that have cysteic acid at positions other than NH$_2$-terminal. Peptides that contain histidine have been excluded. The diagonal lines indicate the positions of the lines of equal charge in Fig. 1. Certain points have the line on which they should lie indicated by arrows. All other details are the same as in Fig. 1. Reproduced by permission of the publishers from R. E. Offord, *Nature* (*London*) **211**, 591 (1966).

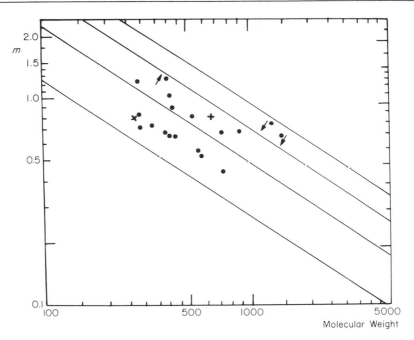

Fig. 4. Mobilities at pH 6.5 of peptides that have cysteic acid at the NH₂ terminus. Peptides that contain histidine have been excluded. +:2 cysteic-acid residues, one of them NH₂-terminal. X:1 cysteic acid residue, MetSO₂ at the NH₂ terminus. All other details are the same as in Fig. 1. Reproduced by permission of the publishers from R. E. Offord, *Nature (London)* **211**, 591 (1966).

low pK of the α-amino group is given by staining with the ninhydrin–cadmium reagent of Barrolier.[6] Such peptides are very frequently those that give a yellow, orange, or brown color, as opposed to the normal red. The principal exceptions to this rule are peptides with NH₂-terminal glycine, which give a yellow color but have normal pK values, and those with basic residues NH₂-terminal, which give the red color but have abnormal pK values. Peptides with low pK values of the α-amino group contribute much of the raggedness to the lines on Fig. 1, but are unlikely to lead to serious errors in interpretation.

Conversely, peptides with MetSO₂ at the COOH-terminus move slightly more to the cathode than anticipated, probably because of the raising of the pK of the α-carboxyl group. Also, Ambler and Brown[8] have found that peptides in which aspartic or glutamic acids are adjacent to a lysine show a similar effect. Such a peptide, which should have been neutral, had $m = +0.05$. The effect is much more noticeable at pH 6.0 than at pH 6.5. As with the perturbations of the α-amino group, these effects do not seem likely to lead to errors of interpretation at pH 6.5.

Accuracy in the Determination of Molecular Weight

The 1966 paper[3] pointed out that the spread of values in Fig. 1 was such as to call for more caution when estimating molecular weights than when estimating net charge. While a peptide of composition Asp: Leu = 1:1 and $m = -0.65$ is clearly (Asp$_1$, Leu$_1$) and not (Asp$_2$, Leu$_2$) or (Asp$_1$, Asn$_1$, Leu$_2$), there can be instances in which the results are ambiguous. The ambiguity can be lessened by determining m values before and after the carrying out of some operation that will produce a known change in the net charge.[e.g.,17] For instance, consider an N$^\varepsilon$-maleylated peptide of unknown charge and mass that had $m = -1.0$ before treatment for the removal of maleyl groups and $m = -0.4$ afterward. Since the removal of a maleyl group converts a negative charge to a positive one, reference to Fig. 1 indicates that by far the most likely solution is that the peptide had a molecular weight of about 600 and, when demaleylated, a net charge of -1. An analogous approach has been adopted for disulfide-bridged peptides. Measurements of m before and after oxidation with performic acid have been used to provide information on molecular weights (within certain limits of reliability) and amide content.[e.g.,18,19] The opening and closing of the lactone ring in peptides with COOH-terminal homoserine can also produce informative changes in m values.

In spite of the possibility of ambiguity, single measurements have been used, both to give a rough estimate when only very small quantities were available, as in the preliminary work on enkephalin[20] and, on several occasions for more accurate measurements. For instance, di-Donato and D'Alessio[9] carried out replicate measurements on a peptide from the ribonuclease of bovine seminal plasma. The amino acid composition was Met:Cys:Arg = 1:2:1 and m was 0.474 ± 0.004. After reduction m was 0.422 ± 0.011. Carboxymethylation of the reduced peptide changed m to -0.368 while aminoethylation gave $m = 1.210$. The authors deduced that the original peptide was a cyclic, disulfide-bridged dimer of the sequence 30–33 of the protein, Met-Cys-Cys-Arg. A discussion of their further conclusion—that these findings have revealed

[17] B. S. Hartley, *Biochem. J.* **119**, 805 (1970).
[18] B. D. Burleigh and C. H. Williams, Jr., *J. Biol. Chem.* **247**, 2077 (1972).
[19] S. Ronchi and C. H. Williams, Jr., *J. Biol. Chem.* **247**, 2083 (1972).
[20] J. Hughes, T. Smith, B. Morgan, and L. Fothergill, *Life Sci.* **16**, 1753 (1975).

the first example of interchain disulfide bonding in an oligomeric enzyme not derived from a zymogen—is beyond the scope of this review.

A highly successful example of molecular-weight measurement is provided by the work of Wenn,[21] who drew up an analog of Fig. 1 for dansylated glycopeptides run at pH 2.1. He obtained the following molecular weights (with the theoretical values in parentheses) for a related set of products formed by the sequential action of a series of glycosylases on the parent glycopeptide: 1250 (1257), 1085 (1095), 925 (933), 755 (771), 570 (566). It should be mentioned that Wenn gave reasons why glycopeptides should be particularly suitable for this approach.

On the other hand, Ambler[22] had to decide whether a certain peptide was pyroglutamyl-Phe or some such derivative as N^α-acetyl-Gln-Phe. The use of mobilities to solve this problem amounts to a determination of molecular weight. The predicted mobilities for the two compounds were −0.64 and −0.52, respectively, but, although the observed mobility was −0.70, Ambler did not regard the question as having been settled with certainty. Whether or not such caution was justified in this instance, it is certainly true that the use of Fig. 1 for determinations of molecular weight is more likely to lead to unrecognized error than its use for the determination of net charge.

Accuracy of Amide Assignments

It was suggested[3] that the degree of scatter of points in Fig. 1 was such that, while ambiguous mobilities might eventually be found, the lines up to $\epsilon = \pm 3$ were sufficiently far apart for it to be unlikely that a result would be obtained that was actually misleading. D. A. Jenner (private communication) has found that the Science Citation Index lists 410 publications that refer to the original paper in the 10-year period after its publication, of which the majority deal with amide determination. Of the 405 that were accessible in Oxford, none reports that Fig. 1 gave incorrect assignments. The plot has been used in the correction of amide assignments made on the basis of all the other major methods but no method has been used to correct assignments made on the basis of the plot. (Since an additional, random sample of sequence papers showed that the method is used more often than it attracts a citation, the question arose that the Citation Index might be a less comprehensive means of searching for errors than had been hoped.

[21] R. V. Wenn, *Biochem. J.* **145**, 281 (1975).
[22] R. P. Ambler, *Biochem. J.* **135**, 751 (1973).

But it is probably fair to assume any difficulties with the method would have provoked a direct citation of the 1966 paper.[3])

One paper[23] reported the reluctance of the authors to place too much reliance on m values of peptides derived from a proposed sequence -Asp-Glu-Asp-Glu-, but it was not suggested that the assignments, as based on Fig. 1, were ambiguous or in conflict with any other evidence.

The amide groups in certain sequences are especially labile to hydrolysis. No method can hope to succeed if the group is lost before the analysis begins, but the electrophoretic approach involves what are perhaps the mildest conditions of any method. It can be applied during, rather than after, the preparation of peptides and provides an opportunity to observe the progressive loss of amides during a series of experiments.

If a peptide has both side-chain amides and side-chain carboxyl groups, Fig. 1 will not tell which residue has the amides. Some other method will have to be used, or m values must be determined on subfragments of the peptide. In either case the value for the net charge of the original peptide will provide a check that no amides have been lost in the course of the experiments.

The factors mentioned above probably explain the seeming accuracy of amide assignments based on Fig. 1. But I feel there is a danger that its reliability has become self-reinforcing and that an incorrect result might not now be recognized as such. Against this, the way in which workers have discovered new amino acids as a result of following up apparently spurious mobilities (see below) suggests that the results of the method still receive a measure of critical attention.

Mobilities at pH 1.9

As at pH 6.5, peptides should have integral charges in the region of pH 2. Since this value is well below the pK of carboxyl groups there will be no distinction between the mobilities of peptides with side-chain amides and those with side-chain carboxyl groups. Nonetheless the plot for pH 1.9 published in the 1966 paper[3] has been used from time to time as a check on the proposed amino acid composition of a peptide. Electrophoresis at this pH had tended to be a separation method of last resort and there were correspondingly fewer data to plot. The situation has been considerably improved by the work of Bailey and Ramshaw,[4] who prepared or purchased 104 peptides and constructed a plot of m related to the mobility of N^α-dansylarginine at this pH (Fig. 5). Both

[23] P. R. Milne, J. R. E. Wells, and R. P. Ambler, *Biochem. J.* **143**, 691 (1974).

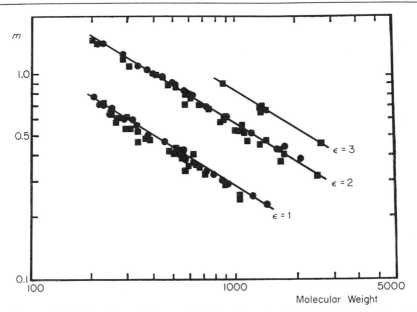

Fig. 5. Mobilities of peptides at pH 1.9 relative to N^α-dansylarginine. Peptides indicated by squares were run in formic acid/acetic acid/water (9:12:179, by volume) cooled by White Spirit 100. Peptides indicated by circles were run in formic acid/acetic acid/water (1:4:45, by volume) cooled by passage of water through electrically insulated metal plates. Reproduced by permission of the authors and publishers from C. J. Bailey and J. A. M. Ramshaw, *Biochem. J.* **135**, 889 (1973).

the original plot and Fig. 5 are corrected for bulk flow of solvent. There were too few points on the original plot to have justified a numerical measurement of the average slope, but it was noticeably less than in Figs. 1 and 5. The average slope in Bailey and Ramshaw's plot is −0.62, and so all the plots that have been constructed for peptides and for some of their derivatives (see below) have been brought into a pleasingly uniform accord with Eq. (1). Bailey and Ramshaw do not give a factor for the conversion of values of m relative to serine into values of m relative to dansylarginine. The factor is approximately 0.65 (K. Rose, private communication). Given this value, it appears that the only points that are significantly discrepant between the original pH 1.9 plot and Fig. 5 are the four of highest molecular weight on the $\epsilon = 1$ line of the former.

Molecules Other than Simple Peptides

Numerical values of mobilities have been determined for a number of peptide derivatives. Some have been given logarithmic plots of their own, while some have been fitted on one or other of the plots for simple peptides.

Gray[24] has constructed plots for dansylated peptides as a means of checking on sequential Edman degradations (the slopes are about −0.6). The principle is to dansylate small samples after each cycle and to determine the m values. The changes in m should be in accord with the sequence as proposed by identification of the PTH, by end-group determination, or by the subtraction of the amino acid compositions of successive cycles. The m values should move up the appropriate line as the molecular weight falls until a charged residue is removed, when the m value should join the appropriate new line. Dansyl peptides of above about 8 residues tend to trail badly.

The plot constructed by Wenn[21] on dansyl glycopeptides has already been mentioned. It has a slope of −0.6. Other studies on glycopeptides include that of Dixon,[25] who used Fig. 1 to deduce a pK value in glucosyl-Val-His.

It is tempting to try to use Fig. 1 (or perhaps Figs. 2 or 3) to try to characterize serine phosphate peptides. The 1966 paper[3] warned that the pK of the ionization of the phosphate group was dangerously close to 6.5, and that anomalous results might be obtained. Daile et al.[26] had a particularly good opportunity to study the problem as they used m values to follow the addition and removal of phosphate groups in a synthetic heptapeptide. They found that the addition of a phosphate group at sites within the sequence caused a charge change of −1.35, and that phosphorylation of the NH_2-terminal residue caused a change of −1.6. Shaw and Wells[27] also found that the charges of serine-phosphate peptides were not always integral, but Carnegie[28] was able to use m values to locate phosphate groups within peptides. Also, Sanders and Dixon[29] found that the first three cycles of sequential Edman degradation of a peptide which they proposed to be Ser(P)-Ser(P)-Ser(P)-Arg-Pro-Val-Arg led to changes of mobility in dansylated samples consistent with changes of charge of +0.98, +1.02 and +0.90. We shall see later that m values for active-site peptides that are covalently modified are interpretable, even if the substituting group carries a phosphate moiety.

The discovery of γ-carboxyl glutamic acid residues (which result from a postsynthetic modification of particular importance in the blood-

[24] W. R. Gray, this series Vol. 11, p. 469 (1967); Vol. 25, p. 333 (1972).
[25] H. B. F. Dixon, Biochem. J. 129, 203 (1972).
[26] P. Daile, P. R. Carnegie, and J. D. Young, Nature (London) 257, 416 (1975).
[27] D. C. Shaw and J. R. E. Wells, Biochem. J. 128, 229 (1972).
[28] P. R. Carnegie, Nature (London) 249, 147 (1974).
[29] M. M. Sanders and G. H. Dixon, J. Biol. Chem. 247, 851 (1972).

clotting pathway) was brought about as a result of the observation that the m values of some fragments of prothrombin indicated that there were more negative charges than could be accounted for by the proposed sequence.[30] The anomalous m values were not found in material isolated after certain forms of anticoagulant therapy. Once attention had been drawn to these peculiarities, mass-spectrometry was used to identify their structural basis. γ-Carboxylglutamic acid, like glutamine, reverts to glutamic acid during acid hydrolysis and so was missed in the original sequence determination. It will also revert without any other serious effect simply on heating, and the changes in m consequent on heating are now being used as an additional means of characterization.[31,32]

Measurements of m values have been used to help identify residues at the active sites of enzymes. These experiments have involved the characterization of peptides that were covalently modified either as a result of the normal mechanism of the enzyme (such as the examination of N^ε-phosphopyridoxal-lysine peptides[33,34]) or by an active-site-directed inhibitor. Examples of the latter application are the work on a peptide of alcohol dehydrogenase inhibited by an analog of NAD[35]; on a peptide isolated from L-isoleucine–transfer-RNA ligase after inhibition by L-isoleucine bromomethyl ketone[36]; and on a peptide of triose phosphate isomerase isolated after inhibition with bromohydroxyacetone phosphate.[37]

That the above peptide derivatives give interpretable m values suggests that the logarithmic plots might be applicable to a wide variety of modified peptides. Measurements of m have been used to characterize the site of action of iodinating reagents—e.g., the work of Thomas and Harris,[38] in which the m values were used to show that the modified residues were mono- rather than diiodotyrosine. Nitration by tetranitro-

[30] S. Magnusson, L. Sottrup-Jensen, T. E. Petersen, H. R. Morris, and A. Dell, *FEBS Lett.* **44**, 189 (1974).

[31] P. Fernlund, J. Stenflo, P. Roepstorff, and J. Thomsen, *J. Biol. Chem.* **250**, 6125 (1975).

[32] H. R. Morris, A. Dell, T. E. Petersen, L. Sottrup-Jensen, and S. Magnusson, *Biochem. J.* **153**, 663 (1976).

[33] O. Avromoric-Zikic, K. G. Welinder, S. Shechosky, and J. Sodek, *Can. J. Biochem.* **51**, 21 (1973).

[34] M. Campos-Cavieres and C. P. Milstein, *Biochem. J.* **147**, 275 (1975).

[35] H. Jörnvall, C. Woenckhaus, E. Schättle, and R. Jeck, *FEBS Lett.* **54**, 297 (1975).

[36] P. Rainey and E. Holler, *Eur. J. Biochem.* **63**, 419 (1976).

[37] A. F. W. Coulson, J. R. Knowles, J. D. Priddle, and R. E. Offord, *Nature (London)* **227**, 180 (1970).

[38] J. O. Thomas and J. I. Harris, *Biochem. J.* **119**, 307 (1970).

methane has been followed in a similar way, and in one instance m values have helped to show that some peptides are cross-linked to others by the action of the reagent.[39] Mobilities have been used to help follow protein modification by 4-sulfophenylisothiocyanate[40] and by 7-chloro-4-nitrobenzo-2-oxa-1,3-diazole.[41] N-Acyl peptides also behave in a straightforward manner, and m values have been determined for numerous N^{α}-acetylated peptides[e.g.,42] and for succinyl, maleyl, glutaryl, citraconyl and sarcosyl peptides,[43] as well as for penicilloyl peptides.[43a] Equation (1) was not found suitable for the characterization of a peptide so highly substituted by maleyl groups that it lay above all the lines in Fig. 1.[44] The existence of pyroglutamyl residues at the NH_2-termini of peptides has been inferred partly on the basis of m values (e.g., in an immunoglobulin [45]) as has the conversion of peptides to diketopiperazines.[46]

Two observations have been made that so far lack an explanation. In one,[47] a peptide of *Chlorella* plastocyanin had an m value corresponding to a charge of -2 when first run, but this dropped to -1 after elution and storage. The amino acid composition (which did not change when the m value changed) could only explain a charge of -1, and the authors tentatively suggest that a labile, acidic carbohydrate moiety had been attached to the peptide to begin with. In another instance[48] mobility and other data agree that the NH_2-terminal peptide of a protozoal cytochrome has some sort of amino substitution, but its nature is not known.

It should not be forgotten that it is sometimes just as important to be assured that there has *not* been a covalent modification. For instance, Carraway and Leeman[49] used Fig. 1 and the 1.9 plot from the 1966 paper[3] to show that the fragments obtained in the course of sequencing bovine neurotensin contained only the normal amino acids that had been suggested by their compositions after acid hydrolysis.

[39] J. Williams and J. M. Lowe, *Biochem. J.* **121**, 203 (1971).
[40] U. Gallwiltz, L. King, and R. N. Perham, *J. Mol. Biol.* **87**, 257 (1974).
[41] N. C. Price and G. K. Radda, *Biochim. Biophys. Acta* **371**, 102 (1974).
[42] H. Jörnvall, H. Ohlsson, and L. Philipson, *Biochem. Biophys. Res. Commun.* **56**, 304 (1974).
[43] J. W. Payne, *Biochem. J.* **123**, 245 (1971).
[43a] P. H. Corran and S. G. Waley, *Biochem. J.* **149**, 357 (1975).
[44] L. King and R. N. Perham, *Biochemistry* **10**, 981 (1971).
[45] J.-C. Jaton and D. G. Braun, *Biochem. J.* **130**, 539 (1972).
[46] H. Jörnvall, *FEBS Lett.* **38**, 329 (1974).
[47] J. Kelly and R. P. Ambler, *Biochem. J.* **143**, 681 (1974).
[48] G. W. Pettigrew, J. L. Leaver, T. E. Meyer, and A. P. Ryle, *Biochem. J.* **147**, 291 (1975).
[49] R. Carraway and S. E. Leeman, *J. Biol. Chem.* **250**, 1907 (1975).

TABLE

Examples of the Use of m Values for the Characterization of Peptides Protected for Use in Semisynthesis[a]

Peptide	Assumed state of protection						Mobility at pH 6.5	
	α-NH$_2$	ϵ-NH$_2$	α-COOH	Side-chain COOH	Side-chain imidazole	—SH	Observed	Predicted
Trp-Trp-Cys-Asn-Asp-Gly-Arg	Free	—	Free	Free	—	S—SO$_3^-$	−0.25	−0.26
							(−0.45)[c]	(−0.51)[c]
	Boc[b]	—	Free	Free	—	S—SO$_3^-$	−0.42	−0.38
							(−0.65)[c]	(−0.72)[c]
	Boc	—	Benzyl	Benzyl	—	S—SO$_3^-$	Does not run	0
	Boc	—	Free	Benzyl	—	S—SO$_3^-$	−0.22[d]	−0.22
							(−0.42)[c]	(−0.44)[c]
Lys-Val-Phe-Gly-Arg	Free	Free	Free	—	—	—	0.9	0.9
	Maleyl	Maleyl	Free	—	—	—	−0.63	−0.65
	Boc	Boc	Free	—	—	—	0	0
His-Gly-Leu-Asp-Asn-Tyr-Arg	Free	—	Free	Free	Free	—	0.07	—
	Boc	—	Free	Free	Boc-NH—CH(CF$_3$)—	—	−0.25	−0.28
	Boc	—	Anisyl	Anisyl	Boc-NH—CH(CF$_3$)—	—	Insoluble	—
	Boc	—	Free	Anisyl	Boc-NH—CH(CF$_3$)—	—	0	0

[a] Data from A. R. Rees and R. E. Offord, Biochem. J. **160**, 467 (1976).

[b] Boc = tert-butyloxycarbonyl-.

[c] Small quantities of material were observed at these mobilities, which were taken to be des-amido forms of the major components at each stage.

[d] The spot trailed, probably owing to low solubility.

Applications to Peptide Synthesis

Some authors have used Fig. 1 to help check the characterization of their synthetic products,[26,50,51] and to follow certain side reactions during syntheses.[52] Figure 1 has been particularly useful in protein semisynthesis,[53] in which fragments of proteins are used as ready-made intermediates in the chemical synthesis of protein analogs. These fragments have to be protected on at least their amino, carboxyl, sulfhydryl, and imidazoyl side chains. They then have to be specifically deprotected at the α-amino or carboxyl group, depending on which is required for the coupling to the next fragment in the sequence. All these operations are very conveniently studied by mobility charges, so long as the protected fragment remains soluble (see the table).

Conclusion

By reporting the great variety of peptides and peptide derivatives that have given interpretable m values on logarithmic plots, it is hoped that this review will stimulate their application to a still wider range of problems.

Acknowledgments

I thank Mr. D. A. Jenner for carrying out the literature survey for this review, and Christ Church, Oxford, for a contribution to the costs of the exercise. I would like to renew my thanks to those colleagues named in the 1966 paper[3] for supplying me with data and many valuable ideas, to the Editorial Board of the *Biochemical Journal* and to Drs. C. J. Bailey and J. A. M. Ramshaw for permission to reproduce Fig. 5; to the Editor of *Nature* for permission to reproduce Figs. 1, 3, and 4; and to the Editorial Board of the *Biochemical Journal* for permission to reproduce the data contained in the table.

[50] E. Wünsch, J. E. Brown, K.-H. Deiner, F. Drees, E. Jaeger, J. Musiol, R. Schouf, H. Stockur, P. Thamm, and G. Wendlberger, *Z. Naturforsch. Teil C* **28**, 235 (1973).
[51] B. Gutte, *J. Biol. Chem.* **250**, 889 (1975).
[52] C. C. Yang and R. B. Merrifield, *J. Org. Chem.* **41**, 1032 (1976).
[53] R. E. Offord, *Nature (London)* **221**, 37 (1969).

Section II
End-Group Methods

[7] Carboxypeptidase C

By HARALD TSCHESCHE

Carboxypeptidases are enzymes that remove (L) amino acids one after another from the COOH terminus of a peptide chain. Carboxypeptidases A and B from bovine and porcine pancreatic glands[1] have been widely used in COOH-terminal residue and sequence determinations. The methods have been reviewed.[2-4] Carboxypeptidase A removes COOH-terminal aromatic and aliphatic amino acids with long side chains most rapidly, but glycine and acidic residues only slowly, while lysine, arginine, proline, and histidine block the action of the A enzyme. The B enzyme has a more restricted specificity and cleaves lysine and arginine most rapidly, but neutral amino acids only slowly and in rare cases.

A carboxypeptidase C(CPC)[5] has been isolated from the flavedo of citrus fruits,[6-8] from orange leaves,[8,9] and French beans.[10,11] The enzymes from citrus fruit and orange leaves have the same substrate specificity and pH optima at acid pH of 5.3,[9] but differ in stability against pH changes and lyophilization. The C enzyme from orange leaves seems to be more stable. It is not inhibited by diisopropyl phosphorofluoridate (DFP),[9] whereas the citrus fruit enzyme looses 36% of its activity[6,7] upon treatment with DFP. Acid carboxypeptidases of this type seem to be a common constituent of many higher angiosperm plants.[8] Procedures for the isolation and characterization of the enzymes have been reviewed in this series (Volume 45 [47]). Carboxypeptidase C removes COOH-terminal lysine, arginine, and proline, as well as all other neutral, aliphatic, aromatic, *and* the acidic protein amino acids. The C enzyme combines the specificity of the carboxypeptidases A and B and in addition

[1] H. Neurath *in* "The Enzymes" (P. D. Boyer, H. Lardy, and K. Myrbäck, eds.), Vol. IV, p. 11. Academic Press, New York, 1960.

[2] J. L. Bailey, "Techniques in Protein Chemistry," p. 222. Elsevier, Amsterdam, 1967.

[3] R. P. Ambler, this series Vol. 19, pp. 155, 436 (1970).

[4] R. P. Ambler, this series Vol. 25, pp. 143, 262 (1972).

[5] The designation C originally refers to its occurrence in "citrus fruits."[6]

[6] H. Zuber, *Nature (London)* **201**, 613 (1964).

[7] H. Zuber, *Hoppe-Seyler's Z. Physiol. Chem.* **349**, 1337 (1968).

[8] H. Zuber and P. Matile, *Z. Naturforsch. Teil B* 663 (1968).

[9] B. Sprössler, H.-D. Heilmann, E. Grampp, and H. Uhlig, *Hoppe-Seyler's Z. Physiol. Chem.* **352**, 1524 (1971).

[10] J. R. E. Wells, *Biochem. J.* **97**, 228 (1965).

[11] The enzyme from French beans was originally named phaseolain.[10]

TABLE I
APPROXIMATE RELATIVE RATES OF RELEASE OF THE PROTEIN
AMINO ACIDS BY CARBOXYPEPTIDASE C[a]

Rapid release:	Phe, Tyr, Trp. Leu, Ile, Val, His
Good release:	Ser, Thr, Met, Ala, Asp, Asn, Glu, Gln, Lys, Arg, Pro, S-carboxymethylcysteine
Slow release:	Gly
No release:	Hydroxyproline, D-amino acids

[a] The rate of release of the COOH-terminal amino acid is strongly influenced by the penultimate residue. Residues with "slow" or "no" release generally significantly decrease the rate of release of the COOH-terminal amino acid (see text).

cleaves prolyl peptides on both the NH_2- and COOH-terminal sides of the imino acid proline.[6,7,9,12,13] Yeast contains a carboxypeptidase Y[14] that has a similar specificity to carboxypeptidase C (this volume 47 [8]).

Specificity of Carboxypeptidase C

Carboxypeptidase C requires a free α-carboxyl group for its action on peptides. Only L-amino acids are cleaved. COOH-terminal amino acid amides are not released.[7,13] The hydrolysis of methyl ester derivatives of peptides has been reported.[6,7] Whether this is due to an inherent property of the enzyme or a minor impurity (such as acetylcholinesterase, a common contaminant of commercial CPC preparations) at present remains unknown. Table I serves as a guide to the approximate relative rates of release of the protein amino acids by carboxypeptidase C.

The rates of hydrolysis of an amino acid A from a hypothetical peptide R-X-A are strongly influenced by the nature of the penultimate amino acid X and vary by factors between 1 and 20. Rapid release of the A amino acid occurs if X is an aromatic (Trp, Tyr, Phe, His) or aliphatic amino acid with a bulky side chain (Leu, Val).[9,13] When X is an acidic residue (Glu, Asp) good release rates are still obtained. With proline in position X the rates are significantly decreased, but COOH-terminal proline and glycine are released.[9,13] However, glycine is not cleaved from glycyl-glycyl-glycine,[9] nor is proline from R-Gly-Pro.[9,12,13] Hydroxyproline in the COOH-terminal position blocks the action of the enzyme.[12] The relative release rates of the A amino acid from several

[12] A. Nordwig, Hoppe-Seyler's Z. Physiol. Chem. 349, 1353 (1968).
[13] H. Tschesche and S. Kupfer, Eur. J. Biochem. 26, 33 (1972).
[14] R. Hayashi, S. Moore, and W. H. Stein, J. Biol. Chem. 248, 2296 (1973).

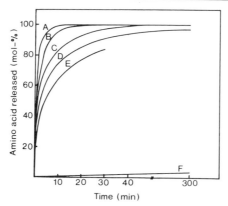

Fig. 1. Time-release plot of the COOH-terminal amino acid from synthetic benzyl-oxycarbonyl dipeptides in 0.1 M sodium citrate, pH 5.3, with 1 mol-% carboxy-peptidase C. Enzyme:substrate ratio ~ 1:100. Curve A, Z-Pro-Val; B, Z-Pro-Met; C, Z-Pro-Phe; D, Z-Pro-Leu; E, Z-Phe-Pro; F, Z-Pro-Pro.

R-prolyl-A peptides are obvious from Fig. 1. Polyproline is degraded by the C enzyme.[12] D-Amino acids in position X seem to inhibit the release of the amino acid from the carboxyl end.[6] Carboxypeptidase C cleaves no dipeptides.[6,7,12,13] The degradation of peptide chains comes to a halt when the stage of a dipeptide is reached. However, further degradation of N-protected dipeptides with bulky N-acyl groups, e.g., benzyloxycarbonyl (Z group), butyloxycarbonyl (Boc group), takes place. The free COOH-terminal amino acid A plus the N-protected residue X are obtained.[6,7,9,13]

Preparation of Enzyme

The C enzyme used should be free of endopeptidase and aminopeptidase activity, which can lead to very misleading results with peptide and protein substrates. This is especially the case if the COOH terminus is formed by a residue that is not rapidly released and the peptide sequence contains a highly susceptible bond for the contaminating peptidase activity. The commercial preparations of carboxypeptidase C, which generally are isolated from orange leaves, contain varying amounts of a trypsinlike and a chymotrypsinlike enzyme, and an aminopeptidase.[15,16]

---

[15] H. Tschesche, R. Klauser, and S. Kupfer, *Hoppe-Seyler's Z. Physiol. Chem.* Submitted for publication.

[16] CPC prepared from freshly collected orange leaves is free of any endopeptidase and aminopeptidase activity. These contaminating activities seem to develop as soon as the leaves wither (H. Uhlig, personal communication).

TABLE II
Contaminations of a Commercial Preparation of Carboxypeptidase
C by a Trypsinlike and a Chymotrypsinlike
Endopeptidase and an Aminopeptidase[a]

Enzyme activity	Substrate	Preparation[b] (mU/mg CPC preparation)	
		I	II
Carboxypeptidase C	Z-Leu-Phe	74	65
Aminopeptidase	Leu-p-nitroanilide	0.5	2.2
Trypsinlike	Benzoyl-Arg-p-nitroanilide	0.7	1.3
Chymotrypsinlike	Carboxypropionyl-Phe-p-nitroanilide	0.2	0.45

[a] For the assay, the procedures described have been used and CPC from Röhm GmbH.

[b] The commercial preparation is not a pure protein but contains sodium citrate from lyophilization of a pH 4.7 solution (Röhm GmbH instructions, Darmstadt, Germany).

Two typical examples of contaminations of a commercial sample of carboxypeptidase C by these peptidase activities are given in Table II. The assay procedures for the different enzyme activities are given below. The pH optima are different for the individual activities and are at pH 7.0 for the aminopeptidase and chymotrypsinlike activity and at pH 7.8 for the trypsinlike activity.[15]

Assay of Carboxypeptidase C Activity

The enzyme activity is determined by hydrolysis of the substrate Z-Leu-Phe and (a) spectrophotometric assay at 340 nm of the amount of Phe released after reaction with picrylsulfonic acid (trinitrobenzenesulfonic acid). An alternative method (b) uses the determination of the ninhydrin color for assay of the CPC activity.[9]

Method a. The lyophilized enzyme preparation is dissolved in water (5 mg/2 ml H_2O[17]), and 0.1 ml is mixed with 0.3 ml of 0.1 M sodium citrate, pH 5.3, and 0.4 ml of the solution of substrate (8.24 mg of Z-Leu-Phe dissolved in 0.3 ml of ethanol made up to 10 ml by addition of 0.1 M sodium citrate, pH 5.3). The mixture is incubated at 30° for 30 min, and then the reaction is stopped by addition of 0.2 ml of acetone.

[17] The commercial, sodium citrate-containing preparation of CPC has about 70 mU/mg dry weight, the purified, citrate-free enzyme about 2500 mU/mg, Röhm GmbH, Damstadt, Germany. The preparation of Boehringer Mannheim, Mannheim, Germany has about 800 mU/mg.

Aliquots of 0.5 ml each are mixed with 1.5 ml of 4% sodium bicarbonate and 0.5 ml of 0.1% picrylsulfonic acid in water. The mixture is allowed to react for 1 hr at 40° in the dark, then is acidified by addition of 0.5 ml of 1 N HCl; the absorbance is measured at 340 nm against a substrate blank without enzyme. One enzyme unit corresponds to the hydrolysis of 1 μmol of substrate per minute at 30° and causes a change in absorbance of $\Delta A^{1\,cm}_{340\,nm} = 1.78$.

Method b. From the solution of enzyme (5 mg/2 ml H$_2$O) 0.4 ml is mixed with 1.5 ml of 0.1 M sodium citrate, pH 5.3, and 1.5 ml of the solution of substrate (8.24 mg of Z-Leu-Phe made up to 10 ml in 0.1 M sodium citrate, pH 5.3, as above). The mixture is incubated at 30°. After 1 min and 21 min, two aliquots of 1 ml each are mixed with 1 ml of a solution of ninhydrin[18] and heated for 15 min at 100°. The mixture is cooled with ice and diluted by ethanol/water (1:1, v/v) to 20 ml. The absorbance at 578 nm is read. The molar absorbance coefficient is $\epsilon^{1\,cm}_{578nm} = 18,200\ M^{-1}\ cm^{-1}$.

Assay of Aminopeptidase Activity (Contaminant)[9]

The aminopeptidase activity in preparations of CPC is measured by hydrolysis of the substrate leucine-p-nitroanilide and spectrophotometric determination at 405 nm of the p-nitroaniline released. An aliquot of 0.2 ml of the CPC solution (5 mg/2 ml of H$_2$O; 350 mU of CPC) is mixed with 2.4 ml of 0.2 M sodium phosphate, pH 7.0, and 0.4 ml of a 5 mM solution of leucine-p-nitroanilide (1.26 mg/ml) in 0.2 M sodium phosphate, pH 7.0. One enzyme unit corresponds to the hydrolysis of 1 μmol of substrate per minute at 25° and causes a change in absorbance of $\Delta A^{1\,cm}_{405\,nm} = 3.32$/min.

Assay of Chymotrypsinlike Activity (Contaminant)

The chymotrypsinlike activity in preparations of CPC is measured by hydrolysis of the substrate carboxypropionyl-phenylalanine-p-nitroanilide and spectrophotometric determination at 405 nm of the p-nitroaniline released. An aliquot of 0.2 ml of the CPC solution (5 mg/2 ml of H$_2$O; 350 of mU CPC) is mixed with 1.8 ml of 0.2 M sodium phosphate, pH 7.0, and 1 ml of 25 mM carboxypropionyl-phenylalanine-p-nitroanilide (5 mg/ml) in 0.2 M sodium phosphate, pH 7.0. One enzyme unit corresponds to the hydrolysis of 1 μmol of substrate per minute at 25° and causes a change in absorbance of $\Delta A^{1\,cm}_{405\,nm} = 3.32$/min.

[18] S. Moore, *J. Biol. Chem.* **243**, 6281 (1968).

Assay of Trypsinlike Activity (Contaminant)[19]

The trypsinlike activity in preparations of CPC is measured by hydrolysis of the substrate N-benzoyl-DL-arginine-p-nitroanilide and spectrophotometric determination at 405 nm of the p-nitroaniline released. An aliquot of 0.2 ml of the CPC solution (5 mg/2 ml of H_2O; 350 mU of CPC) is mixed with 1.8 ml of 0.2 M triethanolamine/HCl, pH 7.8, and 1 ml of 23 mM N-benzoyl-DL-arginine-p-nitroanilide (1 mg/ml H_2O). One enzyme unit corresponds to the hydrolysis of 1 μmol of substrate per minute at 25° and causes a change in absorbance of $\Delta A^{1\,cm}_{405\,nm} = 3.32$/min.

Assay of Acetylesterase Activity

An acetylesterase activity has been reported as a property of highly purified CPC.[6,7,9] This activity is measured by the hydrolysis of p-nitrophenyl acetate and spectrophotometric determination of the increase in absorbance at 405 nm. A solution of 2 ml of 2 mM p-nitrophenyl acetate in 0.1 M sodium barbital/HCl, pH 7.0, or sodium citrate, pH 5.3, is mixed with 1 ml of CPC solution (100–300 mU of CPC) and incubated at 30°. After 1 min and/or 10 min the change in absorbance is read at 405 nm. One enzyme unit corresponds to the hydrolysis of 1 μmol of substrate per minute at 25° and causes a change in absorbance at pH 7.0 of $\Delta A^{1\,cm}_{405\,nm} = 3.31$/min.

Pretreatment of CPC with DFP[15]

The contaminating aminopeptidase and endopeptidase (trypsinlike and chymotrypsinlike) activities of commercial preparations of CPC can effectively be inhibited by preincubation in 10^{-3} M DFP in buffer of pH 5.3–8 as described below. This concentration of DFP is without significant effect on the activity of CPC from orange leaves.[9,15] It has been reported[7] that the enzyme from citrus fruit under these conditions is inactivated to almost 40%. However, this inhibitory effect might well be caused by the organic solvent (propanol) used for preparation of the DFP solution (1 g of DFP/20–50 ml of propanol or isopropanol), since CPC from orange leaves is highly sensitive to low concentrations of organic solvents.[9,15] The activity is decreased by 35–40% in a 0.4 M aqueous solution of isopropanol. This is the usual concentration of alcohols if 10^{-3} M solutions of DFP are prepared for enzyme treatment from organic stock solutions.

[19] H. Fritz, I. Trautschold, and E. Werle, *in* "Methoden der Enzymatischen Analyse" (H. Bergmeyer, ed.), p. 1021. Verlag Chemie, Weinheim/Bergstr., 1970.

A preparation of carboxypeptidase C suitable for investigation of peptide sequences is obtained by dissolving 25 mg of the commercial CPC + sodium citrate preparation ($\sim$350 mU) in 4.75 ml of 0.05 M sodium citrate, pH 5.3, mixing it with 0.25 ml of freshly prepared aqueous solution of DFP (1 ml in 50 ml of H_2O or if unavoidable in 10–20 ml of isopropanol), and allowing the mixture to react at 20–30° for 30 min.

Preparation of Sample

The peptide or protein to be treated should be denatured and preferably in a random coil state. Peptides and proteins containing disulfide bridges are most conveniently subjected to performic acid oxidation, as has been described earlier in this series (see this series Vol. 11 [19]). Reduction by thiol reagents in denaturing agents, e.g., 8 M urea or 6 M guanidine-HCl, followed by S-carboxymethylation (see Vol. 11 [20] and [62] and Vol. 25 [34a]), S-carboxamidomethylation, S-aminoethylation (see Vol. 11 [33]), or sulfitolysis (see Vol. 11 [22]) are all suitable procedures for preparing peptides and proteins for treatment with CPC. The sample is most conveniently stored frozen or handled as a lyophilized, dry material, in which form it is obtained after performic acid oxidation or after gel filtration on Bio-Gel P-2, or Sephadex G-10 or G-25, or Merckogel 2000 or a similar gel in 10% acetic acid or volatile buffer followed by lyophilization.

Choice of Reaction Conditions

The choice of reaction conditions is the same whether a peptide or a protein is used as a substrate. The amount of enzyme and the period of digestion to be used are best determined by one or two trial experiments with sufficient amounts of peptide or protein for two or more samples to be applied to the amino acid analyzer (see below). Alternative methods for the separation of the products of carboxypeptidase digestion of peptides have been reviewed by R. P. Ambler (see Vol. 11 [14]).

Lyophilized peptide (0.02–0.1 μmol) is dissolved in 0.1 ml of 0.05 M sodium citrate, pH 5.3, and mixed with 35 mU of CPC in 0.05 ml of 0.05 M sodium citrate, pH 5.3. The molar enzyme:substrate ratio is about 1:100 and the peptide substrate concentration depends on the molecular weight but about 0.1–1% (w/w). The mixture is incubated at 30°. Aliquots of 0.01–0.05 ml are withdrawn after 10 and 60 min containing 2–50 nmol of peptide and released amino acids. The suitable amount depends on the sensitivity of the amino acid analyzer. The pH of the samples is immediately adjusted to pH 2.5–3 by addition of 0.06–0.1 ml of 0.2 N

HCl. The samples are frozen and stored frozen until deproteinized and prepared for application to the amino acid analyzer, as described below. The amount of COOH-terminal amino acid(s) released from the peptide within the chosen periods of 10 or 60 min should correspond to the amount of peptide treated. Otherwise more drastic conditions with higher enzyme: substrate ratio and/or longer incubation periods should be selected. The temperature of incubation can be raised to 40–50°, which is the optimum for the hydrolysis of Z-Leu-Phe.[9] If no amino acid is released, hydrazinolysis (see Vol. 25 [9]), preferably after previous oxidation of the protein by performic acid (see Vol. 25 [31]), may reveal the nature of the C-terminal residue and indicate why the protein is not degraded by CPC.

Separation and Detection of Released Amino Acids

Several methods, e.g., the amino acid analyzer, paper electrophoresis, and gel filtration, have been used for separation and detection of the products of the reaction of carboxypeptidases and a peptide or protein, and have been reviewed in other volumes of this series (see Vol. 11 [49] and Vol. 25 [10]). Separation of the whole reaction mixture by an amino acid analyzer has the advantage of being (1) highly sensitive, (2) quantitative, and (3) relatively unaffected by quite large amounts of salts in the sample. The major disadvantages are (1) the need for an amino acid analyzer; (2) little information is gained about the residual peptide,[20] and (3) the necessity to perform additional investigations to identify the amino acid amides, asparagine and glutamine (see below). But, since CPC generally releases more than one or two COOH-terminal residues, isolation of sufficient quantities of homogeneous residual peptide for further studies from small-scale experiments is difficult. The use of an amino acid analyzer is highly recommended for the study of reaction mixtures[20] obtained with CPC, which is characterized by its broad specificity. This method provides quantitative data accurate enough to perform kinetic studies of the amino acid release if internal standards are incorporated in the reaction mixture (see below). However, careful deproteinization of the samples applied to the analyzer is necessary in order to avoid contamination and/or clogging of the analyzer column. A poor

---

[20] In the case of small peptides, which are not precipitated with trichloroacetic acid or by heating (see deproteinization), these have to be removed by appropriate gel filtration. Otherwise the peptides show up in the amino acid chromatogram and might complicate the interpretation of the results. However, the positions of peptides on the effluent curves are generally different from those of the amino acids. In the case of coincidences, the peaks given by peptides on the (highly cross-linked) resins are not as sharp as those given by amino acids.

resolution of the amino acid peaks on the analyzer results from contamination of the column with denatured CPC.

Preparation of the Reaction Mixtures for the Amino Acid Analyzer

The aliquots of the reaction mixture, containing the appropriate amount of internal standard (see below), are mixed with an equal volume of 6% trichloroacetic acid and heated for 1 min to 80–100°. The precipitate of enzyme (and protein substrate) is centrifuged, and the supernatant is extracted several times by three times its volume of ethyl ether in order to remove excess of trichloroacetic acid.[20] Separation of the ether extract and the aqueous solution by decantation is greatly facilitated by freezing the aqueous solution. Finally the aqueous solution is brought to dryness by rotary evaporation or lyophilization and then taken up in the appropriate amount of the first amino acid analyzer buffer and applied to the top of the analyzer column.

Sequence Determination from Release Rates Using Internal Standardization

The COOH-terminal amino acid sequence of a peptide or protein substrate can be deduced from the subsequent release of the individual amino acids. A careful determination of the exact molar quantity of each amino acid released at a given time is a prerequisite to obtain reliable time-release plots that allow deduction of the sequential order of residues (see Fig. 2). Determination of the exact quantity of each amino acid is greatly facilitated by adding a known amount of internal standard. The standard should be added to the enzyme-substrate mixture on a mole/mole basis prior to the start of the reaction. The amount of amino acid released is then calculated by comparison to the amount of internal standard found in each sample of the reaction mixture. Evaluation is based on the ratio of the amount of amino acid released and the amount of standard in the sample [Eq. (1)] and is, thus, independent of losses and inaccuracies during sample processing:

$$[\text{AA, released}]_{\text{theor.}} = \frac{[\text{Standard}]_{\text{theor.}}}{[\text{Standard}]_{\text{found}}} \cdot [\text{AA released}]_{\text{found}} \qquad (1)$$

As an internal standard any nonprotein amino acid can be used that is chromatographically well separated on the amino acid analyzer from all protein-constituent amino acids. Norleucine has proved to be well suited as a standard, since it is commercially available, highly stable, and generally well separated from leucine and isoleucine.

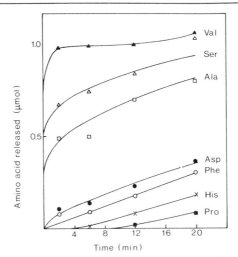

FIG. 2. Rate of release of amino acids from performic acid-oxidized bovine ribonuclease by carboxypeptidase C. Protein concentration 1% (enzyme:substrate ~ 1:100) in 0.1 M sodium citrate, pH 5.3, 30°. The molar amounts were calculated on the basis of norleucine added as an internal standard (see text).

The results from quantitative experiments such as those described below are best interpreted from plots of amino acids released (as mole/mole of peptide or protein) against time (Fig. 2). For unambiguous results, the substrate should be a single-chain peptide or protein. The different types of time-release plots occurring from such experiments and their interpretation have been discussed by R. P. Ambler in this treatise (see Vol. 11 [14]).

Example of Quantitative Experiment.[13] Protein or peptide (1 μmol; e.g., 15 mg of performic acid-oxidized ribonuclease) is dissolved in 0.5 ml of 0.5 M sodium citrate, pH 5.3. An internal standard of 1 μmol[21] of norleucine is added in 0.5 ml of the same sodium citrate buffer. The reaction is started by addition of 1000–5000 mU of CPC dissolved in 0.5 ml of the 0.05 M citrate, pH 5.3. The final concentration of substrate (ribonuclease) is 1% (w/w). The molar ratio of enzyme/substrate based on $M_r = 179,000$[9] for CPC thus approximates 1:100 for a specific activity of 2500 mU/mg. The mixture is most conveniently incubated at room temperature, but higher temperatures of 30° or 40° can be used if the initial experiment has indicated slow release of the first amino acid residues. In this case, however, substrate and enzyme solution should both

[21] The amount of substrate and internal standard were chosen for instruments with normal sensitivity range of 30–80 nmol of amino acid. If the sensitivity of detection is higher, the amount of aliquot withdrawn or the amount of substance in the reaction mixture can be reduced correspondingly.

be preequilibrated to the higher temperature before mixing in order to obtain accurate time-release plots. At convenient time intervals (e.g., 0, 2, 6, 12, 20, 40, 100, and 240 min) 0.15-ml aliquots are withdrawn from the reaction mixture. If the enzyme concentration is reduced to one-tenth of the above amount, the time intervals have to be increased correspondingly (for example see Vol. 25 [21])[22] The reaction is stopped immediately by addition of 0.6 ml of 0.2 N HCl, adjusting the pH to 2.5–3. The samples are frozen unless not immediately subjected to deproteinization as described above. If only a few samples are analyzed on a column it is sufficient to heat the aliquots for 1 min to 70° and to remove the denatured enzyme by centrifugation. The supernatant is then either lyophilized and stored or adjusted to the proper pH for application to the amino acid analyzer column. A blank containing the same amount of enzyme and internal standard as the sample, but no substrate, should be treated in the same way and analyzed for release of amino acids from the enzyme by autodigestion. Generally, it is not necessary to correct the amount of amino acids released from the substrate for quantities generated by autolysis of the enzyme.

Identification of Asparagine and Glutamine

Differentation between serine, asparagine, and/or glutamine on the amino acid analyzer, in general, is impossible without a special program. The problem can be overcome if two samples are withdrawn from the reaction mixture at the same time. The first is analyzed on the amino acid analyzer as described above. The other is freed from residual protein or peptide by gel filtration (e.g., on a column 1 × 100 cm of Sephadex G-10 or G-25 (fine beads), or Bio-Gel P-2, or Merckogel 2000 in 10% acetic acid), and the fraction containing the amino acids is subjected to 2–6 hr of acid hydrolysis in 2 N HCl at 100°. This sample containing serine, aspartic, and/or glutamic acid is then resolved on the amino acid analyzer in the usual way. The amides are then identified by comparison of the two chromatograms.

Qualitative identification of amides is also possible by electrophoresis (see this series Vol. 25 [21]), thin-layer chromatography (see Vol. 11 [5]), preparation of the DNP-derivatives (see Vol. 11 [9] and [14]), or preparation of the amino acid phenyl- or methylthiohydantoins (see Vol. 25 [24] and [25]) followed by the recommended chromatographic or mass spectral identification procedures. In the mass spectrum, asparagine and glutamine give rise to molecular ion peaks with 1 mass unit less than

[22] R. Nehr, B. Riniker, H. Zuber, W. Rittel, and F. W. Kahnt, Helv. Chim. Acta 51, 917 (1968).

the corresponding dicarboxylic amino acids.[23-29] Quantitative determination of the methylthiohydantoins is possible by adding a known amount of ^{15}N-labeled amino acids to the reaction mixture prior to the methylthiohydantoin derivatization (see Vol. 25 [25]).[23] A less expensive procedure is conversion of the amino acids into the corresponding p-bromophenylthiohydantoins and addition of a tetradeuterated p-bromophenylthiohydantoin prior to the conversion and extraction.[29] Quantitative identification is then easily obtained by comparing the heights of the molecular ion peaks of the sample [or quasi molecular ion peaks $(m + 1)^+$ in chemical ionization mass spectrometry] with that of the known amount of tetradeuterated-p-bromophenylthiohydantoin occurring 4 mass units higher in the mass spectrum.[29]

Acknowledgments

    This article is based on work carried out with the skillful assistance of Mrs. S. Kupfer, Miss C. Frank, and Dr. R. Klauser and was supported by the Otto Röhm Gedächtnis-Stiftung and the Deutsche Forschungsgemeinschaft (Bonn-Bad Godesberg, Germany). The author is indebted to Dr. Uhlig and Dr. Sprössler from Röhm GmbH (Darmstadt, Germany) for generous supplies of carboxypeptidase C isolated from orange leaves.

[23] T. Fairwell, W. T. Barnes, F. F. Richards, and R. E. Lovins, *Biochemistry* **9**, 2260 (1970).
[24] H. Hagenmaier, W. Ebbinghaus, G. Nicholson, and W. Vötsch, *Z. Naturforsch. Teil B* **25**, 681 (1970).
[25] F. Weygand and R. Obermeir, *Eur. J. Biochem.* **20**, 72 (1971).
[26] H. Tschesche, E. Wachter, S. Kupfer, and K. Niedermeier, *Hoppe-Seyler's Z. Physiol. Chem.* **350**, 1247 (1969).
[27] H. Tschesche and E. Wachter, *Eur. J. Biochem.* **16**, 187 (1970).
[28] H. Tschesche, G. Reidel, and M. Schneider, in "Proteinase Inhibitors," *(2nd Int. Res. Conf.–Bayer Symp. V)* (H. Fritz, H. Tschesche, L. J. Greene, and E. Truscheit, eds.), p. 235. Springer-Verlag, Berlin and New York, 1974.
[29] H. Tschesche and M. Schneider, *Hoppe-Seyler's Z. Physiol. Chem.* **357**, 1339 (1976).

[8] Carboxypeptidase Y in Sequence Determination of Peptides

By RIKIMARU HAYASHI

    Carboxypeptidase Y was obtained from bakers' yeast and characterized as an enzyme of broad specificity.[1-3] The ability of this enzyme

[1] R. Hayashi, S. Aibara, and T. Hata, *Biochim. Biophys. Acta* **212**, 359 (1970).
[2] R. Hayashi and T. Hata, *Biochim. Biophys. Acta* **263**, 673 (1972).
[3] R. Hayashi, S. Moore, and W. H. Stein, *J. Biol. Chem.* **248**, 2296 (1973).

to release proline is especially useful for structural studies of proteins and peptides. The active site of this enzyme is also unique. The enzyme has no essential metals,[4] but has an active serine at the active site[4,5] and, thus, differs from the pancreatic carboxypeptidases A [EC 3.4.12.2] and B [EC 3.4.12.3]. Carboxypeptidase Y is also unique in its strong esterase activity toward the substrates of chymotrypsin [EC 3.4.21.1], i.e., N^{α}-acetyl-L-tyrosine ethyl ester, in contrast to the pancreatic enzymes which hydrolyze only ester substrates with a free carboxyl group, i.e., Bz-Gly-β-phenyllactate for A, and Bz-Gly-arginate for B. Thus, carboxypeptidase Y appears to be quite similar to chymotrypsin both in mechanism and in active site,[6] although carboxypeptidase Y is an exopeptidase, whereas chymotrypsin is an endopeptidase.

Despite these differences, carboxypeptidase Y is an enzyme that removes amino acids one residue at a time from the carboxyl termini of proteins and peptides. Therefore, in using carboxypeptidase Y for sequence studies of peptides and proteins, methods described for carboxypeptidases A and B can be principally applied.[7] The properties of carboxypeptidase Y have also been reviewed in detail.[8]

Assay Methods

Carboxypeptidase Y hydrolyzes both peptide substrates of the pancreatic carboxypeptidases A and B, i.e., N-acylated dipeptides, and also the synthetic ester and amide substrates of chymotrypsin. Therefore, the activity of carboxypeptidase Y can be determined by methods similar to those for determining carboxypeptidases A[9] or B[10] and chymotrypsin.[11]

In order to evaluate the activity of the enzyme solution, peptidase activity toward benzyloxycarbonyl-L-phenylalanyl-L-leucine (Z-Phe-Leu) or esterase activity toward Ac-Tyr-OEt is recommended.

One unit of activity is defined as the amount of enzyme that produces 1 μmol of the product, or the amount of enzyme that hydrolyzes 1 μmol of substrate per minute in the respective assay.

[4] R. Hayashi, Y. Bai, and T. Hata, *J. Biochem.* **77**, 1318 (1975).
[5] R. Hayashi, S. Moore, and W. H. Stein, *J. Biol. Chem.* **248**, 8366 (1973).
[6] R. Hayashi, Y. Bai, and T. Hata, *J. Biol. Chem.* **250**, 5221 (1975).
[7] See this series, Vol. 25 [10] and [21].
[8] R. Hayashi, Addendum Volume on Proteolytic Enzymes (Vol. 45).
[9] P. H. Petra, this series, Vol. 19.
[10] J. E. Folk, this series, Vol. 19.
[11] P. E. Wilcox, this series, Vol. 19; see also G. W. Schwert and Y. Takenaka, *Biochim. Biophys. Acta.* **16**, 570 (1955).

Assay of Peptidase Activity[12]

Principle. This method is based on the rate of the enzymic hydrolysis of benzyloxycarbonyl-L-phenylalanyl-L-leucine (Z-Phe-Leu). The rate of the reaction can be measured either with the colorimetric ninhydrin method of Moore and Stein for the estimation of liberated leucine, or spectrophotometrically by the decreases in absorbance at 224 nm.

Reagents

Substrate: Z-Phe-Leu (Fluka AG), 0.5 mM in 0.05 M sodium phosphate buffer, pH 6.5

Enzyme, Dilute the enzyme with water or 0.01 M sodium phosphate buffer, pH 7.0, to obtain a solution containing approximately 1 μg/ml for the ninhydrin method or 300 μg/ml for the spectrophotometric method.

Procedure for the Ninhydrin Method. Enzyme solution (0.1 ml) is added to 0.9 ml of substrate solution, which has been prewarmed in a test tube (1.5 $\times$ 17 cm) in a water bath at 25°. After the mixture has been incubated at 25° for 10 min, the reaction is terminated by the addition of ninhydrin reagent (1 ml),[13,14] and color development is performed immediately. The zero-time sample serves as a blank. Under these conditions, approximately 30% hydrolysis occurs.

For samples of unknown activity, the reaction curve should be linear up to an absorbance of about 1.0.

Procedure for the Spectrophotometric Method. Two 1-cm cuvettes, one containing 2 ml of substrate solution and the other 2 ml of buffer, are placed in a double-beam spectrophotometer and allowed to equilibrate at 25°. The wavelength used is 224 nm. The reaction is initiated by adding 10 μl of the enzyme solution to the reaction cuvette, and the change in absorbance is recorded for 3 min. The hydrolysis of 1 μmol of substrate causes a decrease in absorbance of 0.2.

Assay of Esterase Activity[12,15,16]

Principle. This method is based on the titrimetric measurement of the release of protons or upon the spectrophotometric measurement of the change in ultraviolet absorbancy that occurs as a result of the enzymic hydrolysis of Ac-Tyr-OEt.

[12] R. Hayashi, Y. Bai, and T. Hata, *J. Biochem.* 77, 69 (1975).
[13] S. Moore and W. H. Stein, *J. Biol. Chem.* 211, 907 (1954).
[14] S. Moore, *J. Biol. Chem.* 243, 628 (1968).
[15] Y. Bai, R. Hayashi, and T. Hata, *J. Biochem.* 77, 81 (1975).
[16] Y. Bai, R. Hayashi, and T. Hata, *J. Biochem.* 78, 617 (1975).

Titrimetric Method

Reagents

Substrate: Ac-Tyr-OEt, 12.5 mM in H$_2$O
KCl, 1.25 M
Base, 0.01 N NaOH, standardized
Enzyme, 20–30 μg per ml of H$_2$O

Procedure. The assay is carried out in a pH-stat equipped with a thermostated vessel and a syringe containing the standardized base solution. A reaction mixture, which consists of 2 ml of substrate solution and 0.2 ml of KCl solution, is allowed to equilibrate at 25° for 5 min, then is adjusted to pH 8.0. The reaction is initiated by adding 50–200 μl of enzyme solution. During titration, the reaction vessel is covered with a nitrogen gas stream. The rate of the reaction is followed using the rate of consumption of the standardized alkaline solution.

Spectrophotometric Method

Reagents

Substrate: Ac-Tyr-OEt, 1 mM, in 0.05 M sodium phosphate, pH 7.5
Enzyme, 1.2 mg/ml of H$_2$O or 0.01 M sodium phosphate buffer, pH 7.0

Procedure. Two 1-cm cuvettes, one containing 2 ml of substrate solution and the other 2 ml of buffer (0.05 M sodium phosphate, pH 7.5) are placed in a double-beam spectrophotometer and allowed to equilibrate at 25° for 5 min. The wavelength used is 237 nm. The reaction is initiated by adding 10 μl of the enzyme solution to the reaction cuvette, and the change in absorbance is recorded. The hydrolysis of 1 μmol of Ac-Tyr-OEt causes a decrease in absorbance of 0.1.

Purification Procedure

The procedure for a laboratory scale has been previously described.[8] Pure carboxypeptidase Y can be easily obtained from commercial compressed bakers' yeast after 2 weeks of purification procedures.

Yeast protease A[17] and aminopeptidases[18] are the enzymes that could contaminate the purified carboxypeptidase Y. Protease A and aminopeptidases are inactivated by pepstatin and EDTA, respectively. Therefore, when using carboxypeptidase Y for sequence studies, with utmost care,

[17] T. Hata, R. Hayashi, and E. Doi, *Agric. Biol. Chem.* **31**, 150 (1967).
[18] T. Masuda, R. Hayashi, and T. Hata, *Agric. Biol. Chem.* **39**, 499 (1975).

a third chromatography on DEAE-Sephadex A-50 under conditions described previously,[8] or gel-filtration chromatography on Sephadex G-100 using 0.01 M sodium phosphate buffer, pH 7.0, containing 0.1 M NaCl and 10^{-3} M EDTA as the eluent, is recommended. To avoid the action of protease A and aminopeptidases in prolonged digestion of peptides, the addition of pepstatin and EDTA to the digestion mixture may be effective.

Properties

Stability. A thick suspension of the enzyme in saturated $(NH_4)_2SO_4$ can be stored at $-20°$ indefinitely. For use, a portion is dissolved in a minimum volume of H_2O and dialyzed against H_2O or 0.01 M sodium phosphate buffer, pH 7.0, to give an approximately 1% aqueous solution of the enzyme. About 1-ml portions of the solution are stored in vials at $-20°$. The frozen solutions show no loss of activity for at least 2 years. The activity of greatly diluted solutions (below 0.1 mg/ml) is quickly lost; therefore, these should be prepared just before use.[3] Repeated freezing and thawing of solutions of the enzyme, or prolonged storage at room temperature, could lead to autodigestion with the liberation of free amino acids, primarily the amino acids in the COOH-terminal portions of the enzyme. In that case, the enzyme is usable after dialysis as described above.

The salt-free, pure enzyme can also be lyophilized with the loss of approximately 20% of its activity, based on the assay of activity just after the lyophilized powder is dissolved in water. However, almost full activity is restored by leaving the solution overnight at 5° or $-20°$. Therefore, the lyophilized enzyme should be dissolved in water or in 0.01 M sodium phosphate buffer, pH 7.0, 1 day before use.[4]

The enzyme is stable at least for 8 hr in the pH range of 5.5–8.0 at 25°. At higher temperatures, the enzyme is most stable at pH 7.0. The activity is quickly lost in incubation below pH 3 or above 60°.[19]

The enzyme is relatively stable in the presence of protein denaturants or certain solvents,[20,21] where the substrates of the enzyme are either soluble or are in the denatured form so as to render their COOH-termini available for the enzyme. About 80% of the activity is retained after incubation of the enzyme with 6 M urea at 25° for 1 hr. In the presence of 10% methanol, the enzyme is completely stable at pH 5.5–8.0 for 8 hr at 25°. Even in 20% methanol, only 20% activity is lost after incubation of the enzyme at pH 7 for 24 hr.[6] For prolonged incubation of the

[19] E. Doi, R. Hayashi, and T. Hata, *Agric. Biol. Chem.* **31**, 160 (1967).
[20] R. Hayashi, Y. Minami, and T. Hata, *Agric. Biol. Chem.* **36**, 621 (1972).
[21] R. Hayashi, and T. Hata, *Agric. Biol. Chem.* **36**, 630 (1972).

enzyme, therefore, the presence of 10% methanol (or ethanol) would protect against microbial contamination. In the presence of 30% dioxane, 10% 2-chloroethanol, or 60% ethylene glycol, the enzyme is stable at pH 7.0 for at least 15 min.

Physical and Chemical Properties. The enzyme is an acidic protein with a molecular weight of approximately 61,000.[22] The molar concentration of the enzyme is calculated using $A_{280\ nm}^{1\ cm}$ of a 1% solution, which is 15.0, and a molecular weight of 61,000.

The NH_2-terminal residue is lysine[3] and the NH_2-terminal sequence is Lys-Ile-Lys-Asp-Pro-Lys-Ile-Leu-Gly-Ile-Asp-Pro.[23] The COOH-terminal sequence is -Asp-Phe-Ser-Leu.[3] The enzyme is a glycoprotein having about 16 residues of glucosamine and about 15% hexose. The sugar content varies slightly depending on the lot of the enzyme preparation.

Carboxypeptidase Y has one SH-group per molecule as determined with *p*-hydroxymercuribenzoate and by measurement of S-carboxymethylcysteine. The SH-group is not available to iodoacetate, to iodoacetamide, or to the Ellman reagent, 5,5′-dithiobis(2-nitrobenzoic acid), unless the protein has been denatured.[5]

Inhibitors. Carboxypeptidase Y is a diisopropylphosphorofluoridate (DFP)-sensitive enzyme.[5] Its inactivation by [^{32}P]DFP is accompanied by the formation of 1 mol of labeled serine per mole of enzyme. The enzyme is also inhibited stoichiometrically and irreversibly by phenylmethane sulfonylfluoride.[4] Site-specific reagents, i.e., a chloromethyl ketone derivative of benzyloxycarbonyl-L-phenylalanine (ZPCK), inactivate both the peptidase and esterase activities of the enzyme in a much slower reaction than for chymotrypsin.[6] In these respects, carboxypeptidase Y is a serine enzyme having a reactive serine that links with a histidine as in chymotrypsin.

Carboxypeptidase Y is also inhibited by *p*-hydroxymercuribenzoate, probably by its reaction with a single SH-group of the enzyme.[5] The SH-group is assumed to be located near or at the substrate binding site.[24]

Carboxypeptidase Y is sensitive to metal ions.[4] Cu^{2+}, Ag^+, or Hg^{2+} results in a complete loss of activity. Further, Cu^+, Mg^{2+}, Ca^{2+}, Ba^{2+}, Cr^{2+}, Mn^{2+}, Fe^{2+}, Fe^{3+}, Co^{2+}, or Ni^{2+} cause a partial loss in the activity. Therefore, one must take the general precaution of preventing contact with metal ions in purifying and handling carboxypeptidase Y. EDTA and *o*-phenanthroline have no effect on enzymic activity. Trypsin inhibitors from the soybean and the lima bean do not inhibit the activity. All the above inhibitions are seen with both peptidase and esterase activities.

[22] S. Aibara, R. Hayashi, and T. Hata, *Agric. Biol. Chem.* **35**, 658 (1971).
[23] R. W. Kuhn, K. A. Walsch, and H. Neurath, *Biochemistry* **13**, 3871 (1974).
[24] Y. Bai and R. Hayashi, *FEBS Lett.* **56**, 43 (1975).

Product and substrate analogs act as reversible inhibitors of carboxypeptidase Y.[16] L-Amino acids and NH_2-blocked L-amino acids show the competitive type of inhibition. Some phenylalanine analogs, e.g., β-phenylpropionate and *trans*-cinnamate, are also reversible inhibitors of the enzyme. The type and constant (K_i) for these inhibitors are generally parallel for both the peptidase and esterase activities.

Specificity. Carboxypeptidase Y has the ability to remove most amino acid residues, including that of proline, from the COOH termini of proteins and peptides at pH 5.5 to 6.5.[3] In general, however, when the penultimate and/or terminal residues have aromatic or aliphatic side chains, catalysis is most effective. When glycine is placed in the penultimate position, the release of the terminal amino acid is slow. The release of terminal histidine, arginine, and lysine is relatively slow.[12] The cleavage of tripeptides is difficult and dipeptides are completely resistant to hydrolysis.

The K_m values for peptides containing glutamic acid at the penultimate position or the COOH-terminal position increase markedly with a slight change in k_{cat} when the pH of the assay is altered from pH 5.5 to more than pH 6.5.[12] Therefore, peptide bonds involving the carboxyl and/or the amino group of glutamic acid are well hydrolyzed at, or below, pH 5.5 rather than at pH 6.5, where peptides composed only of neutral amino acids are most susceptible.

In contrast to the carboxypeptidases A and B, terminal proline and β-alanine are rather good substrates. However, the rate of splitting of the peptide bond on either side of proline depends extensively on the type of the adjacent amino acid; proline placed on the COOH-terminal side of glycine is not likely to be released.[12] The release of S-carboxymethylcysteine, homoserine, and methioninesulfone has been shown.

Carboxypeptidase Y rapidly hydrolyzes poly-α-L-glutamic acid at pH 4.2, liberating only glutamic acid; poly-L-lysine and poly-L-proline are never hydrolyzed.[2] Poly-α-L-aspartic acid is slowly hydrolyzed by the enzyme.

Carboxypeptidase Y hydrolyzes ester and amide substrates of chymotrypsin.[16] The properties of the amidase action could be applied to the sequence analyses of peptides with amidated COOH-terminal groups, such as oxytocin and vasopressin.

Application to Sequence Studies

Actions of carboxypeptidase Y on peptides and proteins are essentially the same as with other carboxypeptidases. In using the enzyme for sequence studies, one may consult Volume 25 of this series for methods concerning preparation of substrates, separation and identification of the products, and interpretation of the results.

Determination of COOH-Terminal Sequences of Proteins

Carboxypeptidase Y is appropriate to determine the carboxyl-terminal sequences of proteins, because the enzyme acts more strongly on proteins than do pancreatic carboxypeptidases, as has been shown in the hydrolysis of the amyloid A protein[25] and pancreatic deoxyribonuclease.[26] The enzyme is particularly useful at the low pH (pH 4.5) necessary to solubilize the amyloid A protein. For proteins that are difficult to dissolve in a denatured form, digestion in 6 M urea has been used.[3] However, when carboxypeptidase Y is incubated with proteins and large peptides, which are resistant to attack by the enzyme, amino acids, such as leucine and phenylalanine, unexpectedly appeared during prolonged incubation. Experimental conditions in determining carboxyl-terminal sequences of proteins are described below.

Protein (10 mg) is dissolved in 3 ml of 0.05 M pyridine-acetate buffer or sodium acetate buffer, pH 5.5, containing 1.5 μmol of norleucine. Norleucine is used as an internal standard for quantitative determination of amino acids released. Carboxypeptidase Y (50–100 μl) in water or 0.01 M sodium phosphate buffer, pH 7.0 (1 mg/ml), is added and the mixture is incubated at 25°. Aliquots (0.2 ml) are withdrawn at zero time and at appropriate time intervals and mixed with 1 ml of 10% trichloroacetic acid to stop the reaction. The precipitate is then removed by centrifugation. The supernatant is extracted with ether and dried. The sample is then dissolved in 1 ml of 0.2 M sodium citrate buffer, pH 2.2, and amino acid analysis is performed.

For denaturation of protein substrates, buffer for digestion may be made 6 M in urea.[3] Instead of adding trichloroacetic acid, the digestion mixture may be boiled for 5 min to terminate the reaction, and the precipitate that appears can be removed by centrifugation; the supernatant is adjusted to pH 2 with 1 N HCl and analyzed directly for free amino acids.

Sequence Studies of Peptides

Carboxypeptidase Y may be applied to sequence analysis of all peptides. However, it must be kept in mind that the enzyme sometimes fails

---

[25] M. A. Hermodson, R. W. Kuhn, K. A. Walsh, H. Neurath, N. Eriksen, and R. Benditt, *Biochemistry* **11**, 2934 (1972).
[26] T. E. Hugli, *J. Biol. Chem.* **248**, 1712 (1973).

to release aspartic acid and basic amino acids.[3] In particular, hydrolysis of peptides in which basic amino acids are clustered will be limited and no basic amino acids released. Experimental conditions for sequence studies of peptides are described below.

Peptides (1–5 nmol) are hydrolyzed with 5–10 μg of carboxypeptidase Y at 25° or 37°. The buffer used is 0.1 M pyridine–acetate or 0.1 M ammonium acetate, pH 5.5. The hydrolyzate is lyophilized; the residue is dissolved in citrate buffer, pH 2.2, and applied to an amino acid analyzer.

Typical examples of release of amino acids from peptides submitted to hydrolysis with carboxypeptidase Y are shown below.

Example 1

	Val-Val- Ser -Glu-Pro[27]
10 min	0.54 0.83
60 min	0.20 0.91 1.00

Example 2

Lys-CySO$_3$H-Gly-Met(SO$_2$)-Gln-Asn-Pro-Met(SO$_2$)-Arg[28]

4 hr	0.6 0.6 0.7 0.9 1.0

Example 3

Glu-Ala-Phe-His-Pro-Met-Tyr-Gly-Pro -Asp- Ala -Met-Phe- Ser -Gly[29]

3 min	0	ND	0	0	0	0.16 0.56 0.64 0.65 0.74
15 min	0	ND	0	0	0	0.58 0.71 0.91 0.79 0.91
120 min	0	ND	0.15	0	0	0.67 1.00 0.80 0.66 0.94
10 hr	0	ND	0.58	0.73 0.62 0.86 1.54 1.12 1.03 1.42

Other Uses of Carboxypeptidase Y

Carboxypeptidases Y, A, and B can be used in a mixture to hydrolyze peptides that are resistant to hydrolysis by the individual enzymes.

Since the specificity of these carboxypeptidases is different, successive digestion of large peptides by these enzymes provides useful information for a large part of the carboxyl-terminal sequence. The successive digestions are made in a volatile buffer. For example, the first digestion by carboxypeptidase Y is performed in 0.1 M pydidine-acetate buffer, pH 5.5. Then, the reaction mixture is lyophilized and subjected to a

[27] J. Salnikow, T.-H. Liao, S. Moore, and W. H. Stein, *J. Biol. Chem.* **248**, 1480 (1973).
[28] T. E. Hugli, *J. Biol. Chem.* **250**, 8293 (1975).
[29] A. Tsugita, *Proc. Natl. Acad. Sci. U.S.A.*, in press.

second digestion by carboxypeptidase A in 0.1 M N-ethylmorpholine-acetate buffer, pH 7.6. Digestion by carboxypeptidases A and B may be made in ammonium bicarbonate buffer, pH 8.5.

Example 1

Ala-Ser-His-Leu-Gly-Leu-Ala-Arg[28]

Carboxypeptidase Y digestion	
24 hr	1.0 1.7 1.0 0.9
Successive digestion by carboxypeptidase B	
4 hr	0.9
followed by carboxypeptidase Y	
2 min	0.5 0.6 1.0 0.9

Example 2

His-Ile-Ala-Gly-Glu-Ser-Tyr-Ala - His - Gly - Tyr- Ile - Pro- Val -Phe[5]

Carboxypeptidase Y	
digestion	
25°, 10 min	0.05 0.35 0.56 0.56 0.57
25°, 60 min	0.06 0.88 0.99 0.82 0.97
25°, 16 hr	0.77 0.70 0.83 0.82 0.94 1.09 1.02 0.96
Carboxypeptidase A	
digestion	
37°, 60 min	0.07 1.05
Successive digestion by	
carboxypeptidase Y	
25°, 60 min	0.10 0.42 0.90 0.98 0.99
followed by carboxy-	
peptidase A	
37°, 60 min	0.03 0.25 0.18 0.94 1.02 1.07 1.00 1.05
37°, 3 hr	0.25 0.71 0.75 1.07 1.01 0.89 1.05 0.95

The optimum pH for the release of COOH-terminal neutral and basic amino acids is 6–7, while that for the release of COOH-terminal acidic amino acids is at, or below, pH 5.5.[12] Therefore, around pH 7, the enzyme hydrolyzes peptides through the neutral and basic amino acids, but not through the acidic amino acids, whereas at pH 5.5, the enzyme hydrolyzes peptides through acidic amino acids. In addition to these properties, using the difference in specificity of carboxypeptidases A, B, and Y, it is possible to remove selectively a limited part of the COOH-terminal sequence of a protein or peptide. For example, ten COOH-terminal amino acid residues of pancreatic ribonuclease A have been removed.[30] The modified RNase A was reactivated by mixture with synthetic peptides containing 9 or 10 residues of the COOH-terminal sequence.

[30] R. Hayashi, S. Moore, and R. B. Merrifield, *J. Biol. Chem.* **248**, 3889 (1973).

Section III

Chain Separation

[9] Gel Filtration

By C. H. W. HIRS

General. The present article is intended to serve as a guide to those whose primary interest in gel permeation chromatography is in the use of the techniques as a separation tool in protein chemistry. Applications concerned with molecular weight determination and the investigation of interactions between proteins and ligands are dealt with in other contributions to this series (see Volume 26, pp. 28–42).

Permeation chromatography as applied in enzyme and protein chemistry is usually performed with hydrophilic matrices based on cross-linked polysaccharides (e.g., Sephadex, Agarose) or polyacrylamide (e.g., Bio-Gel) and is often termed gel filtration for simplicity.[1] Equilibrated with the eluent, the matrix typically is in the form of relatively fine particles, either spherical or irregular in shape. In gel filtration, as in all chromatographic processes, separations occur when solutes have different distribution coefficients[2] between the stationary gel matrix and the moving eluent. Although the distribution coefficient in an ideal gel permeation separation should reflect the extent to which a solute can permeate unit volume of equilibrated matrix, in actuality additional effects are frequently involved, notably adsorption of solute on the matrix, as in conventional adsorption chromatography, and ion exchange caused by charged groups formed on the matrix as a result of chemical modification by impurities in the eluents employed. In consequence, the distribution coefficient, and thus the extent of retention of the solute during chromatography, is not necessarily a straightforward function of those quantities most important in determining the extent of gel permeation, viz., molecular volume and shape. While this circumstance is unsatisfactory from the theoretical standpoint, the experimentalist interested in obtaining separations should recognize that the existence of absorptive

---

[1] An excellent review of the theoretical background to gel filtration has been provided by G. K. Ackers *in* "The Proteins," 3rd ed. (H. Neurath and R. L. Hill, eds.), pp. 1–92, Academic Press, New York, 1975.

[2] In gel filtration the term distribution coefficient (or partition coefficient) is used to describe the fraction of the internal volume of the gel accessible to the solute. Operationally this coefficient is equivalent to the distribution coefficient employed in conventional chromatography and defined as the ratio of the concentrations at equilibrium of the solute in the stationary and moving phases. However, the former can have values only between 0 and 1, while the latter can have values anywhere between 0 and infinity.

and ion-exchange contributions may on occasion be exploited to advantage.

Selection of Matrix. To a first approximation most of the molecules with which a protein chemist deals are either globular or exist in solution as random coils. In consequence, those hydrodynamic properties important in determining the extent to which members of these two classes of molecules permeate into the gel filtration matrix are generally comparable functions of the molecular weight. It is customary, therefore, to determine the choice of the matrix in a gel filtration experiment on the basis of the molecular weights of the solutes to be fractionated, to the extent these are known or can be estimated.

If the solutes to be separated are of similar molecular weight, a single matrix may prove sufficient for adequate separations. However, when a mixture of solutes of widely differing molecular weights must be dealt with, it may be necessary to use a succession of matrices to obtain a satisfactory separation of the components (see Fig. 1 for an example). The performance to be expected of a given matrix can be estimated from specifications provided by the manufacturer. The most significant quantity is the exclusion limit, or the molecular weight for a given class of solutes (spherical molecules or random coils) at which the distribution coefficient just reaches zero. Molecules with molecular weights in excess of the exclusion limit may be anticipated to pass through a column of the matrix unretarded. For molecules with molecular weights below the exclusion limit there is a molecular weight range in which the distribution coefficient is essentially directly proportional to the logarithm of the molecular weight. This range usually corresponds to values of the distribution coefficient between approximately 0.1 and 0.7. Outside this range the relationship is nonlinear. In general it is an advantage to confine separations to the linear range. For a given pair of closely similar solutes, the separation efficiency will obviously increase the larger the distribution coefficient can be set by proper choice of matrix.

In selecting a matrix, careful attention should be given to the stability of the matrix in the eluent needed to handle the solutes being fractionated. Matrices of the polysaccharide type are sensitive to solvents with low pH values and undergo hydrolytic breakdown under these conditions. In addition, polysaccharide matrices are quite sensitive to oxidation by oxygen dissolved in the eluent, which leads to the introduction of carboxyl groups and thereby confers ion-exchange properties on the matrix. Polyacrylamide matrices are sensitive to hydrolytic attack at extremes of pH and undergo deamidation. The carboxylate functions formed in this fashion likewise confer ion-exchange properties on the matrix. These sensitivities underline the importance of using, wherever

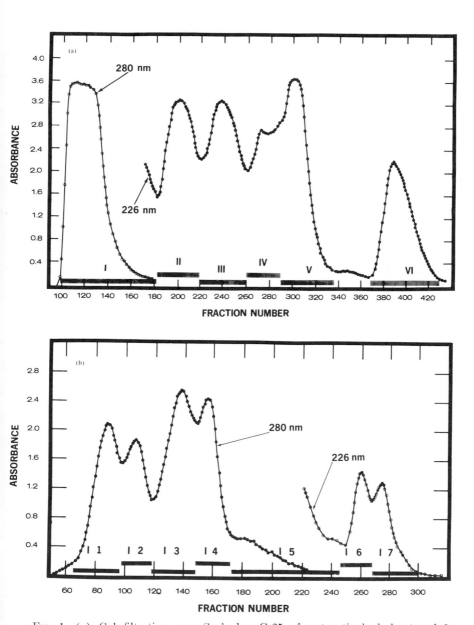

Fig. 1. (a) Gel filtration over Sephadex G-25 of a tryptic hydrolysate of 9 μmol of maleylated, reduced, S-aminoethylated bovine carboxypeptidase B. The column, 2.5 × 200 cm, was equilibrated with 0.2 M ammonium bicarbonate and operated at 25° at a flow rate of 5 cm/hr (25 ml/hr throughput). The progress of the fractionation was monitored by absorbance measurements at 226 and 280 nm on 2.7-ml fractions accumulated in a fraction collector. The fractions combined for further analysis are shown by bars. (b) Gel filtration on Sephadex G-50 of the peptides in fraction I of (a). The column, 2.5 × 200 cm, was equilibrated with 0.2 M ammonium bicarbonate and operated at 25° at a flow rate of 5 cm/hr (25 ml/hr throughput). Other details as in (a). From J. J. Schmidt and C. H. W. Hirs, *J. Biol. Chem.* **249**, 3756 (1974). Reproduced by courtesy of the *Journal of Biological Chemistry.*

possible, eluents with pH values in the neutral range. If a column is to be used repeatedly, thorough deaeration of the eluent will minimize oxidative modification of the matrix.

Resolution in gel filtration is determined by the degree to which equilibrium conditions are approached in the operation of the column. The rate of equilibration is diffusion controlled and will increase with temperature and state of subdivision of the matrix. Although technological improvements have permitted the development of stable matrices of extremely small particle size, an important limitation is the deformation which many matrices undergo when subjected to pressure. This deformation restricts the flow rate which may be used with finely divided matrices and thus offsets the advantages to be gained by reduction of particle size. In practice, therefore, choice of the particle size to be used in a given separation must involve a compromise between the opposing requirements for practicable flow and maximum subdivision of the matrix. An important new development has been the introduction of gel filtration matrices suitable for use in high-pressure liquid chromatography. These matrices are prepared by coating such materials as porous glass with a layer of cross-linked polysaccharide. The products are deformation-resistant and permit gel filtration at the high flow rates characteristic of high-pressure liquid chromatography. Obviously, of course, such matrices do not possess the capacity of isotropic matrices, and their principal use will be in analytical applications.

Swelling the Matrix. Commercially available matrices are usually sold as dry powders and must be swollen in the eluent to be used in the separation before they are poured into a chromatograph tube. Adherence to recommended procedures is of critical importance in assuring that the performance of the column will accord with specifications. Since swelling can be a time-consuming step, particularly with matrices which have a low degree of cross-linking and swell extensively, it is advisable to schedule the swelling well in advance of an experiment.

The quantity of matrix to be taken for an experiment can be determined from the water regain value provided by the manufacturer. In many instances the volume of the packed gel matrix per dry gram is also given. The requisite quantity of dry matrix is placed in a filter flask of such size that the swollen matrix will occupy about one-quarter of the total volume. Sufficient eluent to approximately half fill the flask is then added, and the contents are thoroughly mixed by shaking. The flask is stoppered, leaving the sidearm open, and placed in a boiling water bath for the period of time specified for that particular type of matrix by the manufacturer. The contents should be agitated thoroughly at intervals of 15 min during the heating step. After cooling, a magnetic stirring bar is introduced and the suspension is stirred gently for at least 1 hr while the flask is evacuated with a water aspirator. Vigorous stirring should be avoided as it tends to generate fines by breaking down particles of the matrix. In general, spherical matrices exhibit greater mechanical resilience than those obtained by crushing bulk polymer. Deaeration of the matrix will occur under gentle boiling of the liquid and will minimize the tendency for bubble formation when the matrix is packed into a column and percolated with eluent.

Assembly of Column. To take full advantage of the resolution afforded by a gel filtration column it is important not to compromise that resolution by allowing liquid to accumulate in pockets and dead zones on the effluent side of the column. Mixing will take place in these regions and obviate the separations obtained to extents directly related to the volume of fluid held up. This problem can be minimized by careful attention to the design of the fittings used on gel filtration columns. A number of commercially available systems perform very satisfactorily (e.g., columns manufactured by Altex, Laboratory Data Control, Glenco, Pharmacia, etc.) and are characterized by effluent fittings designed to minimize hold-up and mixing. Entirely comparable performance may be attained on columns fabricated in the laboratory from readily available commercial materials. We have found this to be the preferred course when large, preparative-scale columns are needed with diameters between 7.5 and 18 cm. Such columns can be fabricated relatively inexpensively to satisfy the needs of a particular separation.[3] Connections between the

---

[3] Our experience has shown that Lucite is a convenient material for the fabrication of such equipment. Its chemical properties are compatible with most applications involving aqueous solvents, and it can be machined by unskilled workers. Chromatograph tubes can be made by closing the end of a segment of tubing with a Lucite disk cut to the correct diameter, using methylene chloride for cement. The disk is drilled and tapped at the center to accept a standard Chromatronix tube fitting. The column support is cut from ¼-inch porous polyethylene stock to just fit snugly in the tube and is held about 1 mm from the bottom plate with small inserts of sheet Lucite cemented in a symmetrical pattern to the plate. The chromatograph tube is closed at the top with a cylindrical plunger fabricated from 1-inch Lucite stock and cut to give sufficient clearance between its outer surface and the inside of the tube to accommodate O-rings under moderate compression. To provide proper alignment the plunger is grooved to accept two O-rings at a separation of about ¾ inch. Manipulation of the plunger is made possible with the aid of an extension of suitable diameter cemented to the plunger. A 1.5-mm hole is drilled through the plunger and into the extension for a sufficient distance to meet a hole of similar diameter drilled at right angles to the axis of the extension. The latter hole is also tapped to accept a standard Chromatronix tube fitting. The depth to which the plunger is inserted is fixed with the aid of a yoke attached to a collar cemented around the top of the chromatograph tube. If the tube requires

column and pumps, detectors, etc., are made with $\frac{1}{16}$- or $\frac{1}{8}$-inch Teflon tubing, depending on the size of the column and the rate at which eluent is passed through.

Columns should always be poured at a temperature equal to or higher than the temperatures at which they are to be used. Failure to observe this precaution can lead to bubble formation when the columns are used subsequently.

For pouring, the exit line from the chromatograph tube is closed off and a careful check made that the tube is precisely vertical. The well-equilibrated, swollen matrix is now carefully suspended in a volume of eluent twice that of the settled matrix, preferably at the same temperature as the tube. Then, as much of the slurry as possible is poured into the tube. Some authorities recommend pouring all the matrix at once, using an extension on the tube equipped with a stirrer; others recommend pouring the column in sections without stirring. In our experience comparable performance is obtained with columns poured by either of these procedures, although more care is usually required if the columns are poured in sections. The latter method will be described here because it requires less equipment and is the only method really suitable for very large columns.

After the slurry has been poured into the tube and all air bubbles have risen to the surface, the exit line is opened and effluent is diverted to a waste line. As the column begins to accumulate in the tube, a careful check should be made to see that the upper surface is completely flat. If it is not and there is a tendency for buildup to form on one side, the tube almost certainly has not been supported completely vertically and the column should be started again. If a concave surface develops initially, this may be ignored if it does not persist. It is usually due to the slurry having been poured into the tube too precipitously. Pouring should be slow and the slurry should be directed at the wall of the tube close to the top. In this fashion there will be little or no mixing as the slurry descends in a film to the liquid surface.

As the column accumulates, the surface should be checked periodically to be certain it remains perfectly flat and to determine when it no longer continues to sink. When this point is reached the exit tube is

a jacket for water circulation, this can readily be fabricated and cemented into place.

In use such tubes function most satisfactorily if a diffuser screen is inserted between the plunger and the matrix. Such a screen can conveniently be made by cutting a disk of suitable diameter from $\frac{1}{4}$-inch coarse polyethylene stock. Hold-up in the end fittings of such tubes is not greater than is found with commercial equipment.

closed again and supernatant fluid in the tube is removed by aspiration until only 2–3 cm remains above the surface of the column. The next lot of slurry is now poured, again filling the tube to the top and taking great care not to disturb the surface of the column. A convenient way to control addition of slurry at this critical point is to pipette it in along the walls of the tube in small portions. The speed of addition can then be increased as the liquid head grows. Settling is continued as before, the entire operation being repeated with progressively smaller additions of slurry until the desired, nominal column height has been established. The tube is now provided with an inlet fitting and eluent is pumped through the column at a rate of flow between 2 and 6 cm/hr.[4] Further settling of the column usually occurs. Eluent flow is maintained until inspection reveals that the matrix bed has settled fully.

Final adjustment of the height of the column can now be made. If matrix needs to be withdrawn, a 4–5 cm layer of eluent should be added over the top of the column and stirred gently with a glass rod so as to resuspend the matrix in progressively larger quantities until the desired level is reached. Stirring is stopped, and the suspension is quickly aspirated off the top of the column. Fresh eluent is added and the stirring is reinitiated and continued until 1 or 2 cm of matrix have become suspended. The matrix is now allowed to settle to produce a flat column surface. The top fitting is reintroduced, and fresh eluent is pumped through the column for a final settling.

Application of Sample. In gel filtration the volume of sample applied to the column determines the resolution attainable. For critical separations it is essential to keep the sample volume at less than 2% of the internal volume of the column.[5] Obviously, therefore, it is important to use eluents in gel filtration in which the solubility of the components to be separated is high; otherwise, the capacity of the column will be too small to be useful.

Because resolution is sensitive to sample volume, there are distinct practical advantages to the use of special column inlet fittings which permit sample application with a minimum of dilution or rinsing. When

---

[4] In specifying column flow rates it is often advantageous to use the flow rate per unit area of column, with dimensions of centimeters per unit of time. It should be realized, however, that the velocity thus expressed is the rate at which the solvent front would move if the tube were empty. Obviously, in a packed chromatograph tube, the actual rate of movement would be larger in proportion to the fraction of the total volume of the packed bed occupied by impermeable matrix.

[5] For critical separations the internal volume of the column should be measured using a solute of low molecular weight. We have found it convenient to use NaCl, which is readily detected in the effluent by measurement of the conductivity.

properly adjusted, such fittings reduce the quantity of free liquid at the top of the column to a negligible level. The sample may then be applied through a suitable sample loop (see Volume 25, pp. 1–9) or from a syringe. In either case, no rinsing is required. The rate at which sample is added is best controlled with a pump and in any event should not exceed the rate at which the column will be operated subsequently.

Pumps. Although gel filtration columns can be operated by gravity from suitably placed eluent reservoirs (Mariotte bottles are convenient because a constant pressure head can be maintained), the use of pumps offers a number of significant advantages in terms of reproducibility of flow rates and conveniences of controlling the columns.

Since gel filtration columns packed with isotropic matrices cannot be operated successfully under significantly elevated pressures, peristaltic pumps are entirely satisfactory for such columns. Our experience has shown that pumps of the type distributed by LKB Industries or Pharmacia perform very satisfactorily over prolonged periods.

The pump may be placed on the inlet or outlet side of the column, or both if it is a multiple-channel pump. Care should be taken to be certain that the tubing used in the pump does not constitute a significant source of mixing in the fluid lines to and from the column. This is best achieved by the use of tubing of nearly the same internal diameter in the pump as the tubing making up the fluid lines.

In most applications the preferred location for the pump is on the outlet side of the column. In this configuration the pump draws liquid through the column and there is less tendency for the release of gas bubbles in the lower parts of the column.

If prolonged use of a column is contemplated, precautions should be taken to deaerate the eluent and remove dust particles from it. As indicated earlier, deaeration is essential if oxidation of the matrix is to be minimized, particularly when this is of the polysaccharide variety. Dust particles in eluents ultimately end up in the top of the column and are one of the principal factors contributing to the increase in back-pressure observed during prolonged use of a column.

Deaeration is conveniently achieved by placing the eluent in a serum bottle of such capacity that only about one-third of the volume is used and stirring the liquid magnetically while the bottle is evacuated with a water aspirator. After 1.5–2 hr the eluent is transferred to a reservoir bottle and stored under nitrogen.

Dust is satisfactorily removed from eluents by careful filtration through a suitable grade of analytical filter paper, such as Whatman No. 12 paper. Filtration can be expedited by fluting the paper.

Increasing column back-pressure can also be caused by penetration of the filter disk at the end of the column by particles of matrix. This is particularly frequently found with matrices that swell extensively on exposure to the eluent and are more prone to deform under pressure. The problem can be minimized if the bottom 2–3 cm of the column are filled with the same matrix, but more highly cross-linked. The different extent of cross-linking will make this segment of the column readily distinguishable visually, so that the two forms of the matrix can be separated mechanically when the time comes to dismantle the column.

Monitoring Separations. A variety of procedures have been found to have a relatively broad range of application in protein chemistry. Of these the nonconsumptive clearly are to be preferred because no material is lost in detection. Perhaps the most widely used of the nonconsumptive procedures is measurement of the absorbance of the effluent at 226 or 230 nm, a region in which light absorption by the peptide bond becomes significant and thus allows considerable sensitivity in detection and measurement of peptides and proteins. The procedure is limited to eluents which transmit light well at these low wavelengths.[6] A much larger variety of eluents transmit light well at 280 nm. Proteins and peptides containing aromatic amino acid residues can be detected at this wavelength, but generally with less sensitivity than is possible at 226 or 230 nm.

Although the fractionation achieved can be monitored entirely satisfactorily by making absorbance measurements on the individual effluent fractions accumulated in a collector, much time is saved, particularly when many experiments must be performed, by continuously recording the absorbance of the effluent at a suitable wavelength with an ultraviolet monitor. A number of commercially available instruments (e.g., LKB, Laboratory Data Control, Altex) are designed for such application. They are equipped with micro cells in which there is negligible mixing of effluent. Variable wavelength as well as fixed wavelength (filter) instruments are offered. For most purposes the less expensive, filter-equipped monitors are entirely satisfactory. Sensitivity and baseline stability in these instruments are excellent, and continuous operation is possible for days without significant baseline drift.

Among consumptive procedures for monitoring the progress of gel filtration experiments, frequent use has been made in the past of the ninhydrin reaction (see Volume 11, pp. 325–329) and the Lowry pro-

---

[6] An eluent which has proved useful in a wide range of applications is ammonium acetate buffer at acetate concentrations between 0.05 and 0.1 M. This buffer is volatile and is readily removed during lyophilization. If added ionic strength is required NaCl may be used, but recovery of solutes will then require a suitable desalting step.

cedure (see Volume 3, p. 448). These procedures are being replaced by fluorescence procedures based on fluorescamine (see this volume [26]) and o-phthaladehyde (see this volume [1, 26]), which afford significantly higher sensitivity.

Recycling. The separation of two solutes in gel filtration is directly proportional to the volume of the column. In principle, therefore, separations can be improved by increasing the dimensions (usually the length) of the column. In practice, however, there are mechanical problems when substantial increases in column dimensions are attempted. For example, to increase separation, it is possible to connect several identical columns in series, but difficulties are encountered when this is done because of development of gas bubbles at the inlet fittings of the columns farther down line. An alternate and more practicable solution is to recycle the effluent repeatedly through the same column thereby gaining the same effect as increasing column dimensions.

Operation of a column in the recycling mode requires that there be provision for continuous monitoring of the absorbance at 226 or 280 nm of the effluent from the column. Suitable valving must also be incorporated to permit flow of eluent from a reservoir into the column at those times when effluent is being diverted to the fraction collector. As the components in the mixture being separated emerge from the column for the first time, a careful check should be made of the width (in volume units) of the entire effluent profile. It should be recognized that this width will increase in direct proportion to the number of passages through the column. Regardless of the initial width, therefore, the effluent profile will eventually attain a width larger than the internal volume of the column, at which point the most rapidly moving components in the original mixture will begin contaminating the slowest. The wider the initial band, the sooner this contamination will set in. For this reason components moving either much more slowly or rapidly than the components desired should be diverted to the fraction collector as soon as possible. In this regard it is important for the correlation of fraction number with recorder position to have an accurate calibration of the volume of fluid in the effluent lines between the flow cell in the detector and the delivery tip of the fraction collector.

A further factor to be taken into consideration is the diminution in height of the elution profiles as recycling occurs. This height diminishes by a factor of $1/\sqrt{2}$ for each passage. When the initial height is relatively low, this effect during recycling can occasion difficulty in detecting the desired component unless a detector with a high sensitivity range is employed.

Maintenance of Aseptic Conditions. Many of the eluents used in gel filtration and the matrices themselves are susceptible to bacterial and fungal attack. The consequences of such attack can be serious because it is usually impossible to recover a matrix satisfactorily after such contamination. Although additional effort is required, routine use of sterile distilled water in the preparation of solutions and the addition of various antibacterial and fungal agents to all prepared solutions is strongly recommended. In our experience the only really effective preservatives are sodium azide and a product marketed by ICI (America) under the name chlorhexidine.

Recovery of Matrix after Use. If proper precautions have been observed to keep a column aseptic during use, the matrix may be recovered and stored for subsequent use. The savings thus achieved can be substantial. The following procedure has been used routinely for matrices operated in aqueous solvents. The matrix is transferred from the column to a beaker by flushing the column with a stream of distilled water administered through a piece of flexible plastic tubing. After the matrix has settled, the supernatant fluid is removed by aspiration. The matrix is now suspended in an equal volume of 0.1 N NaOH made 1 N in NaCl. The suspension is stirred gently overnight. We have found this treatment to be effective in removing small quantities of denatured protein adhering to the matrix. The matrix is then washed copiously with distilled water, using decantation, until the washings are neutral. At this point dehydration is begun by washing the matrix with 20% ethanol, again by decantation. This, in turn, is followed by successive washes with 40%, 60%, 80%, and finally 100% ethanol. The dehydrated matrix can then be stored dry in the cold.

If the matrix has been used in eluents which are predominantly organic, e.g., 50% acetic acid, the treatment with sodium hydroxide may be omitted.

Section IV

Cleavage of Disulfide Bonds

[10] Reductive Cleavage of Cystine Disulfides with Tributylphosphine

By URS TH. RÜEGG and JOSEF RUDINGER*

General Considerations

Disulfides in proteins are customarily reduced with mercaptans, such as 2-mercaptoethanol, thioglycolic acid, cysteine.[1] Molar excesses of 20- to 1000-fold are used in order to drive the equilibrium reaction

$$R_1\text{-S-S-}R_2 + (R_3\text{-SH})_n \rightleftarrows R_1\text{-SH} + \text{HS-}R_2 + R_3\text{-S-S-}R_3 + (R_3\text{-SH})_{n-2}$$

toward the desired products.

Dithioerythritol and dithiothreitol[2] favor the reaction to go to the right because their oxidized products form a thermodynamically stable 6-membered ring. However, even these reagents have to be used in an approximately 20-fold molar excess in order to get close to 100% reduction of a protein.[3]

After the above equilibrium is established, the mixture containing the rather labile mercapto protein is, in most cases, treated with an alkylating agent, which has to be used in an even larger excess than the thiol reagent. Apart from the risk of reoxidation, this "double excess procedure" can lead to undesirable (side) products, because most—if not all—alkylating agents will react to a certain extent with nucleophilic groups other than sulfhydryl in proteins. In order to avoid these drawbacks, more potent reducing agents with a nonequilibrium behavior have been sought.

Simple organic disulfides may be reduced in the presence of water by derivatives of trivalent phosphorus (for references see Grayson and Farley).[4] This finding has been applied to wool,[5] a complex and insoluble protein with still unknown structure. In particular, tri-n-butylphosphine was shown to be a potent and specific agent for the cleavage of wool

* Deceased April 30, 1975.
[1] R. Cecil, in "The Proteins" (H. Neurath, ed.), 2nd ed., Vol. 1, p. 394. Academic Press, New York, 1963.
[2] W. W. Cleland, Biochemistry 3, 480 (1964).
[3] T. A. Bewley and Choh Hao Li, Int. J. Pept. Protein Res. 1, 117 (1969).
[4] M. Grayson and C. E. Farley, in Chimie Organique du Phosphore, Colloq. Int. C.N.R.S., No. 182, p. 275 (1969).
[5] B. J. Sweetman and J. A. Maclaren, Aust. J. Chem. 19, 2347 (1966).

cystine.[6] Kinetic studies additionally indicate trialkylphosphines to be among the most reactive phosphines.[4] Tris (2-carboxyethyl)phosphine and tris(hydroxymethyl)phosphine have been used for the partial reduction of a γ-globulin.[7] These phosphines, however, are not commercially available and appear to be much less reactive than tributylphosphine.

In order to study the applicability of the tributylphosphine reduction to general protein chemistry, we have carried out some work on highly purified preparations of soluble proteins with known structure.[8,9] Under the conditions described below, we find the reduction of protein disulfides with tributylphosphine to be as follows:

1. Stoichiometric: One mole of tributylphosphine reduces 1 mol of disulfide, as can be expected from the postulated mechanism[5]:

$$R_1-S-S-R_2 \quad \longrightarrow \quad \left[\begin{array}{c} R_1-\overset{\ominus}{S}-S-R_2 \\ Bu_3\overset{\oplus}{P} \quad \overset{\ominus}{OH} + H^{\oplus} \end{array} \right] \quad \longrightarrow \quad R_1-\overset{\ominus}{S} + R_2-\overset{\ominus}{S}$$
$$Bu_3P + H_2O \qquad \qquad \qquad \qquad Bu_3PO + 2H^{\oplus}$$

2. Quantitative: All peptides and proteins studied ([Lys^8]-vasopressin, insulin,[8,9] human serum albumin,[10] bovine ribonuclease) can be fully reduced with a 5–20% molar excess of tributylphosphine. No strong denaturing agent such as urea or guanidinium chloride seems to be necessary, except when (partially) reduced proteins precipitate.

3. Rapid: From experiments with a model compound[11] under essentially the same conditions as for proteins, we found $t_{1/2}$ to be about 1 min. Preliminary experiments with insulin[12] indicate that the reaction goes to completion in 30–40 min.

4. Specific for disulfides: With large (e.g., 10-fold) excess of PBu$_3$ to protein disulfide, no modification of other residues than cystine could be detected, as judged by electrophoresis and amino acid analysis of the hydrolyzates. Elimination of sulfur ("desulfurization") occurs to some extent with other phosphines at more acidic pH.[5] However, even

---

[6] J. A. Maclaren and G. J. Sweetman, *Aust. J. Chem.* **19**, 235 (1966).

[7] M. E. Levison, A. S. Josephson, and D. M. Kirschenbaum, *Experientia* **25**, 126 (1969).

[8] J. H. Seeley, U. T. Rüegg, and J. Rudinger, *Proc. Eur. Pept. Symp. 12th*, p. 86 (1973).

[9] U. T. Rüegg and J. Rudinger, *Isr. J. Chem.* **12**, 391 (1974).

[10] U. T. Rüegg and J. Rudinger, *Int. J. Pept. Protein Res.* **6**, 447 (1974).

[11] The reduction of (Boc-Cys-OH)$_2$ (1 mM) with PBu$_3$ (1.1 mM) in 0.1 M NaHCO$_3$/propanol (1:1) at 25° was followed spectrophotometrically by measuring the decrease of the absorbance at 249 nm.

[12] Determined by following the uptake of base during the reduction of insulin (1 mM) with PBu$_3$ (3.6 mM) in H$_2$0/propanol (1:1) in an autotitrating device at pH 8.3.

when reducing insulin with PBu_3 at acid pH values, we detected no desulfurization.[13]

5. Compatible with many alkylating agents: Phosphine and the resulting phosphine oxide are both unreactive toward ethyleneimine,[6,8,9] epoxypropane,[6] chloroacetate,[6] 2-vinylpyridine,[14] 4-vinylpyridine,[15] and 1,3-propane sultone.[10] Therefore, reducing and alkylating agents can be added together. "Trapping" of the newly generated protein thiol with one of these reagents should allow selective partial reductions.[16] Iodo- and bromoacetic acid and their amides react with PBu_3 to form a quaternary phosphinium salt. This is much less reactive than PBu_3, and it is therefore advisable to carry out reduction and alkylation in two steps.[15]

Procedures

The reduction is carried out best at room temperature at slightly alkaline pH in Tris buffer (0.1 M) or in bicarbonate (0.5 M). Reduction at acid pH can be performed but is much slower,[13] and the resulting thiols are unreactive in respect to alkylation under acidic conditions. 1-Propanol[6] (20–50%, v/v) has to be added in order to dissolve the phosphine reagent. In our hands, addition of propanol improved the solubility of proteins quite often, and the solvent can easily be removed by rotary evaporation. For proteins with poor solubility, urea, guanidinium chloride, or dimethylformamide can be used (e.g., 8 M urea in 0.1 M Tris·HCl (pH 8.2)/propanol 1:1) together with the conditions described below.

All examples given are reductions and *in situ* alkylations. Reduced proteins are very often poorly soluble and therefore difficult to handle, and—more important—their tendency to be reoxidized renders proof of complete reduction difficult.

Reduction and S-alkylation with ethyleneimine[17] of insulin and human serum albumin exemplifies the "one-batch process,"[6] i.e., addition of tributylphosphine and alkylating agent at the same time. A similar pro-

---

[13] Insulin (5 mg/ml) was reduced with PBu_3 (5-fold excess) in 8 M urea in 0.1 M acetate buffer, containing 20% propanol (pH 4.8) and in 0.1 M AcOH, 8 M urea containing 20% propanol (pH 3) for 24 and 48 hr, respectively. The reduced material was desalted by gel filtration in acid media (50% AcOH), but partial reoxidation (10–20%) could not be prevented.

[14] M. Friedman and A. T. Noma, *Text. Res. J.* **40**, 1073 (1970).

[15] J. A. Maclaren, *Text. Res. J.* **41**, 713 (1971).

[16] Partial reduction has been investigated with insulin[9]; however, the product was heterogeneous, most likely owing to rapid thiol–disulfide interchange.

[17] M. A. Raftery and R. D. Cole, *Biochem. Biophys. Res. Commun.* **10**, 467 (1963); *J. Biol. Chem.* **241**, 3457 (1966); and R. D. Cole, this series, Vol. 11, p. 315 (1967).

cedure, reduction and S-sulfopropylation,[10] is described elsewhere.[18] Reduction and S-carboxymethylation[19] has to be carried out in a "sequential one-batch process"[6]; i.e., iodoacetate is added only after termination of the reduction step.[15] The reduction and S-carboxymethylation of RNase is given as an example of this method.

If reductions are to be performed with a very small excess of PBu₃ (5–20%), as described for insulin and serum albumin below, it is advisable to exclude oxygen completely. This can be achieved as follows: The protein is dissolved in a sealed flask connected with a vacuum pump and a nitrogen cylinder. The flask is evacuated with stirring or shaking and then filled with nitrogen. This procedure is repeated two or three times, and reagents are added under positive nitrogen pressure.

Owing to the high selectivity of PBu₃, larger excesses (e.g., 2–10-fold) of the reagent can be used. Traces of oxygen present initially in the reaction mixture can then be tolerated.

Tributylphosphine. This colorless liquid with a fairly unpleasant smell is very toxic. We strongly advise performing all manipulations involving more than an analytical amount (a few micromoles) in the hood.

Since PBu₃ is readily oxidized by air, commercial samples are only 70–85% pure, most of the remainder being the unreactive oxide (OPBu₃) and some phosphinite esters of the type BuOPBu₂.[20] Purification is therefore not necessary, but it is advisable to keep the reagent bottle and all solutions under nitrogen. For small-scale work, we have been using 1–5% (v/v) solutions of PBu₃ in 1-propanol. The "reducing capacity" can be determined as described below and remains constant over a period of many months. As an approximation, a 2% solution of PBu₃ in propanol has a "reducing capacity" of 55–70 μmol/ml.

Titration. An aliquot of the stock solution, 0.1–1 ml, is added with stirring to 10 ml of 10 mM iodine in propanol. Unreduced iodine is titrated with 0.1 M aqueous $Na_2S_2O_3$ until the solution becomes colorless. The phosphine content of the sample added can be calculated as follows: Reducing capacity (μmol) = 50 (2-x); x = ml 0.1 M $Na_2S_2O_3$ needed for titration.

General Procedure for Reduction and S-Alkylation

The protein (2–10 mg/ml) is dissolved in a 1:1 mixture of 0.5 M $NaHCO_3$/1-propanol. The pH is checked and, if necessary, adjusted to 8.0–8.3 with 1 M NaOH or Na_2CO_3. A stream of nitrogen is blown over the

[18] See this volume [11] for details.
[19] C. H. W. Hirs, this series, Vol. 11, p. 199 (1967); and F. R. N. Gurd, Vol. 11, p. 532.
[20] S. A. Buckler, *J. Am. Chem. Soc.* 84, 3093 (1962).

surface of the stirred solution and PBu$_3$ (1–5 μmol or 15–90 μl of the 2% stock solution per milligram of protein, respectively) is added.

For S-aminoethylation,[17] ethyleneimine (4–40 μmol or 0.2–2 μl per milligram of protein, respectively) can be added together with the phosphine reagent. After 2–4 hr, isolation is started.

For S-carboxymethylation,[19] sodium iodoacetate (2–10 μmol or 2–10 μl per milligram of protein of a 1 M solution of iodoacetic acid in 1 M NaOH) is added in the dark after 2 hr reducing time. The pH is checked after about 5 and 15 min and, if necessary, adjusted to 8–8.3 with 1 M NaOH or Na$_2$CO$_3$. After 30 min, excess sodium iodoacetate can be destroyed by addition of a small amount of 2-mercaptoethanol (0.2–1 μl/mg protein), and after a further 15 min, isolation is begun. Alternatively, the solution can be acidified to render the remaining sodium iodoacetate unreactive toward sulfhydryl groups.

Isolation of the Alkylated Protein. For large peptides and for proteins, the reaction mixture can be dialyzed directly against (volatile) buffer (e.g., 0.01 M NH$_4$HCO$_3$). If dialysis is carried out against acidic buffers or dilute acetic acid, the reaction mixture should be acidified first. With large amounts of proteins (>10 mg), "hollow fiber dialyzers" (Dow Chemical Co.) have been used successfully, their main advantage being speed (2–3 hr for complete desalting in a 100-ml "beaker dialyzer," type b/HFD-1[10]). The product is recovered by lyophilization.

Alternatively, the reaction mixture can be decreased to ~5–30% of the initial volume by rotary evaporation and applied to a column of appropriate size of Sephadex G-25 (medium) or G-15 equilibrated with 0.05 M NH$_4$HCO$_3$. After elution, the protein fractions (A_{280}) are pooled and the product is recovered by lyophilization. If the product becomes insoluble after removal of propanol, acetic acid can be added and desalting is carried out as described above but in 1 M or 50% (v/v) AcOH.

Reduction and Aminoethylation of Insulin

Bovine insulin (144 mg, 25 μmol) is dissolved in 20 ml of 0.5 M NaHCO$_3$/propanol (1:1). After evacuating and flushing with nitrogen, ca. 80 μmol of PBu$_3$ is added as an aliquot of a titrated 2% solution of PBu$_3$ in propanol. Ethyleneimine (250 μl, 4.8 mmol) is added together with the phosphine. After 5 hr, the solution is reduced to about 6 ml by rotary evaporation (bath temperature ~25°), the same volume of AcOH is added, and the solution is applied to a column (4 × 32 cm) of Sephadex G-25 (medium) equilibrated with 50% AcOH. Elution is carried out with the same solvent at 40–80 ml/hr. The protein fractions are pooled, reduced to about 10 ml by rotary evaporation (bath temperature

$<35°$), diluted with ca. 50 ml of water, and lyophilized to afford a mixture of aminoethylated insulin A and B chains in quantitative yield. The chains can be separated by gel filtration to give $A(SAet)_4$ and $B(SAet)_2$ in yields over 90%.[9]

Reduction and Aminoethylation of Serum Albumin

Human serum albumin, (6.7 mg, 0.1 μmol), is dissolved in 1 ml of 0.5 M NaHCO$_3$/propanol (1:1) and treated under nitrogen with an aliquot of the titrated PBu$_3$ stock solution (to give at least 1.8 μmol of reducing capacity) and 7 μl (136 μmol) ethyleneimine. After 3 hr, AcOH is added to pH 3, and the clear solution is desalted on a Sephadex G-25 column (1.7 $\times$ 43 cm) in 1 M AcOH. Elution is performed at about 20 ml/hr; protein fractions are pooled and lyophilized to afford fully S-aminoethylated serum albumin in quantitative yield.

Reduction and Carboxymethylation of Ribonuclease

If ribonuclease is treated with PBu$_3$ under the conditions used for insulin and serum albumin, partially reduced RNase will precipitate. For complete reduction and *in situ* alkylation, the reaction is therefore carried out in the presence of urea as a co-solvent.

Bovine pancreatic ribonuclease (64 mg, 5 μmol) is dissolved in 7.5 ml of 7 M urea in 0.1 M Tris·HCl, pH 8.2/1-propanol (2:1). Under nitrogen and stirring, 50 μl of undiluted PBu$_3$ (130–180 μmol, 6- to 9-fold excess) is added. After 2 hr, 140 μl of 1 M sodium iodoacetate (1 M iodoacetic acid in 1 M NaOH; 3.5-fold excess) is added in the dark; the pH is adjusted after 5 min to 8.2 with 1 M Na$_2$CO$_3$ and after another 30 min 0.5 ml of AcOH is added. The solution is then dialyzed in the cold against three changes of 2 liters of 0.1% AcOH, and the material is recovered quantitatively by lyophilization.

[11] Alkylation of Cysteine Thiols with 1,3-Propane Sultone

By URS TH. RÜEGG and JOSEF RUDINGER*

An investigation of the structure of a protein generally requires derivatization of the labile cysteine residues. Such a modification reaction should allow specific S-alkylation under mild conditions, and the cysteine derivative formed should be stable to more drastic treatments, such as

*Deceased April 30, 1975.

acid hydrolysis or Edman degradation. Optimally, the product should have good solubility properties as well as distinctive analytical characteristics.

The sulfonic acid group with its strongly hydrophilic behavior gives rise generally to good solubility. Owing to its very low pK_a value (about 1–2) it serves as an analytical (e.g., electrophoretic) marker.

A classical procedure introduces sulfo groups into protein by oxidation of cysteine and cystine with peracids to form cysteic acid.[1,2] However, other residues (Met, Trp, Tyr) are also affected by the strong oxidant.[2] Similarly, thiols and disulfides can be treated by oxidative sulfitolysis yielding S-sulfocysteine,[3] but this residue is destroyed under the conditions of acid hydrolysis.

The most common procedures used presently are S-alkylation with iodoacetic acid (or its amide)[4,5] and with ethyleneimine.[6] The resulting thioether derivatives, carrying carboxyl (or carboxamide) and amino substituents are stable to most reactions used in protein chemistry. However, these substituents do not differ drastically from functional groups found commonly in proteins, and, in general, they do not enhance product solubility. Furthermore, light-catalyzed oxidation of iodide ion liberated upon carboxymethylation can cause side reactions (oxidation of Met, Trp, Tyr, His)[4] and NH_2-terminal S-carboxymethylcysteine is liable to a side reaction (cyclization to the thia analog of pyrohomoglutamic acid)[7].

S-Sulfoalkyl derivatives of cysteine combine the advantages of the stability of the S-alkyl substituent with the solubilizing effect of the sulfo group. Iodoalkanesulfonates have been investigated as S-alkylating reagents, but their reactivities were found to be 2–3 orders of magnitude lower than iodoacetate.[8] Recently, one compound of this group, 2-bromoethanesulfonate has been used to modify protein sulfhydryl groups.[9] The low reactivity was counterbalanced by the use of a large excess of the reagent.

We have investigated the reaction of 1,3-propane sultone (1,2-oxathiolane-2,2-dioxide or 3-hydroxypropanesulfonic acid γ-sultone) with

---

[1] F. Sanger, *Nature (London)* **160**, 295 (1947).
[2] C. H. W. Hirs, this series Vol. 11, pp. 59, 197.
[3] R. D. Cole, this series Vol. 11, p. 206.
[4] C. H. W. Hirs, this series Vol. 11, p. 199.
[5] F. R. N. Gurd, this series Vol. 11, p. 532.
[6] R. D. Cole, this series Vol. 11, p. 315.
[7] A. F. Bradbury and D. G. Smyth, *Biochem. J.* **131**, 637 (1973).
[8] M. A. Jermyn, *Aust. J. Chem.* **19**, 1293 (1966).
[9] V. Niketic, J. Thomsen, and K. Kristiansen, *Eur. J. Biochem.* **46**, 547 (1974).

several nucleophiles present in proteins.[10] The sulfonyl grouping of 1,3-propane sultone (I) serves the dual purpose of activating the terminal methylene group and, after reaction, providing the desired side-chain substituent in the S-3-sulfopropyl derivative (II):

$$
\begin{array}{ccc}
\underset{\text{(I)}}{\ce{H2C-CH2}\\ \ce{O2S} \quad \ce{CH2}\ \overset{\ominus}{\ce{S}}-\ce{CH2}-\overset{\displaystyle|\atop\displaystyle CO}{\underset{\displaystyle|\atop\displaystyle NH}{\ce{CH}}}} & \longrightarrow & \underset{\text{(II)}}{\ce{O3S}-\ce{CH2CH2CH2}-\ce{S}-\ce{CH2}-\overset{\displaystyle|\atop\displaystyle CO}{\underset{\displaystyle|\atop\displaystyle NH}{\ce{CH}}}}
\end{array}
$$

S-3-Sulfopropylcysteine can readily be obtained by reaction of propane sultone (PS) with cysteine at room temperature. The compound is completely stable to acid hydrolysis and very water soluble, and is not retained by sulfonic acid ion-exchangers.

Kinetic experiments with cysteine at pH 8.3 showed that PS reacts about 12 times slower than iodoacetate.[10] Alkylation of methionyl and histidyl residues occurs with both reagents at pH 8.3 to about the same extent, if a 12-fold correction is made for the reaction of PS. N-Alkylation of α- and ϵ-amino groups occurs to a measurable degree only at higher pH values (>9). From these model experiments, it appears that the selectivity of both reagents is about the same. PS is hydrolyzed slowly at pH 8.3, the rate of hydrolysis being about 1% of that of the S-alkylation.

It should be kept in mind, however, that under inappropriate conditions propane sultone will react with nucleophiles in proteins. Reactivities of chemically identical functional groups in undenatured proteins often vary up to 100-fold, and side reactions might therefore occur. Even though the solvent used for carrying out reduction-sulfopropylation (20–50% propanol) has only a weak denaturing effect, we have not been able to detect any products of such a side reaction under the conditions described below by means of electrophoresis and amino acid analysis.

For derivatization of cysteine in proteins, alkylation generally has to be preceded by reduction of the cystine disulfides. Quantitative reduction of proteins with tributylphosphine (PBu₃) is described elsewhere.[11] Under the conditions described below, PS and PBu₃ do not interfere with each other. They can therefore be added together and both reactions occur simultaneously. This "one-batch" procedure not only requires less handling, but also permits the use of only relatively slight excesses of

[10] U. T. Rüegg and J. Rudinger, *Int. J. Pept. Protein Res.* **6**, 447 (1974).
[11] See this volume [10] for details.

the reagents. Therefore, the overall reaction is more easily controlled and is much less liable to side reactions (alkylation of His, Met, Lys, etc.) than reduction with the common thiols (e.g., 2-mercaptoethanol) followed by *in situ* reaction with large excesses of alkylating agent.

The generally excellent solubility of S-sulfopropylated peptides and proteins in aqueous media prevent losses due to "wash out" in automated Edman degradation: Bis-S-sulfopropylinsulin B-chain gave higher repetitive yields than the corresponding carboxymethyl derivative.[12] Detailed work on sequencing has been carried out with proteins carrying the S-sulfoethyl substituent.[9]

It seems that S-sulfopropylation meets the requirements of protein chemistry as outlined above. In comparison with carboxymethylation, the main drawback is the lower reaction rate, but the advantages lie in the compatibility with the tributylphosphine reduction, the solubility of the products, the improved analytical properties and the absence of a side reaction which can occur with S-carboxymethylcysteine.[7]

Procedures

The reagent, 1,3-propane sultone (PS) is a carcinogen and mutagen[13] and should therefore be handled with great care. The commercial crystalline solid is fairly hygroscopic. Some of the sultone can hydrolyze even in the crystalline state to form 3-hydroxypropanesulfonic acid whose content can be determined by titration. If purity is being checked by gas chromatography at elevated temperature, the acid can cyclize back to the sultone under elimination of water,[14] and an apparently higher degree of purity will be found. Samples (Fluka A. G., Buchs, Switzerland) that were labeled >95% pure contained 10–12% acid by titration. Essentially pure propane sultone can be obtained with some losses by pulverization of the crystals, extraction with small portions of water, filtration, and fast drying in a vacuum desiccator. Alternatively, the sultone can be distilled.[14] For small-scale work, 250 mM (306 mg/10 ml) solutions of PS in water/propanol (1:1) can be used; these should always be prepared just prior to use. Precautions when handling the reducing agent, *tributylphosphine*, its titration and the method used for excluding oxygen are described elsewhere.[11]

[12] Dr. K. Wilson, Institute of Biochemistry, University of Zurich, personal communication.
[13] J. McCann, E. Choi, E. Yamasaki, and B. N. Ames, *Proc. Natl. Acad. Sci. U.S.A.* **72**, 5135 (1975).
[14] J. H. Helberger, *Ann. Chem.* **588**, 71 (1954).

Preparative Procedures for Reduction and S-Sulfopropylation

The preparative procedure for reduction-sulfopropylation is technically very simple, reliable, and reproducible: the protein is dissolved under a nitrogen atmosphere, PBu₃ and PS are added, the solution is set aside and desalted, and lyophilized.

Under the conditions given below the alkylation is always the slower and therefore rate-limiting step. Generally, we found a 2-fold excess of PS requires 16 hr reaction time, a 10-20-fold excess 2 hr. Kinetic experiments with insulin have been performed.[10]

General Procedure. The protein is dissolved to give a concentration of 5–10 mg/ml in a suitable, buffered solvent at pH 8–8.5 (0.1 M Tris·HCl or 0.5 M NaHCO₃) containing 20–50% (v/v) 1-propanol. If urea has to be used (in order to dissolve the protein or the intermediate mercaptoprotein), Tris is the buffer of choice or, alternatively, a primary amine (∼0.1 M methylamine) should be added in order to trap cyanate ions formed during exposure of urea to alkali.[15] The solution is deaerated by flushing with nitrogen, or better, repeatedly evacuated with stirring and flushed with nitrogen.[11]

Both reagents, PB₃ and PS are added together with stirring and under positive nitrogen pressure. Molar excesses of PBu₃ of 10% to 5-fold over protein disulfide can be used.[11] For proteins with unknown cystine content, 1–3 μmol of PBu₃ per milligram of protein is sufficient. Irrespective of the amount of PBu₃ used, PS should be added in excess over protein thiol. For convenient overnight reaction (16 hr) at room temperature, a 2-fold excess over released sulfhydryl is used (4-fold over disulfide, respectively) and 2 μmol of PS per milligram of protein are added for proteins with unknown cyst(e)ine content. Alternatively, a 10- to 20-fold excess of PS can be used, requiring reaction times of 2 hr.

Isolation of the S-Sulfopropylated Protein. Any method used for desalting can be applied. Gel filtration on Sephadex G-25 (medium) in 0.05 M NH₄HCO₃ is optimal for peptides and relatively small sample volumes. Dialysis (against water or 0.1% acetic acid) is recommended with larger amounts and proteins. Rapid desalting (2–3 hr),[10] can also be achieved in hollow-fiber devices (100-ml Beaker Dialyzer b/HFD-1, Dow Chemical Co.). The products are recovered by lyophilization.

Two examples of the method are given.

Reduction and Sulfopropylation of Serum Albumin. A sample of human serum albumin (17 mg, 0.25 μmol) is dissolved in 1.8 ml 0.5 M NaHCO₃/ propanol (5:4) in a vial. The flask is evacuated and the solution saturated with nitrogen, 4.7–5.1 μmol (10–20% excess) PBu₃ solution in

[15] G. R. Stark, this series Vol. 11, p. 590.

propanol (as determined by titration[11]) and 70 μl (17.5 μmol) of a 250 mM solution of PS in H_2O/propanol (1:1) are added and the vial set aside. After 16 hr, the solution is applied to a column (ca. 2.6 $\times$ 34 cm) of Sephadex G-25 (medium) equilibrated with 0.05 M NH_4HCO_3. Elution (30–50 ml/hr) is carried out with the same buffer, the fractions containing protein are collected and lyophilized to yield quantitatively human serum albumin substituted with ^{35}S-sulfopropylcysteine residues.

Reduction and Sulfopropylation of Ribonuclease. Bovine pancreatic ribonuclease (64 mg, 5 μmol) is dissolved in 7.5 ml of 7 M urea in 0.1 M Tris·HCl, pH 8.2/propanol (2:1). The solution is deaerated and saturated with nitrogen. PBu_3 (22–40 μmol, 1.1- to 2-fold excess) as a 5% titrated solution in propanol[11] and 320 μl (80 μmol) of a 250 mM solution of PS in propanol/H_2O (1:1) (a 2-fold excess) are added. The flask is set aside for 16 hr at room temperature, the solution is acidified by addition of 0.5 ml of AcOH and dialyzed in the cold against three changes of 0.1% AcOH. S-Sulfopropylated RNase is recovered by lyophilization of the retentate.

Analytical Procedure: Reduction-Sulfopropylation and Direct Hydrolysis

This method permits reduction, S-sulfopropylation, and acid hydrolysis in a one-batch procedure. Removal of salts and reagents is not necessary after termination of the modification reaction and hydrolysis is begun in the presence of residual amounts of PBu_3, $OPBu_3$, and PS. There is no indication that any of these compounds react irreversibly under the conditions of acid hydrolysis with any of the amino acids. Residual phosphine might even help to trap traces of oxygen present at the initial stage of hydrolysis and thus prevent oxidation of thioethers (methionine, S-sulfopropylcysteine).

The hydrolysis of propane sultone, originally thought to be a disadvantageous side reaction during S-alkylation, turns out to be an advantage in this context: the hydrolysis is pH independent,[16] and residual sultone is converted quickly into inert 3-hydroxypropanesulfonic acid. This prevents undesirable side reactions, such as alkylation of methionine.

All the products formed from the reagents upon hydrolysis are neutral or acidic and therefore pass through the ion-exchange resin of the amino acid analyzer unretarded. They do not interfere with the standard ninhydrin colorimetric detection method.

[16] A. Mori, M. Nagayama, and H. Mendai, *Kogyo Kagaku Zasshi* **74**, 715 (1971); A. Mori, M. Nagayama, and H. Mendai, *Bull. Chem. Soc. Jpn.* **44**, 1669 (1971).

The following procedure has been applied to four model proteins and peptides: [Lys[8]]vasopressin, bovine insulin,[10] bovine pancreatic ribonuclease, and human serum albumin.[10] In all cases, amino acid analysis indicated complete reduction and sulfopropylation, not a trace of cystine or cysteine was left and all other values, with one exception,[17] agree well ($\pm 5\%$) with the control experiment (analysis of the hydrolyzate of unmodified material). RNase that precipitates partially after reduction under the conditions used redissolves almost completely during the course of the reaction.

General Procedure. A thick-walled Pyrex hydrolysis tube (1.5×15 cm or 0.8×15 cm) serves as receptacle for all operations. A sample (0.2–3 mg, depending on the amount required for amino acid analysis) of the protein is dissolved in 0.5 M NaHCO$_3$/1-propanol (1:1 or 2:1) at 1–5 mg/ml. Nitrogen is blown through the solution via a glass tube or pipette and, after a few seconds, over the surface of the solution. PBu$_3$ (2- to 10-fold excess over protein disulfide, i.e., ca. 1–5 μmol/mg protein or 6–35 μl of a 5% stock solution in propanol[11] per milligram of protein) and PS (10- to 20-fold excess over sulfhydryl, i.e., ca. 10–20 μmol/mg protein or 40–80 μl of a 250 mM solution in H$_2$O/propanol (1:1) per milligram of protein[18]) are added together without interruption of the nitrogen stream. The tube is stoppered and kept at 20–25°.

After 2 hr, constant-boiling (5.7 M) HCl is added (0.1–0.5 ml) carefully (to prevent excessive foaming due to the release of carbon dioxide gas), the tube is connected to a rotary evaporator via a rubber stopper with a hole, and the solution is brought to complete dryness. More 5.7 M HCl (0.5–1 ml) is added, the solution is deaereated under vacuum, the tube is sealed, and hydrolysis is carried out as usual.[19]

Amino Acid Analysis of Samples Containing S-3-Sulfopropylcysteine. Owing to its high acidity, S-sulfopropylcysteine [Cys(Sp)] is eluted from the long column of the amino acid analyzer[20] well ahead of Asp, at about the position of cysteic acid. With column dimensions of 0.9×49 cm, the elution volume of Cys(Sp) is 22 ml as compared to Cys(O$_3$H) with 21 ml and Asp with 53 ml. The color constant of Cys(Sp) can be determined with an analytically pure sample; we found it to be 0.96 times the constant of Asp.

---

[17] The value of Lys was decreased by 8% in the analysis of sulfopropylated [Lys[8]]-vasopressin. No attempts were made to identify the possible side product, N-sulfopropyllysine.

[18] For proteins and peptides with known cyst(e)ine content, it is advisable to take the molar amount (10–20-fold excess).

[19] S. Moore and W. H. Stein, this series Vol. 6, p. 819.

[20] D. H. Spackman, this series Vol. 11, p. 3.

[12] Reductive Cleavage of S-Sulfo Groups with Tributylphosphine

By URS TH. RÜEGG

The S-sulfo or "Bunte salt" group can generally be described as an intermediate thiol protecting group in peptide and protein chemistry. In synthetic peptide chemistry, for example, labile sulfhydryl peptides generated upon removal of the protecting groups are commonly converted into stable S-sulfo derivatives. This can best be done by oxidative sulfitolysis.[1-4] Following purification and characterization, the S-sulfo group is removed and the material is oxidized to form the desired disulfide derivative (see, e.g., ref.[5]).

The S-sulfo group has now proved to be more stable than expected.[1,4] In semisynthetic approaches with insulin chains, S-sulfo groups were used successfully when carrying out active ester couplings in the presence of triethylamine[6] and they were even found to be stable to both the basic and strongly acidic conditions employed during the Edman degradation.[7] Owing to its very low pK_a value (about 1–2), the S-sulfo group remains charged over a wide pH range. Therefore, S-sulfo peptides generally exhibit good solubility properties in aqueous solution. In separations based on charge they behave distinctively different from nonsubstituted peptides.

The S-sulfo group can be removed reductively: sodium in liquid ammonia, a reduction method known to cause a great number of side reactions, has been employed.[8,9] Most often, reduction with excess thiol is used.[3] However, before reoxidation of the reduced protein can be carried out, excess reagent has to be removed completely under anaerobic conditions. If oxygen is not excluded during all manipulations of this removal—such as gel filtration, dialysis, or protein precipitation—formation of undesired mixed disulfides between reagent and protein is possible.

[1] J. M. Swan, *Nature (London)* **180**, 643 (1957).
[2] J. L. Bailey and R. D. Cole, *J. Biol. Chem.* **234**, 1733 (1959).
[3] R. Cecil, *"The Proteins,"* 2nd ed. (H. Neurath, ed.), Vol. 1, p. 438. Academic Press, New York, 1963.
[4] R. D. Cole, this series, Vol. 11, p. 206.
[5] B. Gutte and R. B. Merrifield, *J. Biol. Chem.* **246**, 1922 (1971).
[6] D. Brandenburg, *Hoppe-Seyler's Z. Physiol. Chem.* **353**, 263 (1972) (cf. ref. 1–3).
[7] D. Brandenburg, M. Biela, L. Herbertz, and H. Zahn, *Hoppe-Seyler's Z. Physiol. Chem.* **356**, 961 (1975).
[8] C. L. Tsou, Y. C. Du, and G. J. Xu, *Sci. Sinica* **10**, 332 (1961).
[9] Y. Shimonishi, H. Zahn, and W. Puls, *Z. Naturforsch. Teil B* **24**, 422 (1969).

Obviously, these difficulties arise only because the reagent used to remove the S-sulfo group is chemically the same as protein sulfhydryl groups. Therefore, the reagent of choice should be chemically different and hence unable to interfere with the formation of protein disulfides.

Tributylphosphine was found to be a highly specific reagent that reduces protein disulfide bonds stoichiometrically.[10] This finding has been extended to the cleavage of the sulfenyl sulfite bond of the S-sulfo group.[11] The reaction mechanism can be expected to be similar to the one suggested for the reduction of disulfides,[10] the nucleophilic attack of the phosphine taking place at the divalent sulfur:

$$R-S-SO_3^{\ominus} \quad \longrightarrow \quad \left[R-S + SO_3^{2\ominus} + H^{\oplus} \atop Bu_3P^{\oplus} \quad {}^{\ominus}OH \right] \quad \longrightarrow \quad R-S^{\ominus} + SO_3^{2\ominus}$$
$$Bu_3P \quad H_2O \qquad\qquad\qquad\qquad\qquad\qquad\qquad\qquad Bu_3PO + 2H^{\oplus}$$

The advantages in using tributylphosphine for reduction of disulfides[10] hold true for the reduction of S-sulfo groups: the reaction is highly specific, stoichiometric, rapid, and compatible with most alkylating agents.

Since the S-sulfocysteine residue is not stable to acid hydrolysis,[2,4] reduction of an S-sulfo peptide with PBu$_3$ and *in situ* alkylation followed by hydrolysis and amino acid analysis serves as a quick method for the quantitation of cysteine and cystine present.[11] It should be kept in mind, however, that PBu$_3$ reduces both S-sulfo groups and disulfides.[10] Therefore, the method cannot be used to determine the extent of prior sulfitolysis.

Another application of the method is reduction followed by air oxidation to form the corresponding disulfide directly. The whole operation can be performed in two simple steps: the S-sulfo protein is first reduced at high concentration with PBu$_3$ and then diluted into air-saturated reoxidation buffer. The products of the reduction step are present during the reoxidation and do not seem to interfere with the correct formation of disulfides: phosphine oxide is inert and therefore harmless, whereas sulfite could reverse the reaction by sulfitolysis.[1-4] This point was not investigated, but the high yield of insulin obtained[11] with this method indicates that the concentration of sulfite is probably too low to cause any significant interference. Preliminary results with ribonuclease indicate[12] that the same procedure might be applied to more complex single-

[10] See this volume [10] for details.
[11] U. T. Rüegg and H.-G. Gattner, *Hoppe-Seyler's Z. Physiol. Chem.* **356**, 1527 (1975).
[12] Octa S-sulfo RNase (obtained by sulfitolysis in the presence of cupric ions)[4] was treated at 10 mg/ml for 40 min in 4 M urea in 0.5 M NaHCO$_3$/propanol (1:1) with a 2-fold excess of PBu$_3$, diluted to a final concentration of 0.02 mg/ml into 0.1 M

chain proteins which preferentially refold into their native conformation.[13] However, since not enough data are available presently and since refolding conditions vary from protein to protein, no detailed general procedure is given here.

Methods

Tributylphosphine reagent. As outlined,[10] all manipulations with larger amounts of PBu$_3$ should be performed in the hood and under nitrogen. For carrying out S-sulfo group reductions, 2–5% (v/v) solutions of PBu$_3$ in 1-propanol can be used. The "reducing capacity" is determined iodometrically, as previously described.[10] Because of the strictly stoichiometric and therefore quantitative course of the reaction, we have generally used only a slight (i.e., 10%) excess of the reagent. If desirable, larger excesses of PBu$_3$ can be used. The reaction is carried out at slightly alkaline pH in a buffered solution containing 1-propanol (20–50%, v/v). Urea and guanidinium chloride are both compatible with PBu$_3$.[10]

Reduction and Direct S-Aminoethylation of S-Sulfo Peptides

This method converts S-sulfocysteine directly into S-aminoethylcysteine and allows quantitative determination of the sum of cysteine and cystine present in a protein. S-Aminoethylation[14] was chosen as a modification procedure because the change from the negatively charged sulfo function to the positively charged amino function can be followed electrophoretically. As an example, procedures for the reduction and aminoethylation of the tetra S-sulfo derivative of bovine insulin A-chain,[15] insulin-A(SSO$_3$H)$_4$, are given[11]:

Insulin-A(SSO$_3$H)$_4$ (6.7 mg; 2.5 μmol) is dissolved in 1.5 ml of 0.5 M NaHCO$_3$/propanol (2:1). The pH is adjusted, if necessary, to 8.3 with 1 M Na$_2$CO$_3$ and the flask is evacuated and the solution flushed with nitrogen. An aliquot of the titrated 2–5% stock solution of PBu$_3$ in

Tris·HCl, pH 8.0, and oxidized for 20 hr.[13] The product had the same activity (i.e., 30% of native RNase) as the material from the control experiment (reduction of native RNase with PBu$_3$ and reoxidation as described above).

[13] E. Haber and C. B. Anfinsen, *J. Biol. Chem.* **237**, 1839 (1962); F. H. White, Jr., this series, Vol. 11, p. 481.

[14] M. A. Raftery and R. D. Cole, *Biochem. Biophys. Res. Commun.* **10**, 467 (1963); *J. Biol. Chem.* **241**, 3457 (1966); R. D. Cole, this series, Vol. 11, p. 315.

[15] Insulin-A(SSO$_3$H)$_4$ was obtained by sulfitolysis[2,4] of insulin and was separated from Insulin-B(SSO$_3$H)$_2$ and purified as described by Rüegg and Gattner.[11]

propanol is added,[10] to give at least 11 μmol (10% excess) of PBu$_3$. Ethyleneimine (10–15 μl; 200–300 μmol) is added simultaneously. The vial is stoppered, and the reaction is left to proceed for 3 hr at room temperature. The solution is acidified to about pH 3 by addition of AcOH and applied to a column (ca. 1.7 × 43 cm) of Sephadex G-10 equilibrated with 1 M AcOH. Elution is carried out with the same buffer (20–30 ml/hr), the fractions containing the protein are collected and the solution is lyophilized to give insulin-A(SAet)$_4$ in quantitative yield. This product migrates at neutral pH in the opposite direction to insulin-A(SSO$_3$H)$_4$.

Reduction and carboxymethylation[16] are carried out in two steps: an example is given below.

Reduction and Carboxymethylation of S-Sulfo Ribonuclease

RNase(SSO$_3$H)$_8$ (10.8 mg, 0.8 μmol)[4] is dissolved in 1.0 ml of a 1:1 mixture of 8 M urea in 0.5 M NaHCO$_3$ and propanol (final concentration of urea 4 M). Nitrogen is passed through the solution and an aliquot of a titrated 2–5% solution of PBu$_3$ in propanol[10] is added with a reducing capacity of about 6–15 μmol (2–5-fold excess). After 40 min at room temperature, 50 μl (50 μmol, ca. 8-fold excess) of 1 M sodium iodoacetate is added in the dark. Under these conditions, the pH should remain constant at 8.3. After 30 min, 50 μl AcOH is added to stop S-alkylation. The solution is dialyzed against 0.1% AcOH in the cold and the fully S-alkylated material is recovered by lyophilization.

[16] C. H. W. Hirs, this series, Vol. 11, p. 199; F. R. N. Gurd, Vol. 11, p. 532.

Section V
Selective Cleavage by Chemical Methods

[13] Cleavage at Cysteine after Cyanylation

By George R. Stark

Principle. The SH groups of denatured proteins can be converted quantitatively to SCN groups with 2-nitro-5-thiocyanobenzoate (NTCB)[1,2] or with other reagents (see below). NTCB is readily synthesized and can be labeled with [14C]cyanide.[3] Exposure to pH 8–9 results in cleavage at the amino group of the modified cysteine residue in excellent yield.[1]

At this time there is no complete method for removing the iminothiazolidine carboxylyl (ITC) group from the amino termini of the cleavage products, so that sequential Edman degradation can be performed directly only on the peptide containing the original amino terminus. Since disulfides do not react with NTCB, selective cleavage at cysteine can be obtained in the presence of cystine. Many other variations are possible. For example, the SH groups of a denatured protein can be blocked, followed by reduction and cyanylation of the disulfides, or reactive SH groups of a native protein can be cyanylated selectively, followed by cleavage, or these SH groups can be blocked, followed by cyanylation and cleavage in denaturant at the unreacted groups.

Procedure.[1] Dissolve the protein in 6 M guanidinium chloride–0.2 M Tris acetate buffer, pH 8. If there are no disulfides, add a slight excess of dithiothreitol (DTT) and allow the solution to stand at room temperature for 0.5 hr. To reduce disulfides, add DTT to 10 mM and heat the solution to 37° and 1–2 hr. Add NTCB[3] in 5-fold excess over total thiol, readjust the pH to 8 with NaOH if necessary, and allow the reaction to proceed for 15 min at 37°. Acidify the reaction mixture to pH 4 or below, cool it to 4°, and separate the modified protein from small molecules, for example, by dialysis or by gel filtration into 50% acetic acid or another

[1] G. R. Jacobson, M. H. Schaffer, G. R. Stark, and T. C. Vanaman, *J. Biol. Chem.* **248**, 6583 (1973).
[2] Y. Degani and A. Patchornik, *Biochemistry* **13**, 1 (1974).
[3] Y. Degani and A. Patchornik, *J. Org. Chem.* **36**, 2727 (1971).

volatile solvent of low pH. The sample may be stored at $-20°$ in 50% acetic acid without cleavage.[1] For cleavage, take the protein to dryness to remove the acidic solvent thoroughly and redissolve it in 6 M guanidinium chloride—0.1 M sodium borate buffer, pH 9, for 12–16 hr at $37°$.

The Cyanylation Reaction. Degani and Patchornik[2] found that modification of small peptides with NTCB is incomplete at high concentrations of reactants and recommended that the total concentration of SH not exceed 0.5 mM. However, Jacobson *et al.*[1] used much higher concentrations of DTT in the reduction and modification of proteins in denaturants, with nearly quantitative incorporation of [14]CN from labeled NTCB, and the experience of some other investigators has been similar (see below). If there is difficulty in achieving complete reaction in a particular case, it may be wise to do the cyanylation reaction at low total concentration of SH. Alternatively, one can use the two-step method of Vanaman and Stark,[1,4] in which the protein-thionitrobenzoate mixed disulfide is isolated first and then treated with cyanide. Recently, 1-cyano-4-dimethylamino pyridinium salts were shown to be effective for cyanylation of protein SH groups.[5] With this reagent, the side reactions which may make the use of low concentrations of SH desirable are not possible. Fontana[6] has found that 2-nitrophenylsulfenyl chloride reacts with the SH groups of proteins extremely rapidly and quantitatively under acidic conditions, where the oxidation of SH groups is slow. Subsequent treatment with cyanide results in cyanylation of the protein. The reagent also modifies tryptophan, but this should not interfere with sequence analysis of the peptides. Also, the thiosulfonate derivative of 5,5′-dithiobis (2-nitro-benzoate) (DTNB), prepared with peracetic or performic acid, is a more reactive alternative to DTNB itself.[6]

Cleavage or β-Elimination? Degani and Patchornik[2] found that thiocyanate and dehydroalanyl residues were formed in variable amounts from the β-thiocyanoalanyl residues of model compounds at pH 8–10. With Z-Phe-CySCN-Gly-OH, the model compound closest to the thiocyanoalanyl residues of proteins, 11% of the theoretical amount of thiocyanate was released after 20 hr at pH 9, $37°$, without denaturant. In most cases with proteins (see below), cleavage seems to go nearly to completion, and it is often not clear whether incomplete cleavage results from β-elimination or from incomplete modification.

[4] T. C. Vanaman and G. R. Stark, *J. Biol. Chem.* **245**, 3565 (1970).

[5] M. Wakselman, E. Guibejampel, A. Raoult, and W. D. Busse, *J. Chem. Soc., Chem. Commun.* **1976**, 21 (1976).

[6] Angelo Fontana, Institute of Organic Chemistry, University of Padova, Padova, Italy, personal communication.

Properties of Native Proteins after Cyanylation. Specific SH groups of undenatured papain,[7] aspartate transcarbamylase,[4] isocitrate dehydrogenase,[8] aspartate aminotransferase,[9] NAD-specific glutamate dehydrogenase,[10] glyceraldehyde-3-P dehydrogenase,[11] and myosin[12,13] have been modified with NTCB or with DTNB plus cyanide. Often, SH groups thought to be essential on the basis of inactivation following modification with other, more bulky, reagents can be converted to SCN groups with little or no loss of activity. In most cases, there is evidence that the proteins retain the SCN groups without cleavage, so long as the native structure is retained. This is understandable in terms of the mechanism of the reaction, in which the carbon atom of the SCN group must be oriented appropriately for nucleophilic attack by the amide nitrogen atom of the thiocyanoalanine residue.[1] Such orientation may be constrained by the structure of the native protein, or it may be too difficult for hydroxide ion to attack the scissile carbonyl atom, to provide the required anionic tetrahedral intermediate.

Of course, specific cleavage at cysteine residues specifically modified in native proteins provides a convenient method for locating such residues. Use of disulfides formed from the 6-thio analog of ATP followed by cyanide, or direct cyanylation of protein SH groups with a 6-thiocyano analog of ATP has led to incorporation of label and specific cleavage in several cases.[12–14]

Experience of Other Workers with Cleavage. In two cases the extent of cleavage has been evaluated quantitatively. Cowgill[15] used the DTNB-KCN method with paramyosin and achieved cleavage at the single internal cysteine (per chain of $M_r = 110,000$) with a yield of nearly 90%. This work has been extended to separation of the fragments of molecular weight 140,000 and 60,000 generated by cleavage at the two internal cysteine residues of the paramyosin dimer.[16] Casey and Lang[17] have achieved cleavage yields of 85–95% at cysteine 104 of the human hemoglobin α-chain with the NTCB procedure of Jacobson et al.[1] The method

[7] Y. Degani, H. Neumann, and A. Patchornik, *J. Am. Chem. Soc.* **92**, 6969 (1970).
[8] A. E. Chung, J. S. Franzen, and J. E. Braginsky, *Biochemistry* **10**, 2872 (1971).
[9] W. Birchmeier, K. Wilson, and P. Christen, *J. Biol. Chem.* **248**, 1751 (1973).
[10] Y. Degani, F. M. Veronese, and E. L. Smith, *J. Biol. Chem.* **249**, 7929 (1974).
[11] L. D. Byers and D. E. Koshland, Jr., *Biochemistry* **14**, 3661 (1975).
[12] P. D. Wagner and R. G. Yount, *Biochemistry* **14**, 1900 (1975).
[13] P. D. Wagner and R. G. Yount, *Biochemistry* **14**, 1908 (1975).
[14] R. G. Yount, *Adv. Enzymol.* **43**, 1 (1975).
[15] R. W. Cowgill, *Biochemistry* **13**, 2467 (1974).
[16] R. W. Cowgill, *Biochemistry* **14**, 4277 (1975).
[17] R. Casey and A. Lang, *Biochem. J.* **145**, 251 (1975).

has also been applied to myosins,[18,19] troponin I,[20] H-2 and HL-A antigens,[21] and basement membrane proteins[22] with evidence in each case of extensive cleavage but without more detailed quantitation.

Removal of ITC Groups. Since ITC-peptides are not degraded sequentially with the Edman reagent,[1] most peptides resulting from specific cleavage at cysteine residues can not be degraded from their amino termini and further specific cleavages will usually be required if sequence information is to be obtained from them. Therefore, a method for selective removal of ITC residues would be most useful. Our preliminary tests with pyrrolidone carboxylyl peptidase were not successful,[1] but this approach deserves further exploration. Some hope is provided by the preliminary results of Schaffer and Stark[23] in which a reduced nickel catalyst was effective in reducing iminothazolidine carboxylate (ITC) to alanine and ITC-Gly to Ala-Gly. A complicating factor is that the catalyst itself appears to cause specific cleavage at additional sites in unmodified proteins, possibly at Phe-Thr and Phe-Ser sequences. Furthermore, its effectiveness with large ITC-peptides is untested.

[18] A. G. Weeds and K. Burridge, *FEBS Lett.* **57,** 203 (1975).
[19] K. Burridge and D. Bray, *J. Mol. Biol.* **99,** 1 (1975).
[20] H. Syska, J. M. Wilkinson, R. J. A. Grand, and S. V. Perry, *Biochem. J.* **153,** 375 (1976).
[21] P. A. Peterson, L. Rask, K. Sege, L. Klareskog, H. Anundi, and L. Östberg, *Proc. Natl. Acad. Sci. U.S.A.* **72,** 1612 (1975).
[22] R. Alper, and N. A. Kefalides, *Biochem. Biophys. Res. Commun.* **61,** 1297 (1974).
[23] M. H. Schaffer and G. R. Stark, *Biochem. Biophys. Res. Commun.* **71,** 1040 (1976).

[14] Cleavage at Asn-Gly Bonds with Hydroxylamine

By PAUL BORNSTEIN and GARY BALIAN

The use of nucleophilic compounds, such as hydroxylamine, as esterolytic agents is well known. However, the ability of hydroxylamine to cleave asparaginyl-glycyl peptide bonds, or more precisely, the cyclic imide derivative of this bond, is not generally appreciated. Early attempts to convert amide to hydroxamate groups by hydroxylaminolysis ($-CONH_2 + NH_2OH \rightleftarrows -CONHOH + NH_3$) employed relatively severe conditions (3 M NH_2OH, 60°, pH 6.5–7.5, 24 hr) and were attended by an appreciable degree of cleavage of peptide bonds of the X-Pro type in which X was a variety of amino acids.[1,2] Gallop, Seifter, and co-

[1] G. Braunitzer, *Biochim. Biophys. Acta* **19,** 574 (1956).
[2] L. K. Ramachandran and K. Narita, *Biochim. Biophys. Acta* **30,** 616 (1958).

workers[3,4] observed that a small but significant number of bonds in collagen were cleaved by hydroxylamine using milder conditions (1 M NH_2OH, 40°, pH 10, 2–3 hr). A number of other proteins yielded positive tests for protein-bound hydroxamate after treatment with NH_2OH, but in general the level was less than 20% of that observed for collagen.[4] These investigators then examined the formation of hydroxamates of model compounds and of gelatin, their modification by dinitrophenylation and the rearrangement of the resulting dinitrophenylhydroxamates by the Lossen reaction.[5-7] In the case of the model compound polyanhydroaspartic acid (polysuccinimide), it was shown by analysis of the products of the Lossen rearrangement that both α- and β-hydroxamates were initially formed.[6] It was therefore concluded that α- and β-carboxyl groups of aspartyl residues participated in the formation of hydroxylamine-susceptible bonds in collagen. Although it was emphasized that hydroxylaminolysis did not distinguish between ester and imide linkages,[3,4] results were interpreted in terms of a subunit model for collagen α chains in which individual α chains were composed of lower molecular weight subunits linked by nonpeptide bonds.[8,9]

In subsequent efforts to examine the primary structure of collagen α chains, Butler[10] and Bornstein[11] established that well defined CNBr-produced fragments of the α1 chain of rat collagen were cleaved with a high degree of specificity at Asn-Gly bonds. Eventually, the elucidation of the entire sequence of the 1052 amino acids in the α1 chain of collagen[12] demonstrated that only α-amino, α-carboxyl bonds linked the amino acids in the polypeptide chain. It was therefore clear that the increased susceptibility of collagen to hydroxylamine could not be ascribed to the presence of nonpeptide bonds in this protein. A mechanism involving the formation of the cyclic imide derivative of an Asn-Gly bond, which accounts for most aspects of the hydroxylamine cleavage reaction has been proposed.[13] The characteristics of the cleavage of the single Asn-Gly bond

[3] P. M. Gallop, S. Seifter, and E. Meilman, *Nature* (*London*) **183**, 1659 (1959).

[4] O. O. Blumenfeld, M. Rojkind, and P. M. Gallop, *Biochemistry* **4**, 1780 (1965).

[5] S. Seifter, P. M. Gallop, S. Michaels, and E. Meilman, *J. Biol. Chem.* **235**, 2613 (1960).

[6] P. M. Gallop, S. Seifter, M. Lukin, and E. Meilman, *J. Biol. Chem.* **235**, 2619 (1960).

[7] O. O. Blumenfeld and P. M. Gallop, *Biochemistry* **1**, 947 (1962).

[8] S. Seifter and P. M. Gallop, *in* "The Proteins," 2nd ed. (H. Neurath, ed), Vol. IV, p. 153. Academic Press, New York, 1966.

[9] D. Volpin, H. Hörmann, and K. Kuhn, *Biochim. Biophys. Acta* **168**, 389 (1968).

[10] W. T. Butler, *J. Biol. Chem.* **244**, 3415 (1969).

[11] P. Bornstein, *Biochem. Biophys. Res. Commun.* **36**, 957 (1969).

[12] P. P. Fietzek and K. Kuhn, *Int. Rev. Connect. Tissue Res.* **7**, 1 (1976).

[13] P. Bornstein, *Biochemistry* **9**, 2408 (1970).

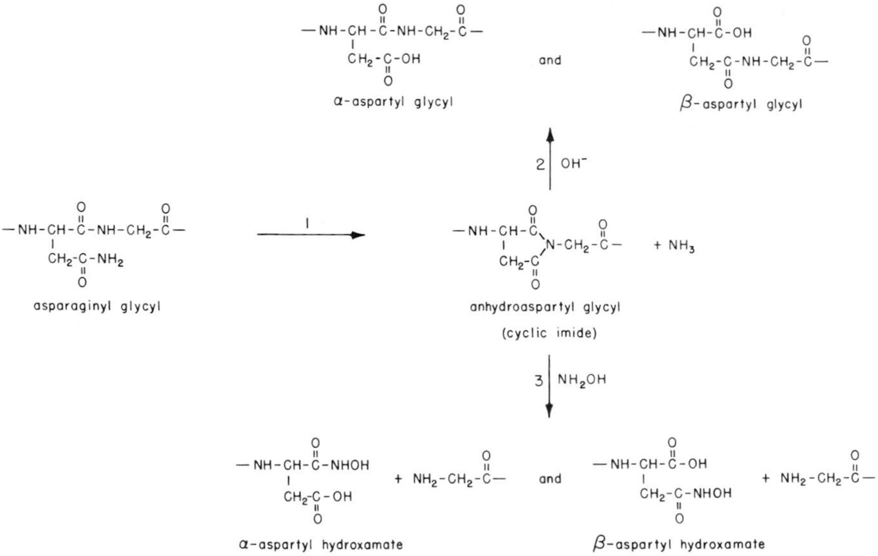

FIG. 1. Proposed structure of the hydroxylamine-sensitive bond in proteins. Cyclization of an asparaginyl-glycyl bond gives rise to the cyclic imide, anhydroaspartylglycine (reaction 1). Cleavage with hydroxylamine yields (through the intermediary of the dihydroxamate) a mixture of α- and β-aspartyl hydroxamates and a new NH_2-terminal glycine (reaction 3). Alternatively, nucleophilic attack by hydroxide ion leads to a mixture of α- and β-aspartyl bonds (reaction 2).

in bovine pancreatic ribonuclease by hydroxylamine were consistent with this mechanism.[14] These studies suggested that hydroxylaminolysis could be adapted as a relatively specific means for the nonenzymic cleavage of Asn-Gly bonds in proteins generally.

Rationale

A structure for the proposed hydroxylamine-sensitive derivative of an Asn-Gly bond is provided in Fig. 1 and a mechanism for the nucleophilic addition of hydroxylamine to the cyclic imide, resulting in chain cleavage, is suggested in Fig. 2.

The factors that promote cyclization of the asparaginyl side chain (Fig. 1, reaction 1) are not known in detail, but it seems likely, based on the following findings, that low pH fosters cyclic imide formation. Difficulties in amino acid sequence determination, attributed to the formation of β-aspartyl peptide bonds, were partially alleviated when anhydrous

[14] P. Bornstein and G. Balian, *J. Biol. Chem.* **245**, 4854 (1970).

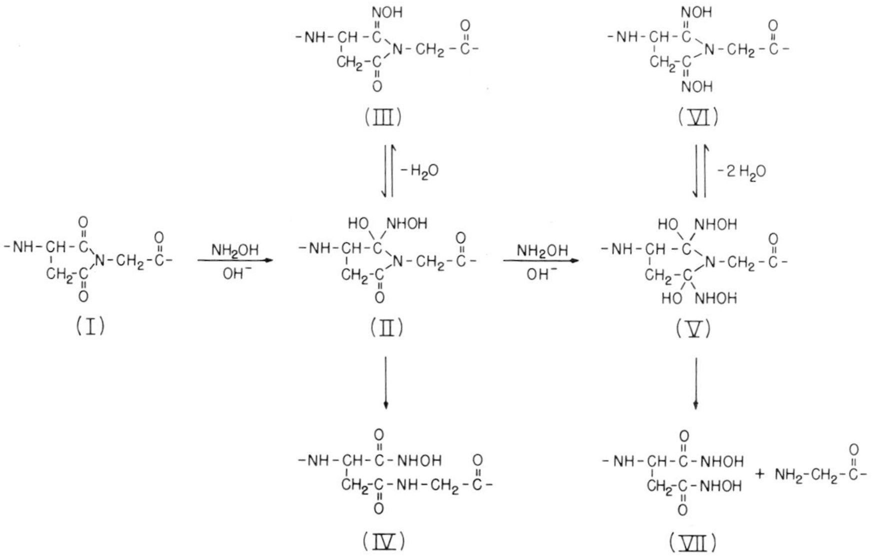

FIG. 2. Proposed mechanism for the nucleophilic addition of hydroxylamine to the cyclic imide, anhydroaspartylglycine (I). Rearrangement of the hydroxylamine adduct (V) results in chain cleavage. See text for further details. Reproduced, with permission, from P. Bornstein, *Biochemistry* 9, 2408 (1970).

trifluoroacetic acid at 25° was substituted for glacial acetic acid and HCl gas at 100° in the cyclization step of the Edman reaction.[15,16] Nevertheless, a reduction in the yield of NH_2-terminal asparagine when asparagine preceded glycine in a thermolysin sequence was still observed during automatic sequenator analysis.[17] During synthesis of aspartyl peptides, removal of protecting groups with HBr in acetic acid at 60° was also shown to lead to aspartimide formation.[18] Finally, Swallow and Abraham[19] observed formation of ε-(aminosuccinyl)lysine from ε-aspartyllysine by heating at 80° in concentrated HCl.

The possibility of base-catalyzed cyclization also exists,[20,21] but competing hydrolysis of the cyclic imide leading to replacement of the amide

[15] D. G. Smyth, W. H. Stein, and S. Moore, *J. Biol. Chem.* 237, 1845 (1962).

[16] D. G. Smyth, W. H. Stein, and S. Moore, *J. Biol. Chem.* 238, 227 (1963).

[17] M. A. Hermodson, L. H. Ericsson, K. Titani, H. Neurath, and K. A. Walsh, *Biochemistry* 11, 4493 (1972).

[18] P. M. Bryant, R. H. Moore, P. J. Pimlott, and G. T. Young, *J. Chem. Soc. London* 3868 (1959).

[19] D. L. Swallow and F. P. Abraham, *Biochem. J.* 70, 364 (1958).

[20] E. Sondheimer and R. W. Holley, *J. Am. Chem. Soc.* 76, 2407 (1954).

[21] A. J. Adler, G. D. Fasman, and E. R. Blout, *J. Am. Chem. Soc.* 85, 90 (1963).

by a carboxyl group excludes base pretreatment of the protein as a practical means of enhancing cleavage of Asn-Gly bonds.

It is clear that cyclic imide formation can be a common occurrence in aspartyl or asparaginyl peptides and that esterification of the β-carboxyl group greatly enhances the rate of cyclization. A detailed mechanism for hydrolysis of β-esters of aspartylpeptides which involves substituted succinimide intermediates was presented by Bernhard et al.[22] In view of the chemical similarity between the ester and amide groups it is reasonable to assume that asparaginyl bonds would be more susceptible to imide formation than aspartyl peptide bonds.

Strong support for the preferential tendency of asparaginylglycyl sequences, in comparison with Asn-X sequences, to undergo cyclic imide formation is provided by the work of DeTar et al.[23,24] These authors investigated the "active" ester method of synthesizing sequential peptide polymers and noted that the synthesis of the p-nitrophenyl ester of HBr-Asp(OCH$_3$)-Gly-Gly posed serious difficulties related to imide formation whereas the synthesis of the p-nitrophenyl ester of HBr-Asp-(OCH$_3$)-Ser-Gly proceeded far more readily.

Ample precedent exists for the base-catalyzed hydrolysis of the cyclic imide of peptidylaspartic acid (Fig. 1, reaction 2).[19,25-27] Saponification results in preferential formation of the β-aspartyl peptide bond, since the stronger acidic character of the α-carbonyl group favors nucleophilic attack at this point.

Alternatively, chain cleavage occurs in the presence of alkaline hydroxylamine (Fig. 1, reaction 3). A mechanism that entails nucleophilic addition of hydroxylamine first to one and then to a second carbonyl-carbon of the substituted succinimide is illustrated in Fig. 2. The hydroxylamine adducts (II) and (V) are in equilibrium with their respective oximes (III) and (VI). Rearrangement of (II) and (V) gives rise to the aspartyl mono- and dihydroxamates, (IV) and (VII), respectively. The formation of the dihydroxamate entails cleavage of the peptide chain and liberation of a new NH$_2$-terminal amino acid. Rearrangement of II

[22] S. A. Bernhard, A. Berger, J. H. Carter, E. Katchalski, M. Sela, and Y. Shalitin, *J. Am. Chem. Soc.* **84**, 2421 (1962).

[23] D. F. DeTar, M. Gouge, W. Honsberg, and U. Honsberg, *J. Am. Chem. Soc.* **89**, 988 (1967).

[24] D. F. DeTar, F. F. Rogers, Jr., and H. Bach, *J. Am. Chem. Soc.* **89**, 3039 (1967).

[25] A. R. Battersby and J. C. Robinson, *J. Chem. Soc.* p. 259 (1955).

[26] W. D. John and G. T. Young, *J. Chem. Soc.* p. 2870 (1954).

[27] J. Kovacs, H. Nagy Kovacs, I. Konyves, J. Csaszar, T. Vadja, and H. Mix, *J. Org. Chem.* **26**, 1984 (1961).

to the monohydroxamate can account for the incorporation of hydroxylamine into a polypeptide without chain cleavage.[28] The dihydroxamate is likely to be unstable, hydrolysis yielding a mixture of α- and β-hydroxamates. Presumably hydrazinolysis of the phthalimide protecting group during peptide synthesis, to form phthalyl hydrazide and a free amino group,[29] represents an analogous reaction to the cleavage of a cyclic imide by hydroxylamine.

Procedure

Hydroxylamine Cleavage

A solution, 6 M in guanidine and 2 M in hydroxylamine, is prepared as follows. Guanidine-HCl, 23 g, and 5.5 g of hydroxylamine-HCl are weighed in a 100-ml beaker. Then 4.5 M LiOH (filtered through glass fiber paper before use) is added slowly with vigorous stirring in an ice bath. The pH of the solution is monitored and maintained at 7–8 until all solute has dissolved and a total volume of about 35 ml has been reached. At this time the pH is adjusted to 9.0 with LiOH and water is added to a total volume of 40 ml. The solution is transferred to a reaction vessel maintained at 45°. The protein or peptide, reduced and alkylated to break disulfide bonds, is added to a concentration of 1–5 mg/ml and the pH is readjusted to 9.0 in a pH-stat using 4.5 M LiOH as titrant. A reaction time of 4 hr is usually optimal, since longer times generally result in an increased frequency of nonspecific cleavages. If the volume of solution precludes the use of a pH-stat, cleavage can be performed in a solution containing 2 M hydroxylamine and 2 M guanidine, buffered with 0.2 M K_2CO_3, pH 9.0.

The reaction is terminated by addition of concentrated formic acid to a pH of 2–3. The mixture is then desalted on a column of Sephadex G-25 or Bio-Gel P2, developed with 9% formic acid. Peptides with molecular weights in excess of 2500 can be expected to appear in the void volume of the column.

[28] Compound (II) may rearrange as indicated or may undergo internal nucleophilic attack by the —NHOH group on the β-carbonyl group to produce a bicyclic intermediate that can decompose to liberate a free NH_2-terminal group and a labile hydroxylimide. The latter would be rapidly hydrolyzed to give a mixture of α- and β-hydroxamates. We thank Professor Theodor Wieland for suggesting this mechanism.

[29] E. Schröder and K. Lübke, "The Peptides," Vol. 1, p. 9. Academic Press, New York, 1965.

Hydroxamate Determination

Hydroxamates can be detected colorimetrically after oxidation by iodine. The resulting nitrite is used to diazotize sulfanilic acid which can then form an azo dye with α-naphthylamine.[30] Since hydroxylamine itself undergoes oxidation to nitrite, preparations must be free of hydroxylamine. The following procedure is based on that described by Bergman and Segal[30] and Seifter *et al.*[5]

Reagents

Sodium acetate, 6%, in water
Sulfanilic acid, 1%, in 25% acetic acid
Iodine, 1.3%, in glacial acetic acid
Sodium thiosulfate, 0.4 N in water
α-Naphthylamine, 0.6%, in 30% acetic acid

Procedure. The sample (0.3 ml) containing no more than 50 nmol of hydroxamate is placed in a glass tube and to it is added 1.3 ml of sodium acetate solution. Next is added 0.2 ml of sulfanilic acid followed by 0.1 ml of iodine solution. The mixture is allowed to stand for 10 min and 0.05 ml of thiosulfate solution is added, sufficient to decolorize the iodine. Naphthylamine reagent, 0.04 ml, is then added, and the solution is mixed and allowed to stand at room temperature for 60 min to permit complete and reproducible color development. Optical density at 520 nm is determined in 1-ml cuvettes using reagent blanks. In this fashion quantities of hydroxamate as low as 5 nmol can be determined accurately. Equimolar quantities of sodium nitrite, hydroxylamine, and benzhydroxamic acid yield the same degree of color and can be used to produce standard curves which are linear to 50 nmol.

Analyses for peptide-bound hydroxamate are not essential but serve to identify the fragment proximal to the cleaved bond. In this regard it should be recalled that a fraction of the homoserine in CNBr-produced peptides may be in the form of the lactone and that this lactone is susceptible to hydroxylamine cleavage. Consequently, both fragments produced by hydroxylamine cleavage of a CNBr peptide will contain hydroxamate.[13] In addition, a certain level of hydroxylaminolysis of amide groups may occur.

Applications

Hydroxylamine cleavage of Asn-Gly bonds has been performed successfully in a substantial number of cases in both proteins and large pep-

[30] F. Bergman and R. Segal, *Biochem. J.* **62**, 542 (1956).

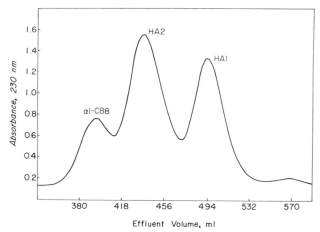

FIG. 3. Elution pattern of fragments obtained by cleavage of rat collagen α1-CB8 with hydroxylamine. Chromatography performed on 8% agarose in 1 M CaCl$_2$, 0.05 M Tris-HCl, pH 7.5. HA-1 and HA-2 represent the two hydroxylamine-produced fragments. Reproduced, with permission, from G. Balian, E. M. Click, and P. Bornstein, *Biochemistry* **10**, 4470 (1971).

tide fragments; this experience is summarized in the table. In addition, Hormann, Heidemann, and co-workers have used hydroxylamine cleavage in a series of studies on fragmentation of collagen chains.[9,31,32] While the bonds split in the latter studies were not identified, the fragments were characterized by molecular weight and amino acid analysis and by electron microscopy of long-spacing segments. When reviewed in the light of the known sequence of collagen α chains[12] it seems likely that most of these cleavages also occurred at Asn-Gly bonds.

As seen in the table, there appears to be no preference for other amino acids adjacent to asparagine and glycine. Presumably, therefore, neither chemical nor steric effects influence the cleavage reaction at that distance.

An example of the extent of cleavage that can be expected at an Asn-Gly bond is shown in Fig. 3. The observed cleavage of the CNBr-produced fragment of a rat collagen α1 chain, about 80% or more, can be obtained by use of the procedure described above. When the substrate was pretreated with alkali (0.2 M K$_2$CO$_3$, pH 10.5, 35°, 90 min) the fragment was almost totally resistant to cleavage.[13] This finding is thought to result from nucleophilic attack by hydroxide ion, leading to opening of the cyclic imide (Fig. 1, reaction 2).

[31] E. Heidemann and W. Heinrich, *Eur. J. Biochem.* **14**, 61 (1970).
[32] H. Schlebusch and H. Hörmann, *Biochim. Biophys. Acta* **221**, 370 (1970).

APPLICATIONS OF THE HYDROXYLAMINE CLEAVAGE REACTION TO PROTEINS AND PEPTIDES

Protein or peptide	Sequence	NH$_2$OH (M)	Solvent	Temp. (°C)	Time (hr)	Extent of cleavage (%)	References[d]
Collagen, rat, α1-CB3	Asn-Asn-Gly-Ala	0.72	1 M K$_2$CO$_3$, pH 9.5	37	7–24	NR[a]	1
Collagen, rat, α1-CB8	Ala-Asn-Gly-Ala	2	0.2 M K$_2$CO$_3$, pH 9.0	45	2	70–80	2, 3
Collagen, chick, α1-CB6A	Lys-Asn-Gly-Asp	1	0.5 M K$_2$CO$_3$, pH 10.5	45	1.5	NR	4
Collagen, bovine, α2-CB4	His-Asn-Gly-Leu Glu-Asn-Gly-Thr Pro-Asn-Gly-Leu	2	0.2 M K$_2$CO$_3$, pH 9.0	30	4	NR	5
Ribonuclease,[b] bovine pancreas	Lys-Asn-Gly-Gln	2	0.2 M K$_2$CO$_3$, pH 9.0	45	2	70	6
ACTH, porcine	Pro-Asn-Gly-Ala	2	0.2 M K$_2$CO$_3$, pH 9.0	NR	NR	NR	7
Glutamate dehydrogenase, bovine liver	Ala-Asn-Gly-Pro	2	0.2 M K$_2$CO$_3$, pH 9.0	45	2	30–50	8
Thermolysin, *Bacillus thermoproteolyticus*	Trp-Asn-Gly-Ser Asp-Asn-Gly-Gly	1	0.1 M K$_2$CO$_3$, pH 10.5	35	2	NR	9
Trypsin,[c] porcine	Phe-Asn-Gly-Asn Cys-Asn-Gly-Gln	2	6 M Guanidine, 0.2 M K$_2$CO$_3$, pH 9.0	45	4	30 50	10
Factor X (Stuart factor),[c] bovine	Lys-Asn-Gly-Ile Asp-Asn-Gly-Gly	2	6 M Guanidine, 0.2 M K$_2$CO$_3$, pH 9.0	45	4	45 70	11

Phosphorylase, rabbit muscle	Gly-Asn-Gly-Gly, Val-Asn-Gly-Val, Thr-Asn-Gly-Ile, Leu-Asn-Gly-Ala	2	6 M Guanidine, 0.2 M K$_2$CO$_3$, pH 9.0	45	4	} 60	12
Glyceraldehyde-3-phosphate dehydrogenase,[b] *Bacillus stearothermophilus*	-Asn-Gly-	2	Guanidine, 0.2 M K$_2$CO$_3$, pH 9.0	45	2	70	13
Superoxide dismutase, bovine	Lys-Asn-Gly-Val	2	4 M Guanidine, 0.2 M K$_2$CO$_3$, pH 9.0	45	3	VR	14

[a] Not reported.
[b] Reduced, carboxymethylated.
[c] Reduced, S-pyridylethylated.
[d] References:

1. W. T. Butler, *J. Biol. Chem.* **244**, 3415 (1969).
2. P. Bornstein, *Biochemistry* **9**, 2408 (1970).
3. G. Balian, E. M. Click, and P. Bornstein, *Biochemistry* **10**, 4470 (1971).
4. S. N. Dixit, J. M. Seyer, A. O. Oronsky, C. Corbett, A. H. Kang, and J. Gross, *Biochemistry* **14**, 1933 (1975).
5. F. W. Rexrodt, P. P. Fietzek, and K. Kuhn, *Eur. J. Biochem.* **59**, 105 (1975).
6. P. Bornstein and G. Balian, *J. Biol. Chem.* **245**, 4854 (1970).
7. L. Gráf, S. Bajusz, A. Patthy, E. Barat, and G. Cseh, *Biochim. Biophys. Acta* **6**, 415 (1971).
8. K. Moon and E. L. Smith, *J. Biol. Chem.* **248**, 3082 (1973).
9. K. Titani, M. A. Hermodson, L. H. Ericsson, K. A. Walsh, and H. Neurath, *Nature (London) New Biol.* **238**, 35 (1972).
10. M. A. Hermodson, L. H. Ericsson, H. Neurath, and K. A. Walsh, *Biochemistry* **12**, 3146 (1973).
11. D. L. Enfield, L. H. Ericsson, K. A. Walsh, H. Neurath, and K. Titani, *Proc. Natl. Acad. Sci. U.S.A.* **72**, 16 (1975).
12. A. Koide, K. Titani, L. H. Ericsson, K. A. Walsh, and H. Neurath, *Fed. Proc., Fed. Am. Soc. Exp. Biol.* **35**, 1622 (1976).
13. J. E. Walker and J. I. Harris, personal communication.
14. H. M. Steinman, V. R. Naik, J. L. Abernathy, and R. L. Hill, *J. Biol. Chem.* **249**, 7326 (1974).

Cleavage at Asn-Gly bonds can produce relatively large peptide fragments that are amenable to automated analysis by sequenator. Since the probability of occurrence of an Asn-Gly bond is of the order of 1/400 for most proteins, no more than one or a few Asn-Gly bonds can be expected. When used in conjunction with CNBr cleavage at methionyl residues the method offers a means of determining the position of some CNBr-produced fragments in the polypeptide chain and provides a new starting point for sequential Edman degradation.

Application of the method to determination of the structure of glutamate dehydrogenase was responsible for correction of a previously published incorrect sequence.[33] In the cases of bovine factor X and thermolysin the use of hydroxylamine cleavage was instrumental in the determination and proof of structure.[34] In special cases the method may be used to distinguish between Asn-Gly and Asp-Gly bonds.[35] To our knowledge, under conditions in which cleavage of Asn-Gly bonds is effected Asp-Gly bonds appear to be essentially resistant.

The disadvantages of the method of hydroxylamine cleavage at Asn-Gly bonds include incomplete cleavage of susceptible bonds and a low level of cleavage at Asn-X bonds. Incomplete cleavage is in most cases not a serious problem since the products are not numerous and can often be separated from the original fragment by molecular sieve chromatography. Only a low level of cleavage at Asn-X bonds has been encountered. In most cases the susceptible bonds have not been identified. Extensive nonspecificity of the reaction, suggested by the findings of Deselnicu et al.[36] with the bovine collagen α2 chain was not encountered by Rexrodt et al.[37] in analyses of the smaller CNBr fragment, α2-CB4. Bornstein and Balian[14] noted cleavage to the extent of 5–10% at the Asn-Leu bond (residues 34–35) in bovine pancreatic ribonuclease and Hermodson[38] observed a low degree of cleavage of an Asn-Met bond (residues 165–166) in porcine trypsin. Cleavage at an Asn-Ala bond in bovine superoxide dismutase was also reported.[39] Such extraneous cleavages, if abundant, could complicate the isolation of fragments sufficiently pure for sequencer analysis.

[33] K. Moon and E. L. Smith, *J. Biol. Chem.* **248**, 3082 (1973).
[34] K. A. Walsh, personal communication.
[35] L. Gràf, S. Bajusz, A. Patthy, E. Barat, and G. Cseh, *Biochim. Biophys. Acta* **6**, 415 (1971).
[36] M. Deselnicu, P. M. Lange, and E. Heidemann, *Hoppe Seyler's Z. Physiol. Chem.* **354**, 105 (1973).
[37] F. W. Rexrodt, P. P. Fietzek, and K. Kuhn, *Eur. J. Biochem.* **59**, 105 (1975).
[38] M. A. Hermodson, personal communication.
[39] H. M. Steinman, V. R. Naik, J. L. Abernathy, and R. L. Hill, *J. Biol. Chem.* **249**, 7326 (1974).

Discussion

It is now firmly established that the increased susceptibility of collagen to hydroxylamine cleavage results from a relatively high incidence of Asn-Gly bonds in this protein and cannot be attributed to the existence of special nonpeptide bonds. Since glycine is present as every third amino acid in collagen and asparaginyl residues precede glycine in five instances in the α1 chain,[40] one would expect six fragments after cleavage with hydroxylamine. This number is roughly consistent with the number of fragments observed during cleavage of collagen, particularly when account is taken of the apparently greater incidence of Asn-Gly bonds in the α2 chain.[12] These observations together with the now proven susceptibility of other proteins to hydroxylamine cleavage establish the general validity of the method.

The original method for hydroxylamine cleavage was adapted from that described by Blumenfeld et al.[4] Modifications that have been introduced include a higher NH_2OH concentration (2 M), slightly higher temperature (45°), a lower pH (9) and the use of 6 M guanidine as a solvent and LiOH as titrant.[41] A lower pH may be expected to reduce competing nucleophilic attack by hydroxide ion which leads to opening of the cyclic imide without chain cleavage (Fig. 1, reaction 2). However, cleavage of Asn-Gly bonds is substantially reduced below pH 9.0,[13] even though the pK_a of hydroxylamine is 6.02. The use of 6 M guanidine as solvent enhances the solubility of large peptide fragments and is likely to increase the exposure of a given Asn-Gly bond to the nucleophile. Marked dependence of the degree of cleavage on conformation of the substrate was observed in the case of ribonuclease.[14] Substitution of LiOH for NaOH takes advantage of the increased solubility of LiCl and avoids the formation of a salt precipitate which appears when NaOH is used to neutralize the guanidine·HCl and $NH_2OH·HCl$.

There is substantial evidence for the prevention of hydroxylamine cleavage of Asn-Gly bonds by base pretreatment.[13] As expected, proteins subjected to such treatment acquire an additional negative charge due to loss of the amide group.[13]

The specificity for cleavage at Asn-Gly bonds, as opposed to other asparaginyl links, appears to be a function of the relative ease with which the asparaginyl side chain can approach the amide bond formed with the succeeding amino acid in the chain when the latter amino acid lacks a side chain. This is readily demonstrable with space-filling models. The

[40] G. Salem and W. Traub, *FEBS Lett.* **51**, 94 (1975).
[41] We thank Drs. K. Titani, D. L. Enfield, and K. A. Walsh for acquainting us with the unpublished procedure for hydroxylamine cleavage in use in their laboratory.

addition of any side chain to the residue following asparagine results in unacceptable contact distances between atoms of the side chain and the oxygen of the β-carbonyl group for most conformations of the polypeptide chain. Some conformations do, however, permit cyclic imide formation even in the presence of substituents on the α-carbon. Possibly, the unusual chemical properties of the X-Gly peptide bond may contribute to the enhanced susceptibility of Asn-Gly bonds to hydroxylamine cleavage.

Although some γ-glutaminyl esters in peptides are also known to form cyclic compounds,[42] cleavage of Gln-Gly bonds by hydroxylamine has not been observed.

It is of interest that β-aspartyl glycine has been identified in the urine of normal individuals and that the quantity of the excreted compound can be increased by feeding a gelatin-rich diet.[43,44] The level of β-aspartyl glycine far exceeds that of any other β-aspartyl dipeptide in human urine.[44] Presumably anhydroaspartylglycyl groups preexist in collagen and other proteins or are formed in the course of digestion and metabolism. Preferential ring opening at the α-carbonyl bond and the metabolic inertness of the β-aspartyl bond would account for excretion of the dipeptide.

Conclusions

Nonenzymic cleavage of proteins with hydroxylamine provides a relatively specific means of producing large peptide fragments suitable for further chemical analysis. Cleavage occurs at Asn-Gly bonds and results from the tendency of the asparaginyl side chain to cyclize forming a substituted succinimide that is susceptible to nucleophilic attack by hydroxylamine. The increased susceptibility of Asn-Gly bonds in comparison with other asparaginyl bonds may result from the greater ease with which the asparaginyl side chain can cyclize in the absence of steric hindrance imposed by a side chain on the succeeding amino acid. The extent of cleavage achieved varies with the protein; cleavage is enhanced by complete denaturation of the protein and by the use of 6 M guanidine as a solvent during hydroxylaminolysis. A low level of cleavage at Asn-X bonds has been observed in some cases, but aspartyl bonds appear to be resistant under conditions used. The infrequency of Asn-Gly bonds in most proteins results in the production of very large fragments that may

[42] D. W. Clayton, G. W. Kenner, and R. C. Sheppard, *J. Chem. Soc.* p. 371 (1956).
[43] J. J. Pisano, E. Prado, and J. Freedman, *Arch. Biochem. Biophys.* **117**, 394 (1966).
[44] F. E. Dorer, E. E. Haley, and D. L. Buchanan, *Biochemistry* **5**, 3236 (1966).

overlap CNBr-produced fragments and could serve as new start points for sequential Edman degradation.

Acknowledgment

We thank Professor Y. Pocker for many helpful discussions regarding the mechanism of hydroxylamine cleavage of Asn-Gly bonds.

[15] Cleavage at Aspartyl–Prolyl Bonds

By Michael Landon

The particular acid lability of the peptide bond linking aspartyl residues to prolyl residues in proteins was first noticed when primary structure analysis was performed on proteins containing these bonds. Substantial cleavage was occurring at such bonds during the procedures employed in the production and isolation of the peptide fragments required for protein amino acid sequence determination.[1] Work on the primary structure of bovine glutamate dehydrogenase, which contains two aspartyl-prolyl bonds, demonstrated that the bonds were being broken during the preparation of tryptic peptides,[2] tryptic peptides from the maleylated protein,[3] peptides produced by cyanogen bromide cleavage,[4] and peptic peptides.[5] In view of the known acid lability of peptide bonds involving aspartyl residues,[6] it was not surprising to have found splitting occurring at such bonds during the acidic conditions employed in hydrolysis with pepsin and in chemical cleavage using cyanogen bromide. However, two other observations were less expected. First, that the milder conditions, albeit still acidic, that were used at that time for peptide purification (and indeed are still used) should cause splitting and, second and more important, that cleavage under all conditions was substantially confined to Asp-Pro bonds.

The lability of Asp-Pro bonds was, therefore, initially seen as anomalous cleavage under the acid conditions routinely used in specific peptide bond scission and subsequent peptide separation procedures. The apparent high specificity of the reaction and the yields obtained, even

[1] D. Piszkiewicz, M. Landon, and E. L. Smith, *Biochem. Biophys. Res. Commun.* **40,** 1173 (1970).

[2] M. Landon, M. D. Melamed, and E. L. Smith, *J. Biol. Chem.* **246,** 2360 (1971).

[3] W. J. Brattin and E. L. Smith, *J. Biol. Chem.* **246,** 2400 (1971).

[4] T. J. Langley and E. L. Smith, *J. Biol. Chem.* **246,** 3789 (1971).

[5] D. Piszkiewicz, M. Landon, and E. L. Smith, *J. Biol. Chem.* **248,** 3067 (1973).

[6] J. Schultz, this series Vol. 11 [28].

under conditions that were appropriate for the intended procedure and not necessarily for the anomalous cleavage that was observed, suggested that the lability of peptide bonds linking aspartyl residues to prolyl residues would lend itself to a specific chemical cleavage procedure.

The limitation of specific chemical cleavage at Asp-Pro bonds is the rarity of these bonds, but this is also an inherent advantage of the method.

The mechanism of selective cleavage at Asp-Pro bonds has been discussed by Piszkiewicz et al.[1] Selective hydrolysis of aspartyl peptide bonds occurs under mildly acidic conditions for periods of 4 to 24 hr at elevated temperatures.[6] Mechanistically this involves intramolecular catalysis by aspartate carboxylate anion displacement of the protonated nitrogen of the peptide bond. The properties of proline are clearly influential with regard to the greatly increased lability of the bond linking an aspartyl residue to a prolyl residue. Piszkiewicz et al.[1] suggested that in peptide linkage there is a greater basicity of the prolyl nitrogen compared with that of other amino acid residues. This greater basicity would, by analogy with the hydrolysis of anilides,[7] enhance the rate of hydrolysis of the peptide bond by increasing protonation of the leaving group. It is pertinent to note that specific cleavage at bonds preceding prolyl residues in the peptide antibiotic alamethicin was obtained under acid conditions (12 M HCl, 37°, 30 min).[8] Piszkiewicz et al.[1] also commented on other factors that might influence the reaction.

Procedure

The protein should be dissolved at a concentration in the range 0.5–10 mg/ml in the appropriate acid solution. Table I lists the various acid solutions that have been used for specific cleavage at Asp-Pro bonds. The final choice of conditions will depend on the particular protein and will have to be determined by trial procedures. In general, the temperature should not exceed 40° and the duration of the procedure can be as long as 120 hr in the milder conditions, but should not be extended beyond 48 hr in the case of the more vigorous conditions (a formic acid concentration above 70%), where significant nonspecific peptide bond splitting may occur if the exposure to acid is prolonged. The extent of cleavage can be determined by procedures involving the separation and quantitation of peptide products or by quantitative end-group analysis.

[7] M. L. Bender and R. J. Thomas, J. Am. Chem. Soc. 83, 4183 (1961).
[8] J. W. Payne, R. Jakes, and B. S. Hartley, Biochem. J. 117, 757 (1970).

TABLE I

METHODS FOR SPECIFIC CLEAVAGE AT ASPARTYL-PROLYL BONDS

	Conditions		
Method	Acid solution	Temp. (°C)	Duration (hr)
A[a]	10% (v/v) acetic acid, adjusted to pH 2.5 with pyridine	40	24–120
B[a]	10% (v/v) acetic acid, adjusted to pH 2.5 with pyridine, in 25% (v/v) 1-propanol	40	24–96
C[a]	10% (v/v) acetic acid, adjusted to pH 2.5 with pyridine, in 7 M guanidinium chloride	40	24–96
D[b]	70% (v/v) formic acid	37	24–72
E[b]	70% (v/v) formic acid, in 7 M guanidinium chloride	37	24–48
F[b]	75% (v/v) formic acid	37	24–48
G[b]	75% (v/v) formic acid, in 7 M guanidinium chloride	37	24–48
H[b]	90% (v/v) formic acid	37	24

[a] K. J. Fraser, K. Poulsen, and E. Haber, *Biochemistry* **11**, 4974 (1972).
[b] J. Jauregui-Adell and J. Marti, *Anal. Biochem.* **69**, 468 (1975).

Discussion of Applications

The first reported use of specific chemical cleavage at Asp-Pro bonds was by Poulsen *et al.*,[9] who applied the procedure to rabbit antibody light chain. The conditions employed were 10% acetic acid, adjusted to pH 2.5 with pyridine, in 7 M guanidinium chloride, at 40° for 96 hr. This system gave cleavage at the single susceptible bond in yields of 90%. The effects of different conditions on the splitting of rabbit antibody light chain were investigated by Fraser *et al.*,[10] who showed that the effectiveness of the acid employed (as above) was substantially increased by the presence of 7 M guanidinium chloride. However, in the case of a carboxypeptidase inhibitor from potato, splitting was obtained in yields of 84% using acetic acid (10%, at pH 2.5) alone.[11]

More vigorous conditions were required with other proteins. Mare milk lysozyme required 75% formic acid in 7 M guanidinium chloride, at 37° for 48 hr for quantitative cleavage of the single Asp-Pro bond,[12]

[9] K. Poulsen, K. J. Fraser, and E. Haber, *Proc. Natl. Acad. Sci. U.S.A.* **69**, 2495 (1972).
[10] K J. Fraser, K. Poulsen, and E. Haber, *Biochemistry* **11**, 4974 (1972).
[11] G. M. Hass, H. Nau, K. Biermann, D. T. Grahn, L. H. Ericsson, and H. Neurath, *Biochemistry* **14**, 1334 (1975).
[12] J. Jauregui-Adell and J. Marti, *Anal. Biochem.* **69**, 468 (1975).

while with tobacco mosaic virus protein (vulgare strain) maximum yields of only 38% at the more labile of two bonds were obtained using 90% formic acid, at 37° for 24 hr.[12] In the latter case the exposure to acid could not be extended as nonspecific cleavage was observed. Studies on the protein from the other strains of tobacco mosaic virus showed that even small changes in overall primary structure significantly changed the lability at Asp-Pro bonds.[12] Variation in susceptibility to cleavage of Asp-Pro bonds is therefore apparent. A summary of the results of procedures applied to particular proteins is given in Table II.

TABLE II
APPLICATIONS OF SPECIFIC CLEAVAGE AT ASPARTYL-PROLYL BONDS

Protein	Method[a]	Percentage cleavage after hours of reaction					Protein con-centration (mg/ml)
		24	48	72	96	120	
Rabbit antibody light chain[b]	A	15	27	24	26	28	0.5–5
Rabbit antibody light chain[b]	B	18	24	29	44		0.5–5
Rabbit antibody light chain[b]	C	30	56	68	87		0.5–5
Rabbit antibody light chain[c]	C[d]				64		8–16
Rabbit antibody light chain[e]	C[d]				70		16
Carboxypeptidase inhibitor from potato[f]	A				84		5
Mare milk lysozyme[g]	C[d]		30				2
Mare milk lysozyme[g]	D	35	40	75			2
Mare milk lysozyme[g]	E		35				2
Mare milk lysozyme[g]	F	30	80				2
Mare milk lysozyme[g]	G	70	100				2
Tobacco mosaic virus protein[g,h]	D	12					2
Tobacco mosaic virus protein[g,h]	G		19				2
Tobacco mosaic virus protein[g,h]	H	38					2
Rubredoxin from *Desulfovibrio vulgaris*[i]	D	40					2

[a] The letter refers to the method in Table I.

[b] K. J. Fraser, K. Poulsen, and E. Haber, *Biochemistry* **11**, 4974 (1972).

[c] J. Jaton. *Biochem. J.* **141**, 1 (1974).

[d] Procedure carried out at 37°.

[e] J. Jaton, *Biochem. J.* **141**, 15 (1974).

[f] G. M. Hass, H. Nau, K. Biemann, D. T. Grahn, L. H. Ericsson, and H. Neurath, *Biochemistry* **14**, 1334 (1975).

[g] J. Jauregui-Adell and J. Marti, *Anal. Biochem.* **69**, 468 (1975).

[h] Vulgare strain: figures are for cleavage at the more labile Asp-Pro bond; cleavage at the other Asp-Pro bond occurred only in Method H and did not exceed 10%.

[i] M. Bruschi, J. Bonicel, G. Bovier-Lapierre, and P. Couchoud, *Biochim. Biophys. Acta* **434**, 4 (1976). Almost equal cleavage occurred at each of the two Asp-Pro bonds.

Summary

Specific cleavage at Asp-Pro bonds of a protein can be effected by exposure to acid at moderate temperature for periods up to 120 hr. The effectiveness of cleavage is usually increased by incorporating a denaturing agent in the acid solution. Some Asp-Pro bonds are resistant to cleavage probably because of retention of protein folding even under the most vigorous conditions that have been employed, and under these circumstances some nonspecific peptide bond cleavage may occur, particularly if the exposure is prolonged.

It is very likely that cleavage at Asp-Pro bonds will occur when acid procedures are employed in peptide bond scission or peptide separation and purification. With the larger peptides produced by cyanogen bromide cleavage it has been observed that partial cleavage at Asp-Pro near the termini of the authentic peptide may yield fragments that copurify with the larger component, thus complicating subsequent sequence determinations.[13]

The relative paucity of Asp-Pro bonds in proteins is both a limitation and an advantage of this specific cleavage procedure. The limitation is that many proteins do not contain any Asp-Pro bonds but as the Asp-Pro bond is never abundant the number of fragments obtained by application of the procedure can be advantageously small, which will simplify subsequent fractionation processes.

[13] K. Titani, M. A. Hermodson, L. H. Ericsson, K. A. Walsh, and H. Neurath, *Biochemistry* **11**, 2427 (1972).

[16] Phthalylation of Amino Groups

By J.-F. Pechère and R. Bertrand

The reversible chemical modification of amino groups of proteins and peptides by acylating reagents is a widely used technique in protein chemistry because of its vast range of applications in both active-site and sequence studies. Comparative studies performed on lysozyme[1] and myoglobin[2] have shown that maleic anhydride[3,4] and citraconic anhydride[5,6] are the most satisfactory reagents for such modifications, and

[1] A. F. S. A. Habeeb and M. Z. Atassi, *Biochemistry* **9**, 4939 (1970).
[2] R. P. Singhal and M. Z. Atassi, *Biochemistry* **10**, 1756 (1971).
[3] P. J. G. Butler, J. I. Harris, B. S. Hartley, and R. Leberman, *Biochem. J.* **112**, 679 (1969).
[4] P. J. G. Butler and B. S. Hartley, this series Vol. 25, p. 191.
[5] H. B. F. Dixon and R. N. Perham, *Biochem. J.* **109**, 312 (1968).
[6] M. Z. Atassi and A. F. S. A. Habeeb, this series Vol. 25, p. 546.

these two reagents are indeed those whose use is the most frequently reported in the literature. However, the reversible blocking of amino groups by maleic or citraconic anhydrides is not always free of problems, which often belong to a given type of polypeptide. Generally, the unblocking of maleylated proteins requires rather drastic or lengthy conditions and is not always quantitative, while that of citraconylated proteins can occur so easily that the derivative is not stable under the conditions which are required for its examination or its selective splitting (e.g., tryptic digestion). This last situation is encountered, among others, in the case of proteins containing a large proportion of acidic amino acid side chains, probably because some of these can participate locally in the catalysis[3] of the unblocking reaction. There is thus a need for a reagent yielding derivatives of stability intermediate between those of maleylated and citraconylated polypeptides.

A successful step in this direction was made by Riley and Perham,[7] who described the interesting properties and applications of exo-*cis*-3, 6-endoxo-Δ^4-tetrahydrophthalic anhydride. Their work suggested that the more easily obtainable phthalic anhydride could perhaps behave similarly and this possibility was therefore explored. It has thus been found[8] indeed that phthalic anhydride reacts rapidly and specifically at pH 8.5 with the free amino groups of proteins and peptides, thereby converting them quantitatively to *o*-carboxybenzoyl derivatives. This reaction is also quantitatively reversible by incubation of the phthalyl derivative at pH 3.5 and 50° for 48 hr.

Reaction of Phthalic Anhydride with Proteins

It has been found that complete blocking of amino groups is achieved (Fig. 1) after a protein, at concentrations of 5–10 mg/ml, has been allowed to react at pH 8.5–9.0 and at room temperature, with 5–6 aliquots of a 2-fold excess of phthalic anhydride over the number of free amino groups present. The reaction is most conveniently conducted in a pH-stat using dry dioxane to introduce the anhydride in solution. These conditions are, as expected, very similar to those prevailing for the reaction of maleic anhydride with proteins.[3,4] Carbonate buffers are useful here also. Urea or guanidinium hydrochloride can be incorporated in the reaction mixture when an insoluble protein is being handled or when the presence of "buried" amino groups is suspected.

There is no doubt that, like maleic and citraconic anhydrides, phthalic anhydride also reacts to some extent with seryl, threonyl, and tyrosyl hydroxyl groups, with histidine imidazole groups and cysteine

[7] M. Riley and R. N. Perham, *Biochem. J.* 118, 733 (1970).

[8] R. Bertrand, P. Pantel, and J.-F. Pechère, *Biochimie* 56, 515 (1974).

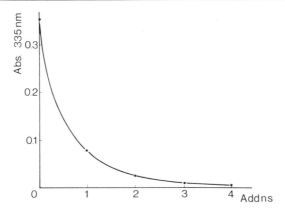

FIG. 1. Reaction of phthalic anhydride with carp 4.47 parvalbumin at pH 8.5 (1% sodium bicarbonate). The abscissa represents the number of additions of a 2-fold molar excess of phthalic anhydride over the number of free amino groups initially present. The absorbances at 335 nm of the ordinate refer to the application of the 2,4,6-trinitrobenzenesulfonic acid procedure according to T. Okuyama and K. Satake, *J. Biochem.* (*Tokyo*) **47,** 454 (1960) on an aliquot of the reaction mixture after each addition, and is proportional, after blank subtraction, to the number of free amino groups still present. Results reproduced, with permission, from R. Bertrand, P. Pantel, and J.-F. Pechère, *Biochimie* **56,** 515 (1974).

thiol groups. However this presents no practical problems, as the ester bonds thus produced can be quantitatively cleaved by a treatment with hydroxylamine. On the other hand, the absence of an activated double bond in phthalic anhydride, as in exo-*cis*-3,6-endoxo-Δ^4-tetrahydrophthalic anhydride, is the reason why no alkylation of thiol groups can occur as it does with the maleic and citraconic anhydrides. The existence of isomeric derivatives, which are formed with the last reagent,[5] is also excluded in the present case. Phthalic anhydride can thus rightly be considered to be entirely specific for free amino groups.

Spectral changes occur during the reaction of phthalic anhydride with free amino groups (Fig. 2), allowing the conclusion[8] that open *o*-carboxybenzoyl derivatives, and not cyclic imides, are formed exclusively under the above reaction conditions (slightly alkaline pH). The important contribution to the absorption, at 270 nm, of the incorporated benzene rings provides a sensitive way of ascertaining the extent of the reaction.

Procedure

The desired amount of protein is dissolved, in a pH-stat cell, in 1% sodium bicarbonate (pH 8.5) so that a 5–10 mg/ml solution is obtained. Possibly, 1025 mg of guanidinium hydrochloride are added to each

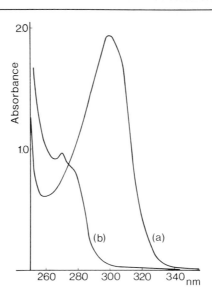

FIG. 2. Ultraviolet spectra of phthalylglycylglycine imide (7.5×10^{-4} M) (a) in acid medium (0.01 M HCl) and (b) in alkaline medium (0.1 M NaOH or after 3 hr in 1% NaHCO₃, pH 8.5). Results reproduced with permission from R. Bertrand, P. Pantel, and J.-F. Pechère, *Biochimie* **56**, 515 (1974).

milliliter of this solution in order to render it 6 M in the last reagent (with a concomitant 1.8-fold increase in volume). The pH is adjusted to 8.5 with 2 M NaOH contained in the pH-stat syringe and the machine is set up to keep this pH automatically. An aliquot (usually 20–50 μl) from a 250 mg/ml (1.69 M) solution of phthalic anhydride in dry dioxane, corresponding to a 2-fold molar excess over the number of free amino groups present, is added to the reaction vessel under vigorous stirring, resulting in an immediate consumption of base. When this has leveled (ca. 10 min), a second and four further aliquots are added in the same way so that a total of 5–6 additions is made over about 1 hr. To the contents of the pH-stat cell, hydroxylamine hydrochloride (12 mg/mg original protein) is now added, the pH is adjusted to 9.0, and the mixture is left standing at room temperature for about 20 hr. The contents of the pH-stat cell are subsequently transferred to the top of a column of Sephadex G-25 (cross section: 0.125 cm²/mg protein; length: 15 diameters, e.g., 2 × 30 cm for 25 mg of protein) equilibrated with 0.5% ammonium bicarbonate (pH 8.5), where they are freed of unreacted material (flow rate: 17.5 ml/hr/cm² cross section, e.g., 55 ml/hr for a 2 × 30-cm column). The protein peak (front) is collected and the solution is lyophilized.

Two methods have been found useful for checking that the reaction has indeed taken place quantitatively.[9,10] They are most conveniently applied to a sample of the pool of the protein fractions from the Sephadex G-25 desalting column, a sample of the earlier fractions providing for an adequate blank when necessary. The first method relies on the spectral changes associated with the reaction. The spectra of the starting and of the derivatized proteins, both dissolved in 0.1 M NaOH, being 6 M in guanidinium hydrochloride, are recorded between 240 and 300 nm and the difference between the specific extinction coefficients of the two materials at 271 nm is computed. By comparison with the value $\epsilon = 1208$ M^{-1} cm^{-1}, corresponding to the addition of one carboxybenzoyl group (Fig. 2), the amount of incorporated phthalyl groups can be estimated. The second method makes use of the reaction of dansyl chloride with free amino groups.[11] This is conveniently performed according to Sodek and Hofmann[2] on an aliquot of derivatized protein corresponding to 300 nmol of free ϵ-lysyl groups initially present in the original protein. A blank of similar volume is treated in parallel. At the end of the procedure, the whole acid hydrolysate is spotted on a silica gel thin-layer plate together with appropriate standards of 1–2 nmol of ϵ-DNS-lysine. The plate is developed with the solvent methyl acetate/isopropanol/conc. NH$_3$ 45:35:20 (v/v) described by Crowshaw et al.[13] The absence, in the sample lane, of an ϵ-DNS-lysine spot stronger than that of the 1 nmol standard ensures that less than 0.5% of the free lysines of the original protein have been left unreacted.

Cellulose acetate electrophoresis (in 0.075 M sodium bicarbonate buffer, pH 8.6, 100 V for 15 min) is helpful in ascertaining that the material obtained is homogeneous.

Stability of Phthalyl-Protein Bonds

At a pH above 6.5, and in particular at pH 8.5 and 40°, phthalyl-protein bonds are perfectly stable (Fig. 3). In contrast, at a lower pH, they become labile, the blocked amino groups being progressively regenerated into their free form, a process that is completed in 48 hr at pH 3.5 and 50°.

[9] Abbreviations: TNBS, 2,4,6-trinitrobenzenesulfonic acid; DNS, 5-dimethylaminonaphthalene-1-sulfonyl; PTH, phenylthiohydantoin.
[10] The progress of the reaction may be followed with TNBS (Fig. 1) only provided no urea or guanidinium chloride are present in the reaction mixture.
[11] W. R. Gray, this series Vol. 25, p. 121.
[12] J. Sodek and T. Hofmann, Can. J. Biochem. 48, 425 (1970).
[13] K. Crowshaw, S. J. Jessup and P. W. Ramswell, Biochem. J. 103, 79 (1967).

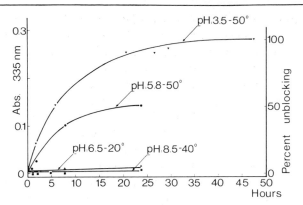

Fig. 3. Stability as a function of time (hours) of the phthalyl-protein bonds in fully phthalylated carp 4.47 parvalbumin under different conditions of pH and temperature. The absorbances at 335 nm of the left ordinate are as in Fig. 1. For convenience, the values have also been expressed as the corresponding percentages of unblocking on the right ordinate. Results reproduced with permission from R. Bertrand, P. Pantel, and J.-F. Pechère, *Biochimie* **56**, 515 (1974).

This situation is particularly favorable to the use of phthalic anhydride as a reversible blocking agent, permitting the selective tryptic cleavage of peptide chains at the level of their arginyl bonds.[8,14] The fragments thus generated can be fractionated, preferably in their phthalylated form which ensures their solubility, at least in nonacid media, by chromatography on either molecular sieving supports or anion exchangers. The peptides which are obtained at sufficient purity can be directly subjected[14] to Edman degradation[15] under particularly favorable conditions, as they will generally have a COOH-terminal arginine and as solubility problems often associated with the phenylthiocarbamylation of the ε-amino groups of lysine are avoided. This amino acid, when present, is recovered in the form of α-PTH, ε-phthalyllysine which can be readily recognized by thin-layer chromatography. Alternatively, isolated phthalylated peptides can be unblocked and be exposed to further tryptic digestion at the level of their regenerated lysyl side chains, while mixtures of peptides that resist fractionation on the basis of charge in their blocked form can often be separated in this way after they have been unblocked.

[14] J.-P. Capony, J. Demaille, C. Pina, and J.-F. Pechère, *Eur. J. Biochem.* **56**, 215 (1975).

[15] S. Iwanaga, P. Wallèn, N. J. Cröndahl, A. Henschen, and B Blombäck, *Eur. J. Biochem.* **8**, 189 (1969).

Unblocking of Phthalylated Peptide Chains

The phthalylated material is dissolved in distilled water so as to provide a solution at a maximum concentration of 4 mg/ml. Subsequently, 0.200 ml of 1 M NaOH is added for each milliliter of solution under vigorous stirring, followed by 0.560 ml of 1 M formic acid. Under these conditions, the phthalylated derivative precipitates as a very fine suspension while a strongly buffered medium of pH ca 3.50 is obtained. This suspension is incubated at 50° for 48 hr under constant gentle stirring. The insolubility of the material throughout the process does not impair its unblocking. It can be brought back into solution at the end of the incubation period by addition, under vigorous stirring, of 5 M NaOH (0.120 ml/ml initial solution). The unblocked product can be recovered free of contaminants by desalting on a molecular sieving column of appropriate porosity equilibrated with 0.5% ammonium bicarbonate (e.g., as described above for the recovery of the phythalylated protein).

Comments

The above procedures have been used with constant success in work dealing mainly with muscular parvalbumins.[14,16] where neither maleylation nor citraconylation have been found adequate. Other proteins, however, have also been handled, such as frog troponin-C[17] yeast phosphoglycerate kinase,[18] horse and dogfish myoglobins,[19] and mouse hemoglobin.[20] All these proteins have been observed to behave in a way very similar to muscular parvalbumins. The only significant difference noted relates to the occasional insolubility of some of the hemoproteins considered under the coupling conditions (1% bicarbonate, pH 8.5). When this is the case, one proceeds as described above for the unblocking reaction, the phthalylation being effected on a fine suspension of the protein obtained from its acid solution by increasing the pH. The suspension progressively dissolves during the progress of the coupling reaction.

It is probable that diagonal methods, in particular those described in conjunction with the use of maleic anhydride,[4] would also be of application when this is replaced by the phthalic homolog. However, no experience has yet been gained about this possibility.

[16] J.-P. Capony, C. Pina, and J.-F. Pechère, *C. R. Acad. Sci.* Ser. D **280**, 1615 (1975).
[17] J.-P. Capony, J.-P. van Eerd, and J.-F. Pechère, unpublished (1976).
[18] J.-P. Sallei and F. Martin, unpublished (1976).
[19] P. Pantel and G. Roseau, unpublished (1975).
[20] J.-P. Sallei and E. Zuckerkandl, *Biochimie* **57**, 343 (1975).

[17] Reversible Blocking at Arginine by Cyclohexanedione

By EMIL L. SMITH

Because of its relatively specific cleavage at arginyl and lysyl residues, trypsin remains one of the most valuable enzymes for hydrolysis of polypeptides and proteins in studies of amino acid sequences. The use of trypsin has been greatly enhanced by the development of methods for reversible blocking at lysyl residues, e.g., by maleylation,[1] citraconylation,[2] etc., thus limiting the hydrolysis to peptide bonds involving arginyl residues. It obviously became desirable to find a reagent and conditions which would similarly permit reversible blocking at arginyl residues to allow hydrolysis only at lysyl peptide bonds.

Any reagent that reversibly binds to guanidinium groups should also be useful in probing the functional roles of specific arginyl residues in peptides and proteins, e.g., in binding of substrates or cofactors, and in the active sites of enzymes. Until recently, there have been few such reagents available.

It has been known for some time that bifunctional aldehydes and ketones react with the guanidinium groups of arginyl residues; however, in earlier studies the reactions were performed with a variety of reagents under conditions that resulted either in irreversible changes of the arginyl residues, modification of other residues, incomplete substitution of the arginyl residues, or spontaneous regeneration of the arginyl residues in neutral or slightly alkaline solutions.[3] This suggested a systematic investigation[3] with a previously studied reagent,[4] 1,2-cyclohexanedione, to find conditions that would permit its use for the blocking of arginyl residues while avoiding, as much as possible, the limitations mentioned above. The methods described here are derived from the papers of Patthy and Smith.[3,5]

Treatment with cyclohexanedione of a mixture of all of the usual amino acids found in proteins, under the specified conditions, showed complete recovery, except for arginine, of all of them, as judged by recovery on the amino acid analyzer. The compound formed by arginine and cyclohexanedione is stable under mildly acidic conditions. Similarly,

[1] P. J. G. Butler, J. I. Harris, B. S. Hartley, and R. Leberman, *Biochem. J.* **112**, 679 (1969).
[2] H. B. F. Dixon and R. N. Perham, *Biochem. J.* **109**, 312 (1968).
[3] L. Patthy and E. L. Smith, *J. Biol. Chem.* **250**, 557 (1975).
[4] K. Toi, E. Bynum, E. Norris, and H. A. Itano, *J. Biol. Chem.* **242**, 1036 (1967).
[5] L. Patthy and E. L. Smith, *J. Biol. Chem.* **250**, 565 (1975).

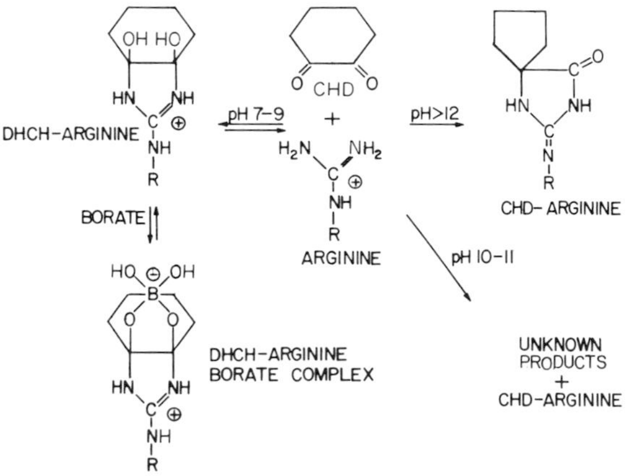

FIG. 1. The reaction of arginine with 1,2-cyclohexanedione (CHD) at various pH values is shown.

after reaction of lysozyme and oxidized ribonuclease with cyclohexanedione, followed by acid hydrolysis under the usual conditions (6 N HCl at 110° for 24 hr), quantitative recoveries of all amino acids were obtained, except for arginine, which is largely destroyed under these conditions.

The use of several reagents for arginine modification has already been reviewed in an earlier volume of this series by Yankeelov.[6]

Principle

As shown in Fig. 1, 1,2-cyclohexanedione (CHD) reacts with arginine (or arginyl residues) at pH 8–9 in sodium borate buffer to form a single product N^7-N^8-(1,2-dihydroxycyclohex-1,2-ylene)-L-arginine (DHCH-arginine). At pH values above 9 the product is converted first to substances of unknown structure and, finally and irreversibly, to CHD-arginine, N^5-(4-oxo-1,3-diazaspiro[4,4]non-2-ylidene)-L-ornithine.[4] As previously demonstrated by Toi et al.,[4] CHD-arginine is the sole product in strongly alkaline solutions (above pH 12).

The structure of DHCH-arginine was proved by examination of its nuclear magnetic resonance and infrared spectra which were consistent with the proposed structure. The presence of vicinal cis-hydroxyl groups was indicated by the formation of a complex with borate and by cleavage on oxidation with periodate to yield N^7-adipyl-L-arginine (Fig. 2), presumably because the expected cyclic intermediate is unstable. Hydrolysis of adipylarginine in acid (or weak alkali) yielded adipic acid and

[6] J. A. Yankeelov, Jr., this series Vol. 25, p. 566.

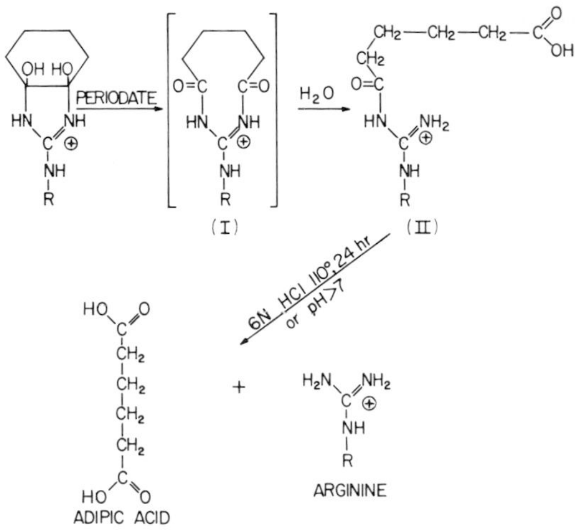

Fig. 2. The cleavage of DHCH-arginine by periodate to yield N^7-adipylarginine (II) is shown. The latter gives adipic acid and arginine on hydrolysis under the conditions specified.

arginine. The formation of a specific red complex with Ni^{2+} indicated the dioxime structure of DHCH-arginine.

Formation of the complex of DHCH-arginine with borate serves both to accelerate the reaction and to stabilize the product, thus driving the reaction to completion in a relatively short time. Under these conditions there is no detectable reaction with lysyl residues, a reaction which occurs much more slowly and at a much higher pH optimum (near pH 11).

Conditions of Reaction and Characterization of Product

Investigations with free arginine demonstrated that only DHCH-arginine is formed in the presence of cyclohexanedione between pH 7.5 to 9.0 in the presence of borate buffer.[7] The reaction is first order with respect to both reactants and the rate is increased almost 10-fold between 25° and 40°. Higher temperatures were not studied since secondary changes may occur in peptides and proteins.

DHCH-arginine is stable indefinitely in 30% acetic acid and, indeed, is stable below pH 7.0, thus permitting hydrolysis of proteins and peptides by proteolytic enzymes under acidic conditions, and isolation of the resulting peptides without regeneration of arginyl residues. DHCH-

---

[7] Tris or other amine buffers should not be used since complexes may form slowly with cyclohexanedione.

arginine is stable under the usual conditions of amino acid analysis and is eluted from the short column of the analyzer between ammonia and arginine. In paper chromatography with 1-butanol–pyridine–acetic acid–water (15:10:3:12), the R_f of DHCH-arginine is 0.34–0.35 as compared to 0.23 for arginine.[3] On paper impregnated with boric acid, the R_f is 0.27.[5] On paper electrophoresis at pH 1.9, DHCH-arginine migrates between glycine and alanine with an R_f of 0.76 relative to arginine.[3,8]

Hydrolysis in 6 N HCl at 110° for 24 hr destroys most of the DHCH-arginine but there may be considerable (15–20%) regeneration of arginine. If the hydrolysis is performed in the presence of excess mercaptoacetic acid (20 μl), DHCH-arginine is converted to a neutral product of unknown structure but with no regeneration of arginine[3]; this product migrates on paper electrophoresis at pH 1.9 between alanine and valine. The last procedure permits estimation by difference analysis of the number of modified arginine residues in a peptide or protein.

At pH 8–9, arginine is regenerated slowly when borate is removed by dialysis or by gel filtration. This is accelerated by amines that react with cyclohexanedione, such as Tris, hydrazine, ammonia, etc. The process is also accelerated by strong nucleophiles such as hydroxylamine. pH has little effect since these reagents serve to trap the liberated cyclohexanedione. At pH 7.0 and 37° the half life of DHCH-arginine in 0.5 M hydroxylamine is 100 min. The reaction is performed in the absence of oxygen for 6–7 hr in evacuated, sealed tubes under nitrogen. Solutions should be prepared with deionized water or in the presence of EDTA to bind Cu^{2+}, since this metal ion can catalyze secondary reactions with proteins in the presence of hydroxylamine. Note also that hydroxylamine may produce cleavage of asparaginylglycine peptide bonds when these are present in the peptide or protein.[9,10]

Prevention of Tryptic Hydrolysis at Arginyl Residues

A model experiment with oxidized ribonuclease is described below to illustrate the utility of the method in specific cleavage by trypsin at lysyl residues.[3]

Performic acid-oxidized bovine pancreatic ribonuclease A (2 μmol/ml) was treated with 0.15 M cyclohexanedione (a 75-fold excess) in 0.25 M sodium borate buffer at pH 9.0 in a sealed vial at 35° for 2 hr. An equal volume of 30% acetic acid was then added and the protein was dialyzed

---

[8] The electrophoretic mobility of DHCH-arginine relative to arginine at various pH values is given in Fig. 3 of the paper by Patthy and Smith.[3]

[9] W. T. Butler, *J. Biol. Chem.* **244**, 3415 (1969).

[10] P. Bornstein and G. Balian, *J. Biol. Chem.* **245**, 4854 (1970).

in the cold successively against 15%, 7.5%, and 1% solutions of acetic acid. After freeze-drying, the sample was dissolved in 0.1 M borate at pH 8.0 at a concentration of 1 μmol/ml and hydrolyzed with trypsin for 4 hr. The hydrolysis was terminated by adding 30% acetic acid.

Each of the peptides in the hydrolyzate was isolated in pure form by the conventional methods of gel filtration on Sephadex G-50 in 30% acetic acid followed by paper chromatography in 1-butanol–pyridine–acetic acid–water or by electrophoresis at pH 1.9, without regeneration of arginyl residues. No peptide was found that was produced by tryptic action at arginyl residues, and all of the peptides expected from hydrolysis at lysyl residues were isolated in pure form.

The method has also been used in our laboratory on a large peptide obtained after cyanogen bromide cleavage of the NADP-specific glutamate dehydrogenase of *Neurospora crassa*.[11] In this instance, it was essential to obtain an overlap of a tryptic peptide terminating in arginine. Use of the above method limited the tryptic action to hydrolysis at lysyl residues and permitted isolation of the peptide necessary for proving the overlapping sequence.

Recently, the method has been used for completing the sequence of ribonuclease St from *Streptomyces erythreus*.[12] After treatment with 1,2-cyclohexanedione by the method of Patthy and Smith, hydrolysis with trypsin was limited to the two lysyl residues present in this protein of 102 residues.

Identification of Specific Arginyl Residues

Treatment of native proteins with cyclohexanedione permits ascertaining the effect of modifying arginyl residues on enzymic activity. First studies were made with pancreatic ribonuclease and egg white lysozyme.[5]

With small proteins it is possible to identify peptides containing modified arginyl residues by comparing peptide patterns or by specific reaction with Ni^{2+} on paper chromatograms.[5] The latter yields the typical red color of the Ni^{2+}–dioxime complex. Unfortunately, however, the color of the complex is usually weak against the light green background of the color given by Ni^{2+} salts on paper. Washing the paper to remove the Ni^{2+} after staining the peptides will elute the soluble peptides; however, peptides containing Ni^{2+} bound to the dioxime are sparingly soluble and will remain on the paper.

[11] K. M. Blumenthal, K. Moon, and E. L. Smith, *J. Biol. Chem.* **250**, 3644 (1975).
[12] N. Yoshida, A. Sasaki, M. A. Rashid, and H. Otsuka, *FEBS Lett.* **64**, 122 (1976).

Recently, [¹⁴C]cyclohexanedione has become available from a commercial source.[13] This now permits convenient identification of peptides containing arginyl peptides specifically labeled with ¹⁴C. As an example, the NADP-specific glutamate dehydrogenase of *Neurospora* is rapidly inactivated by reaction with cyclohexanedione.[14] The most rapidly reacting arginyl residue in the protein was specifically labeled and a pure peptide containing a single residue of [¹⁴C]DHCH-arginine was isolated after enzymic hydrolysis, first with pepsin and then with trypsin at a slightly acid pH.[15,16]

[13] New England Nuclear, 549 Albany Street, Boston, MA 02118.
[14] K. M. Blumenthal and E. L. Smith *J. Biol. Chem.* **250**, 6555 (1975).
[15] B. M. Austen and E. L. Smith, *J. Biol. Chem.* **251**, 5835 (1976).
[16] The initial hydrolysis was done with pepsin since the reaction of cyclohexanedione and the enzyme could be terminated rapidly by acidification, in order to avoid further reaction with other arginyl residues. The presence of borate does not interfere with peptic action.

Section VI

Selective Cleavage with Enzymes

[18] Cleavage at Arginine Residues by Clostripain

By William M. Mitchell

Clostripain [clostridiopeptidase B (EC 3.4.22.8)] is a sulfhydryl protease with a narrow specificity range limited to the carboxyl peptide linkage of positively charged amino acids. Unlike trypsin, however, clostripain markedly prefers arginine over lysine residues.[1] Thus, under conditions of controlled hydrolysis peptide bond cleavage can be limited to arginine sites, including trypsin-resistant arginylprolyl bonds.[2] Although this enzyme could be expected to be a widely used tool for the sequence chemist, the lack of a commercial source has greatly limited its usage.[3] The assay, purification, and physical–chemical properties of clostripain have been reviewed previously.[4,5] Only the enzymic properties relevant to its specificity and potential utility will be considered.

Active Site Properties

Clostripain possesses both esterase-amidase as well as protease activities. The enzyme is easily assayed by its ability to hydrolyze N-benzoylarginine ethyl ester (BAE) in the presence of a reducing agent.[4] As illustrated in the table, trypsin and clostripain differ markedly in their active site kinetic parameters, although their molar rates of hydrolysis are comparable.[6] Clostripain owes its speed of hydrolysis of BAE to a remarkably high k_{cat} while trypsin is dependent on a better binding affinity of substrate to its active site. The specificity for the arginine moiety in naturally occurring proteins is illustrated in Fig. 1. In these experiments the indicated protein was exposed for various lengths of time to clostripain (100:1, w/w). After proteolysis, peptide fragments were further digested with carboxypeptidase B [treated with diisopropylphosphorofluoridate (DFP) and soybean trypsin inhibitor] to liberate carboxyl-terminal arginine and lysine residues with subsequent

[1] W. M. Mitchell and W. F. Harrington, *J. Biol. Chem.* 243, 4683 (1968).

[2] W. M. Mitchell, *Science* 162, 374 (1968).

[3] Worthington Biochemical Corporation (Freehold, NJ) recently introduced clostripain as a standard stock enzyme. It is too early to assess the general quality of this product for standard sequence studies as reported by users.

[4] W. M. Mitchell and W. F. Harrington, this series Vol. 19 [45].

[5] W. M. Mitchell and W. F. Harrington, Clostripain *in* "The Enzymes," 3rd ed. (P. D. Boyer, ed.), Vol. 3, p. 699.

[6] W. H. Porter, L. W. Cunningham, and W. M. Mitchell, *J. Biol. Chem.* 246, 7675 (1971).

COMPARISON OF KINETIC CONSTANTS OF CLOSTRIPAIN, TRYPSIN, AND
PAPAIN-CATALYZED HYDROLYSIS OF N-BENZOYLARGININE ETHYL ESTER AT $25°$

Enzyme	$K_{m(app)}$ (mM)	k_{cat} (sec^{-1})	$k_{cat}/K_{m(app)}$ (mM^{-1} sec^{-1})
Trypsin[a]	0.0043	14.6	3395.0
Papain[b]	12.2	15.3	1.25
Clostripain[c]	0.235	643.3	2738.0

[a] Data from T. Inagami and J. M. Sturtevant, *J. Biol. Chem.* **235**, 1019 (1960); 0.025 M CaCl₂, pH 8.0.

[b] Data from J. R. Whitaker and M. L. Bender, *J. Am. Chem. Soc.* **87**, 2728 (1965); 0.2 M sodium acetate, pH 5.79.

[c] Data from W. H. Porter, L. W. Cunningham, and W. M. Mitchell, *J. Biol. Chem.* **246**, 7675 (1971); 0.1 M sodium phosphate, pH 7.8, 5 mM DTT.

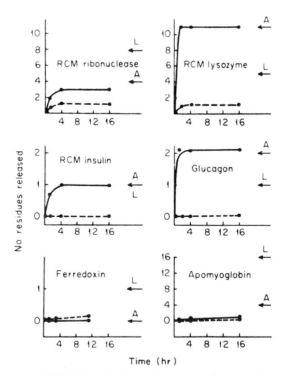

FIG. 1. Number of COOH-terminal residues exposed per molecule during timed hydrolysis of various polypeptides in the presence of clostripain. COOH-terminal arginine and lysine exposed during clostripain digestion are released as free amino acids by carboxypeptidase B. The arrows indicate the arginine and lysine residues susceptible to tryptic hydrolysis. RCM, reduced carboxymethylated. Illustration from W. M. Mitchell and W. F. Harrington, *J. Biol. Chem.* **243**, 4683 (1968).

analysis by the Spackman *et al.*[7] short-column method. The solid line plots as a function of time for arginine liberation and the dashed line indicates lysine liberation. In each case the rapid appearance of arginine with respect to lysine confirms the general specificity of clostripain. The solid arrows represent the tryptic release of arginine (A) and lysine (L), respectively. One residue of lysine was released in clostripain-treated RCM (reduced carboxymethylated) lysozyme, but this substrate contains a lysylarginine sequence that would be attacked by carboxypeptidase B following clostripain hydrolysis at arginine. The only apparent exception is RCM ribonuclease, where one residue of lysine was found per three residues of arginine. Figure 1 further illustrates that not all potential substrates are susceptible to clostripain hydrolysis. Apomyoglobin contains four arginine residues per molecule, with less than one arginine bond split in 16 hr. A similar phenomenon has been observed with arginine-rich histone, in which less than 10% of the total arginine sites were split. Thus not only is the enzyme highly selective in its side chain acceptance (i.e., arginine) but also it is probably sensitive to the local environment of the potential substrate site as well.

Sequences immediately adjacent to the arginine specificity site are important in the relative binding affinity of substrate to clostripain. Rainey[8] uses a variety of defined synthetic polypeptides to determine their competitive k_1 on BAE hydrolysis. The presence of secondary binding sites distant to the anionic guanidinium binding site was demonstrated over a minimum distance of six amino acid residues. There are at least three binding subsites for side chains in the active site, plus the major arginyl specificity site. Immediately adjacent to the major arginyl specificity site toward the substrate amino terminus, there is an anionic binding site that binds either lysine or arginine. No influence of amino acids in residue No. 2 was found, although a second anionic binding site was discovered at residue No. 3. Proceeding toward the carboxyl terminus of the substrate, a single hydrophobic site which markedly prefers tryptophan residues was found immediately adjacent to the major arginyl specificity site. The contribution of these subsites to the binding affinity of substrate to enzyme may account for the anomalous situation in which neither polyarginine nor polylysine is hydrolyzed by clostripain, although both polymers inhibit BAE hydrolysis (i.e., polyarginine two orders of magnitude greater than polylysine).[1]

The length of the guanido side chain is critical for maximal binding affinity of substrate to enzyme. L-Homoarginine with one additional carbon in the side chain is a very effective inhibitor of BAE hydrolysis.[6] The most

---

[7] D. H. Spackman, W. H. Stein, and S. Moore, *Anal. Chem.* **30**, 1190 (1958).
[8] J. M. Rainey, Doctoral Dissertation, Vanderbilt University, 1971.

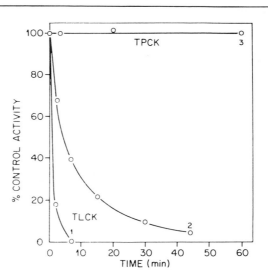

Fig. 2. Effect of tosyllysinechloromethyl ketone (TLCK) and tosylphenylalanine-chloromethyl ketone (TPCK) on the enzymic activity of clostripain. Clostripain is at a concentration of 37.5 nM. Reactions were carried out in 0.05 M sodium phosphate, pH 6.5, in the presence of 0.11 mM DTT. Curve 1, 60 nM TLCK; curve 2, 15 nM TLCK; curve 3, 15 nM TPCK. Illustration from W. H. Porter, L. W. Cunningham, and W. M. Mitchell, *J. Biol. Chem.* **246**, 7675 (1971).

effective competitive inhibitors, however, are the alkylguanidines, which also serve as competitive inhibitors for trypsin. The most effective competitive inhibitor found to date is butylguanidine with approximately 2.5 orders of magnitude better binding affinity for clostripain than for trypsin.[9] As the alkyl side chain decreases in length, the effectiveness as an inhibitor of BAE hydrolysis substantially decreases.

Although the active site of clostripain shows a marked preference for binding guanido groups, tosyllysinechloromethyl ketone (TLCK) is an effective active site label.[6] Figure 2 illustrates the loss of clostripain activity against BAE, a phenomenon that is easily prevented when the enzyme is incubated in the presence of a competitive inhibitor such as benzamidine or tosylhomoarginine methyl ester. Moreover, tosylphenylalaninechloromethyl ketone (TPCK) is without effect except at high concentrations. Although the site of attack has not been elucidated to date, a sulfhydryl reactive group appears to be involved in view of clostripain's free sulfhydryl requirement, the known sites of attack by these reagents (i.e., histidine and cysteine) in other systems, a pronounced lability of labeled site in clostripain to acid hydrolysis, and an

[9] P. W. Cole, K. Murakami, and T. Inagami, *Biochemistry* **10**, 4246 (1971).

observed $pK_a \simeq 8.25$ in log V_{max} vs pH plots.[9] The rapid attack by TLCK on the active site of clostripain would appear to contradict the general phenomenon of restricted arginine specificity. However, in retrospect, the rapid and specific affinity of TLCK for the active site of clostripain is reasonable. Although the rate of TLCK inactivation is faster in clostripain than trypsin, the expected greater reactivity of a sulfhydryl in the active site of clostripain over the histidyl-active nucleophile in trypsin is compatible with the known major specificity for guanidine groups by clostripain. Unfortunately, no arginine homolog (i.e., tosylarginine chloromethyl ketone) has been available to test this hypothesis.

Utility as a Sequence Tool

The lack of a commercial source for clostripain until recently has severely limited the use of this unique protease. The initial production of large fragments from proteins under conditions of limited hydrolysis has not been used to date, but it offers substantial advantages to investigators of primary sequence. Ideally, hydrolysis rates should be followed by pH or chemical titration[10] of liberated α-NH_2 groups in order to avoid secondary cleavage at lysine residues. In addition, screening of clostripain preparations for adventious nonclostripain protease activity can be easily accomplished by the following techniques: (1) Hydrolysis of BAE should be completely eliminated by H_2O_2 and by equimolar quantities of TLCK. (2) Polyarginine should be resistant to hydrolysis and should inhibit hydrolysis of BAE.

An example of a sequence problem in which clostripain was used to solve an overlap is provided by structural studies on a lysozyme produced by a species of *Chaloropsis*. Edman degradation of a cyanogen bromide fragment (CNBr III) resulted in a tentative assignment of a lysine residue to the No. 5 position. However, loss of a substantial percentage of lysine at the first cycle necessitated an independent assignment of lysine to residue No. 5. Figure 3 illustrates CNBr III and its clostripain and tryptic peptides.[11] Isolation of a unique peptide (Cl-2) produced by clostripain which contained an arginine COOH terminus plus an internal lysine provided the correct peptide placement when com-

---

[10] pH-stat kinetics offers the advantage of immediate knowledge of the rate of hydrolysis but suffers from the relatively large quantities of substrate required. Chemical titration of liberated α-NH_2 groups with reagents such as fluorescamine or *o*-phthalaldehyde offers the advantage of small substrate quantities required for detection but suffers a moderate time lag in analysis.

[11] J. W. Shih and J. H. Hash, *J. Biol. Chem.* **246**, 994 (1971).

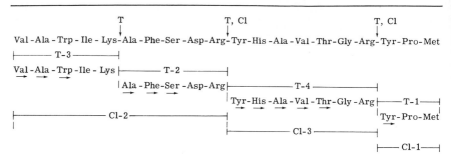

Fig. 3. Sequence of CNBr-III peptide from *Chaloropsis* lysozyme. Tryptin (T) and clostripain (Cl) cleavage sites as indicated by ↓. Edman degradation steps of individual tryptic peptides indicated by →. Illustration from J. W. Shih and J. H. Hash, *J. Biol. Chem.* **246**, 994 (1971).

pared to sequenced tryptic peptides. Clostripain has been used also as an aid in the primary structure determination of tuftsin[12] and as a tool in structure–function studies on glucagon.[13]

[12] K. Kishioka, A. Constantopoulos, P. S. Satah, W. M. Mitchell, and V. A. Najjar, *Biochim. Biophys. Acta* **310**, 217 (1973).
[13] P. W. Felts, N. E. C. Ferguson, K. A. Hagey, E. S. Stiff, and W. M. Mitchell, *Diabetologia* **6**, 44 (1970).

[19] Specific Hydrolysis by Trypsin at Alkaline pH[1]

By Lewis J. Greene and Diana C. Bartelt

Trypsin hydrolysis at pH 7–8 is extensively used in amino acid sequence determinations to hydrolyze the peptide bond at the carboxyl terminus of arginine, lysine, and S-2-aminoethylcysteine residues.[2] Specific hydrolysis at arginine by trypsin can be achieved by acylation of the ε-and ω-amino groups of lysyl and S-2-aminoethylcysteinyl residues, respectively. When a reversible modification reaction, such as maleyla-tion,[3] is employed, subsequent hydrolysis at both lysyl and S-2-amino-

[1] Research carried out at Brookhaven National Laboratory under the auspices of the U.S. ERDA. By acceptance of this article, the publisher and/or recipient acknowl-edges the U.S. Government right to retain a nonexclusive royalty-free license in and to any copyright covering this paper.
[2] D. G. Smyth, this series Vol. 11, p. 214 (1967).
[3] P. J. G. Butler and B. S. Hartley, this series Vol. 25, p. 191 (1972).

ethylcysteinyl residues can be obtained by a second exposure to trypsin at pH 7–8 after removal of the acyl blocking groups. In this way, overlapping peptides may be obtained through the use of only one proteolytic enzyme.

Hydrolysis at pH 11.0 is an alternative method to achieve specific cleavage by trypsin which does not require chemical modification of the substrate. A significant advantage over the acylation procedure is that hydrolysis at alkaline pH permits selective cleavage of lysyl residues in the presence of S-2-aminoethylcysteinyl peptide bonds. However, the discrimination between arginyl and lysyl bonds at alkaline pH is not as precise as that obtained by the acylation method.

Principle

The procedure is based on the kinetic studies of Wang and Carpenter.[4,5] Selective hydrolysis by trypsin at alkaline pH is based on the requirement of the enzyme for a positively charged distal amino group adjacent to the site of hydrolysis. The differences in the acid dissociation constants (pK_a) of the guanidine group of arginine (ca. 12.5), the ϵ-amino group of lysine (10.3) and the ω-amino group of S-2-aminoethylcysteine (9.4) provide the basis for the selectivity when hydrolysis is carried out at pH 10.7 to 11.0. However, the relative rates of hydrolysis of these peptide bonds also depend on the amino acid sequence of the substrate adjacent to the scissile bond, because both ionization of the distal amino groups and trypsin catalysis are influenced by adjacent amino acids.

Procedure

Apparatus. Trypsin hydrolysis is carried out in a pH-stat to maintain the pH at 11.0 and to monitor the rate and extent of reaction.[2,6] An automatic burette with a small total delivery volume (0.25 ml) and a combination electrode with a narrow diameter (Radiometer GK 2322 C, or equivalent) is recommended for reaction volumes of 5–8 ml. The reaction vessel (8–10 ml total volume) is fitted with a jacket for circulating water at a constant temperature (25°) and mixing is achieved with a plastic-coated magnetic stirring bar. A slow stream of nitrogen freed of

[4] S. S. Wang and F. H. Carpenter, *Biochemistry* **6**, 215 (1967).
[5] S. S. Wang and F. H. Carpenter, *J. Biol. Chem.* **243**, 3702 (1968).
[6] C. F. Jacobsen, J. Léonis, K. Linderstrøm-Lang, and M. Ottesen, *Methods Biochem. Anal.* **4**, 171 (1957).

CO_2 by passage through KOH is directed over the surface of the reaction solution during hydrolysis to reduce the adsorption of $CO_2 + NH_3$ from air.

Trypsin. Trypsin (bovine, crystalline, salt-free) is treated with L-(1-tosylamido-2-phenyl)ethyl chloromethyl ketone (TPCK) as described by Carpenter[7] in order to inactivate chymotrypsin, which is a frequent contaminant of commercial bovine trypsin preparations. Stock solutions of trypsin (1 mg/ml) are prepared in 1 mM HCl and stored at 5°. The concentration of TPCK-treated trypsin is given in terms of the weight of the lyophilized powder. These preparations usually contain only 50–60% active trypsin.

Trypsin is sufficiently active and stable at pH 10.7–11.0 to be used at concentrations of 1–4% w/w relative to the substrate for periods of 1–2 hr. For example, a 4% w/w ratio corresponds to a molar ratio of 1:600 for active trypsin: substrate molecular weight 2000. There is, however, a significant decrease in trypsin activity when going from pH to 10.7 to 11.0. This is due to an ionizable group (pK_a 10.4) in trypsin which must be protonated in order to have a fully active enzyme.[5] TPCK-treated trypsin (100 μg/ml) at pH 11.0, 25°, for 1 hr loses 15 and 25% of its activity, respectively, in the presence and in the absence of $CaCl_2$.

Trypsin Hydrolysis at pH 11.0. The substrate (1–5 μmol) dissolved in 4 ml of 0.1 M KCl is transferred to the reaction vessel of the pH-stat and equilibrated at 25° under a stream of nitrogen. The pH of the solution is manually titrated to 10.5 with 0.1 N NaOH and then automatically to pH 11.0 with the autoburette containing 0.033 N NaOH. The reaction is started by the addition of 0.1–0.3 ml of trypsin. Since the trypsin solution is acidic, it is necessary to determine the amount of base required to neutralize the trypsin solution before hydrolysis of the peptide or protein is carried out, so that the value for total base uptake can be corrected. The reaction is stopped by adjusting the pH to 2.0 with normal HCl when base uptake has ceased, when the theoretical number of bonds has been hydrolyzed, or after a predetermined amount of time which is based on a preliminary experiment.

Examples

1. Selective Hydrolysis at Lysine in the Presence of S-2-Aminoethylcysteine. Residues 6 through 44 of *S*-2-aminoethylcysteinyl porcine pancreatic secretory inhibitor I has arginine at the carboxyl terminus and

---

[7] F. H. Carpenter, this series, Vol. 11, p. 237 (1967).

contains three residues of lysine and five residues of S-2-aminoethyl-cysteine per molecule.[8] The peptide (0.3 μmol/ml) was hydrolyzed with 4.5%, w/w, TPCK-treated trypsin at pH 11.0, 25° for 20 min.

Glu-Ala-Thr-$Cys(Ae)$- Thr-Ser-Glu-Val-Ser-Gly-$Cys(Ae)$-Pro-Lys-Ile- Tyr-Asn-Pro-
6   10   15   18   20
Val-$Cys(Ae)$-Gly-Thr-Asp-Gly- Ile -Thr-Tyr-Ser-Asn-Glu-$Cys(Ae)$-Val-Leu-$Cys(Ae)$-
   25   30   35
Ser-Glu-Asn-Lys-Lys-Arg
   40   42   43   44

Glu —— 42% —— Lys   Ile ——————— 45% ——————— Lys   Lys ——— 68% ——— Arg
16   18   19   42   43   44

Glu ———————————— 44% ———————————————— Lys
6   42

                    Ile ——————————————— 7% ——————————————————— Arg
                    19   44

The partially hydrolyzed products, residues 6 through 42 and residues 19 through 44, were redigested under the same conditions for 40 and 50 min, respectively. The overall recoveries of the products relative to the amount of the starting peptide, residues 6 through 44, were: residues 6 through 18, 80%; residues 19 through 42, 89%, and residues 43 and 44, 74%. No hydrolysis at any of the five residues of S-2-aminoethyl-cysteine was detected. Hydrolysis at the Lys_{42}-Lys_{43} peptide bond occurred more rapidly than at the Lys_{18}-Ile_{19} peptide bond by a factor of at least 2.

2. *Selective Hydrolysis at Arginine in the Presence of Lysine.* Residues 40 through 56 of performic acid-oxidized bovine pancreatic secretory trypsin inhibitor contain one arginine and two lysine residues.[9] The peptide (0.5 μm/ml) was hydrolyzed with 4%, w/w, TPCK-treated

Glu-Asn-Lys-Glu-Arg-Gln-Thr-Pro-Val-Leu-Ile-Gln-Lys-Ser-Gly-Pro-CySO$_3$H
40   42   44   50   52   56

Glu ——— 76% ——— Arg Gln ———————————— 60% ———————————————— CySO$_3$H
40   44   45   56

                    Gln ———————————— 24% ———————————— Lys Ser ——— 25% ——— CySO$_3$H
                    45   52   53   56

trypsin at pH 11.0, 25°C. The reaction was terminated after 7 min when the initial rapid release of acid had stopped (cf. Fig. 3 of Greene and Bartelt.[9]

[8] D. C. Bartelt and L. J. Greene, *J. Biol. Chem.* **246**, 2218 (1971).
[9] L. J. Greene and D. C. Bartelt, *J. Biol. Chem.* **244**, 2646 (1969).

The ratio of the relative rates of hydrolysis of Arg-Gln compared with Lys-Ser is at least 4:1. Selectivity was reduced when the duration of hydrolysis was extended. After 13 min of hydrolysis, the following recoveries were obtained: residues 40 through 44, 82%; residues 45 through 52, 75%; residues 53 through 56, 68%; and residues 44 through 56, 19%. The Lys-Glu peptide bond (residues 42 and 43) was not hydrolyzed in either experiment.

Comments

1. The effect of neighboring amino acids on the relative rates of hydrolysis is illustrated by the more rapid hydrolysis of the Lys-Lys bond than the Lys-Ile bond in Example 1 and by the hydrolysis of the Lys-Ser bond, but not the Lys-Glu bond in Example 2. The experimental conditions given in the examples provide some guidelines, but it is generally necessary to carry out some preliminary experiments before applying the procedure. In Example 2 the duration of hydrolysis was selected on the basis of a preliminary experiment in which the kinetics of the total hydrolysis at pH 11.0 were monitored in a pH-stat (cf. Fig. 3 of Greene and Bartelt.[9]

2. The relative rates of hydrolysis of lysyl and S-2-aminoethylcysteinyl bonds by trypsin at alkaline pH are sufficiently different to permit complete hydrolysis of lysyl bonds before S-2-aminoethylcysteinyl bonds are attacked. At pH 11.0 this may require 60–80 min for some peptides. Since exposure of peptides to this pH may lead to deamidation (especially at Asn-Gly bonds), it may be advisable to attempt to carry out the reaction at pH 10.7, where trypsin is more active, thereby reducing the incubation time.

3. The discrimination between arginyl and lysyl bonds at alkaline pH is not as precise as obtained with the acylation method. The use of short incubation periods to limit the extent of hydrolysis increases the effective selectivity, provides additional overlap information, but also reduces the yields of products and makes their purification more complex. In view of this limitation, it may be advisable in some cases to apply both methods: acylation and hydrolysis at alkaline pH to obtain selective trypsin hydrolysis in three stages.[8,10] The reduced aminoethylated protein may first be maleylated to obtain specific cleavage at arginine and then, after fragments are separated and demaleylated, trypsin hydrolysis at alkaline pH is employed to obtain specific hydrolysis at lysine. Finally, the products are treated at pH 7–8 with trypsin to hydrolyze to S-2-aminoethylcysteine residues.

[10] D. C. Bartelt, R. Shapanka, and L. J. Greene, *Arch. Biochem. Biophys.* **179**, 189–199 (1977).

[20] Applications of Thermolysin in Protein Structural Analysis

By ROBERT L. HEINRIKSON

One prominent class of proteolytic enzymes is composed of neutral metalloendopeptidases, largely bacterial in origin, which share a common specificity and exhibit maximal activity in the pH range from 7 to 8.[1] These enzymes are inactivated by metal chelators, such as EDTA and 1,10-phenanthroline, but are refractive toward the reagents and polypeptide proteinase inhibitors usually employed to inactivate serine and sulfhydryl proteinases. Thermolysin is a thermostable, extracellular metalloendopeptidase isolated from culture filtrates of the thermophilic organism *Bacillus thermoproteolyticus*. At present, it is both the best understood member of this class of neutral proteases and the most widely employed in protein structural analysis. Thermolysin is a single polypeptide chain of 316 residues with a molecular weight of 34,600. Its amino acid sequence[2] and tertiary structure[3] have been elucidated, and each molecule of enzyme has been shown to contain a single atom of Zn^{2+}, which is essential for catalytic activity.

The purification and assay of thermolysin have been described by Matsubara[4] in an earlier volume of this series. The following discussion will be concerned with selected aspects of the enzyme specificity and examples in which thermolysin has been applied to the hydrolysis of native and denatured polypeptides.

Specificity

The hydrolysis of peptide bonds by thermolysin proceeds optimally in cases in which the amino group is donated by an amino acid residue with a bulky hydrophobic side chain. Thermolysin cleavages are therefore predominant at leucine, isoleucine, phenylalanine, and valine residues, but the enzyme displays rather broad specificity with protein substrates and hydrolysis of bonds involving methionine, histidine, tyrosine, alanine, asparagine, serine, threonine, glycine, lysine, and acidic

[1] M. Matsubara and J. Feder, *in* "The Enzymes," 3rd ed. (P. D. Boyer, ed.), Vol. 3, p. 721. Academic Press, New York, 1971.
[2] K. Titani, M. A. Hermodson, L. H. Ericsson, K. A. Walsh, and H. Neurath, *Nature (London) New Biol.* **238**, 35 (1972).
[3] B. W. Matthews, L. H. Weaver, and W. R. Kester, *J. Biol. Chem.* **249**, 8030 (1974).
[4] H. Matsubara, this series Vol. 19, p. 642.

TABLE I
SPECIFICITY OF THERMOLYSIN IN THE HYDROLYSIS OF PROTEINS

Protein[a]	No. of amino acid residues	Conditions	No. of peptides observed	P_1' residues in bonds cleaved[b]
Native and PAO-azurin from *Pseudomonas fluorescens*[c]	128	0.2 M NH$_4$OAc, pH 8.5, 5 mM CaCl$_2$ 37°, 57°; 1–3 hr, S/E = 100 (w/w)	42	Leu, Ile, Phe, Val, Met[c] (Tyr, ½ Cys, Ala, Asp, Thr, His, Gly)
RCM-nerve growth factor[d]	118	pH-Stat, pH 8.0, 37°, 22 hr, S/E = 200 (w/w)	60	Leu, Ile, Phe, Val (Ala, Tyr, Ser, Thr)
RAE αA$_2$ chain of bovine crystallin	173	0.1 M NH$_4$HCO$_3$, pH 8.6, 37°, 15 hr, S/E = 50 (w/w)	10[e]	Leu, Ile, Phe, Val (Ser)
HD-Sulfate binding protein from *Salmonella typhimurium*[f]	~300	pH-stat, pH 8.0, 37°, 6 hr, S/E = 50 (w/w)	41	Leu, Ile, Phe, Val (Ala, His, Tyr, Lys?)
RCM-ferredoxin from *Chromatium*[g]	81	0.1 M Tris, pH 8.0, 5 mM CaCl$_2$ 40°, 2 hr, S/E = 240 (w/w)	24	Leu, Ile, Phe, Val, Glu (Tyr)
Heme-free horseradish peroxidase[h]	~300	Same as[c] except 37°, 2 hr, S/E = 100 (w/w)	120	Leu, Ile, Phe, Val (Met, Ala, Thr, Gly, Glu)
Tobacco mosaic virus protein[i]	158	Protein-buffered, pH 8, 36°, 2 hr, S/E = 200 (w/w)	24	Leu, Ile, Phe Val (Ala, Tyr)
PAO-insulin[j]	51	Same as above except 30 min, S/E = 1200 (w/w)	15	Leu, Phe
PAO-insulin B-chain[j]	30	0.05 M (NH$_4$)$_2$CO$_3$, pH 8.0, 27° 16 hr, S/E = 100 (w/w)	14	Leu, Phe, Val (Tyr, Ala)
PAO-ferredoxin (taro)[k]	97	pH 8.0, 40°, 3 hr, S/E = 100 (w/w)	25	Leu, Ile, Val, Tyr[k]
CNBr F^{III} from carboxypeptidase A[l]	81	pH-stat, pH 8.0, 37°, 2 hr, S/E = 200 (w/w)	22	Leu, Ile, Phe, Val (Ala, Trp)

[a] RCM, reduced and S-carboxymethylated; RAE, reduced and S-aminoethylated; PAO, performic acid-oxidized; HD, heat denatured.
[b] Residues in parentheses indicate sites of minor cleavage.
[c] R. P. Ambler and R. J. Meadway, *Biochem. J.* **108**, 893 (1968). No cleavage observed at methionine sulfone in oxidized protein.
[d] R. H. Angeletti, M. A. Hermodson, and R. A. Bradshaw, *Biochemistry* **12**, 100 (1973).
[e] F. J. van der Ouderaa, W. W. de Jong, and H. Bloemendal, *Eur. J. Biochem.* **39**, 207 (1973). Only peptides necessary to complete the sequence were described.
[f] T. Imagawa, *J. Biochem. (Tokyo)* **72**, 911 (1972).
[g] H. Matsubara, R. M. Sasaki, D. K. Tsuchiya, and M. C. W. Evans, *J. Biol. Chem.* **245**, 2121 (1970).
[h] K. G. Welinder and L. B. Smillie, *Can. J. Biochem.* **50**, 63 (1972).
[i] H. Matsubara, R. M. Sasaki, A. Singer, and T. H. Jukes, *Arch. Biochem. Biophys.* **115**, 324 (1966).
[j] K. Morihara and H. Tsuzuki, *Biochim. Biophys. Acta* **118**, 215 (1966).
[k] K. K. Rao and H. Matsubara, *Biochem. Biophys. Res. Commun.* **38**, 500 (1970). No cleavage was observed for Glu-Phe (15-16) or Ser-Phe (62-63)
[l] R. A. Bradshaw, *Biochemistry* **8**, 3871 (1969).

TABLE II

SITES OF THERMOLYSIN CLEAVAGE IN OXIDIZED INSULIN,
TOBACCO MOSAIC VIRUS PROTEIN, AND CYTOCHROME c^a

P_1	P'_1
Lys, Arg, Ser, Gly, Ala, Met, Leu	Ile
His, Asn, Thr, Ser, Gln, Gly, Ala, Leu, Tyr, Phe	Leu
Lys, Gln, Thr, Phe	Val
Gln, Ser, Gly, Leu, Phe	Phe
Lys, Ser	Ala
Arg	Tyr

[a] Data from H. Matsubara, R. M. Sasaki, A. Singer, and T. H. Jukes, *Arch. Biochem. Biophys.* **115**, 324 (1966).

residues are commonly encountered in greater or lesser degree depending upon the conditions of hydrolysis. The sites of thermolysin cleavage in a variety of proteins are indicated in Table I, together with the experimental conditions employed in each case. Residues listed are those occupying the P'_1 position as defined by the notation of Schechter and Berger[5] (P_4-P_3-P_2-P_1 ↑ P'_1-P'_2-P'_3-P'_4). Although hydrolyses of bonds involving tryptophan are rare, at least one such cleavage in carboxypeptidase A has been documented by Bradshaw.[6] In general, however, the preferred sites of cleavage are those in which the P'_1 residue is leucine, isoleucine, phenylalanine, or valine. A summary of the peptide bonds cleaved by thermolysin in several proteins of known structure is given in Table II.

Exploration of the secondary specificity of thermolysin has been made through analysis of the rates of hydrolysis of model peptides[7-10] and studies of competitive inhibition of substrate binding.[10,11] Several conclusions in addition to those mentioned above relative to the primary specificity are obvious from the results given in Table III. A strict requirement is shown for residues of the L-configuration in both P_1 and P'_1, and D-amino acids in the P'_2, P'_3, P'_4, P_2, and P_3 positions show inhibitory effects, the magnitudes of which are inversely related to the distance

[5] I. Schechter and A. Berger, *Biochem. Biophys. Res. Commun.* **27**, 157 (1967).
[6] R. A. Bradshaw, *Biochemistry* **8**, 3871 (1969).
[7] H. Matsubara, *Biochem. Biophys. Res. Commun.* **24**, 427 (1966).
[8] K. Morihara and T. Oka, *Biochem. Biophys. Res. Commun.* **30**, 625 (1968).
[9] K. Morihara and H. Tsuzuki, *Biochim. Biophys. Acta* **118**, 215 (1966).
[10] K. Morihara and H. Tsuzuki, *Eur. J. Biochem.* **15**, 374 (1970).
[11] J. Feder, L. R. Brougham, and B. S. Wildi, *Biochemistry* **13**, 1186 (1974).

TABLE III
COMPARATIVE RATES OF HYDROLYSIS OF SEVERAL MODEL PEPTIDE
SUBSTRATES BY THERMOLYSIN

$P_4 - P_3 - P_2 - P_1 \downarrow P'_1 - P'_2 - P'_3 - P'_4$	Rate of hydrolysis
Gly–Phe	0^a
Gly–Phe–NH₂	0^a
Z–Gly–Phe	0.025^a
Z–Gly–Phe–NH₂	1.6^a
Z–Gly–Leu–NH₂	18.0^b
Z–Gly–Leu–Gly	7.6^b
Z–Gly–Leu–Ala	76.1^b
Z–Gly–Leu–D-Ala	0.3^b
Z–Gly–Leu–Leu	51.4^b
Z–Gly–Leu–Phe	18.2^b
Z–Gly–Leu–Gly–Gly	11.2^b
Z–Gly–Leu–Gly–Ala	25^b
Z–Gly–Leu–Gly–D-Ala	2.4^b
Z–Gly–Leu–Gly–Phe	18.6^b
Z–Gly–Leu–Gly–Gly – Ala	29^b
Z–Gly–Leu–Gly–Gly–D-Ala	25^b
Z–Gly–Leu–NH₂	6.6^c
Z–Ala–Leu–NH₂	13.6^c
Z–Ala–Gly–Leu–NH₂	15.2^c
Z–D-Ala–Gly–Leu–NH₂	3.0^c
Z–Ala–Gly–Gly–Leu–NH₂	0.8^c
Z–D-Ala–Gly–Gly–Leu–NH₂	5.0^c

[a] Data expressed as μmol/ml/hr [H. Matsubara, *Biochem. Biophys. Res. Commun.* **24**, 427 (1966)].

[b] Data expressed as μmol/min/mg enzyme [K. Morihara and T. Oka, *Biochem. Biophys. Res. Commun.* **30**, 625 (1968)]. Assay conducted in pH-stat at pH 8.0, 4 mM peptide, 0.1 M KCl, 30°.

[c] Same as [b] except assay performed by ninhydrin: 4 mM peptide, 25% dimethylformamide, 50 mM Tris, pH 7.0, 2 mM CaCl₂, 40°.

from the bond cleaved. The endopeptidase nature of thermolysin attack is confirmed by the fact that no hydrolysis occurred with dipeptides possessing a free α-amino group and only marginal cleavage of Z-Gly-L-Phe was observed, even though Z-Gly-L-Phe-NH₂ is a good substrate for thermolysin.[7] The effect on hydrolysis of residues in the P'_2 position was L-Ala > L-Leu > L-Phe ≫ Gly and amide, and in the P'_3 position L-Ala > L-Phe > Gly. Hydrolysis of peptides is favored when aromatic amino acids occupy the P_1 site and hindered when P_1 is an acidic residue.[8] The peptide Z-Pro-Leu-NH₂ was not cleaved by thermolysin,[8] contrary to observations with native peptides and proteins in which

TABLE IV

THERMOLYSIN-CATALYZED HYDROLYSIS OF TRIPEPTIDE SUBSTRATES[a]

Substrate P_2–$P_1 \uparrow P'_1$–P'_2	[S] (mM)	$k_{cat}/K_M \times 10^{-4}$ (sec^{-1} M^{-1})	Relative k_{cat}/K_M
FA–Gly–Ala–Gly	0.4	0.013	1
FA–Gly–Leu–NH$_2$	0.2	2.2	170
FA–Ala–Ala–Ala	0.4	5.5	420
FA–Gly–Leu–Gly	0.2	8.3	640
Ac–Gly–Leu–Gly	1.0	0.064	5
FA–Gly–Leu–Phe	0.05	30	2,300
FA–Gly–Leu–Ala	0.1	87	6,700
FA–Phe–Leu–Gly	0.1	230	18,000

[a] Reactions at 25° in 0.05 M Tris, pH 7.5 containing 0.1 M NaCl and 10 mM CaCl$_2$; k_{cat}/K_M for FA-Gly-Ala-Gly is taken as 1. Data from S. Blumberg and B. Vallee, *Biochemistry* **14**, 2410 (1975).

hydrolysis of bonds with proline in the P_1 position does occur.[12] Cleavage does not take place, however, if proline occupies the P'_2 position.[10,13,14] It should be emphasized, therefore, that if the objective is to generate smaller peptides from a polypeptide fragment, hydrolysis of

$$\text{X-Pro} \overset{\text{Th}}{\underset{\downarrow}{\rule{1cm}{0.5pt}}} \text{Phe} \overset{\text{Cht}}{\underset{\downarrow}{\rule{1cm}{0.5pt}}} \text{Y}$$

may be feasible with chymotrypsin (Cht) or thermolysin (Th), but neither enzyme will cleave at the expected positions in the sequence X-Phe-Pro-Y.

Another approach to defining the specificity of thermolysin has utilized a series of furyl-acryloyl (FA) blocked tripeptides[15] (Table IV). Values of k_{cat}/K_m spanning 4 orders of magnitude emphasize the importance of hydrophobicity not only in the P'_1 residue, but in P_1 and P'_2 as well.

Clearly, both the rate and specificity of thermolysin cleavage depend predominantly on the nature of the P'_1 amino acid and secondarily on the residues in the immediate vicinity of this site. These findings suggest that a number of steric and conformational requirements must be fulfilled

[12] R. Sterner and R. L. Heinrikson, *Arch. Biochem. Biophys.* **168**, 693 (1975).
[13] Y. Ohta and Y. Ogura, *J. Biochem. (Tokyo)* **58**, 607 (1965).
[14] R. P. Ambler and R. J. Meadway, *Biochem. J.* **108**, 893 (1968).
[15] S. Blumberg and B. L. Vallee, *Biochemistry* **14**, 2410 (1975).

for the optimal binding and multiple attachment of the substrate to the active site of thermolysin, although discretion must be exercised in extrapolating the results of model peptide studies to large polypeptides.[10]

Applications of Thermolysin

General Strategies

Before proceeding to examples in which thermolysin has been employed in protein chemistry, it may be helpful to discuss briefly the strategies upon which such implementations have been based. Thermolysin has been applied as a tool in structural analysis both in the elucidation of protein covalent structures and in defining certain elements of the tertiary structures of proteins. Because of its broad specificity, thermolysin is usually employed in the secondary cleavage of tryptic and chymotryptic peptides during the course of the sequence analysis of the protein of interest. A general stratagem in protein sequence analysis is that the first set of fragments from the protein be generated by hydrolytic procedures which are limited and specific so that the resulting peptides are relatively large and easier to align in the final stages of the structural analysis. Subsequent cleavages of these fragments are produced by enzymes which are less and less specific. Clearly, the use of thermolysin in the primary cleavage of most denatured proteins and polypeptides would produce too many small peptides to be of value in the initial phase of the work. Exceptions to this general rule include proteins often found in conjunction with nucleic acids which contain a preponderance of basic residues. Here the roles of trypsin and a proteinase such as thermolysin are reversed, and the latter enzyme is preferable as a primary hydrolytic agent.[16,17] The use of thermolysin in the cleavage of peptides that have already undergone prior hydrolysis with trypsin and chymotrypsin, in effect, increases its specificity. With tryptic fragments, for example, the peptide bond between the NH_2-terminal residue and a hydrophobic amino acid such as leucine in the second position would not be susceptible to cleavage by thermolysin because of the proximity of the α-amino group. In the case of secondary cleavage of chymotryptic peptides, thermolysin hydrolysis may often be restricted to valine and isoleucine residues, since leucine and phenylalanine usually occupy COOH-terminal positions in these parent fragments. The relationship between chymo-

[16] W. S. Kistler, C. Noyes, R. Hsu, and R. L. Heinrikson, *J. Biol. Chem.* **250**, 1847 (1975).
[17] G. Moskowitz, Y. Ogawa, W. C. Starbuck, and H. Busch, *Biochem. Biophys. Res. Commun.* **35**, 741 (1969).

trypsin and thermolysin exemplifies the case in which the primary enzyme can increase the specificity of the secondary protease in the generation of smaller peptides.

The discussion thus far has been based upon applications useful in those phases of sequence analysis in which it is necessary to produce and analyze small fragments. Automated Edman degradative methods, however, are optimally applied to the analysis of large peptides, and the question may be considered of how thermolysin might be employed for limited specific cleavage of proteins. Obviously, such applications would have to be made on native proteins that contain only a few peptide bonds susceptible to cleavage. In fact, as will be discussed later, thermolysin has been used to probe the three-dimensional structure of native proteins, but in most of these cases the sequences of the protein substrates were known. Two properties of thermolysin are important in regard to its potential application in the generation of large polypeptide fragments from a native protein. First, the enzyme does seem to cleave native proteins to a greater extent than most proteases (Table I). The folding of the peptide chain and the consequent stabilization of the native conformation by disulfide bonds, ligands, and other interactions confer upon proteins a great deal of resistance to proteolytic attack.[18] Thermolysin will often produce cleavages in native proteins under mild conditions in which these substrates are refractive toward other proteases. The second property of thermolysin important in this regard is its thermostability. Native proteins unfold as the temperature is increased with T_m values normally in the range of 50–70° at neutral pH. Thermolysin is active at temperatures up to 80°. Preparations of the enzyme maintained at neutral pH in the presence of 2 mM Ca^{2+} show no loss of activity even after incubation for 1 hr at 60° and the half-life of enzyme activity at 80° is about 1 hr.[4] Thermolysin contains no disulfide bonds and the inordinate thermostability of its native conformation relative to that of most proteins is probably due to subtle differences in hydrophobic and ionic interactions, Ca^{2+} complexing, and hydrogen bonding.[3] In any case, the search for limited specific cleavages in a protein by thermolysin may be conducted over a broad temperature range in which the protein substrate may present several structures that differ in susceptibility to thermolysin attack. This provides a capability yet to be generally explored by workers in the field, for producing large fragments from a protein under carefully specified conditions of temperature which may, in turn, be useful for sequence analysis. Thermolysin cleavage at high temperatures has already been applied to the location of disulfide bonds

[18] C. B. Anfinsen and H. A. Scheraga, *Adv. Protein Chem.* **29,** 205 (1975).

in recalcitrant proteins, and its thermostability has permitted delineation of hydrolytically susceptible peptide bonds as a function of temperature for proteins of known sequence.

Experimental Conditions

As is the case for all proteolytic enzymes of broad specificity, optimal results from the application of thermolysin to the hydrolysis of proteins or peptides often require preliminary experimentation with regard to variables such as time, pH, temperature, and enzyme:substrate ratio. Because of its temperature stability, an even wider degree of latitude is possible in the use of thermolysin as a hydrolytic agent and only a general survey of each experimental variable will be given in this section. Details with regard to the conditions of thermolysin hydrolysis have been included in Table I and more will be presented later for other specific applications. The conditions for enzyme assay have been reported earlier.[4]

Thermolysin. Thermolysin may be prepared from culture filtrates of *B. thermoproteolyticus*[4] or it may be purchased as a crystalline preparation from Calbiochem (Los Angeles, California), or from numerous sources in Japan. The commercial product may be recrystallized,[4] and preparations with increased specific activity have been obtained by affinity chromatography.[19] For most purposes, however, the commercially available enzyme is suitable for use directly. Methods for the preparation of active, water-insoluble conjugates of thermolysin and polyacrylamide have been described.[20]

Substrate. The nature of the protein or peptide substrate and any derivatization thereof depend upon the purpose of the application. Solubility at neutral pH is the major requirement. Peptides that do not dissolve under these conditions can often be brought into solution by citraconylation[21] of amino groups. The peptides resulting from subsequent cleavage by thermolysin are often soluble following removal of the citraconyl substituents under acidic conditions. This modification can also be applied to large proteins as long as maintenance of native conformation is not essential to the study at hand. Proteins insoluble at neutral pH may also dissolve with disruption of their native structures by performing the hydrolysis at elevated temperatures or in the presence of 8 M urea or organic solvents, under which conditions thermolysin retains its activity. Of course, if the purpose of the experiment is to probe the

[19] K. Fujiwara and D. Tsuru, *J. Biochem.* (*Tokyo*) **76**, 883 (1974).
[20] R. Epton, J. McLaren, and T. Thomas, *Biochim. Biophys. Acta* **328**, 418 (1973).
[21] J. Y. Tai, A. A. Kortt, T.-Y. Liu, and S. D. Elliott, *J. Biol. Chem.* **251**, 1955 (1976).

tertiary structure of the protein, then conditions must be employed that are consistent with maintenance of the native conformation.

Temperature and pH. From the foregoing discussion it is clear that a broad range of temperatures from 4° to 80° may be employed in digestions with thermolysin, depending, once again, on the purpose of the study. Generally, hydrolysis is carried out at pH 8.0 or at pH values in this vicinity. For special purposes, however, higher pH values near 9 may be employed if solubility is favored under these conditions. Conversely, hydrolysis at pH 6.5 may be preferable in facilitating thermal denaturation[18] and in minimizing disulfide interchange in attempts to locate –S–S– bridges.

Buffer. The major requirement of the buffer is that it maintain the desired pH throughout the hydrolysis and that it contain small amounts (1–10 mM) of Ca^{2+}. Tris and borate buffers and pH-stats have been employed, but for some purposes volatile solvents such as 0.05 to 0.2 M N-ethylmorpholine acetate or 1% NH_4HCO_3 may be preferable. In some cases Ca^{2+} has been omitted from the hydrolysis medium; the enzyme presumably contains enough bound calcium to maintain its activity at moderate temperatures. The presence of added Ca^{2+} is especially important as a deterrent against autodigestion and inactivation at elevated temperatures. Hydrolysis may also be performed in the presence of 8 M urea or organic solvents, although under these conditions thermolability is enhanced. The ratio of enzyme to substrate is also highly subject to the demands of the experiment. In most applications the amount of enzyme is 0.2 to 1% by weight of the substrate (cf. Table I).

In summary, the application of thermolysin to hydrolysis of proteins and peptides is subject to a wide degree of experimental manipulation and, by and large, the intrinsic specificity of the enzyme is not altered significantly over a broad spectrum of conditions. This underscores the versatility of thermolysin as a tool in the exploration of various aspects of protein structure. The enzyme provides a fairly selective method for generating small peptides under conditions that favor extensive hydrolysis. Moreover, a few large fragments may be produced under extraordinary conditions of temperature which favor cleavage of the native protein substrate or partially unfolded derivatives thereof.

Primary Structural Analysis of Proteins

In view of the widespread application of thermolysin in the determination of amino acid sequences of peptides and proteins, a comprehensive review of the literature would be beyond the scope of this article. In many of the examples of specificity given in Table I, thermolysin was employed in the actual sequence analysis of the protein in question. Generally speaking, the overall strategy is usually to generate cyanogen bromide fragments first, followed by subsequent hydrolyses of these peptides with trypsin and/or chymotrypsin. Thermolysin hydrolysis is then used to generate peptides from tryptic or chymotryptic fragments, the structural analyses of which require secondary cleavages. The basis of this strategy has been discussed earlier, and recent examples of its application may be found in the structural analyses of streptococcal proteinase[21] and porcine pepsin.[22] The low content of lysine and arginine in pepsin precluded the use of trypsin except at aminoethylcysteine sites. The strategy was, therefore, to isolate and align the cyanogen bromide fragments and then complete the sequence analysis of these by cleavage with numerous proteinases of broader specificity. Peptides obtained by thermolysin cleavage were entirely in accord with the known specificity of the enzyme toward protein substrates and were crucial to the elucidation of the pepsin sequence.

Other proteins exist in which thermolysin, or an enzyme of similar specificity, is applied with advantage as a primary cleavage agent. For example, the ferredoxin from *Chromatium* contains a single methionine at residue 3 and two residues each of lysine and arginine positioned near the COOH terminus of the 81-residue polypeptide (cf. Table I). Cleavages with cyanogen bromide and trypsin were therefore of little value in the sequence analysis. The use of thermolysin in conjunction with chymotrypsin, however, served to generate the fragments and overlapping sequences necessary to complete the structural determination of this protein. Similar strategies have been successful for proteins with an abundance of basic amino acids. A testis-specific basic polypeptide from the rat was sequenced by a combination of automated Edman degradation and analysis of thermolysin fragments.[16] The protein of 54 residues contains 1 methionine at position 11, and 18 residues of lysine plus arginine distributed uniformly throughout the sequence. The approach was, therefore, to generate thermolysin peptides and to employ trypsin as a secondary cleavage procedure. Hydrolysis was performed in 2 ml of 0.1 M N-ethylmorpholine acetate, pH 8.0, containing 1 mM CaCl$_2$, 7.2 mg of the testis-specific protein, and 40 μg of thermolysin. After incubation for 2 hr at 34°, the mixture was acidified with 0.2 ml of glacial acetic acid and lyophilized. Eight peptides were isolated by a combination of high-voltage paper electrophoresis and descending chromatography. As is often the case with the somewhat apolar thermolysin

[22] P. Sepulveda, J. Marciniszyn, Jr., D. Liu, and J. Tang, *J. Biol. Chem.* **250,** 5082 (1975).

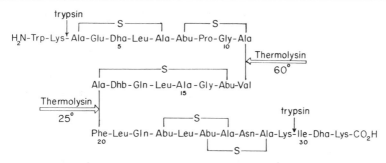

Fig. 1. The covalent structure of the heterodetic pentacyclic peptide subtilin from *Bacillus subtilis*. Sites of cleavage of the peptide by trypsin and by thermolysin at different temperatures are indicated by the arrows. Abu = α-aminobutyric acid; Ala-S-Ala = lanthionine; Abu-S-Ala = 3-methyllanthione; Dha = dehydroalanine; Dhb = 3-methyldehydroalanine. From E. Gross, H. H. Kiltz, and E. Nebelin, *Hoppe-Seyler's Z. Physiol. Chem.* **354**, 810 (1973).

peptides, the latter procedure provided the greatest resolution of the fragments. Cleavages occurred at 2 of the 3 leucine residues (leucine is COOH-terminal in the protein), at both tyrosines, and at the single methionine; the protein substrate is lacking in isoleucine and phenylalanine. A minor cleavage was observed at 1 of the 3 histidine residues, at position 9. Yields of the major peptide products ranged from 50% to almost 90%. It should be pointed out that, owing to the presence of valine and leucine at the NH$_2$ terminus, thermolysin peptides often give somewhat low color yields with ninhydrin.[14] Care must be exercised, especially in detection on paper, to see that no peptides are overlooked.

The temperature stability of thermolysin was exploited in an interesting way in determining the covalent structure of the complex heterodetic polypeptide subtilin from *Bacillus subtilis*.[23] This peptide of 32 residues is highly cross-linked with the thioether function of lanthionine and 3-methyllanthionine. Cleavage with trypsin produced a large-core fragment comprising residues 3–29 and including all of the thioether bridges (Fig. 1). Hydrolysis of this fragment at 25° with thermolysin in 50 mM N-ethylmorpholine acetate, pH 7.9, containing 10 mM CaCl$_2$ produced 2 peptides by a single cleavage between the cross-linked residue 19 and Phe-20. Surprisingly, hydrolysis did not take place between Phe-20 and Leu-21. Subsequent treatment of the product containing residues 3–20 with thermolysin at 60° for 18 hr again resulted in the hydrolysis of a single peptide bond between the cross-linked residue 11 and Val-12. These manipulations were instrumental in the

[23] E. Gross, H. H. Kiltz, and E. Nebelin, *Hoppe-Seyler's Z. Physiol. Chem.* **354**, 810 (1973).

elucidation of the subtilin structure and have been applied with success in studies of similar bacterial cyclic peptides.

Before closing this section on applications to protein structural analysis, reference will be made to an example in which thermolysin has been used to locate disulfide bonds in a protein. In many cases with conventional globular proteins, thermolysin is able to effect considerable or complete hydrolysis under conditions in which other enzymes give only partial cleavage. A dramatic case in point is that of the basic pancreatic trypsin inhibitor (Kunitz) which, although resistant to enzymic degradation, is susceptible to thermolysin hydrolysis both in its free native form and in complex formation with trypsin.[24] This fact served as the basis of a strategy for identifying the location of disulfide bridges in bovine pancreatic secretory trypsin inhibitor (Kazal's inhibitor) by Guy et al.[25] The inhibitor (5 μmol, 31 mg) was incubated for 48 hr at 37° in 10 ml of 0.1 M 2-(N-morpholino)ethane sulfonate buffer, pH 6.5, containing 2 mM CaCl$_2$ and 0.3 mg of thermolysin. Despite the low pH, which was chosen in order to minimize disulfide interchange, extensive hydrolysis of the inhibitor was observed. Two fractions were separated by gel filtration, one of which was composed of peptides similar to the intact inhibitor but having one or two bonds hydrolyzed. The susceptible bonds were Ile-Leu (2–3) and Glu-Val (12–13). These relatively slow cleavages were followed by an opening up of the inhibitor structure which facilitated more rapid and extensive hydrolysis to yield the smaller peptide components of the second fraction. Cleavages occurred at the amino side of isoleucine, leucine, valine, methionine, serine, threonine, and alanine (Fig. 2) and the three disulfide bonds in the inhibitor were easily placed by analysis of the fragments. Thermolysin was particularly useful in that it hydrolyzed between the leucine residues in the sequence Cys-Leu-Leu-Cys; attempts to cleave in this region of the molecule with trypsin and pepsin were unsuccessful. This example illustrates that although pepsin has long been the enzyme of choice for the generation of disulfide-linked peptides, thermolysin may provide a more useful alternative for this purpose in many proteins with highly compact structures.

Studies of Protein Conformation

Some interesting applications of thermolysin have been made in attempts to delineate certain aspects of three-dimensional conformation and structure–function relationships in proteins. Protein–ligand and

---

[24] T.-W. Wang and B. Kassell, *Biochem. Biophys. Res. Commun.* **40,** 1039 (1970).
[25] O. Guy, R. Shapanka, and L. J. Greene, *J. Biol. Chem.* **246,** 7740 (1971).

$$88\% \quad 4\% \qquad\qquad 88\% \qquad\qquad 83\% \qquad\qquad 83\%$$
$$\downarrow \quad \downarrow \qquad\qquad \downarrow \qquad\qquad\quad \downarrow \qquad\qquad\quad \downarrow$$

H₂N-Asn-Ile-Leu-Gly-Arg-Glu-Ala-Lys-Cys-Thr-Asn-Glu-Val-Asn-
10

$$80\% \qquad\qquad 91\% \qquad\qquad 60\% \qquad\qquad 81\%$$
$$\downarrow \qquad\qquad\quad \downarrow \qquad\qquad\quad \downarrow \qquad\qquad\quad \downarrow$$

Gly-Cys-Pro-Arg-Ile-Tyr-Asn-Pro-Val-Cys-Gly-Thr-Asp-Gly-Val-
20

$$25\% \quad 65\% \qquad 11\% \qquad 68\% \quad 55\% \qquad 55\% \quad 40\%$$
$$\downarrow \quad \downarrow \qquad \downarrow \qquad \downarrow \quad \downarrow \qquad \downarrow \quad \downarrow$$

Thr-Tyr-Ser-Asn-Glu-Cys-Leu-Leu-Cys-Met-Glu-Asn-Lys-Glu-
30   40

$$52\% \qquad 89\% \qquad\qquad 91\%$$
$$\downarrow \qquad\quad \downarrow \qquad\qquad\quad \downarrow$$

Arg-Gln-Thr-Pro-Val-Leu-Ile-Gln-Lys-Ser-Gly-Pro-Cys-CO₂H
50

Fɪɢ. 2. Amino acid sequence of bovine PSTI showing peptide bonds hydrolyzed by thermolysin at pH 6.5, 37°, for 48 hr. The extent of cleavage has been calculated on the basis of the yields of peptides recovered from extensively hydrolyzed inhibitor. From O. Guy, R. Shapanka, and L. J. Greene, *J. Biol. Chem.* **246**, 7740 (1971).

protein–protein interactions often result in the masking of peptide bonds, which in the nonassociated protein undergo hydrolysis with thermolysin. Conversely, bonds not susceptible to thermolysin in native protein substrates may become so during the course of limited disruption of the native conformation. Select examples of the use of thermolysin as a probe of conformation and structure–function relationships in macromolecules are given below.

Thermolysin has been employed to generate peptide fragments from calf thymus arginine and glycine-rich histone (GAR-histone), a protein that inhibits transcription of DNA in the DNA-primed RNA polymerase system. Basic residues in GAR-histone are clustered in 4 major regions of the protein, and 4 peptide fragments comprising these regions of high basic amino acid content were tested for inhibitory activity in order to tell if inhibition was due to ionic interactions not dependent upon an intact histone chain.[17] On the basis of the observation that some of the peptides were slightly inhibitory and some stimulated transcription, the authors concluded that inhibition of polymerase action is due to precipitation of a rigid GAR-histone-DNA complex.

The use of thermolysin as a means for cleaving bonds in native proteins which are resistant to hydrolysis by other proteases has been mentioned earlier.[24] By careful manipulation of experimental conditions such cleavages can be limited to varying degrees. This was the case for hydrolysis of Kazal's trypsin inhibitor in which the slow thermolysin hydrolysis of 2 peptide bonds signaled a conformational change that rendered the substrate amenable to rapid and extensive digestion.[25] Thermolysin cleavage of oxidized insulin under two sets of conditions led to quite different sets of cleavage products (Table I). These studies emphasize the importance of a detailed exploration of hydrolytic conditions when the aim of the investigation is to induce limited specific cleavage of the protein substrate.

The resistance of staphylococcal nuclease and nuclease-T to proteolysis in the liganded and free forms was examined by Taniuchi et al.[26] It was found that Ca^{2+} and deoxythymidine protect nuclease and nuclease-T from proteolytic inactivation by thermolysin, chymotrypsin, and subtilisin. The authors concluded that the protective effect of Ca^{2+} and deoxythymidine might be due to physical masking of suceptible bonds necessary for maintaining the active conformation of the nucleases or to decreased flexibility in the conformation induced by ligand binding.

Thermolysin has been applied in the investigation of haptoglobin–hemoglobin interactions. D'Udine et al.[27] observed that while haptoglobin dimers are cleaved by thermolysin, the β-chain of haptoglobin in hemoglobin–haptoglobin complexes is protected against proteolysis. Hemoglobin α-chain dimers alone provide a similar, although somewhat diminished, stabilization in the complex. These findings suggest that the β-chain of haptoglobin and the α-chain of hemoglobin are in close association in the complex and that this association yields a proteolytically resistant conformation.

Finally, it has been shown that at low Ca^{2+} concentrations, the autodigestion of thermolysin at 74° leads to the limited and specific generation of an inactive polypeptide ($M_r = 6000–10,000$) and an active fragment of M_r 28,000.[28] The authors conclude that the thermostability of the enzyme is related to its content of Ca^{2+}, and they propose that slight changes in Ca^{2+} concentration may influence the specificity of thermolysin hydrolysis.

Conclusions

Thermolysin is already a fundamental tool in protein sequence analysis independent of whether or not such studies are based on conventional or automated strategies. Its use in the location of disulfide bridges

[26] H. Taniuchi, L. Morávek, and C. B. Anfinsen, J. Biol. Chem. 244, 4600 (1969).
[27] B. D'Udine and L. F. Bernini, Protides Biol. Proc. Colloq. 22, 603 (1974).
[28] H. Drucker and S. L. Borchers, Arch. Biochem. Biophys. 147, 242 (1971).

in molecules resistant to other proteases is well documented and wider application of thermolysin for this purpose will certainly be forthcoming. The generation of large fragments from native or partially denatured proteins by thermolysin hydrolysis should be possible through a careful analysis of the products formed under varying conditions of hydrolysis. As yet, this application has not been explored in detail, but it is of great potential interest relative to primary structural analysis, to studies of native protein conformation and the molecular events accompanying thermally induced transitions in protein conformations, and to attempts to define the nature of protein–ligand and protein–protein interactions in recognitional processes. It may be of interest in this regard to explore the use of acylated thermolysin derivatives[15] which display "super-activity" toward peptide substrates as a potential tool of greater specificity in approaching these problems of protein structure.

[21] Cleavage at Glutamic Acid with Staphylococcal Protease

By GABRIEL R. DRAPEAU

Proteolytic enzymes catalyzing the hydrolysis of peptide bonds involving exclusively the basic amino acid residues lysine and arginine have been available for many years. Trypsin is by far the best known enzyme exhibiting this high degree of specificity and, for that reason, it has played a central role in studies of the primary structure of proteins. Recently, enzymes which specifically cleave peptide bonds at the carboxyl group of the acidic amino acid residues aspartic acid and glutamic acid have been discovered.[1-3] One of these enzymes, staphylococcal protease, has this specificity and can be further restricted to glutamyl bonds only under certain controlled conditions.[4] During the last few years this enzyme has been used for determination of the amino acid sequences of several proteins and proved to be another valuable tool for such studies.

Preparation of Enzyme and Sample

The staphylococcal protease is available commercially or can be prepared in the laboratory. A detailed procedure for the purification of the enzyme has been given earlier (this series Vol. 45, p. 369).

Digestion of the protein or peptide substrate is allowed to proceed in

[1] G. K. Garg and T. K. Virupaksha, *Eur. J. Biochem.* **17**, 13 (1970).
[2] G. R. Drapeau, Y. Boily, and J. Houmard, *J. Biol. Chem.* **249**, 6468 (1972).
[3] A. C. Ryden, L. Ryden, and L. Philipson, *Eur. J. Biochem.* **44**, 105 (1974).
[4] J. Houmard and G. R. Drapeau, *Proc. Natl. Acad. Sci. U.S.A.* **69**, 3506 (1972).

ammonium bicarbonate, pH 7.8, or ammonium acetate buffer, pH 4.0, at a concentration varying from 50 to 100 mM and also containing 2 mM EDTA.[5] The temperature of incubation is usually 37° for periods varying from 4 to 24 hr. An enzyme-to-substrate ratio of about 1:30 is normally employed.

When the cleavage of both aspartyl and glutamyl bonds is desired, the ammonium bicarbonate or ammonium acetate buffer is replaced by sodium or potassium phosphate buffer, 50–100 mM, pH 7.8.

Discussion

Specificity of the Staphylococcal Protease. Staphylococcal protease cannot degrade casein in which the carboxyl groups have been coupled with glycine ethyl ester in amide linkages. On the other hand, unmodified casein is readily hydrolyzed.[2] Using ribonuclease and lysozyme as substrates, the protease cleaves only glutamyl bonds when digestion is carried out in ammonium bicarbonate or ammonium acetate buffer, with the single exception of Asp-Gly bonds.[4] More recently, the enzyme has been used for the digestion of several proteins, and its specificity for glutamyl bonds has been further confirmed. Cleavage at aspartyl bonds is very infrequent, although it is not restricted to Asp-Gly linkages as originally observed with lysozyme. However, the enzyme shows a marked preference for certain aspartyl bonds when used in ammonium bicarbonate or acetate buffer. For example, in California gray whale myoglobin the Asp-Ala bond at position 109–110 is cleaved while another Asp-Ala bond (position 126–127) is not affected.[6] In the sequences of ribonucleases isolated from three different species the aspartyl bond cleaved is at Asp-121 (Asp-Ala or Asp-Asn), and it occurs in all three enzymes.[7] The NADP-specific glutamate dehydrogenase of *Neurospora* contains four Asp-Ile sequences, only two of which are cleaved to a significant extent (5% or less).[8] These observations suggest that a particular configuration of the substrate molecule around the aspartyl bond may be essential for the cleavage to occur in ammonium bicarbonate.

[5] *Staphylococcus aureus* cells have also been shown to secrete a neutral protease which is inhibited by EDTA [S. Arvidson, T. Holme, and B. Lindholm, *Biochim. Biophys. Acta* **301**, 135 (1973); S. Arvidson, *Biochim. Biophys. Acta* **310**, 149 (1973)]. Although this enzyme does not normally contaminate the glutamyl-specific enzyme preparations, EDTA should be added as a measure of precaution, particularly when high concentrations of the protease are utilized or when the incubation periods are extended.

[6] R. A. Bogardt, Jr., F. E. Dwulet, L. D. Lehman, B. N. Jones, and F. R. N. Gurd, *Biochemistry* **15**, 2597 (1976).

[7] G. W. Welling, G. Groen, and J. J. Beintema, *Biochem. J.* **147**, 505 (1975).

[8] J. C. Wootton, A. J. Baron, and J. R. S. Fincham, *Biochem. J.* **149**, 739 (1975).

The nature of the amino acid residue on the carboxyl-terminal side of glutamic acid may be critical for some protein substrates. In the NADP-specific glutamate dehydrogenase of *Neurospora* a Glu-Glu and two Glu-Asp sequences are not split, or split only with very low efficiency.[8] This suggests that the protease poorly cleaves glutamyl bonds which are followed by an acidic residue. However, an identical Glu-Asp sequence in horse myoglobin and in California gray whale myoglobin is readily cleaved.[4,6] The glutamyl bonds of myoglobin which involved hydrophobic amino acid residues with a bulky side chain are poorly cleaved.[4] This observation appears to be peculiar to myoglobin, since numerous glutamyl bonds found adjacent to such residues are hydrolyzed quantitatively in various proteins. Recently, a detailed account of the action of staphylococcal protease on peptides derived from the NAD-specific glutamate dehydrogenase of *Neurospora crassa* was published.[9] It is reported that glutamyl bonds involving polar, hydrophobic or basic residues are generally hydrolyzed, but it is clear that the type of residues and the length of the peptide in the neighborhood of the potential cleavage site are important factors which influence the rate of hydrolysis. No cleavage of a Glu-Pro sequence has so far been observed and the reported hydrolysis of this bond in the tryptophan synthetase α-chain of *E. coli* probably results from a nonenzymic split.[4] Cleavage of carboxymethylcysteinyl bonds does not seem to occur.

Digestion in Phosphate Buffer. When digestion of proteins and peptides is carried out in sodium or potassium phosphate, both glutamyl and aspartyl bonds are cleaved.[2] These hydrolytic conditions may be used advantageously for proteins that are poorly digested in ammonium bicarbonate. For example, hydrolysis of the glutamyl bonds present in staphylococcal protease is incomplete in ammonium bicarbonate, giving rise to a mixture of peptides, most of which could not be purified with acceptable yields. On the other hand, in phosphate buffer, a quantitative hydrolysis of both glutamyl and aspartyl bonds was obtained.[10] The aspartyl bonds cleaved under these conditions were identified as Asp-Arg, -Thr, -Ala, -Gln, -Asn, and -Leu, but one Asp-Pro, Asp-Lys, and two Asp-Gly sequences were not hydrolyzed.

Finally, it may be worth mentioning that the staphylococcal protease is fully active in the presence of 0.2% sodium dodecyl sulfate and retains 50% of its activity in a 4 M urea solution. Digestion under these conditions could be attempted for proteins or peptides that are not readily attacked by the protease under nondenaturing conditions.

[9] B. M. Austin, and E. L. Smith, *Biochem. Biophys. Res. Comm.* of Vol. **72**, 411 (1976).
[10] G. R. Drapeau, in preparation.

Section VII

Separation of Peptides

[22] Two-Dimensional Thin-Layer Methods[1]

By ROBERT W. GRACY

Thin-layer peptide mapping has the principal advantages of sensitivity, speed, and simplicity, in addition to the high resolving power capable of simultaneously separating large numbers of peptides. Thin-layer, two-dimensional techniques are based on the same fundamental combination of chromatographic and electrophoretic separations originally utilized on large filter papers,[2-5] but are carried out on much smaller sheets coated with thin layers of cellulose or silica gel. Whereas standard, two-dimensional separations require 1–5 mg of digested protein, peptide maps on thin layers can easily be obtained with 10–50 μg (i.e., 0.2–1.0 nmol) of material. Moreover, by using radioisotope-labeled peptides, the level of sensitivity can be extended even further. In addition to the sensitivity and speed of separation, thin-layer methods do not require specialized high voltage supplies or electrophoretic chambers which are capable of dissipating large amounts of heat. A simple, inexpensive thin-layer electrophoresis chamber can be constructed in the laboratory, and power supplies normally utilized for polyacrylamide gel electrophoresis or other routine electrophoreses can be used. Frequently, there are additional advantages of thin-layer methods in particular situations. For example, in the comparison of normal and genetically or chemically modified proteins, it is a distinct advantage to utilize a system in which both the normal and modified protein digests can be simultaneously subjected to electrophoresis or chromatography, thereby assuring optimal comparisons. Moreover, with the recent development of methods for amino acid analysis and sequencing at the subnanomole levels (see Sections I and VIII, this volume), it is now not only possible to obtain peptide maps on small amounts of material (e.g., that recovered from a single polyacrylamide gel), but it is also possible to extend these studies to include structural analyses of the isolated peptides. Although several

[1] This work was supported in part by grants from the U.S. National Institutes of Health (AM14638), the Robert A. Welch Foundation (B-502), a Research Career Development Award from the National Institutes of Health (K04 AM70198), the Alexander von Humboldt-Stiftung, and North Texas State University Faculty Research Funds.
[2] G. Haugaard and T. D. Kroner, *J. Am. Chem. Soc.* **70**, 2135 (1948).
[3] V. M. Ingram, *Nature* (*London*) **178**, 792 (1956).
[4] A. M. Katz, W. J. Dryer, and C. B. Anfinsen, *J. Biol. Chem.* **234**, 2897 (1959).
[5] J. C. Bennett, this series, Vol. 11, p. 330 (1967).

different systems for thin-layer peptide mapping have been described,[6-10] and the precise conditions for optimal separation of a particular mixture of peptides will vary somewhat, the following methods have been successfully utilized in this laboratory for obtaining peptide maps on over a dozen different proteins.

Materials and Equipment

Owing to the sensitivity of the methods employed to detect the peptides, it is essential that all reagents be of the highest purity, and that all reaction vessels be acid-cleaned prior to use. It is also advisable to presiliconize glassware to avoid losses of peptides by adsorption to the glass. Gloves should be worn at all stages to avoid contamination from fingerprints.

The following solutions are prepared from analytical grade reagents. Pyridine is redistilled from ninhydrin, and iodoacetic acid is recrystallized from petroleum ether. Although other purifications are not usually required, it is essential that ultrapure grade guanidinium chloride be used. Electrophoresis buffer is replaced with fresh buffer after approximately ten runs. Chromatography solvent and spray reagents for the fluorescamine reaction are prepared fresh daily.

Solutions

Modification buffer: 0.5 M Tris (Cl), 0.25 M EDTA, 6 M guanidinium chloride, 8 mM 2-mercaptoethanol, pH 8.5; total volume 10 ml

Electrophoresis buffer: Pyridine–glacial acetic acid–water, 1:10:89, pH 3.7; total volume 2.5 liters

Chromatography solvent: 1-butanol–pyridine–glacial acetic acid–water, 50:33:1:40; total volume 50 ml

Fluorescamine spray reagent: 0.1% w/v in acetone; volume 25 ml

Triethylamine spray: 10% v/v in methylene chloride; volume 50 ml

Equipment and Supplies

Thin-layer cellulose sheets[11]: 20 × 20-cm cellulose coated, 100–160 μm thickness, plastic-backed without fluorescent indicator (e.g.,

[6] S. Watanabe and A. Yoshida, *Biochem. Genet.* **5**, 541 (1971).
[7] S. Tawi, N. Koni, and K. Uyeda, *J. Biol. Chem.* **247**, 1138 (1972).
[8] T. H. Sawyer, B. E. Tilley, and R. W. Gracy, *J. Biol. Chem.* **247**, 6499 (1972).
[9] R. M. Waterson and W. H. Konigsberg, *Proc. Natl. Acad. Sci. U.S.A.* **71**, 376 (1974).
[10] D. L. Bates, R. N. Perham, and J. R. Coggins, *Anal. Biochem.* **68**, 175 (1975).
[11] For separation of the peptides on thin layers of silica gel, 20 × 20-cm Polygramsil plastic sheets coated with 0.25-mm silica gel layers can be utilized (Macherey-Nagel and Co., Düren, West Germany).

Eastman Kodak No. 13255 or Polygram Cel 400, Macherey-Nagel and Co., Düren, G. F. R.)

The electrophoretic power supply can be any of the usual types used for routine, low-voltage electrophoresis and should be capable of delivering 20–40 mA at 500 V. The chromatography tank is the standard 22 × 21 × 10-cm glass type used for most thin-layer chromatography (e.g., Kontes Glass No. 416150). The inside of the tank is lined with filter paper to facilitate equilibration of the atmosphere in the tank with the solvent. The chromatography solvent (50 ml) is added to the tank and allowed to saturate the paper wicks, and equilibrate for at least 30 min prior to chromatography.

Although a number of commercially available electrophoresis units can be utilized or modified so that they can be used for thin-layer electrophoresis of peptides, a simple chamber can be constructed from ¼-inch Plexiglas as shown in Fig. 1. The chamber is assembled by fusing the Plexiglas pieces together with methylene chloride. Platinum wire (22 gauge) serves as the electrodes, and cellulose sponges (22 × 10 × 3 cm) serve as buffer wicks. This chamber is designed to permit simultaneous electrophoresis of two thin-layer sheets. The buffer (approximately 1.2 liters) is added to each of the outer buffer chambers, and Varsol (Exxon Oil) is added to both the middle compartment, as well as overlaying the buffers (a total of approximately 7 liters). The nonvolatile, nonconducting Varsol serves not only as a heat sink, but also to prevent the thin-layer sheet from drying and the volatile buffer from evaporating.

Sample Preparation

The initial steps of sample preparation will depend to some degree on the state of the material. For example, if the sample is from a polyacrylamide gel, it must first be eluted from the gel and concentrated.[12,13] A wide variety of proteolytic digestion methods are possible, and these are discussed in Sections IV and V of this volume. However, since one is frequently dealing with less than 1 mg of protein, methods must be adapted to microscale and some techniques modified. If the protein contains either sulfhydryls or disulfides, it is first necessary to allow these groups to react to prevent disulfide interchange of peptides. The protein may be oxidized with performic acid to convert cysteine residues into cysteic acid,[14] or the protein can be S-carboxymethylated with iodoacetic acid.[15] The procedure described below reduces disulfides

[12] D. Bray and S. M. Brownlee, *Anal. Biochem.* **55**, 213 (1973).
[13] P. Dobos and J. Y. Plourde, *Eur. J. Biochem.* **39**, 463 (1973).
[14] C. H. W. Hirs, this series, Vol. 11, p. 199.
[15] A. M. Crestfield, S. Moore, and W. H. Stein, *J. Biol. Chem.* **238**, 622 (1963).

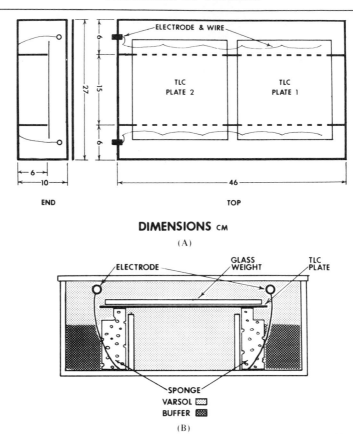

FIG. 1. (A) Construction of a simple electrophoresis chamber for peptide mapping. (B) End view.

and alkylates the cysteines with iodoacetic acid, prior to tryptic digestion. The procedure is given for a total of 100 μg of protein, but can be proportionally adapted to somewhat smaller or larger quantities with no difficulty.

The protein sample (100 μg) is dried with a stream of dry nitrogen in a 1.0-ml thick-walled, conical glass vial fitted with a Teflon-lined rubber septum (e.g., Pierce Reacti-Vial No. 13221). This type of reaction vial is convenient, since it allows the sample to be concentrated in the tip by centrifugation, and also permits the addition or removal of material through the septum while maintaining the sample in an oxygen-free atmosphere. The sample is dissolved in 200 μl of modification buffer (see above), and the vial is centrifuged to compact the sample to the bottom. A small 14 $\times$ 3.5-mm magnetic spin bar is added, and the vial is sealed.

Dissolved oxygen is removed and replaced by dry nitrogen by alternately applying vacuum or the gas through a 18-gauge needle inserted through the septum. The sample is stirred magnetically during this and the following steps. This procedure is repeated several times over a 1-hr period, while denaturation and reduction are allowed to proceed under nitrogen at room temperature. Ten microliters of a freshly prepared solution of iodoacetic acid (29 mg of iodoacetic acid in 1.0 ml of water[16]) are added by injection through the septum. Carboxymethylation of cysteines is then allowed to proceed in the dark at pH 8.5 for 20 min before the reaction is terminated by the addition of 10 μl of 2-mercaptoethanol. The sample is removed and desalted by dialysis in the dark against 0.2 M ammonium bicarbonate[17] returned to the conical vial and lyophilized.

The S-carboxymethylated protein is resuspended in 200 μl of 0.2 M ammonium bicarbonate, pH 8.0 (pH adjusted with either carbon dioxide or ammonium hydroxide). The protein may not be entirely soluble at this stage and should be triturated or placed in an ultrasonic chamber to form a finely divided suspension. A total of 1 μg of trypsin (1% of the total sample) is used to digest the protein, and is added in two stages to minimize autodigestion of the protease. The trypsin solution (0.1 mg per milliliter of ice cold water) is prepared immediately prior to the digestion. A 5-μl aliquot (0.5 μg of trypsin) is added to the protein solution to initiate the digestion, and the remainder of the trypsin solution is stored at 0°. The digestion is allowed to proceed at 37°, while slowly mixing with a magnetic stirrer. After 3 hr an additional 5-μl aliquot of trypsin is added, and the digestion is allowed to proceed for an additional 3 hr. It is important to use trypsin that has been treated to inactivate any traces of chymotryptic activity (e.g., L-1-tosylamido-2-phenylethylchloromethyl ketone-treated). The above conditions are usually adequate to completely hydrolyze most proteins, and any material not in solution at the outset should become solubilized during the digestion.[18] After digestion of the protein, the peptide solution is lyophilized.

---

[16] The amount of iodoacetic acid should not be excessive to avoid carboxymethylation of other residues (see Crestfield et al.[15]). If the cysteine content of the sample is known, this amount should be adjusted accordingly.

[17] Alternatively, the sample can be desalted by gel filtration. However, if the solubility of the S-carboxymethylated protein in the absence of guanidinium chloride has not previously been ascertained, it is advisable to desalt by dialysis, since insoluble protein can still be recovered from the dialysis bag.

[18] If the insoluble "core" material still exists at the end of this period, it can be separated by centrifugation and submitted to a more rigorous digestion (e.g., trypsinization in 2 M guanidinium chloride).

Application of Sample

The lyophilized material is carefully washed from the walls of the vial with 50 μl of 2% ammonium hydroxide and the reaction vial centrifuged to bring the entire sample to the bottom of the tube. The solution is dried with a gentle stream of dry nitrogen and redissolved in 10 μl of 2% ammonium hydroxide. From 0.2 to 1.0 nmol (1–5 μl) of sample solution is then taken up in a small capillary (prepared by drawing out an acid-cleaned melting point capillary in a warm flame). The sample is applied near the corner of the thin-layer sheet 3 cm from either edge. The size of the application spot is important in order to obtain maximal sensitivity and optimal resolution of the peptides, and should not exceed 2 mm in diameter. If necessary, the sample application process can be aided by drying the spot with a stream of dry nitrogen.

Electrophoresis

Prior to electrophoresis, the sponge wicks in the buffer chambers are immersed in the buffer solution, squeezed, and allowed to reabsorb fresh buffer. The cellulose thin-layer sheet is sprayed with a fine mist of electrophoresis buffer. The cellulose should be wet, but not to the point where it will drip. A small vial (approximately 1 cm in diameter) is held over the site of sample application during the spraying so that the buffer does not directly wet it. The buffer is then allowed to slowly wet the site of the sample application by capillary action. In this fashion, the sample is not smeared, or allowed to diffuse. The thin-layer sheet is then inverted (cellulose side down) and placed under the Varsol in the electrophoresis chamber. The edges of the thin-layer plate make a smooth contact wtih the sponge wicks in the buffer reservoirs. Since at pH 3.7 the peptides are positively charged, the thin-layer plate is positioned so that the side nearest the application site is placed nearest the anode. A 20 × 20 cm × 2-mm glass plate is carefully laid over the plastic thin-layer sheet to provide added weight necessary to maintain even contact with the sponge wicks (Fig. 1). Although conditions for optimal electrophoretic separation will vary with the particular peptides involved, electrophoresis at pH 3.7 at 300 V (20 mA) for 2–3 hr is usually sufficient.[19] If the electrophoresis chamber is placed in a cold room or a

---

[19] For optimal separation of some peptides, it may be advantageous to conduct electrophoresis at a higher pH. Such a buffer can be prepared from pyridine–glacial acetic acid–water (10:3:300, pH 6). However, under these conditions, since some peptides may be anionic and migrate toward the anode, the sample application spot should be approximately 7 cm from the edge that will be nearest the anode.

refrigerator at 3–5°, it is not necessary to provide any additional means of cooling. The relatively large volume of Varsol provides an effective heat sink, and the temperature of the buffer does not increase during the electrophoresis. After electrophoresis, the thin-layer sheets are removed from the chamber and dried at room temperature for at least an hour in a dust-free area.

Chromatography

The thin-layer sheets are turned such that the site of application and the peptides separated by electrophoresis are horizontally distributed across the bottom of the thin-layer sheet. The thin-layer sheets are placed in the preequilibrated chromatography tank, and ascending chromatography is carried out until the solvent front is within 2 cm of the top (between 3 and 4 hr). The peptide maps are then removed from the chamber and air dried as before.

Visualization of the Peptides

Although thin-layer peptide maps prepared in this manner can be stained with the usual ninhydrin reagent, if the peptides are to be recovered for further analyses the use of fluorescamine is a distinct advantage. Furthermore, somewhat better sensitivity and clarity are usually obtained using the fluorescamine reagent. The dried peptide map is first sprayed lightly with a mist of the 10% triethylamine solution[20] and allowed to dry for a few seconds. The map is then sprayed with 0.1% fluorescamine solution, followed by a second spraying with the triethylamine and viewed under long wavelength ultraviolet light (336 nm). Although most of the peptides become visible almost immediately, it is advisable to observe the development over a period of time. Thus, it is often possible to decide whether a large fluorescent spot is the result of a single peptide or two peptides with similar mapping coordinates. If the peptides are not resolved into reasonably circular spots, but appear as elongated smears, this is generally an indication of sample overloading or of the presence of salts in the sample solution. A significant amount of fluorescence remaining at the origin is indicative of undigested "core," and redigestion of the sample may be necessary. If background fluorescence or the solvent front can be observed, solvents should be redistilled and the thin-layer sheets prewashed in the chromatography solvent.

[20] A. M. Felix and M. H. Jiminez, J. Chromtogr. 89, 361 (1974).

Analysis of Peptide Maps and Recording of Data

While viewing the peptides, it is desirable to outline them lightly with a pencil on the surface of the cellulose. Even though the peptide maps can be preserved by storage in the dark at 2–5°, the fluorescence still fades within a few days; thus, it is necessary to make permanent records by tracing the two-dimensional coordinates of each of the peptides. Since the thin-layers of cellulose may chip from the surface with repeated handling or prolonged storage, it is also advisable to trace the positions of the peptides on the plastic backing of the thin-layer sheet with a permanent ink marking pen. Composite peptide maps should be prepared from tracings of at least five different maps. It is advantageous if each of the maps is made with varying amounts of the peptide mixture in order to locate peptides of low fluorescence as well as distinguish between highly fluorescent peptides with closely mapping coordinates. Using these two-dimensional thin-layer procedures, up to 50 peptides have been resolved in several laboratories.[6–10,21,22]

Recovery of Peptides

Since the fluorescamine reagent reacts with only a small percentage of the primary amines of the peptide when sprayed on surfaces, and the fluorescent derivatives can be hydrolyzed to free amino acids,[23] peptides can be recovered from the thin-layer plates for amino acid analysis or other structural studies. However, low levels of contaminating protein are present in commercially available celluloses or precoated plates.[24] While the levels of these contaminants are usually low (15–40 nmol per gram of cellulose), they are sufficient to interfere with analysis of peptides recovered from thin-layer plates. The contaminating proteins are not removed from the cellulose by washing with acid, base, or organic solvents, and appear to be firmly bound to the cellulose. Therefore, it is possible to extract the peptides from the cellulose if precautions are taken to remove all cellulose "fines" from the extraction solution. Peptides have been successfully extracted from thin-layer plates in good yields with essentially no contaminants by either electrophoretic elution of the peptides[24] or by solvent extraction and filtration.[24,25] Regardless of the

[21] B. E. Tilley, R. W. Gracy, and S. G. Welch, *J. Biol. Chem.* 249, 4571 (1974).
[22] T. D. Kempe, D. M. Gee, M. Hathaway, and E. A. Noltmann, *J. Biol. Chem.* 249, 4625 (1974).
[23] E. Mendez and C. Y. Lai, *Anal. Biochem.* 65, 281 (1975).
[24] B. E. Tilley, M. Izaddoost, J. M. Talent, and R. W. Gracy, *Anal. Biochem.* 62, 281 (1974).
[25] E. Schiltz, K. D. Schnackerz, and R. W. Gracy, *Anal. Biochem.* 78 (1977).

method utilized to extract the peptides, it is advisable to extract "control regions" of the thin-layer plate where no peptides have been identified and analyze these together with the samples as an indication of background levels.

The cellulose areas identified as containing peptides can be scraped from the plastic support and transferred to 150-mm glass Pasteur pipettes that have been tightly plugged with a membrane filter (e.g., Sartorius 20-mm membrane filter, 1 μm pore size, Catalog No. 13400) and prewashed with 6 N HCl containing 0.02% v/v 2-mercaptoethanol. Then 0.2 ml of 6 N HCl-mercaptoethanol solution is added to the cellulose. After 15 min the HCl is forced through the Pasteur pipette with dry nitrogen and the effluent is collected. This process is repeated twice more, and the effluents containing the extracted peptide are combined. If the peptides extracted in this fashion still contain fines of cellulose, they can be removed by filtering the HCl-extract through a 0.2-μm Millipore Swinney-type filter. If amino acid analysis is desired, the filtered peptide solutions are directly sealed in 10×100 mm glass tubes *in vacuo* and hydrolyzed under standard conditions.

Isotope Methods and Other Techniques

In the method described above, if [14]C- or ^{3}H-labeled iodoacetic acid is utilized to alkylate the cysteines, the S-carboxymethylcysteine-containing peptides can be located by radioautography or by counting the scraped cellulose in a scintillation counter. Specifically labeled peptides such as those obtained by affinity labeling of the catalytic center of an enzyme can similarly be identified. In addition, more general isotope labeling techniques can be employed in order to extend the sensitivity of peptide detection. For example, treatment of the peptide digest with [3H]dansyl chloride[26] or the prior amidination of the protein with [14C] acetimidate[10] permit radiolabeling of most of the peptides prior to thin-layer separation.

Most of the other techniques which were originally developed for conventional two-dimensional peptide separation on large paper sheets can also be utilized in thin-layer systems. For example, diagonal mapping can be employed to locate specific peptides. After the peptide mixture is separated in one direction, specific peptides are modified before separating under identical conditions at 90° to the first separation. Only

[26] M. K. Oskarrson, W. G. Robey, C. L. Harris, P. Fischinger, D. K. Haapala, and G. F. Vande Wou, *Proc. Natl. Acad. Sci. U.S.A.* **72**, 2380 (1975).

the modified peptides(s) will lie off of the diagonal. Likewise, an effective three-dimensional separation can be achieved by first separating the peptides electrophoretically, then scraping and eluting specific regions (e.g., the most anodic, most cathodic, and middle zones). Each of these peptide mixtures can then be applied to separate thin-layer sheets and separated in two dimensions, utilizing a different pH from the first electrophoretic separation.

Thus, while conventional peptide separation on paper continues to provide a fundamental tool to the protein chemist, two-dimensional thin-layer methods in many cases provide a more satisfactory alternative. Not only are two-dimensional thin-layer maps ideally suited for comparative studies, but the growing armament of highly sensitive methods for protein structural studies has extended the utilization of these techniques from the purely analytical to also include the preparative isolation of peptides.

[23] Separation of Peptides on Phosphocellulose and Other Cellulose Ion Exchangers

By CHRISTOPHER C. Q. CHIN and FINN WOLD

The separation of peptides by ion-exchange chromatography on both ion-exchange resins and cellulose derivatives has been discussed in this series (Vols. 11 and 25). In this chapter a rather specific system of complementary cellulose ion exchangers (phosphocellulose and TEAE-cellulose) will be considered with the emphasis on peptide separation with a minimum of special equipment for peptide detection. By taking advantage of the strong ultraviolet absorption by peptide bonds in the range of 200–250 nm, it is possible to detect the peptide peaks by direct monitoring of the column eluates, and this represents the simplest and most economical approach to peptide detection. The method is obviously severely limited by the fact that most common organic buffer ions also absorb strongly in the far ultraviolet region, and it was felt to be of value to explore peptide fractionation by ion-exchange chromatography using only inorganic buffers and salts that are completely transparent in the wavelength region where the peptides absorb maximally. One general procedure which in our hands has given very satisfactory fractionation of the peptides produced from a number of proteins by different cleavage procedures involves chromatography on phosphocellulose in dilute phosphoric acid with a KCl eluting gradient, desalting of individual peptide peaks on Bio-Gel P-2 with dilute NH_4OH as the solvent, and chromatography on TEAE-cellulose in dilute NH_4OH with a $NH_4Cl–NH_4OH$

eluting gradient or in dilute borate with a sodium borate gradient. With all these systems the columns have been monitored simply by following the ultraviolet absorption at 215 nm, either directly with a spectrophotometer equipped with a flow-through cell or by manual determination of the absorbance of the collected fractions.

The major advantages of this peptide fractionation procedure are the ease of peptide detection without any consumption of peptides and the relatively short duration of the chromatographic runs. The slowest step in fractionating 50 mg of peptides is the phosphocellulose column, which requires 15–17 hr. Each of the gel filtration runs and the TEAE-cellulose runs can be completed in 2–4 hr.

The disadvantages of this approach are the usual ones for any chromatographic system with fairly narrowly defined solvents: limitations in terms of applicability. The main problem with the inorganic solvent systems is probably their relatively poor ability to dissociate peptide aggregates. In addition, since the buffer salts are nonvolatile, they have to be removed by gel filtration rather than by the much more convenient and simple method of lyophilization. One major disadvantage directly associated with the detection method is that single amino acids and perhaps also dipeptides may escape detection. This flaw can at least partially be circumvented by determining the elution positions of free amino acids in a separate run, and then monitoring these positions by alternative detection methods. As will be shown below, the acidic and neutral amino acids all elute together early in the chromatogram, and the basic ones also elute together early in the salt gradient. Thus only two narrow regions of the chromatogram require special analytical procedures. A final point worth considering is the fact that because of the nonspecific detection method, peaks may be observed which on subsequent workup are found to be completely free of amino acids. An early identification of these impurities by amino acid analysis after acid hydrolysis, for example, will obviously save time and effort.

As already mentioned, this relatively simple and rapid approach to peptide separation has been found to be quite versatile. Originally designed to separate the 13 tryptic peptides from oxidized pancreatic ribonuclease,[1] both the combined phosphocellulose-TEAE-cellulose chromatography approach and the separate phosphocellulose or TEAE-cellulose systems have subsequently been applied successfully to many other peptide mixtures. The peptic peptides of yeast alcohol dehydrogenase have been fractionated both on phosphocellulose[2,3] and on the phosphocellulose-

[1] C. C. Q. Chin and F. Wold, *Anal. Biochem.* **46,** 585 (1972).

[2] J.-S. Twu and F. Wold, *Biochemistry* **12,** 381 (1973).

[3] C. J., Belke, C. C. Q. Chin, and F. Wold, *Biochemistry* **13,** 3418 (1974).

TEAE-cellulose combination.[4] The active-site tryptic peptides from butyl isocyanate-inactivated chymotrypsin and elastase were also obtained pure by this general approach.[5] The peptides of a bacterial dihydrofolate reductase were successfully fractionated on phosphocellulose as part of the protocol in the determination of the complete sequence of this enzyme,[6] and the 5 tryptic peptides from fully acetylated ribonuclease were obtained pure in a single run on phosphocellulose.[7] In this latter study the 3 thermolysin-produced peptides from the NH_2-terminal tryptic decapeptide were further resolved in a single run on phosphocellulose, and a papain digest of the tryptic peptide containing residues 40–85 were further fractionated on the TEAE-cellulose system. In all these cases the inorganic buffer systems were used and the column eluates were monitored at 215 nm.

It would be misleading to leave the impression that these chromatographic procedures are limited to the completely transparent buffers. By sacrificing some sensitivity and monitoring at slightly longer wavelengths (230 nm), chromatograms can be run with dilute acetate or other ultraviolet light-absorbing buffers. The salt gradient still has to be made up by nonabsorbing compounds, however, since it is generally necessary to go to high salt concentrations to elute all peptides.

It may be useful to list some of the many other uses of cellulose ion exchangers in peptide separation, and thus be able to put them in proper perspective. A survey of the recent literature on peptide separation, mostly in connection with amino acid sequence determination, suggests that DEAE-cellulose[8-20] is the most frequently used of the cellulose de-

[4] J.-S. Twu, C. C. Q. Chin, and F. Wold, *Biochemistry* **12**, 2856 (1973).
[5] W. E. Brown and F. Wold, *Biochemistry* **12**, 835 (1973).
[6] J. M. Gleisner, D. L. Peterson, and R. L. Blakley, *J. Biol. Chem.* **250**, 4937 (1975).
[7] B. Walter and F. Wold, *Biochemistry* **15**, 304 (1976).
[8] M. Tomita, and V. T. Marchesi, *Proc. Natl. Acad. Sci. U.S.A.* **72**, 2964 (1975).
[9] K. C. S. Chen, N. Tao, and J. Tang, *J. Biol. Chem.* **250**, 5068 (1975).
[10] V. B. Pedersen and B. Foltmann, *Eur. J. Biochem.* **55**, 95 (1975).
[11] E. T. Jones and C. H. Williams, Jr., *J. Biol. Chem.* **250**, 3779 (1975).
[12] M. Wade and R. A. Laursen, *FEBS Lett.* **53**, 37 (1975).
[13] F. J. Joubert, *Eur. J. Biochem.* **52**, 539 (1975).
[14] F. J. Joubert, *Biochim. Biophys. Acta* **379**, 345 (1975).
[15] B. A. Cunningham, J. L. Wang, M. J. Waxdal, and G. M. Edelman, *J. Biol. Chem.* **250**, 1503 (1975).
[16] R. S. Shulman, P. N. Herbert, K. Wehrly, and D. S. Fredrickson, *J. Biol. Chem.* **250**, 182 (1975).
[17] C. Y. Lai, *Arch. Biochem. Biophys.* **166**, 358 (1974).
[18] R. Begbie, *Biochim. Biophys. Acta* **371**, 549 (1974).
[19] S. D. Goodwin and W. M. Watkins, *Eur. J. Biochem.* **47**, 371 (1974).
[20] F. J. Joubert, *Hoppe-Seyler's Z. Physiol. Chem.* **356**, 1901 (1974).

rivatives with carboxymethyl cellulose[21-23] and phosphocellulose[24-27] as distant seconds. For situations in which peptide aggregation or unfavorable solubility properties require special solvents, such as 6-8 M urea, both carboxymethyl cellulose[28-32] and DEAE-cellulose[33-34] have been found useful. It should be mentioned that as a corollary to most of these ion-exchange procedures, the most commonly used procedure for desalting is gel filtration, and the references considered in this brief review also present a variety of gels and solvent systems that have been found useful both in desalting and sizing peptides mixtures.

Procedures

In the following, the separation of the 13 tryptic peptides from performic acid-oxidized bovine pancreatic ribonuclease will be used to illustrate a typical application of phosphocellulose chromatography, gel filtration, and TEAE-cellulose, in that sequence, to a typical peptide fractionation using only inorganic buffers and monitoring all chromatograms at 215 nm.[1] It should be noted that there are two dipeptides and one tripeptide in this peptide mixture, and that these peptides were detected in the case illustrated. It should be emphasized, however, that the absorbance of the dipeptides is sufficiently low to make dipeptide detection hazardous at best. The column sizes given in the description below were used for the separation of 50 mg of peptide mixture.

Chromatography on Phosphocellulose. The phosphocellulose is washed with 0.5 M NaOH, and 0.5 M HCl, and then several volumes of water to give an acid-free wash. It is then equilibrated with 0.025 N phosphoric acid (starting buffer) until the wash has the same pH (2.5) and conductivity (1.3 mmho) as the phosphoric acid solution, and a (1.5 × 35 cm)

[21] L. Moroder, B. Filippi, G. Borin, and F. Marchiori, *Biopolymers* 14, 2061 (1975).

[22] A. Henschen and F. Lottspeich, *Hoppe-Seyler's Z. Physiol. Chem.* 356, 1985 (1975).

[23] A. Henschen and R. Warbinek, *Hoppe-Seyler's Z. Physiol. Chem.* 356, 1981 (1975).

[24] R. Scholz and N. Hilschmann, *Hoppe-Seyler's Z. Physiol. Chem.* 356, 1333 (1975).

[25] J. H. Highberger, C. Corbett, A. H. Kang, and J. Gross, *Biochemistry* 14, 2872 (1975).

[26] S. N. Dixit, A. H. Kang, and J. Gross, *Biochemistry* 14, 1929 (1975).

[27] G. Frank and A. G. Weeds, *Eur. J. Biochem.* 44, 317 (1974).

[28] R. P. Ambler *Biochem. J.* 151, 197 (1975).

[29] W. W. de-Jong, E. C. Terwindt, and H. Bloemendal, *FEBS Lett.* 58, 310 (1975).

[30] C. G. Chua, R. W. Carrell, and B. H. Howard, *Biochem. J.* 149, 259 (1975).

[31] K. E. Langley, A. V. Fowler, and I. Zabin, *J. Biol. Chem.* 250, 2587 (1975).

[32] B. M. Austen, J. F. Nyc, Y. Degani, and E. L. Smith, *Proc. Natl. Acad. Sci. U.S.A.* 72, 4891 (1975).

[33] J. Weissenbach, R. Martin, and G. Dirheimer, *Eur. J. Biochem.* 56, 527 (1975).

[34] A. Bourgeois, M. Fougereau, and J. Rocca-Serra, *Eur. J. Biochem.* 43, 423 (1974).

column is finally packed and washed further with the starting buffer while adjusting the flow rate to 120–130 ml/hr. (Used phosphocellulose can be regenerated by the same procedure.) The peptide solution (50 mg of peptides) is applied to the column, and the chromatogram is eluted with 500 ml of starting buffer, followed by a KCl gradient. The gradient is prepared in a 3-chamber gradient mixer with 500 ml of starting buffer in each chamber, containing in addition 0, 0.01 M KCl, and 0.2 M KCl, respectively, in chambers 1, 2, and 3. The flow rate is maintained at about 130 ml/hr, and it is convenient to collect 5-ml fractions. The eluate is monitored for peptide absorption at 215 nm and also for conductivity. The peptide mixture applied to both this and to the TEAE column should be essentially salt-free to prevent the immediate elution of the most weakly bound peptides. Use of volatile buffers in the preparation of peptides and repeated lyophilizations are recommended.

Desalting on Bio-Gel P-2. The lyophilized material from each peak above is dissolved in 5 ml of 0.05 N NH_4OH and subjected to gel filtration on (1.8 × 60 cm) columns of Bio-Gel P-2 equilibrated and eluted with 0.05 N NH_4OH at a flow rate of 90 ml/hr. The peptide peaks, detected by their 215-nm absorption, are pooled and lyophilized several times to remove all ammonia and assure low starting ionic strength. In some instances the gel filtration alone will resolve a peptide mixture, and pure peptides may be obtained at this stage (see below for peptides $P_{1,1}$ and $P_{5,2}$). Sephadex G-10 has also been tried for the desalting step, but gives more pronounced retardation of certain peptides. In some instances this may be desirable, but P-2 has given the most reproducible results in our hands.

Chromatography on TEAE-Cellulose. The TEAE-cellulose is washed with 0.5 M HCl, 0.5 M NaOH, and water and is equilibrated with 5 mM NH_4OH (pH about 10.5) as the starting buffer. A (1.5 × 13 cm) column is packed and the flow rate is adjusted to 150 ml/hr. The lyophilized peptide mixture from the gel filtration is applied to the column, which is eluted first with 150 ml of the starting buffer and then with a 2-chamber linear gradient prepared with 250 ml of starting buffer in the mixing chamber and 250 ml of starting buffer containing 0.15 M NH_4Cl and 1.5 ml concentrated NH_4OH (to give a pH of 9.5) in the reservoir. The eluate is again monitored by the 215-nm absorbance. We have found that borate buffer can be substituted for the ammonia buffers in this step. An elution schedule for the TEAE-cellulose step consisting of 150 ml of 5 mM sodium borate, pH 9 (starting buffer), followed by a 2-chamber linear gradient prepared with 250 ml of starting buffer in the mixing chamber and 250 ml of 0.5 M sodium borate buffer, pH 9, in the reservoir, has been found to give as good resolution as the ammonia buffers described above.

The results obtained with the ribonuclease peptides are illustrated in Fig. 1. The circled peptides in the figure are pure (with the exception of P_7) and correspond to the following segments from the 124-amino acid sequence of ribonuclease: P_3, 40–61; P_6, 8–10; P_7, 32–33 and 38–39; P_8, 1–7; $P_{1,1}$, 67–85; $P_{2,2}$, 92–98; $P_{5,2}$, 34–37; P_2T_1, 62–66; P_4T_1, 99–104; P_4T_3, 105–124; P_5T_1, 86–91; P_5T_2, 11–31. As mentioned above, it is useful to establish where free amino acids elute from the phosphocellulose column, and the results obtained with the columns and gradients used in the illustrative example are included in the figure. It is worth emphasizing that with some attention to experimental procedures, the elution profiles from the phosphocellulose columns are reproducible and are essentially diagnostic of a given protein digest, even if complete separation of the peptides is not achieved.

The particular solvents and gradients used in the separation of ribonuclease peptides are not likely to be universally applicable. We have in fact found that for most peptide mixtures a considerably higher final

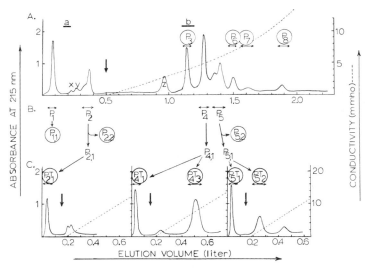

FIG. 1. The separation of the tryptic peptides of oxidized ribonuclease A: (A) chromatography on phosphocellulose; (B) gel filtration (desalting) of individual fractions on Bio-Gel P-2 (elution diagrams are not shown); (C) chromatography of indicated desalted fractions on TEAE-cellulose. Only peaks with a significant amino acid content after acid hydrolysis have been labeled in the figure. Explanation of special symbols: heavy vertical arrows, start of gradients; x, y, z, unidentified UV-absorbing peaks without significant amino acid content; a, elution position for neutral and acidic amino acids; b, elution position of histidine, lysine, and arginine; circled peptides, pure fractions identified by amino acid analysis. Reprinted, with permission, from C. C. Q. Chin and F. Wold, *Anal. Biochem.* **46**, 585 (1972).

KCl concentration (0.5–1.0 M) is required for the elution of all the peptides. It may, however, be convenient to explore the separation of any new mixture of peptides with the phosphocellulose column and the solvent system reported here as the first basic step. At the end of the normal KCl gradient, the column can be eluted further, in stepwise fashion, with increasing concentrations of KCl and finally with strong base to ascertain that all the peptides have been eluted. A similar approach should probably be used for the exploratory chromatograms involving TEAE-cellulose chromatography. With proper selection of salt, acid or base, absorbance at 215 nm can still be used to monitor the elutions, and the collected data will determine how the complete elution schedule for any particular peptide mixture should be designed.

[24] Chromatography on Microbore Columns[1]

By WERNER MACHLEIDT, JOACHIM OTTO, and ELMAR WACHTER

New sensitive techniques for amino acid analysis and sequencing of peptides are described in various articles of this volume. Procedures of equal sensitivity are desirable for the purification of peptides resulting from the cleavage of proteins.

Compared to other separation techniques, column chromatography has several advantages: it is essentially preparative, well suited for automation, and the most versatile in selectivity. The sensitivity of column chromatography is increased by the use of microbore columns (0.1–0.6 cm i.d.).[2,3] On such columns peptides can be purified in amounts between 10 and several 100 nmol. The purification procedure is facilitated by the analytical information obtained from nondestructive detectors. As a consequence of the reduced column diameter, the separations are highly reproducible, and, owing to their small volume, the work-up of fractions is simplified.

The separation of peptides on Dowex 50 microbore columns followed by ninhydrin detection is described by Herman and Vanaman (this volume [25]). The present article is concerned with the use of Sephadex gels or Bio-Gels and Sephadex ion-exchangers in microbore columns monitored by nondestructive detectors.

[1] Supported by the Deutsche Forschungsgemeinschaft, Sonderforschungsbereich 51, "Medizinische Molekularbiologie und Biochemie".
[2] W. Machleidt, W. Kerner, and J. Otto, Z. Anal. Chem. 252, 151 (1970).
[3] W. Machleidt, I. Assfalg, G. Rückl, and E. Wachter, in "Solid-Phase Methods in Protein Sequence Analysis" (R. A. Laursen, ed.), p. 149. Pierce Chemical Co., Rockford, IL, 1975.

Design of the Chromatographic System

Columns

For gel chromatography, 0.3–0.6 × 150-cm columns are used; 0.1–0.3 × 100-cm columns are used for ion exchange. Glass columns of these dimensions with appropriate end fittings and bed supports are commercially available from various sources.

Inexpensive columns can be made by fusing connectors for commercial end-fittings (Altex) to ordinary glass tubes. In 0.6 cm i.d. columns a fused-in glass frit is used as bed support. With all this bench-made equipment, the dead volume at the column end should be kept as small as possible.

Pumps

Any type of pump may be used that is capable of delivering solvents at constant flow rates between 0.05 and 5.0 ml/hr. The flow rates should be independent of column back pressure in the range of 0–50 psi. These requirements are met in an economical way by noncommercial screw-driven piston pumps developed in the authors' laboratory.[4] Up to 6 disposable syringes with a capacity of 30 to 75 ml can be used simultaneously in the same driving unit. Flow rates are selected with the aid of a switchable gear box. Syringes are refilled manually or by mechanical aspiration at increased speed. For gradient forming, special syringes with a built-in magnetic mixing device are used.

Many commercially available solvent delivery systems will not work at the low flow rates needed here. Suitable screw-driven piston pumps are available from Labotron (Kontron, Eching bei München) or may be constructed from medical infusion-withdrawal pumps.

Detectors

According to the properties of the peptides and of the eluent, several detectors are used alternately or connected in series.

Ultraviolet Absorption Detectors. Ultraviolet absorption is continuously monitored in U-shaped flow cells of 1 cm light path and 8 μl volume (Eppendorf Gerätebau, Hamburg). Flow cells with comparable specifications are available from various manufacturers. Absorption at

---

[4] Complete descriptions of noncommercial instrumentation used by the authors have not been published. More detailed information will be personally communicated by request.

225 nm is measured in a spectrophotometer (Gilford 2000), absorption at 280 and 253 nm, or 280 and 289 nm, in a photometer equipped with a mercury lamp and a prism monochromator that can be switched automatically between two preselected wavelength settings (Eppendorf Gerätebau, Hamburg). For economy, both instruments include a 4-sample changer. These sample changers and the monochromator settings are controlled by the two 12-channel recorders (Philips, Hamburg) used for the recording of all absorption profiles.

Instead of this laboratory-assembled instrumentation any set of commercial UV detectors may be used that work at the desired wavelengths and have flow cells with sufficiently low dead volume. The formation of gas bubbles in the flow cells is prevented by maintaining a back pressure of about 3 psi at the outlet. This pressure is preserved by simple flow restrictors made from silicone tubing containing a core (stainless steel or gold) of appropriate length.

Differential Refractometer. For gel chromatography in concentrated organic acids a differential refractometer detector of the deflection type is used (Waters R-401 or Siemens). The heat exchanger of this instrument is held at $20 \pm 0.01°$ by an external thermostated water circulator. The solvents are passed through the reference cell of the refractometer on their way to the column head. A bench-made pressure-switch protects the reference cell from detrimental pressures (higher than 70 psi). If several detectors are used in series, the refractometer is placed closest to the outlet because the flow resistance of its heat exchanger provides the desired back pressure to the entire system.

Radioactivity Detector. Radioactivity from ^{14}C-labeled peptides is detected in a flow cell scintillation counter (Coruflow CMF-101 from ICN Tracerlab). The effluent is passed through a coiled 0.1×20-cm polyethylene tube packed with plastic scintillator spheres (NE 102 A, Nuclear Enterprises, Edinburgh, Scotland) or glass scintillator (NE 901, Nuclear Enterprises).

Connections

Pumps, columns, and detectors are interconnected by 0.5 mm i.d. Teflon or polyethylene tubes via flanged tube end-fittings (Altex).

Alternatively, the following versatile and low-cost technique is recommended[4]: conical tips are fused to the ends of polyethylene tubing by mantling the tube end with a piece of wider tube and pressing it into a heated steel mold while the lumen is kept open by the suction of a water aspirator. Connections between tubes are made by bridging the contacting plane-cut tips with a piece of polyethylene tubing of appropriate diameter. These connections are quickly established without trapping of air bubbles. They are virtually free of dead volume and withstand pressures up to 300 psi. Connections to glass columns are made by direct insertion of the conical tips into appropriate column ends prepared by glassblowing. For special purposes, various fittings and connectors for the conical tube ends can be constructed from Plexiglas or transparent polyvinyl chloride.

Chromatographic Procedures

Packing of Columns

Sephadex gels (superfine; Pharmacia) or Bio-Gels (—400 mesh; Bio-Rad), or Sephadex ion exchangers (QAE-Sephadex A-25, SP-Sephadex C-25; Pharmacia) are preswollen in the solvent or starting buffer according to the recommendations of the manufacturers and used without further fractionation of particles.

Columns of 0.4–0.6 cm i.d. are packed by applying a thick suspension of the preswollen material via a funnel connected to the head of the column, which is prefilled with the solvent. With soft gels of higher porosity the hydrostatic pressure during the packing procedure must be carefully controlled; 0.1–0.3 cm i.d. columns are preferably slurry-packed from a manually operated screw-driven syringe pump mounted vertically above the column head. In this syringe the suspension can be gently stirred by a magnetic bar. The packing of a 0.1 × 100-cm column with Sephadex ion exchangers is usually completed within a few minutes.

Freshly packed columns are equilibrated with several volumes of solvent or starting buffer at the flow rates used for elution. Each ion-exchange chromatography is performed on a freshly packed column. Gel columns are used for several months, with the exception of gels swollen in 80% formic acid, which are renewed every 4–6 weeks.

Application of Samples

In gel chromatography, samples are dissolved in the eluent and applied to the top of the column as usual. For optimal resolution the sample volume should not exceed 100 μl in 0.4 cm i.d. or 200 μl in 0.6 cm i.d. columns (see Comments).

Samples for ion-exchange chromatography are dissolved in 50–200 μl of the starting buffer and are either applied directly to the column bed, or, preferably, pumped onto the column from a sample loop. The filled sample loop is connected to the system manually, or a suitable type of sample injection valve may be used.

Solvents and Buffers

Basically all kinds of solvents and buffers recommended for peptide separations can be used in the microbore system. The choice of developers is restricted, however, by the intention to make full use of the information obtainable from the nondestructive detectors.

With the refractometer detector, gel chromatography may be performed in any type of solvent, including concentrated organic acids which have a strong absorption in the ultraviolet. Virtually all peptides are soluble in 30–80% formic acid.

Typical volatile solvents used for gel chromatography of water-soluble peptides are 0.05–0.1 M ammonium bicarbonate or triethylammonium bicarbonate buffer (pH 8.5–9.0), 0.025–0.05 M ammonium acetate or triethylammonium acetate buffer (pH 4.8), or the corresponding formate buffers (pH 3.8). In all these buffers ultraviolet absorption monitoring is feasible down to 225 nm.

Volatile developers for ion-exchange chromatography on QAE-Sephadex or SP-Sephadex are 0.025 M ammonium bicarbonate (pH 8.5–9.0) or ammonium formate (pH 3.0–4.0) buffer, or the corresponding triethylammonium buffers. Peptides are eluted by concentration gradients of these buffers up to 1 M. Monitoring the absorption of peptides at 225 nm is feasible with the ammonium and triethylammonium bicarbonate, but not with the ammonium and triethylammonium formate gradient. Detection of 225-nm absorption is not impaired by nonvolatile buffers such as 0.1 M Tris-HCl or sodium phosphate with concentration gradients of NaCl. Ultraviolet detectors cannot be used below 280 nm with pyridine-containing developers.

Elution of Columns

For gel chromatography the microbore columns are eluted with a linear velocity[5] of about 10 cm hr^{-1}, corresponding to flow rates of 0.3 ml/hr for 0.3 cm i.d., 0.5 ml/hr for 0.4 cm i.d., and 1.2 ml/hr for 0.6 cm i.d. columns. Usually separations are performed at a room temperature between 20° and 25°; extremes of ambient temperature should be avoided by air conditioning or by the use of temperature-controlled jacketed columns.

Ion-exchange columns are eluted at room temperature with linear velocities near 120 cm hr^{-1}, i.e., 0.4 ml/hr for 0.1 cm i.d., 1.2 ml/hr for 0.2 cm i.d., and 3.6 ml/hr for 0.3 cm i.d. columns. Gradients are produced

[5] For definitions and theoretical background, see B. L. Karger *in* "Modern Practice of Liquid Chromatography" (J. J. Kirkland, ed.), p. 3. Wiley (Interscience), New York, 1971.

by two (or more) syringe pumps operated in parallel, one (or more) of them being used as a magnetically stirred mixing chamber. With the mixing syringe arrested at a constant volume, exponentially shaped gradients are formed. From a pumping mixing syringe with a gradually decreasing volume linear, concave or convex gradients may be obtained. The initial volume of the mixing chamber and the flow rates of starting buffer and limiting buffer are calculated from the equation given by Lakshmanan and Lieberman.[6] Linear gradients frequently used for the separation of complex peptide mixtures are described in the legends to Fig. 2 and Fig. 4.

A typical gel chromatography is completed within 20–40 hr, an ion-exchange chromatogram within 30–60 hr. After application of the sample the system needs no further attention.

The column effluent is collected in 20- or 30-min fractions (0.1–1.8 ml) using disposable polyethylene vessels that can be sealed by caps. The time-based operation of the multichannel fraction collectors (Serva, Heidelberg and noncommercial) is indicated on the recorder charts.

Comments

Performance of Detectors

One major advantage of the microbore technique is the gain in sensitivity that is inversely proportional to the cross-sectional area of the columns. The high concentrations of peptides in the column effluent make it possible to detect nanomole amounts with nondestructive physical methods.

Ultraviolet Absorption Detectors. At 225 nm the molar absorptivity of a single peptide bond is at least 150 M^{-1} cm^{-1}. This value is exceeded up to 200-fold by amino acid residues with an extra absorption (e.g., histidine, phenylalanine, tyrosine, tryptophan). In the complete absence of such residues, about 50 nmol of a dipeptide and, consequently, 5 nmol of a decapeptide can be detected with the described system. This is illustrated by an elution profile of some peptides of known sequence shown in Fig. 2. Small amounts of free amino acids may escape detection.

The absorption of peptide bonds has its maximum between 200 and 210 nm. A wavelength of 225 nm was chosen as a practical compromise to meet the problem of compensating unspecific absorption of solvents and buffers. Compensation of solvent absorption may be improved with a detector containing a reference cell. Moreover, the use of 0.5-mm instead of 10.0-mm cells should allow much more sensitive detection at 206 nm.[7]

[6] T. K. Lakshmanan and S. Lieberman, *Arch. Biochem. Biophys.* **45**, 235 (1953).
[7] See Application Notes No. **77** obtainable from LKB Instrument.

Refractive Index Detector. In concentrated organic acids or pyridine-containing buffers the detection of peptides by their absorption at 225 nm is not possible and is replaced by refractometry. Although widely used in liquid chromatography, until now the differential refractometer has not been applied routinely for monitoring peptide separations.

In the microbore system refractometer detection is about as sensitive as 225-nm photometry of peptides containing no extra-absorbing amino acids (Fig. 1). As a typical bulk property detector, the differential refractometer cannot be used at maximum sensitivity unless all parameters affecting refractive index are carefully controlled or compensated. Temperature can be sufficiently controlled by thermostatting the heat exchanger of the refractometer within ±0.01°. Complete heat exchange is favored by the low flow rates used in the microbore columns. Compensation of flow changes was achieved by inserting the reference cell into the liquid stream from the pump to the column. This arrangement proved superior to passing the solvent through the reference cell from a separate syringe run in parallel to that used for elution. Preliminary experiments indicated that even concentration gradients can be sufficiently compensated by passing the effluent from a dummy column through the reference

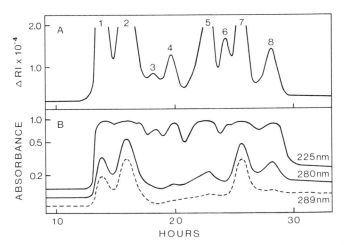

FIG. 1. Elution profiles from different detectors connected in series: differential refractometer (A), ultraviolet absorption detectors at 225 nm, 280 nm, and 289 nm (B). Gel chromatography of cyanogen bromide fragments from glycerol-3-phosphate dehydrogenase (50 nmol) on 0.4 × 150 cm Bio-Gel P-10 in 25 mM ammonium acetate buffer, pH 4.8 (0.5 ml/hr). According to the ratio of absorbances at 280 nm and 289 nm peak 7 contains tryptophan, peak 8 tyrosine. Modified from W. Machleidt, I. Assfalg, G. Rückl, and E. Wachter, *in* "Solid-Phase Methods in Protein Sequence Analysis" (R. A. Laursen, ed.), p. 149. Pierce Chemical Co., Rockford, IL, 1975.

cell. This laborious procedure, however, is not recommended for routine operation.

In a rough approximation, the refractive index of longer peptides is directly related to the number of amino acids and independent of the nature of these residues. Therefore clear-cut refractive index profiles are frequently obtained in cases when the 225-nm absorption detector is overloaded by the exceedingly strong absorption of aromatic residues (see Fig. 1).

The interpretation of overloaded 225-nm absorption profiles is facilitated by simultaneously recorded absorption profiles at 280, 253, or 289 nm. From their absorbances at 280 or 289 nm peptides containing tyrosine and tryptophan can be identified and differentiated from each other (see legend to Fig. 1 for an example). At 253 nm phenylalanine-containing peptides can be detected ($\epsilon = 150$ M^{-1} cm^{-1}) and, with much greater sensitivity, peptides containing citraconylated or maleylated lysine residues (see Fig. 4 for an example).

Radioactivity Detector. With plastic scintillator the average counting efficiency for ^{14}C is 10–12% in the peak maximum. If the same sensitivity was to be achieved by discontinuous liquid scintillation counting, a 15% aliquot would have to be removed from each fraction. In some cases the value of the radioactivity detector is limited by a strong absorption of peptides to the scintillator material, mainly observed at acidic pH. The radioactivity detector is a time-saver in analytical separations of complex mixtures containing ^{14}C- or ^{35}S-labeled peptides.

Reproducibility

The elution profiles obtained from microbore columns are completely reproducible. This holds also for separate ion-exchange columns eluted with the same gradient and (within a reasonable range) for different sample loads (Fig. 2). The high reproducibility of ion-exchange chromatography is a consequence of the noncritical column packing procedure as well as the constancy of gradient formation and elution by the syringe pumps.

About 10–50 nmol of a protein are sufficient for analytical peptide profiles which help to define optimal fragmentation strategy (Fig. 3). Fractions from these separations may be further characterized by TLC, amino acid analysis, and dansylation.

Sample Load

The load of microbore columns can be increased to amounts that will be sufficient for sensitive sequencing techniques in most cases. In gel

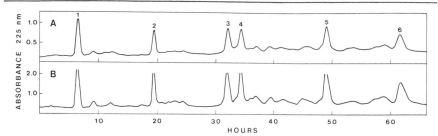

FIG. 2. Reproducibility of ion-exchange chromatography on 0.1 × 100 cm QAE-Sephadex A-25 equilibrated with 0.1 M Tris-HCl buffer, pH 8.6, and eluted with a linear NaCl gradient up to 0.25 M (a mixing syringe with an initial volume of 20.0 ml operating at 0.18 ml/hr is fed with limiting buffer containing 0.5 M NaCl pumped at 0.18 ml/hr). Two separate columns were loaded with tryptic peptides from 100 nmol (A) and 300 nmol (B) of glycerol-3-phosphate dehydrogenase (modified with N-ethylmaleimide) prefractionated on Sephadex G-25. After desalting on Sephadex G-25, the following peptides were obtained: Leu-Pro-Pro-Asn-Val-Val-Ala-Val-Pro-Asp-Val-Val-Lys (from peak 1); Leu-Thr-Glu-Ile-Ile-Asn-Thr-His-Glu-Glu-Asn-Val-Lys (from peak 2); Thr-His-Glu-Glu-Asn-Val-Lys (from peak 4); Leu-Glu-Ser-Cys-Gly-Val-Ala-Asp-Leu-Ile-Thr-Thr-Cys-Tyr-Arg (from peak 6).

chromatography the sample load is mainly limited by the volume of solvent required for complete solubilization of peptides. For difficult separations the width of the starting zone should not exceed 1% of the column length, i.e., 40 μl for a 0.3 × 150-cm column. For this reason

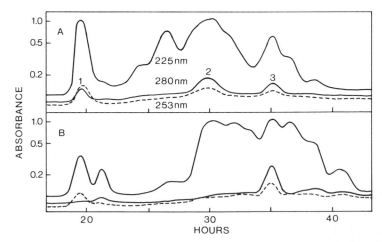

FIG. 3. Analytical peptide profiles obtained on 0.3 × 150 cm Sephadex G-50 in 0.1 M NH₄HCO₃, pH 8.6 (0.27 ml/hr). Chymotryptic digests from 35 nmol of a cyanogen bromide fragment (106 residues) from glycerol-3-phosphate dehydrogenase incubated 10 min (A) and 6 hr (B) at 25°C. The progress of the enzymic cleavage is indicated by the disappearance of the 280-nm peaks 1 and 2 in favor of peak 3.

columns with an i.d. smaller than 0.3 cm are not used in gel chromatography. If gel chromatography is performed for an initial "sizing" of peptides prior to further purification, sample loads up to 10% can be tolerated, e.g., about 1.7 ml on a 0.6 × 150-cm column.

In ion-exchange chromatography the sample volume should not exceed half the void volume to avoid loss of unretarded peptides during sample application. Loss of resolution caused by overloading is dependent on the number and nature of peptides to be separated. Mixtures of small tryptic peptides (up to 15 residues) to be separated in 0.1 × 100-cm columns on Sephadex ion exchangers may contain a total of 2 μmol made up from 5 to 10 components (see Fig. 2 for an example). On the other hand, a 0.3 × 100-cm column had to be used to resolve about 4 μmol of maleylated peptides 20–40 residues in length (Fig. 4). As a rule, complex peptide mixtures should be "sized" by gel chromatography prior to ion-exchange chromatography (see legend to Fig. 4 for a practical example).

Resolution

In typical peptide separations on microbore columns filled with —400 mesh Bio-Gels 10^3 effective plates[5] per 100-cm column length are ob-

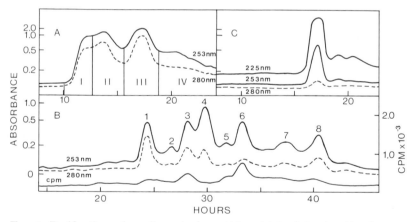

FIG. 4. Purification of medium-sized peptides obtained by tryptic cleavage of ^{14}C-carboxymethylated glycerol-3-phosphate dehydrogenase after maleylation. (A) Gel chromatography of the digest (0.5 μmol) on 0.6 × 150 cm Sephadex G-50 in 0.1 M NH₄HCO₃, pH 8.6. (B) Ion-exchange chromatography of A III on 0.3 × 100 cm QAE-Sephadex A-25 equilibrated with 25 mM NH₄HCO₃, pH 8.8, and eluted with a linear gradient of this buffer up to 1 M (the mixing syringe, initial volume of 80.0 ml, working at 1.8 ml/hr was fed with 1 M NH₄HCO₃ at 1.8 ml/hr). Radioactivity of labeled peptides was monitored with the flow cell scintillation counter. (C) Rechromatography of B 4 on the same column as used in (A). The peak contains 0.3 μmol of a pure peptide 25 residues long.

tained. With this column efficiency two peptides are completely separated when their equilibrium distribution coefficients[5] differ by more than 10%. Even with small differences in molecular size, adequate separations are obtained by repeated runs on 150-cm columns.[3] As an example, peaks 6 and 7 in Fig. 1 were completely resolved after two subsequent steps of gel chromatography and identified as peptides 33 and 32 residues long, respectively.

Due to incomplete equilibrium between stationary and mobile phases, the efficiency of liquid chromatography falls with increasing flow rates.[5] The relatively low flow rates applied here are mainly imposed by the mechanical instability of soft gels. These low flow rates are close to the values leading to optimal resolution. Furthermore, column pressures are always below 30 psi and hence there is no need for expensive high pressure instrumentation.

Per 100 cm of column length, 500–700 effective plates are obtained in peptide separations on Sephadex ion exchangers using the commercially available particles (40–120 μm). Resolution on these ion exchangers is much inferior to that obtainable with finer sized small particles of polystyrene ion exchangers (e.g., 10^4 effective plates in a 0.1 $\times$ 50-cm column filled with Aminex A-6, 17.5 $\pm$ 2 μm, from Bio-Rad). In contrast to polystyrene matrices, however, the derivatized dextrans are fully accessible to larger peptides and do not exhibit irreversible adsorption of peptides. On the whole, microbore chromatography on Sephadex ion exchangers has proved most useful for the purification of medium-sized and large peptides (10 residues and longer). The best separations of these peptides were obtained with nonvolatile buffers at constant pH and elution by a NaCl gradient. Complex mixtures of short peptides may more favorably be separated on polystyrene ion exchangers. Resolution on Sephadex ion exchangers could be improved, however, by the use of fractions composed of smaller particles, but these are not commercially available.

[25] Automated Micro Procedures for Peptide Separations

By ALAN C. HERMAN and THOMAS C. VANAMAN

The increased emphasis in this volume on micromethods for the analysis of protein structure reflects the increasing importance of the application of these techniques to many problems of contemporary molecular biology. Techniques described elsewhere in this volume for performing amino acid analysis [1] and sequence determination [27] on subnanomolar quantities of material make it possible to initiate primary structure analysis on small amounts of protein. Such studies require the separation of peptides derived from nanomolar quantities of protein. Column chromatography utilizing volatile buffer systems continues to be one of the most suitable methods for the preparative-scale fractionation of peptide mixtures. In addition, ion-exchange chromatography performed under rigidly controlled conditions can provide highly reproducible separations of peptides and is therefore suitable for peptide mapping as originally shown in the comparison of abnormal hemoglobins.[1]

A number of recent advances in column chromatography have provided the techniques necessary for purifying peptides on a micro scale. Development of small diameter (9–15 μm) spherical polystyrene-based ion-exchange resins (this volume [2]) has made high-resolution ion-exchange chromatography on microbore columns possible, thereby increasing detection sensitivity (this volume [24]). The use of an appropriately designed system for detection of peptides in column effluents leads to further increases in sensitivity. Simple modifications of the ninhydrin-based monitoring system described by Hill and Delaney (this series, Vol. II [37]) are described below which significantly increase system sensitivity. Procedures for obtaining analytical and preparative peptide separations of nanomolar quantities of material using this modified system also are described.

Procedures

Preparation of Samples. The procedures described in this article are generally applicable to the separation of peptide mixtures from chemical or enzymic cleavage of proteins, discussed in this and preceding volumes of this series. The emphasis in this article is on microscale separations of peptides by column methods on both the analytical scale, for column mapping studies, and on the preparative scale, for use in sequence analysis. Preparing nanomolar amounts of material for microscale separations requires precautions that are not necessary when working on a micromolar scale. First, it is desirable to perform all operations on a single protein sample in one vessel, including, for example, performic acid oxidation, trypsin digestion, lyophilizations, dissolution in chromatography solvents, and centrifugation to remove insoluble material when necessary. This is absolutely essential for 10 nmol of material or less, as prohibitive losses may be incurred from transfers between vessels. Second, the sample must be free of salts. This is extremely important for performing comparative analytical peptide mapping on microbore columns, as elution profiles are significantly altered following elution of large salt peaks.

[1] R. T. Jones, *Cold Spring Harbor Symp. Quant. Biol.* **29**, 297 (1964).

Finally, it is most important to determine (e.g., by amino acid analysis) the exact concentration of the protein digest before applying the sample to chromatographic separation. This information is essential, as it is required for calculating yields of peptides from preparative separations and for standardization of the amounts of samples analyzed in comparative analytical separations.

Buffers. The buffers used in peptide chromatography can be divided into two classes: nonvolatile and volatile. The major disadvantage of using nonvolatile buffers is that there is no simple and efficient method of desalting the purified peptides. The use of a nonvolatile buffer system for strictly analytical peptide mapping is satisfactory; however, we have found it advantageous to use the same volatile buffer system for both analytical and preparative separations. This allows for easy correlation of analytical column profiles and purified peptides from preparative separations. The buffers most frequently used in this laboratory are listed in the table. Generally, in cation-exchange chromatography a single linear gradient using buffer F as starting buffer and buffer J as limit buffer is satisfactory. In some cases, especially with complex mixtures of peptides, the use of two consecutive gradients, buffers C and G and buffers G and J, respectively, will result in better resolution of the early eluting peaks. In unusual cases involving the separation of specific peptides, other gradient systems are used as required. For anion-exchange chromatography we have utilized a simple gradient system consisting of buffer K as starting buffer and buffer J as limit buffer.

COMPOSITION OF PYRIDINE–ACETIC ACID BUFFERS

Buffer	pH	Pyridine concentration (M)	Composition (ml/2 liters solution)	
			Pyridine	Glacial acetic acid
A	2.0	0.01	1.61	620
B	2.2	0.02	3.22	600
C	2.6	0.05	8.06	585
D	2.7	0.10	16.1	576
E	3.0	0.15	24.4	568
F	3.1	0.20	32.2	557
G	3.8	0.50	83.5	568
H	3.9	0.80	128	470
I	4.1	1.00	167	436
J	5.0	2.00	325	285
K	9.1	0.36	60	0

Special attention should be paid to the quality of the pyridine used in buffer preparation. Pyridine should be redistilled after refluxing with ninhydrin prior to use (Hill and Delaney, this series Vol. 11 [37]). Redistilled pyridine still appears to contain ninhydrin-reactive compounds following the alkaline hydrolysis step (see below). Since the gradient systems used employ an increasing pyridine concentration, these contaminants cause a rise in baseline during gradient elution which limits the amounts of recorder chart expansion that can be used. Baseline rise with the normal gradient system can vary from 0.03 to 0.5 absorbance unit depending on the source and preparation of pyridine. We have found J. T. Baker (analyzed reagent grade) pyridine to be consistently low in ninhydrin-positive contaminants.

Reagents. Sodium hydroxide, 5 M (200 g of solid NaOH in deionized water to a total volume of 1 liter) is prepared and stored in plastic containers. This reagent should not be prepared from stock 50% (w/w) NaOH. Acetic acid, 6 M (344 ml of glacial acetic acid to water for a total volume of 1 liter) must be exhaustively deaerated prior to use. Equal volumes of NaOH and acetic acid solutions, when mixed, should have a pH of 5.5 ± 0.2.

Ninhydrin reagent is prepared as follows: 20 g of ninhydrin (Pierce Chemical Co.) and 1.5 g of hydrindantin is dissolved in 650 ml of peroxide-free Methyl Cellosolve (Pierce, Piersolve) in a dark bottle and the solution purged with nitrogen. After 15 min, 350 ml of 4 M sodium acetate (Pierce, pHix buffer), pH 5.5, is added and purging continued for 30 min with vigorous stirring. The resulting solution is diluted to 3 liters with 50% (v/v) Methyl Cellosolve, and nitrogen is continually bubbled through the completed reagent. This reagent should be prepared fresh before each series of peptide analyses.

Effluent Monitor

Column effluents are monitored using a modular Technicon auto-analyzer system similar to that described by Hill and Delaney (this series Vol 11 [37]), the major difference being that the present system is not segmented with nitrogen. A detailed schematic diagram of the monitoring system is shown in Fig. 1. Each reagent line is connected to a 3-way valve to allow the selection of reagent or water. All reagents are pumped with a single proportioning pump at the flow rates shown in the figure legend. Column effluent, pumped at 3–4 ml/hr is mixed with sodium hydroxide in the first of three 3-port mixing manifolds. The alkaline solution is then passed through a Teflon tubing reaction coil immersed in a boiling water bath (this alkaline hydrolysis exposes more amino groups

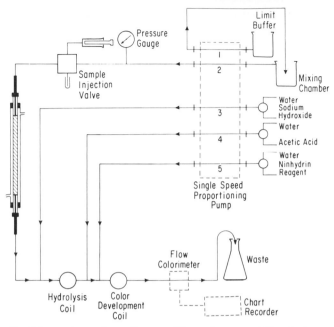

Fig. 1. Peptide analyzer flow diagram. The column and its elution gradient are pumped as described in the text. Column effluent is mixed in a 3-port manifold with 5 M sodium hydroxide and hydrolyzed for 50 min in a boiling water bath. The alkaline solution is then neutralized with 6 M acetic acid, mixed with ninhydrin reagent and color developed for 20 min by passage through the boiling water bath. Absorbance is monitored at 570 nm and recorded on a strip-chart recorder. Pump tube (Technicon) sizes are as follows: 1, 0.03 ml/min (SMA flow-rated, ORANGE-RED); 2, 0.06 ml/min (SMA flow-rated, RED-BLUE); 3 and 4, 0.16 ml/min, 0.02 inch i.d. (Tygon, ORANGE-YELLOW); 5, 0.23 ml/min, 0.025 inch i.d. (Solvaflex, ORANGE-WHITE).

for ninhydrin reaction, thereby increasing system sensitivity). The sample is then neutralized in a second 3-port manifold, mixed immediately with ninhydrin reagent in a third manifold, and passed through a second reaction coil for color development at 100°. It is imperative that a boiling water bath (resin reaction bath, 1000 ml with heating mantle and variable transformer), and *not* an oil bath, be used. The use of an oil bath causes outgassing in the coils. A great deal of variation in the length of the reaction coils does not have an adverse effect on the column profiles. We routinely use a 50-min hydrolysis time and a 20-min color development time. These times are achieved by using 23-m coils of 0.8 mm i.d. tubing or 30-m coils of 0.7 mm i.d. tubing (AWG 22). However, these reaction times can be cut in half with little loss in sensitivity. On the other hand, changes in the reagent or column flow rates listed in the legend to Fig. 1 will reduce peak heights significantly. It is therefore recommended that the flow rates given here be used.

The effluent from the second reaction coil is monitored for absorbance at 570 nm in a flow colorimeter or a flow spectrophotometer and then exits through a waste line to drain. The effluent must exit this waste line at least 1 m above the rest of the system to prevent outgassing in the flow cell.

There are several methods for monitoring at 570 nm. The first is to use the colorimeter available in an amino acid analyzer. This can be done without modification to the analyzer other than connecting the flow cell to the peptide analytical system. With this type monitor and a flow cell with a 6 mm path length, good peak heights are obtained with 2.5 nmol of digest when a recorder setting of 0.1 absorbance units full scale is used.

The second method is to use a flow colorimeter such as that available from Technicon with a 570-nm filter and a Bristol recorder. With this system, recorder sensitivity can be increased by placing a 0-2000 Ω potentiometer in series with the lead from the low side of the colorimeter to the recorder.[2] The application of a resistance in the circuitry at this point results in a nonlinear expansion of the recorder scale. As a greater resistance is placed on the colorimeter output, the lower end of the recorder scale is expanded while the upper end is compressed. A variable resistance, rather than a fixed one, enables the sensitivity of the recorder to be adjusted to match the amount of digested protein being analyzed. With this modification, 5-50 nmol of protein can be analyzed. Five nanomoles (1250 Ω resistance) appears to be a practical lower limit of sensitivity, since greater electronic scale expansion results in an unsteady baseline.

With the recent popularity of high-pressure liquid chromatography, a number of high-quality flow spectrophotometers have become available which are applicable to peptide monitoring. One typical unit is the Altex Model 153 UV monitor which features an 8-μl flow cell, dual-beam optics, and 0.005 absorbance unit full scale sensitivity. We have obtained good elution profiles with 0.5-2.0 nmol of peptide digest using this unit equipped with a 578-nm interference filter and operated at a scale expansion of 0.08 absorbance unit.

Analytical Column Chromatography

Construction. In order to obtain maximum sensitivity, column diameter should be as small as possible since peak height is inversely propor-

[2] A. C. Herman and T. C. Vanaman, *Anal. Biochem.* **63**, 550 (1975).

tional to the cross-sectional area of the column. We have found columns of *approximately* 3 mm i.d. to be readily available, easy to work with, and capable of giving good resolution. Completely assembled, water-jacketed columns fitted with adjustable plungers are available from Beckman Instruments, Inc. and from Durrum Chemical Co. Columns from other sources can be used provided they meet the following criteria:

1. They must be able to withstand at least 300 psi of pressure.

2. The resin support screen or disk must have a small enough pore size to retain a 10-μm resin particle.

3. They must be fitted with adjustable column plungers. (Fig. 2

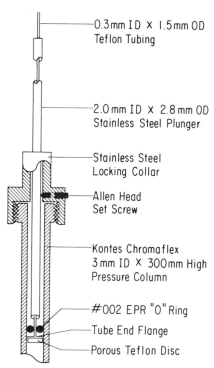

Fig. 2. Construction of adjustable column end fittings for 3-mm Kontes high-pressure columns. The same type of end fitting is used in both ends of the column. The locking collar threads onto the end of the column and holds the adjustable plunger in place by means of an Allen-head set screw. The pressure-tight seal is formed by means of a No. 002 O-ring, which is slipped over the end of the Teflon tubing. This O-ring fits tightly between the Teflon tubing and the inner wall of the column and is held in place on one end by the adjustable plunger and on the other end by a flange in the end of the Teflon tube. The resin support is cut from a standard 9-mm porous Teflon disk (Beckman part No. 331459) with a sharp No. 1 cork borer. For maximum flexibility, the plungers should be about 15 cm long.

illustrates the construction of an adjustable column plunger that can be fitted to a column such as the Kontes Chromaflex 3 mm i.d. column.)

4. They must be water-jacketed.

Columns of the 3 mm size range are suitable for analytical and small-scale preparative runs of 25 nmol or less of protein digest. Larger sample loads result in poor resolution due to overloading.[2]

Packing. The primary column packing used in our laboratory has been an 8% cross-linked spherical sulfonated polystyrene cation-exchange resin with a mean particle diameter of 14 μm (Beckman, type AA-15 or W-1). A similar resin is available from Durrum Chemical Corp. (DC-1A). Greater than 90% recoveries have been obtained for most peptides using this type of resin.

With the advent of high-sensitivity amino acid analysis on microbore columns, Dowex 50-type resins have been developed with very small mean particle diameter. These resins offer increased resolution, sensitivity, and capacity at the expense of higher operating pressures. Although these resins have not been used in this laboratory, peptide separations on a resin with a 9-μm particle size in a 2-mm bore stainless steel column has been described.[3] Resins with a particle size of 11 μm are available from Beckman (AA-20) and Durrum (DC-6A). A 9-μm resin (DC-4A) is also available from Durrum.

Prior to packing the column, resin is conditioned by successive washings in 10 volumes each of 3 M sodium hydroxide, water, 3 M hydrocholoric acid, water, 8 M pyridine, and starting buffer. The conditioned resin is suspended in starting buffer, deaerated, and packed into columns under nitrogen pressure (24 psi). *All columns are poured, maintained, and operated at 55°.* For analytical separations, a 3 $\times$ 200-mm column is satisfactory.

Operation. As shown in Fig. 1, a single multichannel peristaltic pump (Technicon Proportioning pump I) is used to pump the column gradient, the analytical column, and all reagents for subsequent automated analysis. A linear gradient is formed by pumping the limit buffer into the mixing chamber at one half the column flow rate. Since the optimal sampling rate for the monitoring system is 3–4 ml/hr, the column is pumped at 3.6 ml/hr (Technicon, SMA Flow Rated pump tube, 0.06 ml/min) and the limit buffer is pumped at 1.8 ml/hr (Technican, SMA Flow rated pump tube, 0.03 ml/min). This column flow rate results in a column back-pressure of 60–120 psi. We have found that the proportioning pump used in this system can deliver accurate flow rates at pressures in excess of 150 psi *provided the pump tubes are wired to the Teflon*

[3] J. R. Benson, *Anal. Biochem.* **71,** 459 (1976).

tubing as shown in Fig. 3. Teflon tubing 0.8 × 1.5 mm o.d. (the Teflon tubing and necessary fittings and valves are available from most suppliers of liquid chromatography equipment such as Altex, Durrum, LDC, etc.) leads from the pump to the sample injection valve (slider or rotary type with external, interchangeable sample loop). The tubing from the sample injection valve to the column, and the tubing leading from the column to the first mixing manifold of the monitor is 0.3 × 1.5 mm o.d. The microbore tubing is used to reduce both peak spreading and delay time in the system.

Analyses are usually begun about 4:00 PM and run overnight, ending about 10:00 AM the following day. This allows ample time to reequilibrate the column and prepare for a subsequent run so that one analysis per day can be routinely run. Initially, the system is started in the morning so that the baseline has stabilized by late afternoon. The pump is started with all reagent lines, the column pumping line, and the gradient pump line in place. Reagent lines are allowed to pump water for several hours until all air is displaced. Then the sodium hydroxide and acetic acid reagents are introduced, and the system is pumped for another hour prior to introducing the ninhydrin reagent. (This is necessary to preclude the possibility of ninhydrin reagent mixing with sodium hydroxide, which will result in a precipitate that may clog the tubing.) During this time the column is pumped with starting buffer and the limit buffer is recirculated by placing the limit buffer exit tube in the limit buffer chamber. Once all reagents are flowing, the recorder is turned on and the baseline observed until it stabilizes (1–2 hr), at which time the system is ready

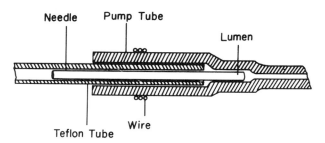

FIG. 3. Wiring pump tubes to Teflon tubing. Stainless steel tubes 1 cm long are cut by rotating a 19-gauge spinal needle, with trocar (or a suitable solid insert wire) in place, between the jaws of a sharp pair of wire cutters. The wire cutters are tightened until the needle, *but not the trocar,* has been cut. The short section of needle is then slipped off the trocar and the process is repeated for the next section of needle. The needle section is placed partially inside the pump tube, and the Teflon tubing is then slipped over the needle, between needle and pump tube. The assembly is held in place by wrapping in several loops of 22-gauge solid, noninsulated wire.

for sample injection. Shortly before initiation of the analysis, both buffer chambers are filled and sample (20–100 μl) is loaded into the sample injector. The sample injection valve is fitted with a sample loop slightly *larger* than the sample size and filled with buffer. The sample is then aspirated into the sample loop leaving a 2-mm air slug separating buffer and sample. The sample is then followed by more buffer (separated by another 2-mm air slug) until the entire sample is contained within the sample loop. The air slugs help to prevent spreading of the sample and make it easy to determine that the entire sample is contained within the sample loop.

The sample is injected into the column, and the time is noted. Starting buffer should be pumped through the column for 30–180 min depending on the nature of the sample being separated and *the dead volume between the mixing chamber and the column* (the resolution of digests containing several early eluting peaks can often be improved by lengthening the amount of time that initial buffer is pumped prior to starting the gradient). To start the gradient, the volume of the mixing chamber is adjusted to 30 ml, the magnetic stirrer is started, and the exit line from the limit chamber is placed in the mixing chamber. The total volume of the gradient is thus 60 ml and, at a flow rate of 3.6 ml/hr, the gradient is completed in 15–18 hr. Although flow rates are constant with a given set of pump tubes, there is some variation from one set of pump tubes to the next. This necessitates performing all comparative analyses with a single set of pump tubes. After completion of the gradient, limit buffer is pumped through the column for at least 2 hr.

After completion of an analysis, the resin is regenerated by forcing 1–2 ml of 8 M pyridine into the packed bed under nitrogen pressure (24 psi), then reequilibrated by pumping starting buffer for 3 hr.

Figure 4 shows the profiles obtained from successive separations performed on 5-nmol samples of a trypsin digest of performic acid-oxidized avian myeloblastosis virus p19 polypeptide using the procedures outlined above.

The shutdown procedure for the monitor system is basically the reverse of the start-up procedure. The ninhydrin and NaOH lines are first flushed with water (1 hr), followed by flushing of the acetic acid line. All tubing should be thoroughly purged with water before the pump is shut off.

Cautionary Notes. All pump tubes should be wired as shown in Fig. 3 and changed after 2 weeks' running time.

Should it become necessary to open the pump platen (for example, to add or remove a pump tube) during operation, the pump tubes should first be clamped off to prevent a pressure drop in the coils. In addition,

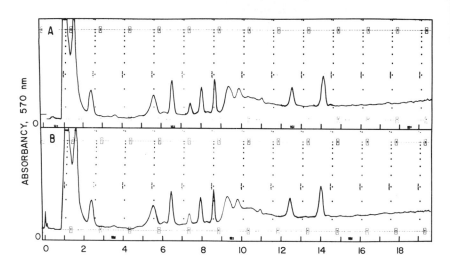

FIG. 4. Reproducibility and sensitivity of the peptide analysis system. Two consecutive separations performed on 6.0 nmol of a trypsin digest of oxidized avian myeloblastosis virus p19 in a 3 × 200-mm column of AA-15. The chromatograms were monitored with a Technicon colorimeter and recorder. Scale expansion was set to 1250 Ω resistance (see text). Reprinted, with permission, from A. C. Herman and T. C. Vanaman, *Anal. Biochem.* **63**, 550 (1975).

the column must be pressurized while the monitoring system is pumping. Loss of back-pressure in the column will cause sodium hydroxide to be forced through the column owing to lower resistance in the column than in the coils.

All O-rings coming in contact with pyridine-containing buffer must be made of EP or red silicone rubber.

Preparative Separations

Column Monitoring. The monitoring system described in Fig. 1 can be used to continuously monitor a portion of the effluent stream from any preparative column separation of peptides or proteins utilizing a suitable nonninhydrin reactive buffer system. Column effluent is divided using a 3-port manifold as shown in Fig. 5. A small aliquot (0.9–3.6 ml) is continuously removed from the effluent stream using a metering tube in the proportioning pump attached to the 3-port manifold (A) as shown. (Flow-rated metering tubes available from Technicon Corporation with flow rates of 0.9 to 3.6 ml/hr are recommended for this purpose.) The remainder of the effluent is passed to a fraction collector where fractions are collected at *timed intervals* (usually 4.5–6 min/fraction). When a

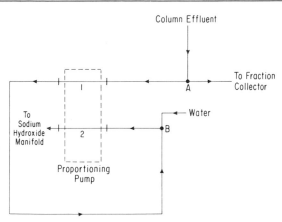

FIG. 5. Sampling system for preparative peptide separations. Column effluent is divided (manifold A), and an aliquot (0.9–3.6 ml/hr) is removed from the effluent by use of a pump tube (1) of appropriate size. If the sample is being removed at 3.0–3.6 ml/hr, this tube is connected directly to the sodium hydroxide manifold. Otherwise, the sample is first diluted with water (manifold B) to obtain a total volume of 3.6 ml/hr with tube No. 2 in the proportioning pump.

sampling rate of less than 3.6 ml/hr is desired, the aliquot taken from the effluent stream is diluted up to 3.6 ml/hr with deionized water in a second 3-port mixing manifold (B) also as shown in Fig. 5. The sample is then introduced into the monitoring system. It should be noted that the percentage of the column effluent being analyzed is determined by the column flow rate and the sampling rate from the divider manifold. These parameters should be chosen so that adequate material is taken by the monitoring system to give suitable peak heights. Specific conditions are given in the following paragraphs for performing and monitoring separations at various ranges of material.

Column Operation. This section deals primarily with separations of peptides by ion-exchange chromatography as described above. Extensive descriptions of automatic gradient elution of columns of this type have been given previously (Vol. 11 [37] and [38]) and will not be repeated here. The description of column operation and monitoring for separating peptides on Beckman AA-15 or W-1 at three different ranges of material are discussed in the following sections. Suitable medium to large-diameter columns are available from several manufacturers (e.g., Altex Scientific, Inc., Beckman Inst., Durrum Chem. Co., and Kontes Glassware). Again, the same criteria must be met for these columns as were listed for the analytical columns. Although adjustable plungers are not essential for occasional preparative separations, they greatly enhance ease of operation, reproducibility, and resolution.

5-25 Nanomoles. The procedures for column size and operation described in the section on analytical peptide mapping may be used to obtain preparative separations of 5–25 nmol. If the more sensitive spectrophotometer system (see above) is available, suitable peak heights are obtained with less than 1 nmol of digest permitting preparative separations of 2–5 nmol of material.

For micro preparative separations, the 0.3 × 20-cm column of Beckman AA-15 (W-1) resin is operated exactly as described in analytical peptide mapping except the column effluent is sent through the divider manifold (Fig. 5), where a suitable aliquot is removed for monitoring.

25–250 Nanomoles. Separation of 25–250 nmol of material should be performed on a 0.5–0.6 × 20-cm column developed at 12 ml/hr. Accurate column flow rates can be obtained using a positive displacement piston pump, such as a Milton-Roy Minipump. Typical gradient elution would be performed with 120 ml each of buffer F and buffer G. Panel A of Fig. 6 shows the elution profile obtained from an analytical separation on a 0.3 × 20-cm column of Beckman AA-15 of 6 nmol of a trypsin digest of performic acid-oxidized p19 polypeptide isolated from avian myeloblastosis virus. Panel B shows the profile obtained when one-

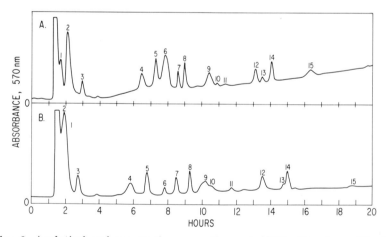

FIG. 6. Analytical and preparative separations of AMV p19 polypeptide. Panel A: 5 nmol of digest chromatographed on a 0.3 × 20-cm column of Beckman AA-15 as described for analytical chromatography in the text. Panel B: Preparative separation of 60 nmol of digest applied to a 0.5 × 25-cm column of AA-15. The column was pumped at 12 ml/hr (total gradient volume = 240 ml) with 7.5% of the column effluent (0.9 ml/hr) being diverted to the monitoring system. Equivalent peaks from the two traces are numbered (peak No. 6 in panel A is primarily ammonia). Both chromatograms were monitored with a Technicon colorimeter and recorder. Scale expansion was set to 1250Ω resistance (see text).

twelfth of the effluent was analyzed from the separation of 60 nmol of the same digest on a 0.5 × 25-cm column of Beckman AA-15 as described in the legend of Fig. 6.

Macro Separations. Satisfactory separation of peptides in amounts of up to 2 μmol can be obtained on a 0.9 × 20-cm column of Beckman AA-15 (W-1) resin operated at a flow rate of 45–60 ml/hr using 250 ml each of starting and limit buffers exactly as described by Hill and Delaney (Vol. 11 [37]). For larger amounts of material (up to 20 μmol), adequate resolution can be obtained on a 2.0 × 25-cm column of Beckman AA-15 (W-1) operated at 80 ml/hr with 500 ml of starting and limit buffers.

Recovery of Purified Peptides. One of the major difficulties encountered in preparing peptides with the automatic monitoring system is the correct assignment of fractions collected to peaks obtained on the column elution profile. This can be easily accomplished by including adequate amounts of a suitable ninhydrin-reactive marker compound which will be observed on the elution profile and also can be detected in the collected fractions by manual methods. For separations on cation exchangers, such as Beckman AA-15, cysteic acid is a suitable marker which is not retained by the resin under the conditions described above. In practice, small aliquots of fractions 1–10 are mixed with 1.0 ml of ninhydrin reagent,[4] heated for 15 min at 100°, and cooled to room temperature; A_{570nm} is read in a spectrophotometer. The ninhydrin color detected in these fractions (usually spanning one or two fractions) is then assigned to the first peak observed in the trace. (It should be noted that there is a considerable delay (2–3 hr) between the time at which sample emerges from the bottom of the column and the time at which its corresponding ninhydrin color is recorded on the trace. This delay time can be minimized by keeping the length of line between the column exit and the monitor system as short as possible.) The abscissa on the trace can then be calibrated directly in units of "fraction number" based on the collection rate and recorder chart speed used for the separation (e.g., at a collection rate of 6.0 min per fraction, each fraction corresponds to 0.2 inch on a trace recorded at 2 inches per hour).

When peaks appear poorly resolved in the elution profile, it is advantageous to examine small aliquots of fractions across the set of peaks by one-dimensional chromatography on thin-layer cellulose plates. This permits more accurate pooling of fractions to obtain homogeneous peptides for further sequence studies.

[4] S. Moore and W. H. Stein, *J. Biol. Chem.* **211**, 907 (1954).

Fractions identified as containing peptides by the techniques described above are combined and evaporated to dryness on a rotary evaporator at reduced pressure (50 mm Hg or less); the dried samples are dissolved in a small known volume of an appropriate solvent (e.g., 50% v/v acetic acid) and stored at −20° until further analyzed. It is highly recommended that prior to pooling the fractions, the pH of every tenth fraction be measured and recorded. This is particularly useful in choosing gradient elution conditions should subsequent rechromatography of any pooled fraction be required. Secondary fractionation of mixtures of peptides is discussed below.

Concluding Remarks

Methods outlined in the previous sections have dealt primarily with fractionation on highly cross-linked sulfonated polystyrene resins such as Beckman AA-15. Such resins are generally best suited for separation of intermediate- and small-sized peptides having moderately acidic to basic isoelectric points. Very acidic and/or large peptides are usually not adsorbed by Dowex 50-X8 resins and hence run as a mixture together with the cysteic acid marker in the initial fractions collected. In addition, large peptides can bind irreversibly to the resin or be eluted slowly throughout the entire gradient elution, contaminating otherwise homogeneous peptides. Separation of these large and/or very acidic peptides can be obtained by column chromatography on a number of different chromatographic media. Gel filtration on columns of Sephadex G-50 or G-75 in 50% (v/v) acetic acid has proved extremely valuable for secondary separations of relatively simple mixtures of peptides. These separations can be monitored using the automated systems described in previous sections. Constant flow rate, though desirable is not essential for chart alignment because fractions are collected as a function of time. N^ε-Dansyllysine added to the sample is an excellent marker for alignment of the elution profile from these separations with collected fractions.

Highly cross-linked spherical polystyrene-based anion-exchange resins such as those described by Benson (this volume [2]) and QAE-Sephadex are ideal for obtaining resolution of mixtures of acidic peptides. Simple linear gradient elution using equal volumes of buffers K and J in the table as starting and limit buffers, respectively, gives excellent resolution using the operating conditions specified in the previous sections for separations on cation-exchange resins. Figure 7 shows the elution profiles obtained with 10 nmol of a trypsin digest of bovine brain modulator

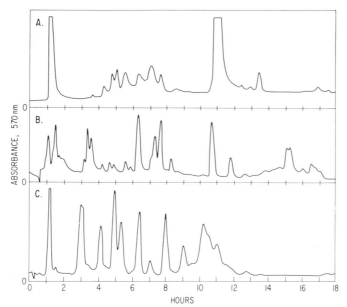

Fig. 7. Comparison of cation- and anion-exchange resins. Analytical separations of a trypsin digest of bovine brain modulator protein were performed as described in text. Panel A, Ten nanomoles of digest separated on a 0.3 × 21-cm column of Beckman AA-15 with 30 ml each of buffers F and J (see table) as starting and limit buffers, respectively (scale expansion 1250Ω). Panel B, Twenty nanomoles of digest on Durrum DA × 8-11 anion-exchange resin (0.3 × 21 cm) with 30 ml each of buffers K and J as starting and limit buffers, respectively (scale expansion 1250Ω). Panel C, Exactly the same as B except that the separation was performed on QAE-Sephadex A-25.

protein,[5] an acidic Ca^{2+} binding protein, applied to a 0.3 × 21 cm column of Beckman AA-15 (panel A) compared to that obtained for 20 nmol of the same digest on 0.3 × 21-cm column of Durrum DA × 8-11 (panel B), an anion-exchanger. Panel C shows the same separation performed in panel B but using QAE-Sephadex A-25. For this particular protein, initial fractionation of tryptic peptides is best accomplished using an anion exchanger. Although better resolution is obtained on the DA × 8-11 than on QAE-Sephadex A-25, the resolution obtained on QAE-Sephadex is suitable for purification of peptides for sequence analysis.

We have routinely used either QAE or SP-Sephadexes with the appropriate volatile buffer systems to obtain secondary separations of peptide mixtures with excellent results and high percent recoveries.

[5] D. M. Watterson, W. G. Harrelson, Jr., P. M. Keller, F. Sharief, and T. C. Vanaman, *J. Biol. Chem.* **251**, 4501 (1976).

Acknowledgments

The studies on which the procedures above are based were supported in part by Grant No. NS 10123 from the National Institutes of Health and Grant No. GB 27597 from the National Science Foundation. We wish to thank Durrum Chemical Corporation and Beckman Instruments Co. for generous gifts of ion-exchange resins.

[26] Detection of Peptides by Fluorescence Methods

By C. Y. LAI

Ninhydrin has long been the reagent of choice for the detection of peptides because of the broad specificity and sensitivity in its reaction with amino compounds. The recent introduction of fluorescamine,[1,2] a novel reagent designed after the fluorogenic reaction of ninhydrin with primary amines in the presence of phenylalanine has now provided a strong alternative to ninhydrin.[3,4] Fluorescamine reacts with primary amino groups of peptides almost instantaneously at room temperature, in aqueous solution at pH 7.5–9, to form a fluorescent compound. The reagent is subsequently hydrolyzed ($t_{1/2} \sim 10$ sec) to yield a water-soluble, nonfluorescent product. Fluorescamine itself is not fluorescent. Furthermore, ammonia in solution yields little or no fluorescence with fluorescamine. These features and the great sensitivity attainable in the fluorescence measurement[5] have made fluorescamine increasingly popular in amino acid and peptide analysis.[6–8]

Another development in recent years is the rediscovery of o-phthal-aldehyde, the fluorogenic reagent used about a decade earlier for the specific detection of cysteine and histidine,[5,9] as a general reagent for amino acids and peptides when used in the presence of 2-mercap-

---

[1] M. Weigele, S. L. DeBernardo, J. P. Tengi, and W. Leimgruber, J. Am. Chem. Soc. 94, 5927 (1972). 4-Phenylspiro [furan-2(3H)-1'-phthalane]-3,3'-dione.

[2] S. Udenfriend, S. Stein, P. Böhlen, W. Dairman, W. Leimgruber, and M. Weigele, Science 178, 871 (1972).

[3] K. Samejima, W. Dairman, and S. Udenfriend, Anal. Biochem. 42, 222 (1971).

[4] K. Samejima, W. Dairman, J. Stone, and S. Udenfriend, Anal. Biochem. 42, 237 (1971).

[5] S. Udenfriend, "Fluorescence Assay in Biology and Medicine." Academic Press, New York, 1964.

[6] See the article by Hare in this volume [1].

[7] S. Stein, P. Böhlen, J. Stone, W. Dairman, and S. Udenfriend, Arch. Biochem. Biophys. 155, 202 (1973).

[8] A. M. Felix and G. Terkelsen, Arch. Biochem. Biophys. 157, 177 (1973).

[9] J. J. Pisano, J. D. Wilson, L. Cohen, D. Abraham, and S. Udenfriend, J. Biol. Chem. 236, 499 (1961).

toethanol.[10] This reagent has recently been applied to the microanalysis of amino acids and peptides.[6,11]

Detection of Peptides in Column Effluents with Fluorescamine[12]

The procedure has in general been adapted from that used with the ninhydrin reagent.[13]

Reagents

Fluorescamine: The crystalline material from a commercial source[14] is dissolved directly in reagent-grade, anhydrous acetone (20 mg/100 ml). The reagent is stable for at least 1 month at room temperature when it is stored in an airtight glass container.

Borate (Na) buffer, pH 8.5: 0.5 M in borate

NaOH, 0.5 M HCl, 0.5 M

Procedure

Reaction after Alkaline Hydrolysis of Peptides. Aliquots containing 0.2–5 nmol of peptides are placed at the bottom of glass tubes,[15] and dried in an oven set at 110°. Aqueous solutions of up to 0.2 ml can be evaporated within 2 hr. Dried samples are hydrolyzed by addition of 0.2 ml of 0.5 M NaOH and heating with steam in an autoclave for 20 min at 120°. After cooling, 0.2 ml of 0.5 M HCl and 1 ml of borate buffer are added successively to each tube. Fluorescamine solution, 0.15 ml, is then added from a dispenser,[16] while vigorously stirring the contents of the tube with a vortex mixer, and the mixing is continued for several seconds. Fluorescence can be measured immediately in the same tube with an Aminco-Bowman spectrophotofluorometer (American Instrument Co., Silver Spring, MD) with excitation set at 390 nm and emission at 475 nm.

[10] M. Roth, Anal. Chem. 43, 880 (1971).
[11] J. R. Benson and P. E. Hare, Proc. Natl. Acad. Sci. U.S.A. 72, 619 (1975).
[12] N. Nakai, C. Y. Lai, and B. L. Horecker, Anal. Biochem. 58, 563 (1974). An improved procedure based on the subsequent experience in the author's laboratory is described here.
[13] C. H. W. Hirs, this series Vol. 11, p. 325 (1967).
[14] Fluram, available from Fisher Chemical Co., Springfield, NJ.
[15] Disposable culture tubes, 13 × 100 mm, available from Scientific Products Inc. (Edison, NJ, Cat. No. T1285-4), have been found satisfactory for this purpose without prior cleanings.
[16] Repetitive dispensers, such as "Repepit" (Fischer 13-687-58, 13-687-60) or "Dispensette" (Brinkhman Instruments Inc., Westbury, NY), are convenient for routine analyses. Reagents may conveniently be stored in the dispenser bottles.

The fluorescence intensity stays constant for about 1 hr, then diminishes slowly with time. The small amount of turbidity formed in the glass tubes on neutralization of the alkali-treated sample has been found not to interfere significantly with the measurement. Aliquots of effluent containing no peptides provide the value of the reagent blank.

Direct Reaction with Peptides. In case a great sensitivity of detection is not essential, and the peptides to be analyzed do not contain proline or an amino acid residue with a blocked α-amino group at the NH_2 terminus, the direct reaction with fluorescamine may be used with convenience and excellent reproducibility.

The sample dried at the bottom of a test tube is dissolved with 1.35 ml of borate buffer, and 0.15 ml of fluorescamine solution is added with vigorous stirring as described previously. A chromatogram using this procedure usually shows a low and smooth base line.

Examples and Discussion

An example of the fluorometric assay of peptide eluted from an ion-exchange column is shown in Fig. 1.[17]

Fluorescamine does not form fluorescent derivatives with secondary, tertiary or quarternary amines, nor with amide groups. Detection of proline (α-amino group is a secondary amine) in an amino acid analyzer[8] is accomplished by injection of N-chlorosuccinimide to the effluent line, during the emergence of proline, to effect its conversion to a primary amine compound.[18] Peptides with proline, pyroglutamic acid, or acylated amino acids at the NH_2 terminus have been found to yield little or no fluorescence with fluorescamine without prior hydrolysis with alkali[12] (Fig. 1); the ϵ-amino group of lysine residues in peptides does not appear to contribute a great deal to the yield of fluorescence. Similarly, the fluorescence yields of peptides containing acidic amino acids at the NH_2 terminus are found to be low in their direct reaction with fluorescamine. Thus, for a general and sensitive detection of peptides, the procedure with alkaline hydrolysis must be used. Unlike the colorimetric method with ninhydrin, small amounts of silicates, which dissolve from glass tubes during the 0.5 M NaOH treatment and precipitate on neutralization, do not significantly interfere with the detection. Although the reagent blank tends to be irregular, and much higher in this procedure than in direct reaction with fluorescamine, all operation may simply be carried out in the same tube.

[17] C. H. W. Hirs, *J. Biol. Chem.* **235**, 625 (1960).
[18] M. Weigele, S. DeBernardo, and W. Leimgruber, *Biochem. Biophys. Res. Commun.* **50**, 352 (1973).

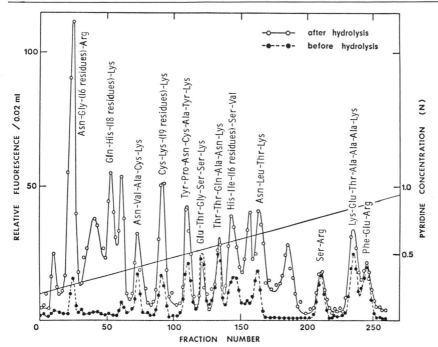

FIG. 1. A chromatogram of RNase A tryptic digest. Tryptic digest of performic acid-oxidized RNase A (21 mg, 1.5 μmol) was chromatographed on a 0.9 × 130-cm column of Dowex 50-X2 at 40° with a gradient of pyridine acetate buffer (0.2 N, pH 3.0 to 2 N, pH 5 in total of 2 liters) as indicated. Each fraction contained 1.8 ml. The effluent was monitored by the fluorescamine procedure for peptides. Amino acid analyses of separated peaks and comparison with the known sequence of RNase A [C. H. W. Hirs, *J. Biol. Chem.* **235**, 625 (1960)] revealed their identity as indicated in the figure. The peak tubes were estimated to contain 2.5–3 nmol of peptides per 0.02 ml of aliquot used for the fluorometric assay. In this experiment (see text footnote 12), the final volume in the assay was 2 ml instead of 1.5 ml as described in this article. Slight increase in the relative fluorescence has been obtained with the latter procedure. The relative fluorescence of 2.5 nmol of norleucine in 1.5 ml is approximately 18 over the blank value.

Detection of peptides by direct reaction with fluorescamine offers an advantage of being simple and rapid, and may conveniently be used for small peptides containing no blocked NH₂ termini or proline. If the sample is in a dilute buffer or at pH 7.5–9, up to 0.5-ml aliquots may be used without prior drying. Most buffer salts, urea, guanidine hydrochloride, and sodium dodecyl sulfate do not react with fluorescamine and thus do not interfere with the analysis. Ammonia, though not forming a highly fluorescent derivative with fluorescamine, does interfere with the reaction when present in a large quantity, probably by consuming and competing with peptides for the reagent. Evaporating the sample to dryness removes most of the ammonia. Urea and guanidine-HCl produce substances that give high blank values on alkaline hydrolysis.

Detection of Peptides on Paper[19] or Thin-Layer Plates

Reagents

Fluorescamine, 10 mg/100 ml solution in acetone
Triethylamine, 1% (v/v) in acetone

Procedure

The paper or thin-layer plates, after electrophoresis or chromatography, are dried in the hood, washed with acetone, and dried. These are then wet with triethylamine solution, dried briefly in the hood (5 min) and treated with the fluorescamine solution. After another acetone wash and drying, peptides may be visualized by their fluorescence under a long-wave (336 nm) UV lamp.[20] The sensitivity of detection is improved by repeating the last acetone wash. Repeating the entire procedure is occasionally helpful in detecting weakly reactive peptides.

The washing and wetting procedure may conveniently be performed by delivering the reagents from plastic wash bottles onto the hanging paper.

After marking the spots that become visible with the above treatments, the paper is placed in an oven and heated at 110° for 3 hr. Peptides with a proline residue at the NH_2 terminus thereupon become detectable as fluorescent spots.

The procedure is applicable without modification to thin-layer plates.

Discussion

For average peptides the limit of detection using the fluorescamine method has been about 0.1 nmol/cm² on Whatman 3 MM paper. The final acetone wash is effective in improving the contrast of fluorescent spots, and is considered essential in attaining the maximum sensitivity. Proline, or peptides containing proline at the NH_2 terminus, become visible on heating for 3 hr or after 2 days at room temperature.[19] The

---

[19] E. Mendez and C. Y. Lai, *Anal. Biochem.* **65**, 281 (1975). Heating the paper prior to the fluorescamine treatment, as originally reported in this paper, has been found to be unimportant in improving the results, and has been omitted from the procedure.

[20] A viewing box such as "Chromato-vue" (Ultra-violet Products, Inc., San Gabriel, CA) is convenient for this purpose.

reaction causing these peptides to form fluorescent derivatives with fluorescamine is unknown. The limit of detection for these peptides is approximately 2 nmol/cm². Occasionally, peptides with a blocked NH_2 terminus, but containing lysine residues, are found to be detectable with fluorescamine, indicating that the ε-amino group can be substantially fluorogenic. The sensitivity of detection on thin-layer plates is generally lower than that on paper.[21]

Rapid and effective separation of peptides may often be achieved by electrophoresis or chromatography on paper. The fluorescamine spray provides a simple and sensitive means of localizing the separated peptides, and does so without chemically altering much of the peptides. In a preparative run dealing with peptides in the amount of 20 nmol or more per square centimeter, less than 5% of the material is found to have reacted with fluorescamine under the conditions described.[19] The peptides may thus be eluted and used in subsequent studies. Furthermore, reaction of fluorescamine with peptides on paper is apparently reversible to some extent. When a peptide, 20% of which is expected to have reacted with fluorescamine, is eluted from paper with 0.1 N NH_4OH, dried, and then hydrolyzed with 5.7 N HCl at 110° for 22 hr, all amino acid residues including the NH_2 terminus are usually recovered, giving the correct amino acid composition of the peptide.[19]

The great sensitivity afforded by fluorescamine can sometimes be the shortcoming of the method. Minor components of a peptide mixture separated on a paper chromatogram may confuse the pattern of major peptides, since the extent of reaction with fluorescamine is greater with smaller amounts of peptides.

Detection of Peptides with the Use of o-Phthalaldehyde

o-Phthalaldehyde has recently been applied successfully to an automated amino acid analyzer.[6,11] Experience in our laboratory indicates that this reagent may be used in the detection of peptides in the same manner as with fluorescamine.[22]

Reagents

o-Phthalaldehyde: (a) aqueous solution containing 0.3 g of o-phthalaldehyde per liter and 0.05 v/v 2-mercaptoethanol; (b) acetone solution containing 0.3 g/l o-phthalaldehyde[23]

---

[21] A. M. Felix and M. H. Jimenz, *J. Chromatogr.* **89**, 361 (1974).

[22] E. Mendez and J. G. Gavilanes, *Anal. Biochem.* **72**, (1976) in press.

[23] In our experience, o-phthalaldehyde obtained from Sigma Chemical Co. (St. Louis, MO) and that from Durrum Instrument Co., [Palo Alto, CA (Fluoropa)] show

Borate (Na) buffer, pH 9.7, 0.5 M in borate
Triethylamine, 1% v/v triethylamine in acetone containing 0.05%
v/v 2-mercaptoethanol

Reagents must be made with highly purified water.[23]

Procedure

Peptide samples are prepared, with or without alkaline hydrolysis, as described above for the fluorescamine procedure. Borate (Na) buffer, pH 9.7, is added to the samples to a final volume of 1.6 ml. *o*-Phthalaldehyde reagent (a), 0.4 ml, is then added with vigorous mixing. Fluorescence intensity is measured in the same tube with excitation set at 340 nm and emission at 455 nm.

For detection of peptides on paper after electrophoresis or chromatography, the paper is processed exactly as described previously for the use of fluorescamine, except that triethylamine solution containing 0.05% mercaptoethanol and *o*-phthalaldehyde in acetone (b) are used.

Discussion

The major advantage for *o*-phthalaldehyde is its solubility and stability in water. The reagent can thus be added to sample solutions without vigorous stirring to ensure rapid mixing, as is necessary for the addition of fluorescamine. In automated amino acid analysis,[11] *o*-phthalaldehyde is directly dissolved in borate (K) buffer, pH 9.7, containing 2-mercaptoethanol so that an extra pump for delivering buffer is eliminated.

Fluorescence intensity produced by the reaction of *o*-phthalaldehyde with amino acids is, on the average, somewhat higher than that obtained with fluorescamine,[11] when measured at a given sensitivity on the fluorometer. With acidic amino acids (Asp and Glu), *o*-phthalaldehyde produces fluorescence values approximately 10 times those obtained by the fluorescamine reaction. With cystine, the ratio is reversed, about 1:5. Like fluorescamine, *o*-phthalaldehyde does not yield fluorescent products with imino acids, such as proline. In the analysis of peptides after alkaline treatment, higher fluorescence is obtained with *o*-phthalaldehyde in general,[22] though also with high backgrounds.

identical sample and blank fluorescence. Water from the ultrapure water system, Hydro Service and Supplies (Durham, NC) and sodium borate from Mallinckrodt (St. Louis, MO) gave satisfactory results.

It should be noted, however, that the sensitivity of detection depends not only on the fluorescence yield, but also on the signal-to-noise ratio, since amplification of the signal also results in the amplification of the background fluorescence. Even with solutions made with carefully selected reagents,[23] o-phthalaldehyde yields background fluorescence severalfold higher than fluorescamine. Thus, sensitivity of detection of peptides with o-phthalaldehyde, even after hydrolysis, is actually not greater than with fluorescamine. Detection of peptides in the picomole range in column effluent has been possible with fluorescamine.[24,25]

Both fluorescamine and o-phthalaldehyde are useful reagents in protein chemistry. The mechanism of reaction between o-phthalaldehyde and primary amines has not yet been elucidated. In contrast, the fluorescamine fluorophors have been characterized,[26] and furthermore, it has been possible to chromatograph them on columns[25] and on thin-layer plates.[27]

[24] P. Böhlen, S. Stein, J. Stone, and S. Udenfriend, *Anal. Biochem.* **67**, 438 (1975).
[25] K. A. Gruber, S. Stein, L. Brink, A. Radhakrishnan, and S. Udenfriend, *Proc. Natl. Acad. Sci. U.S.A.* **73**, 1314 (1976).
[26] S. DeBernardo, M. Weigele, V. Toome, K. Manhart, W. Leimgruber, P. Böhlen, S. Stein, and S. Udenfriend, *Arch. Biochem. Biophys.* **163**, 390 (1974).
[27] K. Imai, P. Böhlen, S. Stein, and S. Udenfriend, *Arch. Biochem. Biophys.* **161**, 161 (1974).

Section VIII

Sequence Analysis

[27] Microsequence Analysis in Automated Spinning-Cup Sequenators

By JACK SILVER

Cell surface molecules are involved in many important biological phenomena which may, in a broad sense, be defined as intercellular communication. This has aroused immense interest in studying proteins that mediate these interactions. However, one of the major obstacles encountered in studying such molecules is obtaining sufficient amounts of material to perform chemical studies, since they are usually present on the cell surface in minute quantities. Sequence analysis of small quantities (<10 nmol) of proteins using the conventional "spinning-cup" automated sequenator has until recently been very difficult because of two basic problems: (i) small amounts of protein tend to wash out of the cup during the liquid phase extraction procedures, and (ii) conventional detection systems which involve reacting ninhydrin with amino acids lack the requisite sensitivity.

This article presents two approaches for circumventing these difficulties and thus permitting the analysis of subnanomole quantities of polypeptides. The first approach involves inclusion of a synthetic "carrier" (succinylated polyornithine) with the protein to be sequenced to reduce protein washout, coupled with the use of either an ultrasensitive amino acid analyzer (Durrum D500) or radioactive phenylisothiocyanate (PITC) to increase the sensitivity of the detection system. This method is critically dependent upon the availability of *homogeneous* protein, since any contaminant will also be sequenced. The second approach involves sequencing radioactive proteins (labeled with tritiated amino acids) by conventional procedures in the presence of a natural protein carrier. Phenylthiohydantoins (PTH) are then resolved by thin-layer chromatography (TLC) or high pressure liquid chromatography (HPLC) and the radioactivity associated with each amino acid measured. This method depends critically upon the ability to radioactively label the protein of interest, but the material to be sequenced need only be *radioactively homogeneous.* Any contaminating proteins introduced either deliberately for purposes of purification (e.g., specific antibodies) or accidentally will not interfere with sequence analysis as long as they are not radioactively labeled. The application of this method to the NH_2-terminal sequence analysis of the transplantation antigens of the mouse is presented.

Materials and Methods

Method 1. Microsequencing of Homogeneous Proteins

Preparation of Succinylated Polyornithine. Poly-L-ornithine-HBR (Research Plus Laboratories, Inc., 200 mg) and Trizma Tris-HCl (15.8 mg) are dissolved in a minimum amount of water and brought to pH 9.5 by the addition of 3 *N* NaOH. Finely ground succinic anhydride (4.0 g) is slowly added to the solution while the pH is maintained at 9.5 by the automatic addition of 3 *N* NaOH. The solution is then dialyzed against distilled water for 2 days with four or five changes of water and subsequently lyophilized. Succinylated polyornithine is also commercially available (Pierce Chemical Co.).

Automated Sequence Analyses. When protein is to be sequenced in the presence of "carrier," 5–7 mg of succinylated polyornithine dissolved in 200 μl of water are placed in a Beckman Model 890 Sequencer, dried under vacuum, and subjected to two complete sequence cycles of the stepwise Edman degradation.[1] Protein samples (50–100 μg) are then dissolved in 200 μl of water, added to the cup, and dried under vacuum prior to commencement of the sequencing cycle. A program similar to that described previously which uses Quadrol as the coupling buffer is employed throughout.[2] Phenylthiohydantoins are hydrolyzed to free amino acids with HI and analyzed by means of a Durrum D500 amino acid analyzer.[3]

When [^{14}C]PITC is employed, a modified Quadrol program is used in which the coupling reaction with unlabeled PITC is preceded by a 30-min coupling reaction with [^{14}C]PITC (Amersham/Searle Corp.). The [^{14}C]PITC (250 μCi, 6 mCi/mmol) is dissolved in 15 ml of heptane and 0.5 ml of this solution is delivered in the initial coupling reaction. After 30 min, the normal amount of unlabeled PITC is added, and the cycle is allowed to proceed as usual. The thiazolinone derivative obtained at the completion of each sequence cycle is cyclized to the phenylthiohydantoin and extracted.[3]

Analysis of [^{14}C]*Phenylthiohydantoins.* The [^{14}C]PTH derivatives are added to a mixture of 20 unlabeled PTH-derivatives and separated by two-dimensional thin layer chromatography on "Cheng-Chin" 15 × 15-cm polyamide plates (Gallard-Schlesinger Chemical Mfg. Corp.). The first-dimension solvent is 45% formic acid, and the second-dimension solvent consists of carbon tetrachloride and acetic acid in a ratio, respectively, of 9:2. This chromatographic system resolves all the PTH-

---

[1] P. Edman and G. Begg, *Eur. J. Biochem.* **1**, 80 (1967).

[2] L. Hood, D. J. McKean, V. Farnsworth, and M. Potter, *Biochemistry* **12**, 741 (1973).

[3] O. Smithies, D. M. Gibson, E. M. Fanning, R. M. Goodfliesh, J. G. Gilman, and D. L. Ballantyne, *Biochemistry* **10**, 4912 (1971).

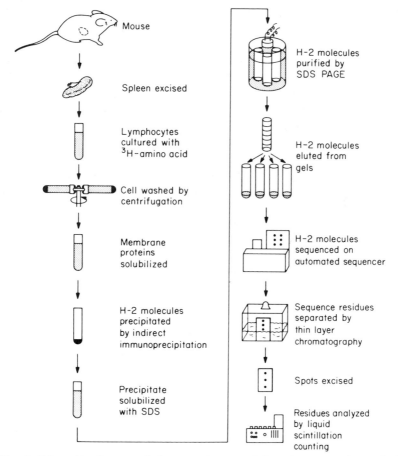

FIG. 1. Schematic diagram of the procedures used for microsequencing technique (see text). SDS PAGE, sodium dodecyl sulfate polyacrylamide gel electrophoresis.

derivatives except those of leucine and isoleucine.[4] The PTH spots are visualized by shortwave ultraviolet light and then cut out of the plate and placed in scintillation vials. Approximately 0.5 ml of methanol is added to each vial. After 5 min, 5 ml of 6.3% Liquofluor (New England Nuclear) in toluene is added to each vial. Samples are counted in a liquid scintillation counter.

Method 2. Microsequencing of Radioactive Homogeneous Proteins (Fig. 1)

Radiolabeling Cell Surface Molecules. One mouse spleen is excised and repeatedly scraped over a wire mesh to release lymphocytes into

[4] J. Silver and L. Hood, *Anal. Biochem.* **67**, 392 (1975).

a Petri dish containing 5 ml of Hanks' balanced salt solution (HBSS) (Grand Island Biological Corp.). The cells are centrifuged, washed once in HBSS, and resuspended in HBSS at a concentration of 4×10^7/ml containing 3–8 tritiated amino acids with the total amount of radioactivity not exceeding 3 mCi. The cell suspension is incubated at 37° in a 2% CO_2 atmosphere for 4 hr with periodic shaking for resuspension. Cell viability at the end of this period is approximately 70%. After centrifugation, the cells are washed once in 10 mM Tris buffer, pH 7.4, containing 0.14 M NaCl (buffer A) and then recentrifuged.

Membrane Solubilization. The cell pellet is resuspended in buffer A containing 0.5% Triton X-100 in a volume equal to that used for the initial cell incubation. Triton X-100 is a nonionic detergent that solubilizes the cell membrane and all its constituent proteins. In addition, proteins retain their specific antigenicity in the detergent solution and subsequent antigen–antibody precipitation reactions are not affected by the detergent. After incubation at 4° for 10 min, insoluble debris (including whole nuclei) is removed by centrifugation in a clinical centrifuge.

Indirect Immunoprecipitation. Specific cell surface molecules are immunoprecipitated by the addition of specific antisera using a double antibody or "sandwich" technique.[5,6] Specific cell surface molecules are culled by the addition of 10 μl of specific mouse antiserum for every 100 μl of soluble membrane preparation, and the solution is incubated at 4°. This permits the formation of soluble antigen–antibody complexes. After 30 min a sufficient amount of rabbit anti-mouse immunoglobulin antibody is added to precipitate all of the mouse antibody along with the cell surface molecules specifically complexed to it. (To avoid coprecipitation of endogenous radioactive mouse immunoglobulin, a preprecipitation with rabbit anti-mouse immunoglobulin antibody is performed prior to addition of specific mouse antisera.) The immunoprecipitate is centrifuged and resuspended in buffer A, and the suspension is transferred to a fresh test tube to avoid undesired proteins that bind nonspecifically to the glass tube. After pelleting the precipitate by centrifugation, it is washed once more as above using buffer A.

Molecular Weight Sieving. The washed immunoprecipitate is dissolved by the addition of 50 μl of Laemmli reducing buffer[7] for every 10 μl of antiserum used in the formation of the precipitate, vortexing, and heating in a boiling water bath for 1 min. The solution is electro-

[5] W. M. Hunter, *in* "Handbook of Experimental Immunology," Vol. 1 (D. M. Weir, ed.), p. 1714. Blackwell, Oxford, 1973.

[6] B. D. Schwartz and S. G. Nathenson, *J. Immunol.* **107,** 1363 (1971).

[7] U. K. Laemmli, *Nature (London)* **227,** 680 (1970).

phoresed (50 μl/gel) on 10% polyacrylamide-sodium dodecyl sulfate (SDS) gels (6 mm $\times$ 10 cm) according to Laemmli.[7] Gels are cut into 1-mm-thick slices, and each slice is incubated overnight at room temperature in 0.5 ml of 0.01% SDS. Aliquots (10%) of each slice eluate are counted in a liquid scintillation counter to locate the desired protein. The gel eluate containing the radioactive membrane protein is filtered through glass wool to remove gel particles and lyophilized. The protein is redissolved in 1 ml of H_2O and dialyzed twice against 0.01% SDS.

Amino Acid Sequence Analysis. Two milligrams of ovalbumin (which has a blocked NH_2-terminal group) is added to the dialyzed sample, which in then lyophilized to a volume of 0.5 ml approximately. The sample is loaded onto a Beckman Model 890 Sequencer and dried under vacuum. Sequential protein degradation is performed using the procedure of Hermodson *et al.*[8]

Analysis of [^{3}H]Phenylthiohydantoins. The thiazolinone derivative obtained at the completion of each sequence cycle is cyclized to the phenylthiohydantoin and extracted.[3] The samples are dried under N_2 and redissolved in 5 μl of ethyl acetate. Each sample is applied as a streak 1 cm long on silica gel plates containing fluorescent indicator (Eastman Kodak). An appropriate amount of unlabeled PTH is also applied to each sample streak to permit visualization. The mixture of PTH derivatives is separated by one-dimensional TLC using an appropriate solvent system (see Fig. 4). The amount of radioactivity associated with each residue is determined by liquid scintillation counting as described above. An alternative method for separating the PTH derivatives is HPLC for which a number of solvent systems are available. Effluent fractions, each containing a different PTH derivative may be collected, evaporated, and analyzed by liquid scintillation counting. The separation of PTH-derivatives is simplified by the fact that only those amino acids included in the incubation mixture in titrated form need to be resolved and analyzed for radioactivity.

Results

Method 1

The effect of "carrier" on the amount of protein washout during microsequencing is reflected in the total nanomoles of all amino acids present at each step in the presence or absence of "carrier" (Tables I and II). In the absence of "carrier," a large amount of amino acids is

[8] M. A. Hermodson, L. H. Ericsson, K. Titani, H. Neurath, and K. A. Walsh, *Biochemistry* 11, 4493 (1972).

TABLE I

AMINO ACID ANALYSIS OF SEQUENATOR RUN IN THE PRESENCE OF "CARRIER"[a]

Amino acid	Lys 1[b]	Val 2	Phe 3	Gly 4	Arg 5	Cys 6	Glu 7	Leu 8	Ala 9	Ala 10
Asp	9.6[c]	7.9	6.8	9.2	11.0	9.4	7.6	6.7	8.0	7.0
Glu	11.6	8.2	6.4	9.0	10.0	8.9	*32.3*	18.9	11.7	8.4
Pro	2.6	1.0	1.2	3.6	5.3	4.5	3.3	3.0	3.0	3.7
Gly	15.6	16.8	13.6	*40.2*	32.9	17.0	12.7	10.6	14.7	10.5
Ala	12.2	9.7	9.8	11.4	14.3	*28.3*	17.2	10.9	*32.8*	*41.7*
Abu	1.0	0.5	0.7	0.8	1.9	3.0	3.0	1.7	1.1	2.7
Val	4.9	*39.7*	5.1	4.5	4.9	5.5	4.6	3.2	3.5	3.4
Ile	3.5	2.7	2.0	3.1	3.3	1.8	1.7	1.6	1.9	1.4
Leu	6.3	4.6	3.0	5.6	5.0	7.0	5.4	*34.8*	15.8	8.1
Tyr	2.2	2.4	2.2	1.8	2.2	4.9	4.8	1.9	1.9	3.6
Phe	3.0	2.7	*46.6*	6.4	3.7	4.6	3.0	2.1	2.1	3.4
Lys	*27.7*	3.9	2.7	4.4	5.5	5.0	4.4	4.7	3.5	4.4
Total nmol in sample	3.4[d]	3.1	3.7	5.2	2.6	3.4	3.7	3.2	4.5	3.7

	Ala 11	Met 12	Lys 13	Arg 14	His 15	Gly 16	Leu 17	Asp 18	Asn 19	Tyr 20
Asp	7.1	9.5	9.4	10.4	9.8	11.0	9.3	*17.1*	*21.5*	17.2
Glu	6.0	9.2	9.8	8.9	9.0	10.4	6.0	8.5	11.0	11.3
Pro	1.5	3.5	4.6	5.6	6.0	0.0	2.9	2.4	2.8	3.1
Gly	10.6	13.4	16.6	14.5	16.7	*33.3*	21.7	17.8	18.8	19.5
Ala	*56.0*	33.3	24.8	23.7	22.6	19.3	22.0	19.6	17.8	18.9
Aba	1.6	3.8	3.1	4.6	4.9	2.6	4.0	3.9	2.3	2.5
Val	3.0	4.8	4.7	5.8	5.2	5.0	5.0	4.9	4.8	4.3
Ile	2.0	1.6	2.1	1.5	2.0	2.9	1.6	1.6	3.4	3.1
Leu	5.4	8.3	7.4	7.7	7.8	5.9	*15.7*	10.6	9.5	7.1
Tyr	1.8	5.0	4.9	6.0	5.5	2.2	4.5	5.1	2.1	*7.1*
Phe	2.0	3.2	3.1	3.8	4.1	3.3	3.0	3.3	2.9	2.6
Lys	3.2	4.6	*9.5*	7.3	6.5	4.2	4.4	5.2	3.1	3.4

	Arg 21	Gly 22	Tyr 23	Ser 24	Leu 25
Asp	18.9	12.1	12.0	11.9	11.3
Glu	8.9	12.0	9.5	8.7	9.8
Pro	2.6	2.6	4.0	1.4	2.7
Gly	18.3	23.3	23.5	24.7	21.8
Ala	18.3	20.1	17.1	23.8	21.7
Aba	2.2	2.2	2.9	2.2	2.5
Val	5.5	4.9	4.6	4.7	5.0
Ile	3.2	3.8	3.2	2.9	3.4
Leu	6.9	7.2	5.7	5.1	*10.5*
Tyr	8.9	4.4	*10.6*	7.6	4.4
Phe	2.7	3.6	2.7	2.8	3.1
Lys	3.5	4.0	4.1	4.2	3.9

[a] Lysozyme (100 μg, 7 nmol) was sequenced in the presence of succinylated polyornithine (5 mg). PTH-derivatives were extracted with ethyl acetate, hydrolyzed, and analyzed on an amino acid analyzer. The aqueous partitioning material was not analyzed, precluding identification of arginine and histidine residues. PTH-cysteine and PTH-serine are degraded to alanine, PTH-asparagine is converted to aspartic acid, and methionine is destroyed by the acid hydrolysis procedure. The correct sequence is presented above the step number and the sequence deduced from the data is italicized.

[b] Step number.

[c] Mole percent of given step.

[d] Corrected against an internal standard of PTH-norleucine.

TABLE II
AMINO ACID ANALYSIS OF SEQUENATOR RUN IN THE ABSENCE OF "CARRIER"[a]

Amino acid	Lys 1	Val 2	Phe 3	Gly 4	Cys 6[b]	Glu 7	Leu 8
Asp	10.7	12.0	10.7	11.6	16.2	10.3	9.1
Glu	8.8	6.8	8.5	8.7	8.1	*22.5*	22.2
Pro	7.7	3.2	1.5	5.3	6.1	1.7	4.5
Gly	19.1	24.2	20.3	*28.0*	22.4	28.3	22.2
Ala	14.1	13.8	12.1	12.7	16.5	13.0	15.5
Abu	1.6	0.6	1.6	1.0	0.0	0.9	0.8
Val	5.1	*16.2*	6.2	6.1	6.4	4.7	5.2
Ile	5.0	5.1	4.0	4.3	4.9	3.2	3.2
Leu	6.7	5.9	5.6	5.5	4.7	4.3	*7.4*
Tyr	3.5	3.9	3.7	5.4	3.8	3.1	2.8
Phe	3.9	3.6	*20.5*	6.1	4.2	3.5	3.1
Lys	*13.8*	4.7	5.3	5.4	6.2	4.5	4.0
Total nmol in sample	15.4[c]	6.3	4.2	3.3	2.3	2.9	5.2

[a] Lysozyme (100 μg) was sequenced and analyzed in a manner identical to that described in Table I except for the omission of succinylated polyornithine.
[b] Step 5 was inadvertently lost.
[c] Corrected against an internal standard of PTH-norleucine.

observed at step 1 (15.1 nmol), mostly due to protein washout (Table II). In contrast, in the presence of "carrier" the total amount of amino acids observed at step 1 is only 3 nmol (Table I) reflecting minimal protein washout. Accordingly, the addition of "carrier" permits the sequential degradation and identification of 19 of 25 amino acid residues from the NH_2 terminus of lysozyme using as little as 7 nmol of protein (ca. $\frac{1}{30}$ the amount normally used) (Table I). In contrast, in the absence of "carrier," protein loss during the sequence cycles makes amino acid identification difficult after step 7 (Table II).

The ability to assign an amino acid residue to a particular step is a function of the change in amino acid mole percent during consecutive steps. The inclusion of succinylated polyornithine has a profound effect on this value and concomitantly the ability to sequence lysozyme further. This is best illustrated by comparing the mole percent values of leucine at steps 7 and 8 (cf. Tables I and II). In the presence of "carrier," the mole percent of leucine rises from 5.4 to 34.8 (544% increase) while in its absence these values change from 4.3 to 7.4 (72% increase). Comparable results are obtained using even smaller amounts of lysozyme (50 μg, 3 nmol) in the presence of "carrier."

Plots of amino acid yields indicate that these differences are due to a drastic decrease in the repetitive yield when "carrier" is omitted, presumably owing to protein washout during the sequence cycle (Fig. 2). The repetitive yield in the presence of succinylated polyornithine calculated from the yields of glycine at step 4 and 16, and from the yields of leucine at steps 8 and 25 are 92 and 91%, respectively (Fig. 2). This compares favorably with sequence runs using macroamounts of protein. Omission of "carrier" reduces the repetitive yield to approximately 67%.

The results of a sequence analysis on 20 μg (1.6 nmol) of lysozyme using [^{14}C]PITC as the coupling reagent is shown in Fig. 3. Succinylated polyornithine (5 mg) is used as "carrier." The number of ^{14}C counts per minute (cpm) associated with each PTH-derivative at each sequence cycle is plotted. An increase in ^{14}C cpm associated with a particular PTH-derivative identifies that amino acid as the sequence residue at that particular position. Therefore, valine, phenylalanine, glycine, glutamic acid, and leucine or isoleucine may be unambiguously identfied as the sequence residues at positions 2, 3, 4, 7, and 8, respectively. The

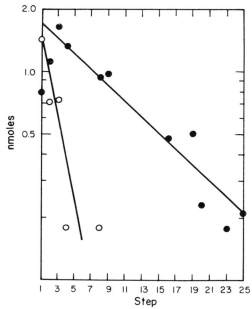

FIG. 2. Amino acid yields of lysozyme sequenator runs in the presence (●) and in the absence (○) of succinylated polyornithine. Background values were subtracted by comparison to a previous step. Losses due to manipulation were corrected against an internal standard of PTH-norleucine.

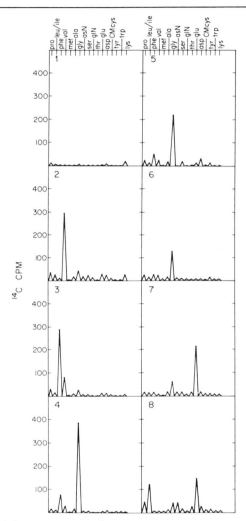

FIG. 3. Radioactivity present at each step on a sequenator run of lysozyme. Radioactive counts per minute are indicated on the ordinate and the 17 identifiable amino acids on the abscissa. Panel 1 indicates the first step in the automatic sequenator run; panel 2 the second; etc.

lysine residue at position 1 does not have any ^{14}C-associated counts for unexplained reasons. The PTH-derivative of the sequence residue at position 5 (arginine) is not extractable from the aqueous phase into the organic phase (ethyl acetate) and therefore is not identified. The PTH-derivative of the sequence residue at position 6 (cysteine) is highly unstable and rapidly degraded, precluding its identification.

Method 2

The results of two typical sequenator runs of a cell surface molecule (transplantation antigen of the mouse) are shown in Fig. 4. Spleen cells

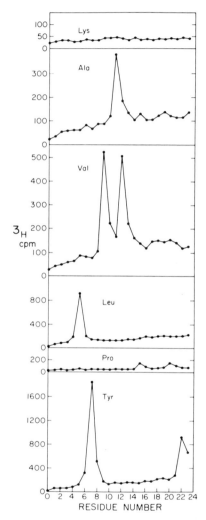

Fig. 4. Amino acid sequence data of a ³H-labeled murine transplantation antigen. The two groups of PTH derivatives (Lys, Ala, Val, and Leu, Pro, Tyr) were separated by one-dimensional thin-layer chromatography on 5 × 5-cm silica gel plates with fluorescent indicator (Eastman Kodak) using 75% heptane, 19% propionic acid, and 6% ethylene dichloride as the solvent system. The amount of radioactivity associated with each of the six incorporated amino acids is plotted against residue number.

TABLE III
AMINO ACID SEQUENCE DEDUCED FROM FIG. 4

Position

2	3	4	5	6	7	8	9	10	11	12	13	14	15	16	17	18	19	20	21	22
—	—	—	Leu	—	Tyr	—	Val	—	Ala	Val	—	—	Pro	—	—	—	—	Pro	—	Tyr

were cultured *in vitro* with groups of tritiated amino acids (i.e., either alanine, lysine, and valine or leucine, proline, and tyrosine) and the radiolabeled cell surface product was isolated and sequenced as described above. Sequence residues are characterized by a sharp rise in radioactivity associated with a particular amino acid followed by a more gradual decline. The partial sequence deduced from these data is shown in Table III; a dash indicates that the corresponding residue is *not* one of the six labeled amino acids (assuming that all six radioactive amino acids have been efficiently incorporated). The data display several characteristics typical of conventional automated sequence analysis. The gradual decline in radioactivity following a sequence residue (known as "lag") is the result of an incomplete Edman reaction and tends to increase throughout the run. In addition, there is a gradually rising background due to random hydrolysis of the protein. Since each amino acid is incorporated to a different extent, the data for each amino acid must be treated independently. Thus, for those residues that appear two or more times (e.g., tyrosine, valine, proline) an average repetitive yield can be calculated. The repetitive yields for these three residues range between 89 and 94%, which is similar to the repetitive yields obtained from conventional runs, although the amount obtained for each step is only 40% of the expected yield (based on the amount of radioactivity loaded on the sequenator).

Discussion

The choice of succinylated polyornithine as "carrier" (Method 1) is based on several considerations. Ornithine, a homolog of lysine, is readily resolved from the other basic amino acids on an amino acid analyzer. Since it is not a naturally occurring amino acid, its presence on amino acid chromatograms may be neglected. The random polymer of ornithine is succinylated in order to render it highly anionic and therefore soluble under alkaline conditions (coupling with phenylisothiocyanate) and insoluble in organic solvents (the ethyl acetate and benzene washes during the sequenator cycles). These are ideal properties for any polypeptide undergoing the three-cycle Edman reaction. Oroszlan *et al.*[9] have reported on the use of succinylated polyornithine and PTH-norleucine in automated sequence analysis of viral proteins. They find that in general succinylated polyornithine as carrier results in less protein washout and more reproducible results than the use of natural proteins which have a blocked NH_2-terminal end (cytochrome *c* and apocytochrome *c*). In addition inclusion of PTH-norleucine or methylthiohydantoin-alanine in the cleaving reagent HFBA, in order to minimize nonspecific losses and degradation of PTH-derivatives during subsequent manipulations, leads to increased repetitive yields. Thus the inclusion of "carrier" with the protein to be sequenced significantly limits protein washout and permits extended automated microsequence analysis.

An alternative microsequencing method which minimizes protein washout is solid-phase sequencing developed by Laursen.[10] This method relies upon the covalent coupling of protein to a solid-phase resin and thus eliminates protein washout during the subsequent Edman degradation.

The use of radioactive PITC as a coupling reagent to increase the sensitivity of the detection system has been reported for both liquid-phase[4,11,12] and solid-phase sequencing[10,13-15] in which as little as 1 nmol of protein was employed. More recently, Bridgen has reported on the N-terminal sequence analysis of 70 pmol of protein in which radioactive PITC was used in conjunction with a solid phase sequencer.[16]

Several recent developments offer alternatives for the detection of sequence residues at the picomole level. These include (1) development of HPLC systems which permit resolution of nearly all 20 PTH amino acid derivatives and which are sensitive to approximately 1 nmol[17] and (2) development of several new reagents, fluorescamine and *o*-phthalaldehyde, for amino acid detection which permit a 50- to 100-fold increase

[9] S. Oroszlan, T. Copeland, M. R. Summers, and G. W. Smythers, *in* "Solid-Phase Methods in Protein Sequence Analysis" (R. A. Laursen, ed.), p. 179. Pierce Chemical Co., Rockford, IL, 1975.

[10] R. A. Laursen, *Eur. J. Biochem.* **20**, 89 (1971).

[11] H. D. Niall, R. T. Sauer, J. W. Jacobs, H. T. Keutmann, G. V. Segre, J. L. H. O'Riordan, G. D. Aurbach, and J. T. Potts, Jr., *Proc. Natl. Acad. Sci. U.S.A.* **71**, 384 (1974).

[12] T. A. Springer, J. Kaufman, C. Terhorst, and J. L. Strominger, in press.

[13] J. Bridgen, *FEBS Lett.* **50**, 159 (1975).

[14] J. Bridgen, D. Snary, M. J. Crumpton, C. Barnstaple, P. Goodfellow, and W. F. Bodmer, *Nature (London)* **261**, 200 (1976).

[15] J. Bridgen, this volume [33].

[16] J. Bridgen, *Biochemistry* **15**, 3600 (1976).

[17] C. L. Zimmerman and J. J. Pisano, this volume [5].

in sensitivity.[18,19] These reagents used in conjunction with a solid-phase sequenator or with synthetic carrier and a liquid-phase sequenator should permit extended automated sequence analysis at the picomole level. In addition, a highly limited method involving the degradation of radioiodinated proteins permits the determination of tyrosine and histidine residues in the sequence of polypeptides available in only picomole quantities.[11]

The second microsequencing method described above is capable of sequencing 100 pmol or less of internally labeled polypeptides isolated in trace quantities from in vitro tissue incubation systems. This is 1000 times less than is required for conventional sequence analysis. The close agreement between results obtained by this method and by conventional microsequencing methods confirms its reliability.[20,21] It should be noted that the presence of SDS (frequently used to solubilize hydrophobic membrane proteins) does not interfere with microsequencing. Similar in vitro techniques have been employed by others for sequencing precursor molecules, such as precursor immunoglobulin[22] and proparathyroid hormone,[23] viral proteins,[20] and other cell surface molecules.[24]

The number and choice of amino acids employed in the incubation mixture is determined by the resolving system employed for the separation of the PTH-derivatives. The groups of 3 amino acids employed above are easily separated by one-dimensional TLC. If an HPLC system is employed, the number of amino acids may be increased since it provides better resolution than TLC.

This microsequencing method suffers from several drawbacks. First, some tritiated amino acids (e.g., aspartic, glutamic) are inefficiently incorporated into protein presumably due to the large cell pool for those particular amino acids. Nevertheless, at least 75% of the NH_2-terminal sequence may be determined by this method.[25] Second, the use of a re-

[18] S. Udenfriend, S. Stein, P. Böhlen, W. Dairman, W. Leimgriber, and M. Weigle, Science 178, 871 (1972).

[19] P. E. Hare, this volume [1].

[20] D. J. McKean, E. H. Peters, J. I. Waldby, and O. Smithies, Biochemistry 13, 3048 (1974).

[21] J. Silver and L. Hood, in "Contemporary Topics in Molecular Immunology" (H. N. Eisen and R. A. Reisfeld, eds.), Vol. 5, p. 35, Plenum, New York, 1976.

[22] I. Schechter, D. J. McKean, R. L. Guyer, and W. Terry, Science 188, 160 (1975).

[23] J. W. Jacobs, B. Kemper, H. D. Niall, J. F. Habener, and J. T. Potts, Jr., Nature (London) 249, 155 (1974).

[24] J. W. Uhr, J. D. Capra, E. S. Vitetta, J. Klein, D. G. Klapper, K. Artzt, E. A. Boyse, D. Bennett, and F. Jacob, Cold Spring Harbor Symp. Quant. Biol. 41, in press.

[25] E. S. Vitetta, J. D. Capra, D. G. Klapper, J. Klein, and J. W. Uhr, Proc. Natl. Acad. Sci. U.S.A. 73, 905 (1976).

stricted number of radioactive amino acids results in gaps in the sequence when the sequence residue is an amino acid which is not radioactively labeled. Thus, mechanical failure of the sequenator or accidental loss of the PTH-derivative may be misinterpreted as the *lack* of an identifiable residue. However, such failures or accidental losses may be monitored by inclusion of a natural "carrier" protein of known sequence for simultaneous sequence analysis.[20,23] Recently, Ballou *et al.* have developed a method for incorporating all 20 radioactive amino acids,[26] thus opening up the possibility of determining entire sequences by this method.

Acknowledgments

I thank Drs. A. Feeney and J. S. Fuhrman for their helpful and critical comments and Dr. L. E. Hood (Division of Biology, California Institute of Technology), in whose laboratory this work was accomplished, for providing intellectual stimulation and the necessary equipment and supplies. This work was supported by grants from NIH and NSF to Dr. Hood. The author has an Established Investigatorship Award from the American Heart Association.

[26] B. Ballou, D. J. McKean, E. F. Freedlender, and O. Smithies, *Proc. Natl. Acad. Sci. U.S.A.* **73**, 4487 (1976).

[28] Peptide Methodology with Braunitzer Reagents

By RAYMOND L. BLAKLEY

The use of the spinning-cup for sequence determinations on peptides of short or moderate chain length has the serious disadvantage that such peptides tend to be lost from the cup in the solvent extractions used during the degradation procedure. This problem is much greater for shorter peptides than for proteins or long peptides.[1,2] Although the problem is less severe in the manual procedures that are available, the speed and ease of the automated Edman degradation makes it desirable to adapt the basic procedure for use with shorter peptides. Various modifications of the degradation procedure have been introduced to minimize washing out of the peptide by solvent.[2] These include changes in the coupling conditions, so as to limit the volume and polarity of the solvents used for extraction. The major modification of this type is the use of a volatile buffer that does not need to be removed by extraction. A modified cleavage procedure is also used which involves a decreased amount of acid and only a single cleavage step.

[1] P. Edman, and G. Begg, *Eur. J. Biochem.* **1**, 80 (1967).
[2] H. D. Niall, this series Vol. 27, p. 942 (1973).

Another procedure for decreasing extractive losses of peptides is the modification of the peptide so as to decrease its solubility, particularly in the chlorobutane used for extraction of the thiazolinone. Loss of peptide from the cup during extraction varies greatly with the composition of the peptide. Limited tryptic peptides do not extract significantly and can usually be sequenced completely. These peptides are produced by tryptic cleavage of proteins or peptides in which lysine residues have been reversibly blocked, and consequently have COOH-terminal arginine. Thus, in the determination of the structure of dihydrofolate reductase of *Streptococcus faecium* all seven of the limited tryptic peptides which contain 14 residues or less could be sequenced completely.[3,4] In the case of two longer limited tryptic peptides, 20 residues out of a total of 40 and 22 of 44 residues could be determined, respectively. Of five other peptides in which arginine was near the amino terminus, four were completely sequenced by the Edman procedure. However, few peptides without arginine at or near the carboxyl terminus could be completely sequenced by the unmodified procedure, particularly if many hydrophobic residues were present.

When lysine is present in the peptide, conversion of the lysine ϵ-amino group to the hydrophobic phenylthiocarbamyl derivative during the first coupling step increases the tendency of the peptide to be extracted by solvents. To reverse this tendency Braunitzer[5-7] introduced the use of isothiocyanates containing one or more sulfonic acid groups. These convert the lysine residue to sulfonated thiocarbamyl derivatives, which are much more polar than phenylthiocarbamyl lysine residues. This causes a significant decrease in the solubility of the peptide in organic solvents with a consequent decrease in extractive losses.

Experimental Procedure

Of the four sulfonated isothiocyanates investigated by Braunitzer, two are available commercially from Pierce Chemical Company, Rockford, Illinois. They are 4-sulfophenylisothiocyanate (SPITC; I) and 3-isothiocyano-1,5-naphthalene disulfonate (II).

[3] J. M. Gleisner, D. L. Peterson, and R. L. Blakley, *J. Biol. Chem.* **250**, 4937 (1975).
[4] D. L. Peterson, J. M. Gleisner, and R. L. Blakley, *J. Biol. Chem.* **250**, 4945 (1975).
[5] G. Braunitzer, B. Schrank, and A. Ruhfus, *Hoppe-Seyler's Z. Physiol. Chem.* **351**, 1589 (1970).
[6] G. Braunitzer, B. Schrank, A. Ruhfus, S. Petersen, and V. Petersen, *Hoppe-Seyler's Z. Physiol. Chem.* **352**, 1730 (1971).
[7] G. Braunitzer, R. Chen, B. Schrank, and A. Stangl, *Hoppe-Seyler's Z. Physiol. Chem.* **353**, 832 (1972).

(I)   (II)

A typical procedure for derivatization of the peptide before proceeding with Edman degradation is as follows. The lysine-containing peptide (0.5–1.0 μmol) is dissolved in 200 μl of DMAA buffer (0.4 M N,N-di-methyl-N-allylamine in pyridine: water, 3:2 v/v, adjusted to pH 9.5 with trifluoroacetic acid). The peptide solution is mixed with 70 μl of freshly prepared aqueous 5% w/v SPITC sodium salt. This represents a 7- to 14-fold excess, assuming the presence of one lysine residue in the peptide. The mixture is heated under nitrogen at 53° for 1.5 hr. The sample is then diluted with a further 800 μl of DMAA buffer and loaded into the sequencer cup. The sample is dried in the sequencer cup by flushing with nitrogen for 3 min, running the vacuum subprogram, followed by the nitrogen subprogram, and finally extraction with 1-chlorobutane. The automated Edman program is then commenced at the first cleavage step. At all subsequent coupling steps the normal phenylisothiocyanate reagent is used. The Quadrol program, which is otherwise used only for longer peptides and proteins, gives satisfactory results for small peptides containing one or more lysine residues under these circumstances,[5] but the DMAA program may also be used.[8]

Results and Limitations

The improvement in the number of identifiable residues and in the yield of amino acid phenylthiohydantoin (PTH) at each step has been reported by many authors. Braunitzer originally reported[5] that whereas Phe-Phe-Tyr-Pro-Thr-Lys-Ala gave only a 5% yield of PTH-Phe in the first step of a Quadrol program without the use of SPITC, use of the reagent permitted 6 steps with the same program, with a 22% yield of free Ala from the cup at step 6. Inman et al.[8] and Powers[9] have also reported greatly improved results with SPITC, though the latter found that the results were still not as good as with the solid phase method. In sequencing the peptide Phe-Gln-Lys-Trp-Gln-Lys-Met-Ser-Lys-Val-Val,

[8] J. K. Inman, J. E. Hannon, and E. Apella, *Biochem. Biophys. Res. Commun.* **46,** 2075 (1972).
[9] D. A. Powers, *in* "Solid-Phase Methods in Protein Sequence Analysis" (R. A. Laursen, ed.) p. 99. Pierce Chemical Co., Rockford, IL, 1975.

we found that with the unmodified Quadrol program only the first three residues could be identified, whereas after initial coupling with SPITC the first 9 residues could be identified.

The method gives the best results if lysine is at the carboxyl terminus. If lysine is present in the interior of the peptide, the modified procedure will, of course, have no further effect on retention of the residual peptide in the cup once the last lysine residue has been cleaved off. The method works equally well for aminoethylcysteine at or near the carboxyl terminus.[8] When SPITC treatment is used, the lysine residues and the amino-terminal residue cannot be identified by GLC because of their low volatility, and if TLC is used for identification authentic sulfoPTH derivatives must be used for reference. If the residues are identified by hydrolysis to the parent amino acid, yields are found to be anomalously low for the first step and those involving lysine, probably because of less efficient extraction of the sulfonated anilinothiazolinones from the cup.

[29] New Supports in Solid-Phase Sequencing[1]

By WERNER MACHLEIDT and ELMAR WACHTER

Since its introduction by Laursen[1a,2] the automatic solid-phase Edman degradation of peptides and proteins has been successfully used in many laboratories (see Laursen[3] and Laursen *et al.*[4] for comprehensive reviews). The growing interest in this method is stimulated by the availability of commercial solid-phase sequenators, new coupling procedures (this volume [30]) and new supports for the immobilization of peptides and proteins.

The solid support originally used by Laursen was a (2-aminoethyl) aminomethyl polystyrene (I) prepared from chloromethylated cross-linked polystyrene (Merrifield resin) by reaction with ethylenediamine[1a] (Fig. 1). The swelling properties of this resin were improved by the addition of aryl amino groups[2] (II). Both supports have been replaced by

[1] Work of the authors reviewed in this article was supported by the Deutsche Forschungsgemeinschaft, Sonderforschungsbereich 51, "Medizinische Molekularbiologie und Biochemie."
[1a] R. A. Laursen, *J. Am. Chem. Soc.* **88**, 5344 (1966).
[2] R. A. Laursen, *Eur. J. Biochem.* **20**, 89 (1971).
[3] R. A. Laursen, *in* "Immobilized Enzymes, Antigens, Antibodies and Peptides" (H. H. Weetall, ed.), p. 567. Dekker, New York, 1975.
[4] R. A. Laursen, A. G. Bonner, and M. J. Horn, *in* "Instrumentation in Amino Acid Sequence Analysis" (R. N. Perham, ed.), p. 73. Academic Press, New York, 1975.

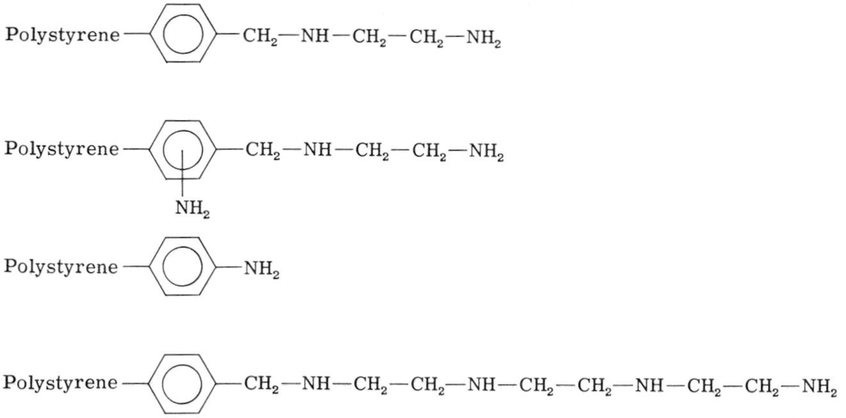

Polystyrene—⟨○⟩—CH_2—NH—CH_2—CH_2—NH_2 (I)

Polystyrene—⟨○⟩—CH_2—NH—CH_2—CH_2—NH_2 (II)
NH_2

Polystyrene—⟨○⟩—NH_2 (III)

Polystyrene—⟨○⟩—CH_2—NH—CH_2—CH_2—NH—CH_2—CH_2—NH—CH_2—CH_2—NH_2 (IV)

FIG. 1. Polystyrene supports.

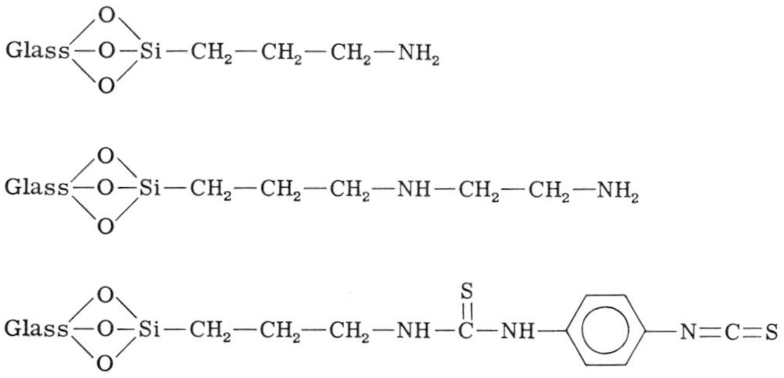

Glass⟨—O—Si—CH_2—CH_2—CH_2—NH_2 (V)

Glass⟨—O—Si—CH_2—CH_2—CH_2—NH—CH_2—CH_2—NH_2 (VI)

Glass⟨—O—Si—CH_2—CH_2—CH_2—NH—$\overset{\overset{S}{\|}}{C}$—NH—⟨○⟩—N=C=S   (VII)

FIG. 2. Porous glass supports.

aminopolystyrene[5] (III) and triethylenetetramine polystyrene[6] (IV), which are now widely used.

Supports with a rigid inorganic matrix are prepared by derivatization of porous glass[7] with aminoalkylsilanes.[8] Supports of this type are 3-aminopropyl glass[9] (V) and *N*-(2-aminoethyl)-3-aminopropyl glass[10] (VI) (Fig. 2). Reaction of these derivatives with *p*-phenylene diisothiocyanate leads to activated isothiocyanato glasses[9] (e.g., VII).

More recently two polyacrylamide-based supports have been applied to solid-phase sequencing, a cross-linked β-alanylhexamethylenediamine polydimethylacrylamide (VIII) designed for solid-phase peptide syn-

[5] R. A. Laursen, M. J. Horn, and A. G. Bonner, *FEBS Lett.* **21**, 67 (1972).
[6] M. J. Horn and R. A. Laursen, *FEBS Lett.* **36**, 285 (1973).
[7] W. Haller, *Nature (London)* **206**, 693 (1965).
[8] P. J. Robinson, P. Dunnhill, and M. D. Lilly, *Biochim. Biophys. Acta* **242**, 659 (1971).
[9] E. Wachter, W. Machleidt, H. Hofner, and J. Otto, *FEBS Lett.* **35**, 97 (1973).
[10] J. Bridgen, *FEBS Lett.* **50**, 159 (1975).

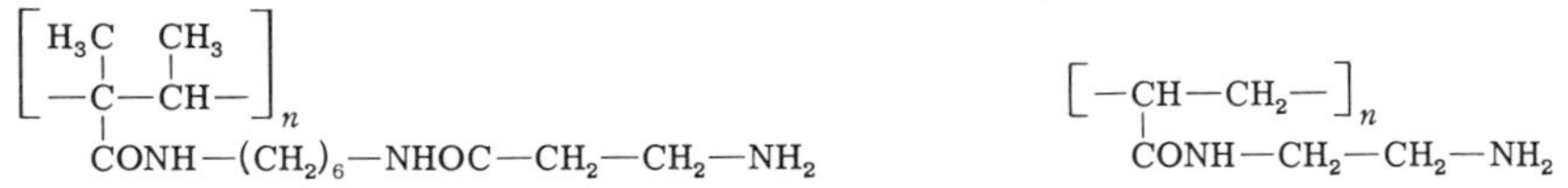

$$\begin{bmatrix} \overset{H_3C}{\underset{\underset{CONH-(CH_2)_6-NHOC-CH_2-CH_2-NH_2}{|}}{-\underset{|}{C}-\underset{|}{CH}-}} \overset{CH_3}{} \end{bmatrix}_n \qquad \begin{bmatrix} -\underset{\underset{CONH-CH_2-CH_2-NH_2}{|}}{CH}-CH_2- \end{bmatrix}_n$$

(VIII)   (IX)

FIG. 3. Polyacrylamide supports.

thesis[11] and an N-aminoethyl polyacrylamide (IX) synthesized from a commercial carboxylic cation exchange resin[12] (Fig. 3).

Synthesis of Supports

Aminopolystyrene[3,5]

The aminopolystyrene resin is prepared by nitration of polystyrene with nitric acid and reduction of the resulting nitropolystyrene with $SnCl_2$.

Fifty grams of 1% cross-linked polystyrene beads (Bio-Beads S-X1, minus 400 or 200–400 mesh, Bio-Rad), are added slowly (over 15 min) with stirring to 600 ml of fuming nitric acid (90%), keeping the temperature below 2° in an ice bath. The suspension is stirred for 1 hr at 0°, then diluted with ice (about 2000 ml) and poured through a fritted-glass filter. The resin is washed several times with dioxane and water. Finally the resin is washed with methanol and dried. Typical yield: 59 g of dry resin.

The nitro resin is suspended in 1000 ml of dimethylformamide (DMF) and heated to 75°. A solution of 420 g of $SnCl_2 \cdot 2H_2O$ in 350 ml of warm DMF is slowly added to the stirred suspension (exothermic reaction). Then the temperature is raised and kept at 140–150°. After 15 min the suspension is cooled and 350 ml of concentrated HCl are added. The mixture is heated again at 100° for 1 hr. Then the resin is washed thoroughly on a fritted-glass filter. Tin salts are removed by washings with water and HCl. Neutralization is achieved by washing the resin with pyridine–water (1:1) followed by DMF–triethylamine (3:1) until the washings do not contain chloride ions. Finally, the yellow-brown product is washed with water and methanol and dried at 60° under vacuum. Typical yield: 44 g of dry resin.

The aminopolystyrene resin is stable and can be stored for years under refrigeration. A ready-to-use aminopolystyrene sequencing resin may be purchased (Pierce).

[11] E. Atherton, J. Bridgen, and R. C. Sheppard, *FEBS Lett.* **64**, 173 (1976).
[12] J.-C. Cavadore, J. Derancourt, and A. Previero, *FEBS Lett.* **66**, 155 (1976).

Triethylenetetramine (TETA) Polystyrene Resin[6]

TETA resin is easily prepared from chloromethyl polystyrene by reaction with triethylenetetramine. Chloromethylated polystyrene may be synthesized from 1% cross-linked polystyrene (e.g., Bio-Beads S-X1, minus 400 or 200–400 mesh, Bio-Rad) according to the procedure given by Laursen.[13] The chloromethylation reaction can be controlled by varying the amount of $SnCl_4$ added or the reaction temperature and time. The chloromethyl derivative should be assayed for its chlorine content.[14] Chloromethyl groups, 1–2 meq per gram of resin, seem to be satisfactory. A higher degree of chloromethylation leads to poor swelling properties resulting from an increased cross-linkage by methylene bridges.[3,15]

Chloromethylated polystyrene beads (Merrifield resin) are commercially available at various degrees of chloromethylation (e.g., Pierce, Bio-Rad) and may be used as starting material for the preparation of TETA resin.

According to Horn and Laursen,[6] 10 g of chloromethyl polystyrene and 125 ml of triethylenetetramine are stirred for 30 min at room temperature and then heated on a steam bath for 1.5 hr. The resin is filtered, washed with methanol, stirred with 20 ml of triethylamine, and filtered again. Then the resin is washed thoroughly with methanol, water, and methanol again and is dried under vacuum at 60° overnight. Typical yield: 11.5 of dry white resin.

TETA resin should be stored in small batches in a freezer. Not more than a few months' supply should be prepared from the stable chloromethyl derivative. Immediately before use, the TETA resin should be washed with DMF to remove traces of nonbound triethylenetetramine which bleeds out of the resin and would react with the peptide.[4]

A TETA sequencing resin prepared from minus 400 mesh polystyrene beads is commercially available (Pierce).

Aminopropyl Glass[9] *and Aminoethylaminopropyl Glass*[10]

Aminopropyl glass (APG) and aminoethylaminopropyl glass (β-APG) are prepared by immersing porous glass beads[16] with a mean pore diameter of 75 Å (CPG 10–75, 200–400 mesh, Electro-Nucleonics Inc.,

[13] R. A. Laursen, this series Vol. 25, p. 344.
[14] J. M. Stewart and J. D. Young, *in* "Solid-Phase Peptide Synthesis." p. 55. Freeman, San Francisco, 1969.
[15] J. A. Patterson, *in* "Biochemical Aspects of Reactions on Solid Supports" (G. R. Stark, ed.), p. 189. Academic Press, New York, 1971.
[16] Cleaning the glass beads with HNO_3–HCl prior to aminosilylation does not lead to significant improvement with respect to capacity or interfering by-products in the degradation.

Fairfield, NJ, distributed by Serva, Heidelberg) in a 2% solution of 3-aminopropyl triethoxysilane or N-(2-aminoethyl)-3-aminopropyl triethoxysilane (Pierce), respectively, in acetone. After evaporation of the acetone at room temperature, the glass beads are washed thoroughly with acetone on a fritted-glass filter and dried under vacuum at room temperature.

The capacity of the aminoalkyl glasses is in the range of 0.2 μmol of amino group per milligram of glass. Both APG and β-APG are commercially available (Pierce). They are perfectly stable when stored in a freezer.

β-Alanyl-hexamethylenediamine-polydimethylacrylamide Resin[11]

The cross-linked t-butyloxycarbonyl (Boc)-β-alanyl-hexamethylene-diamine-polydimethylacrylamide resin is synthesized by persulfate-initiated emulsion copolymerization of a mixture of monomer (dimethyl-acrylamide), cross-linking agent (N,N'-bisacryloylethylenediamine), and functionalizing agent (N-t-butyloxycarbonyl-β-alanyl-N'-acryloylhexamethylenediamine).[17,18]

A solution of 54.7 g of freshly distilled dimethylacrylamide, 11.0 g of N,N'-bisacryloylethylenediamine, and 6.0 g of ammonium persulfate in 500 ml of water is mixed with a solution of 33.0 g of cellulose acetate butyrate and 13.5 g of N-t-butyloxycarbonyl-β-alanyl-N'-acryloylhexamethylenediamine in 1000 ml of 1,2-dichloroethane at 52° under nitrogen and stirred at 500 rpm for 4.25 hr. On this scale of preparation the resin is obtained in a largely amorphous rather than in a beaded form.[11] The β-alanine content of the polymer is 0.24 mmol/g.

Prior to use, the Boc-group is removed by treatment of the support with 1 N HCl-acetic acid.[19]

N-Aminoethyl Polyacrylamide Resin[12]

The N-aminoethyl polyacrylamide support is synthesized from the commercial polyacrylic resin Bio-Rex 70 by reaction with thionyl chloride followed by aminolysis of the chlorinated resin with ethylenediamine.[12]

Ten grams of Bio-Rex 70 (200–400 mesh, Bio-Rad) are purified by heating at 50° for 2 hr in 50 ml of dioxane–2 N NaOH (3:1, v/v) and,

---

[17] E. Atherton, D. L. J. Clive, and R. C. Sheppard, *J. Am. Chem. Soc.* **97**, 6584 (1975).

[18] Several companies have expressed an interest in preparing this support commercially (J. Bridgen, personal communication).

[19] Deprotection was performed by subjecting the resin to steps 1–8 of the solid-phase synthesis cycle described by Atherton *et al.*[17]

after filtering, in 50 ml of dioxane–2 N HCl (3:1, v/v). The filtered resin is washed with water, acetone, and ethyl ether and dried under vacuum.

The purified resin (5 g) is suspended in 25 ml of thionyl chloride containing 5% pyridine and gently refluxed for 3 hr with exclusion of moisture. The chlorinated resin is collected on a fritted-glass filter, washed with benzene and dry ether, and dried under vacuum over KOH pellets overnight. Chlorine content[14] (after hydrolysis in boiling 1 N NaOH): 5 meq/g.

One gram of chlorinated resin is allowed to react with 10 ml of ethylenediamine at 40° for 2 hr with occasional shaking. The aminated resin is filtered, washed successively with water, acetone, trifluoroacetic acid, and ethyl ether, and dried. Complete amination is achieved as judged from the chlorine content in the separated excess of ethylenediamine and water washes.

p-Phenylene Diisothiocyanate (DITC)-Activated Supports

Isothiocyanato Glass (DITC-Glass).[9] Aminopropyl glass is added to a solution of p-phenylene diisothiocyanate (Eastman; 3 mol per mole of amino groups[20]) in DMF (distilled from P_2O_5; 2–3 volumes per volume of glass) and allowed to react for 3 hr at room temperature. The resulting intensely yellow isothiocyanato glass is washed successively with DMF and methanol on a fritted-glass filter and dried under vacuum overnight. Capacity: about 0.03 μmol of accessible isothiocyanato groups per milligram of glass. DITC-glass is stable for months when stored in a freezer.

Isothiocyanato Polydimethylacrylamide.[11] After removal of the Boc-group, 200 mg of the β-alanylhexamethylenediamine-polydimethylacrylamide resin are added in small portions to a gently stirred saturated solution of p-phenylene diisothiocyanate (Eastman) in 10 ml of DMF over a period of 1 hr and stirred for a further 20 min. The activated polydimethylacrylamide resin is washed with DMF and methanol on a sintered-glass funnel and dried under vacuum.

Properties of Supports

Coupling Performance

Capacity. The functional capacity of a support can be evaluated by reaction with an excess of model peptide or protein under optimal

---

[20] The 25-fold excess of DITC used in former experiments to prevent cross-linking of functional amino groups within the support turned out to be unnecessary. On the contrary, a large excess of DITC leads to an increased background in the degradation.

coupling conditions. The amount of covalently bound peptide or protein is determined by amino acid analysis of a hydrolyzed aliquot from the support[21] after careful washes with the coupling medium, methanol, and, finally, trifluoroacetic acid.[22]

By this method, about 30 nmol of reactive isothiocyanato groups per milligram have been found for DITC-activated aminopropyl glass prepared from porous glass with 75 Å pore diameter.[23,24] The functional capacity of this DITC-glass is virtually the same for a variety of models ranging from glycine methyl ester to myoglobin. Hence even the attachment of proteins like myoglobin does not require glasses of higher porosity, which have a lower specific internal surface and do not perform as well as CPG 10–75. Several model proteins have been coupled to DITC-glass and degraded in the solid-phase sequenator, e.g., cytochrome c from *Candida krusei* (35 steps),[25] sperm whale myoglobin (70 steps),[26] and hen egg white lysozyme (50 steps).[26,27]

With polystyrene supports peptide loads of 5–10 nmol/mg have been obtained.[2,4,13] A minimum loading of 8.5 nmol (insulin B chain) per milligram of aminopolystyrene resin and of 5 nmol (insulin A chain) per milligram of TETA polystyrene resin are specified for commercially available products (Pierce).

Cross-linked polystyrene supports have a low capacity for large peptides and proteins, which apparently cannot penetrate the polymer matrix. A practical upper limit seems to be 30–40 residues for aminopolystyrene and about 70 residues for the TETA resin.[4,28]

---

[21] Heterogeneity of peptide loading within the sample may lead to erroneous results, which can be minimized by careful mixing of the beads prior to aliquotation.

[22] In the case of DITC-coupled peptides the trifluoroacetic acid wash effects cyclization of the NH_2-terminal residue and prevents an eventual partial blockage by oxidative desulfurization prior to sequencing.

[23] This functional capacity of 30 nmol/mg resulting from 0.2 μmol/mg of amino groups is remarkably high as compared to the average peptide load of 5 nmol/mg obtained on polystyrene-derived supports with about 1 μmol/mg of amino groups. Apparently only a very small fraction of amino groups is available for peptide attachment on cross-linked polystyrene supports.

[24] E. Wachter, H. Hofner, and W. Machleidt *in* "Solid-Phase Methods in Protein Sequence Analysis" (R. A. Laursen, ed.), p. 31. Pierce Chemical Co., Rockford, IL, 1975.

[25] W. Machleidt, E. Wachter, M. Scheulen, and J. Otto, FEBS. *Lett.* **37,** 217 (1973).

[26] W. Machleidt, H. Hofner, and E. Wachter, *in* "Solid-Phase Methods in Protein Sequence Analysis" (R. A. Laursen, ed.), p. 17. Pierce Chemical Co., Rockford, IL, 1975.

[27] E. Wachter, unpublished results.

[28] This limit does not apply to macroporous polystyrene resins. The authors have performed some preliminary experiments with a macroporous aminoethylenediamine resin (experimental resin provided by E. Merck, Darmstadt; capacity: 0.9 meq amino groups per gram). The myoglobin capacity of the isothiocyanato

Coupling yields of insulin B chain (30 residues) and lysozyme (129 residues) to DITC-activated β-alanylhexamethylenediamine-polydimethylacrylamide resin were estimated to be approximately equal and higher than those obtained with aminopolystyrene and aminopropyl glass in the same laboratory.[11]

Attachment yields between 60% and 80% were reported for proteins like lysozyme, apomyoglobin, and human globin α-chain coupled to the N-aminoethyl polyacrylamide support after activation with DITC. Better results were obtained by DITC-activation of the proteins rather than by activation of the support.[12]

Choice of Coupling Media. A common prerequisite to efficient coupling reactions is the free accessibility of matrix-bound functional groups in the reaction medium.

Polystyrene-derived supports swell sufficiently in DMF or pyridine, but not in aqueous media. In some cases it may be difficult to find solvents that solubilize the peptide and effect a sufficient swelling of the support as well.

Coupling reactions with the polyacrylamide-based supports have been performed in aqueous media such as pyridine–water $(1:1)$[12] or 0.4 M N,N'-dimethylallylamine-trifluoroacetic acid buffer, pH 9.5, in 60% pyridine[11] without any preswelling of the support.

Porous glass support are accessible to all kinds of solvent. The reaction medium can be freely selected according to the solubility of the peptide and the requirements of the coupling chemistry. However, the best attachment yields to porous glass supports have been obtained in completely anhydrous media, such as DMF or nitromethane, provided the peptide is soluble.[24] Most peptides, even larger ones, are rendered soluble in DMF (at least for a short period) by an incubation in anhydrous trifluoroacetic acid (60 min at room temperature) followed by vacuum evaporation (over KOH pellets) of the acid immediately before the coupling reaction. This procedure, originally designed for the lactonization of homoserine peptides,[6] should be generally applied prior to COOH-terminal coupling with carbodiimides and coupling to DITC-glass. Coupling yields between 60% and nearly 100% were obtained with peptides of all sizes.[27,29,30] The widely accepted view that porous glass supports as such are inferior to polystyrene-derived supports with regard

resin prepared by DITC-activation of the macroporous support was somewhat higher than that of DITC-glass.

[29] K. Hochstrasser, G. Bretzel, E. Wachter, and S. Heindl, *Hoppe-Seyler's Z. Physiol. Chem.* **356**, 1865 (1975).

[30] G. Wunderer, H. Fritz, E. Wachter, and W. Machleidt, *Eur. J. Biochem.* **68**, 193 (1976).

to the attachment of smaller peptides (4–10 residues)[24,31] may have to be revised.

The use of DITC-activated glass rather than DITC-activation of the peptide has several advantages. The activated glass is prepared under conditions favoring a maximum degree of substitution with reactive isothiocyanato groups. In the subsequent attachment procedure conditions can be optimized with regard to the properties of the peptide unhindered by the solubility of DITC. The peptide is reacted with a large excess of glass-bound isothiocyanato groups without risk of cross-linking or even polymerization of the peptide. Nonbound peptide may be regained from the washings as it was before.

Other activated porous glass supports which are commercially available (e.g., N-hydroxysuccinimidyl ester CPG, p-nitrophenyl ester CPG, and stable diazonium salt CPG from Pierce) might prove useful in solid-phase sequencing. N-Hydroxysuccinimidyl ester CPG has been used for the attachment of peptides and proteins prior to sequencing from the carboxyl terminus.[32,33]

DITC-activation of the support in a separate reaction prior to peptide or protein attachment has also been used with the β-alanylhexamethylenediamine-polydimethylacrylamide resin.[11]

Choice of the Coupling Reactions. Attachment yields obtained with different coupling reactions are affected by the reactivity of the functional amino groups of the support.

It is general experience that aminopolystyrene is most useful for DITC-coupling (after DITC-activation of the peptide)[5] whereas the TETA resin is the support of choice for all kinds of COOH-terminal coupling.[4] In the case of the TETA resin this may be partially explained by the increased reactivity of the functional amino groups toward activated carboxyls due to the 1,2-diamine structure.[34]

The same effect may favor the coupling of homoserine peptides to β-APG.[10] Previous observations provided evidence that β-APG might be superior to APG for the attachment of large peptides (more than 40 residues).[10,24] From an extended experience this must remain an open question. Various large peptides have been coupled to APG, including a 180-residue fragment from collagen in 70% yield. Aminopropyl glass has been successfully used for COOH-terminal coupling, preferably with

---

[31] B. Wittmann-Liebold and A. Lehmann, *in* "Solid-Phase Methods in Protein Sequence Analysis" (R. A. Laursen, ed.), p. 81. Pierce Chemical Co., Rockford, IL, 1975.

[32] M. J. Williams and B. Kassell, *FEBS Lett.* **54**, 353 (1975).

[33] M. Rangarajan and A. Darbre, *Biochem. J.* **157**, 307 (1976).

[34] M. I. Page and W. P. Jencks, *J. Am. Chem. Soc.* **94**, 8818 (1972).

N-hydroxysuccinimide- and 1-hydroxybenzotriazole-catalyzed carbo-diimide activation.[27]

Degradative Performance

Physical Properties. A sequencing support to be used in a reaction column should be available in small, narrow-sized beads which do not change their volume in the different solvents used in the degradation cycle. The functional groups should be freely accessible to all reagents and solvents.

These requirements are only partially met by the cross-linked polystyrene resins.[35] The polystyrene-derived supports swell appreciably in pyridine, 1,2-dichloroethane, and trifluoroacetic acid, but much less in methanol.[3] The changing bead volume during one degradation cycle would lead to solvent channeling or column blockage. Therefore these resins have to be diluted with glass beads (20-fold excess by weight) in the reaction column. Since the packing quality of the reaction column is mainly determined by the 200–235-mesh glass beads used for dilution, the particle size of the sequencing resin itself seems to be not very critical. Minus 400-mesh and 200–400-mesh particles have been used with equal success. For polystyrene resins diluted with glass beads column back pressures up to 150 psi may be expected at the usual flow rates.[26]

Dilution with glass beads is also necessary with the β-alanylhexa-methylenediamine-polydimethylacrylamide resin which has been used in a largely amorphous rather than in a beaded form.[11]

Porous glass beads retain a virtually constant volume throughout the degradation cycle. They can be employed undiluted in homogeneously packed small reaction columns. Typically a 0.27×7-cm column filled with about 200 mg of aminopropyl glass or DITC-glass prepared from 200–400-mesh CPG 10-75 is used.[26] The column back pressure is always below 10 psi at the maximum flow rate of 0.65 ml/min. The short wetting time (below 1 min) of the small column favors a quick exchange of solvents and effective washes. As a single washing solvent methanol has been used, but impurities can be further reduced by an additional benzene wash.[27]

Following protein attachment in aqueous media, both porous glass supports[26] and N-aminoethyl polyacrylamide resin[12] have been employed in entirely nonaqueous degradation cycles (e.g., DMF/N-ethylmorpholine

---

[35] The macroporous polystyrene resin tested by the authors had mechanical properties similar to those of porous glass beads and was used undiluted in the same short reaction column.

9:1 instead of sequencing buffer[2]). Such properties of sequencing supports may be important when reagents other than phenyl isothiocyanate are used for the degradation.[36,37]

Chemical Stability. Polystyrene-derived supports have been preferred for use in solid-phase sequencing because they are stable to trifluoroacetic acid. Most polyacrylamide-based supports which are successfully utilized for affinity chromatography will be hydrolyzed in the cleavage phase of the Edman degradation. The β-alanylhexamethylenediamine-polydimethylacrylamide resin, however, and the *N*-aminoethyl polyacrylamide support prepared from commercial Bio-Rex 70 resin seem to be reasonably stable in degradations including a trifluoroacetic acid cleavage step.[11,12]

The chemical stability of the support-functional group bond was tested for aminopolystyrene and aminopropyl glass with radioactive glutamic acid attached to the DITC-activated supports.[24] The radioactivity released during one degradation cycle is somewhat higher for the polystyrene resin than for the porous glass support. The maximum release from porous glass is found after the alkaline coupling phase,[38] whereas almost no radioactivity is released during the trifluoroacetic acid phase. With the polystyrene resin, there are about equal losses after the coupling as well as after the cleavage phase. From these experiments a peptide loss of about 0.1–0.2% per degradation cycle is to be expected on porous glass supports. This small value is not significant in terms of repetitive yield.

The instability of supports may lead to interfering backgound in the analytical procedures employed for the identification of phenylthiohydantoins (PTH's). To a varying degree the TETA resin produces an artifact which elutes with PTH-threonine on thin-layer chromatography.[4,39] Using radiolabeled phenyl isothiocyanate for the degradation, radioactive breakdown products of the phenylthiocarbamyl groups from polystyrene resins are observed if the excess amino groups have not been blocked with methyl isothiocyanate prior to sequencing.[2,13]

[36] A. Previero and J.-C. Cavadore, *in* "Solid-Phase Methods in Protein Sequence Analysis" (R. A. Laursen, ed.), p. 63. Pierce Chemical Co., Rockford, IL, 1975.

[37] A. Previero, A. Gourdol, J. Derancourt, and M.-A. Coletti-Previero, *FEBS Lett.* **51**, 68 (1975).

[38] This should be expected from the well-known lability of glass at alkaline pH. Coupling reactions to porous glass supports should be performed at a pH below 10.

[39] An unknown product with the mass of PTH-dehydrothreonine (218) is found in constant amounts in the mass spectra from degradations on aminopropyl glass as well as DITC-glass. This compound runs with PTH-dehydrothreonine in gas chromatography and contains sulfur (as judged from the signal obtained with a sulfur-sensitive detector). It may be a by-product from the degradation rather than a breakdown product from the support.

Blocking the excess amino groups of loaded aminopropyl glass prior to the degradation has been omitted without any adverse effect on the identification of PTH's by gas chromatography and mass spectrometry. The same applies to the blocking of loaded DITC-glass with ethanolamine which had been performed in previous experiments.[25]

Degradative Efficiency. The degradative efficiency of solid-phase sequencing on a certain type of support is characterized by the repetitive yield and overlap of PTH's obtained in quantitated degradations. These parameters may vary considerably within each amino acid sequence. Significant conclusions should be drawn on the basis of mean values derived from extended degradations of longer peptides and proteins or from a sufficient number of experiments with different short peptides.

Both kinds of data are available for porous glass supports. They point very constantly to a mean repetitive yield between 94% and 95% after background subtraction. The mean overlap of 1.0% per cycle is not higher than reported for typical liquid-phase sequenator runs.[40] Apparently the reactions and washings in a porous glass column may reach a similar degree of completeness as when performed in the spinning-cup of a liquid-phase sequenator. The same applies to the so-called nonspecific acidolytic cleavage of peptide bonds which generates background in degradations of large peptides and proteins (Fig. 4).

In degradations of proteins like myoglobin on DITC-glass the polypeptide chain is multiply attached at many lysine residues. There is no evidence, however, that peptides bound via a single lysine side chain are degraded with lower repetitive yield due to an instability of the Si-O bond.[41] Unknown sequences of peptides coupled to porous glass supports have been determined through 30–40 residues.[24,30,41]

A strong overlap has been observed in degradations of a high load of a species of cytochrome *c* on DITC-glass.[25] It may be assumed that at high protein concentrations less favorable sites of the support become occupied and the degradation reactions do not go to completion. In the case of cytochrome *c*, 1–2 nmol of protein per milligram of support seemed to be optimal.

With porous glass supports the initial PTH yield obtained in the degradation varies between 30% and nearly 100% of the amount of peptide or protein bound to the support.[24] The reason for this variation

[40] O. Smithies, D. M. Gibson, E. M. Fanning, R. M. Goodfliesh, J. G. Gilman, and D. L. Ballantyne, *Biochemistry* **10**, 4912 (1971).
[41] W. Machleidt, R. Michel, W. Neupert, and E. Wachter *in* "Genetics and Biogenesis of Chloroplasts and Mitochondria" (T. Bücher, W. Neupert, W. Sebald, and S. Werner, eds.), p. 175. Elsevier/North Holland Biomedical Press, Amsterdam, 1976.

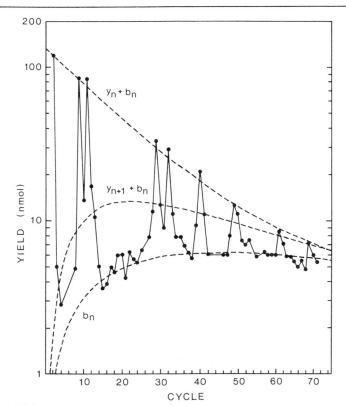

FIG. 4. Yields of PTH-leucine (determined by gas chromatography) from the solid-phase degradation of sperm whale myoglobin on DITC-glass [adapted from W. Machleidt, H. Hofner, and E. Wachter *in* "Solid-Phase Methods in Protein Sequence Analysis" (R. A. Laursen, ed.), p. 17. Pierce Chem. Co., Rockford, Illinois, 1975.]. Peaks emerging from the background indicate positions with a leucine residue in the sequence (2, 9, 11, 29, 32, 40, 49, 61, 69). The broken lines are theoretical curves for the background (b_n), the in-step yield plus background ($y_n + b_n$), and the ($n + 1$)-overlap yield plus background ($y_{n+1} + b_n$) calculated on the assumption of a mean repetitive yield $\alpha = 0.945$, a mean overlap $\beta = 0.01$, and a hydrolytic coefficient $h = 0.00017$ [according to O. Smithies, D. M. Gibson, E. M. Fanning, R. M. Goodfliesh, J. G. Gilman, and D. L. Ballantyne, *Biochemistry* **10**, 4912 (1971)]. An unexplained "prelap" is observed preceding positions 29, 40, and 49.

is still unclear. Losses during automatic sequencing and the following manual workup of samples have been excluded. Since low initial degradation yields have been observed with COOH-terminal-coupled as well as with DITC-coupled peptides, they cannot generally be explained by partial oxidative desulfurization of the NH$_2$-terminal residue prior to sequencing. In carefully quantitated experiments with model peptides attached to polystyrene-derived supports about 10% of the resin-bound peptide was not degraded.[42] The initial degradation yield was unaffected by coupling the peptide with a benzyloxycarbonyl-protected NH_2-terminal amino group. In automated degradations an initial PTH recovery of 60–70% has been found for peptides attached to polystyrene resins,[13] and of 50–75% for proteins coupled to the N-aminoethyl polyacrylamide support.[12]

For degradations of peptides (such as the oxidized insulin chains) on polystyrene resins average repetitive yields between 90% and 94% have been reported.[2,3] The practical limit for solid-phase Edman degradation of peptides on cross-linked polystyrene resins seems to be about 30 cycles.

A great number of unknown peptides has been sequenced by solid-phase Edman degradation on polystyrene supports.[3,4,31,43,44] Large peptides or proteins, if attached to polystyrene resins at all, are degraded with lower efficiency.[45]

Evidence for a high degradative efficiency has been found in experiments with the β-alanylhexamethylenediamine-polydimethylacrylamide support. The basic pancreatic trypsin inhibitor (58 residues) was sequenced through 29 steps.[11]

With the N-aminoethyl polyacrylamide resin proteins like lysozyme, apomyoglobin, and human globin α-chain were sequenced through 10–20 steps; data on the efficiency obtained in these degradations have not been published.[12]

Special Features

In addition to direct sequencing of peptides and proteins, the solid-phase technique offers some new strategies for the fragmentation of polypeptides and the purification of peptide fragments.[3,24,46] These strategies will help in sequencing large peptides with a minimum of effort.

For complete sequencing of large peptides it is often desirable to obtain a smaller fragment containing the COOH-terminal part of the peptide. In the case of a homoserine peptide or a peptide with a single COOH-terminal lysine this may be easily achieved by selective attachment of the COOH-terminal fragment after enzymic cleavage of the peptide with a suitable proteolytic enzyme.[44,47] A more general approach

[42] C. Meek, H. Jeschkeit, and A. Schellenberger, *Eur. J. Biochem.* **66**, 133 (1976).

[43] E. Schiltz and J. Reinboldt, *Eur. J. Biochem.* **56**, 467 (1975).

[44] H.-M. Lee and R. A. Laursen, *FEBS Lett.* **67**, 113 (1976).

[45] J. Bridgen, *Biochem. Soc. Trans.* **2**, 811 (1974).

[46] M. J. Horn, *in* "Solid-Phase Methods in Protein Sequence Analysis" (R. A. Laursen, ed.), p. 51. Pierce Chemical Co., Rockford, IL, 1975.

[47] M. J. Horn, *Anal. Biochem.* **69**, 583 (1975).

is the enzymic cleavage of immobilized peptides bound to the support by any of the available coupling reactions.

Porous glass derivatives or polyacrylamide-based supports should be used in such experiments because they are freely accessible to aqueous solvents. Promising results were obtained with the tryptic digestion of several larger cyanogen bromide fragments (more than 20 residues) bound to aminopropyl glass via their COOH-terminal homoserine lactone. Prior to degradation the loaded support was incubated with trypsin (1:100 molar enzyme:substrate ratio) in 0.1 M NH$_4$HCO$_3$, pH 8.1, at 37°. In one case less than 2.5% of the immobilized peptide remained uncleaved.[48] It seems that aminopropyl glass provides a favorable microenvironment for the selective cleavage of peptide bonds by trypsin or chymotrypsin.[49]

Variable degrees of cleavage were obtained with peptides which were multiply attached via their COOH-terminal and side chain carboxyls. Enzymic cleavage was poor with peptides coupled to DITC-glass. It seems that the hydrophobic microenvironment on this support limits or prevents an effective enzymic attack at peptide bonds.[50] A similar hydrophobic environment is established on aminopropyl glass by the reaction of excess amino groups with phenyl isothiocyanate during the first degradation cycle (or with methyl isothiocyanate prior to the degradation). This should be considered in enzymic cleavage experiments aiming at the creation of a new clean NH$_2$ terminus on a partially degraded peptide.

[48] E. Wachter and G. Rückl, unpublished results.
[49] A tripeptide NH$_2$-terminally attached to N-hydroxysuccinimidyl ester-activated porous glass was successfully degraded with carboxypeptidase A [M. J. Williams and B. Kassell, *FEBS Lett.* **54**, 353 (1975)].
[50] The peptide may be rendered susceptible to proteolytic enzymes by reaction of the excess isothiocyanato groups with alkyldiamines or other primary amines containing polar groups.

[30] Coupling Techniques in Solid-Phase Sequencing

By RICHARD A. LAURSEN

The technique of solid-phase Edman degradation[1] relies on efficient procedures for attaching peptides or proteins to amino-substituted insoluble supports (for a discussion of sequencing resins, see this volume

[1] R. A. Laursen, *Eur. J. Biochem.* **20**, 89 (1971).

[29]; for general reviews, see references 2–6). Since sequencing proceeds from the NH_2 terminus, it is desirable to attach peptides at the COOH terminus. If activation of COOH-terminal carboxyl is used for coupling, the problem then becomes one of selectively activating the COOH-terminal group without simultaneously activating the side chains of aspartic and glutamic acids. In fact, this problem represents the greatest factor limiting the utility of the solid-phase method. Although considerable effort has gone into the search for a completely general method, none has yet been found. Instead, several methods selective for certain types of peptides have been devised. As it happens, this situation has advantages that were not originally appreciated, since such selectivity allows one to remove certain peptides from mixtures and to sequence them without intervening purification steps.

Materials and Methods

Instrumentation. Degradation of peptides is accomplished on an automatic sequencer as described by Laursen.[1,3] These instruments are available commercially from Sequemat, Inc. (Watertown, Massachusetts), LKB Instruments, Inc. (Rockville, Maryland) and Anachem, Ltd. (Luton, U.K.).

Reagents. The reagents and solvents used in peptide coupling procedures are either purified as described earlier[4] or purchased from Pierce Chemical Co. (Sequanal grade). The main contaminants of concern are primary and secondary amines which may react with activated carboxyl groups. For sequencing, reagent grade chemicals are used.[4]

Resins. Three types of resins are commonly used: triethylenetetramine polystyrene (TETA resin),[7] aminopolystyrene,[8] and the aminoglass supports.[9,10] These are described in detail in this volume [29], and are available from Pierce.

Phenylthiohydantoin (PTH) analysis. PTH analysis is possible using

[2] R. A. Laursen, *in* "Immobilized Enzymes, Antigens, Antibodies and Peptides" (H. H. Weetall, ed.), p. 567. Dekker, New York, 1975.
[3] R. A. Laursen, A. G. Bonner, and M. J. Horn, *in* "Instrumentation in Amino Acid Sequence Analysis" (R. N. Perham, ed.), p. 73. Academic Press, New York, 1975.
[4] R. A. Laursen, this series Vol. 25, p. 344 (1972).
[5] R. A. Laursen and M. J. Horn, *in* "Protein Sequence Determination" 2nd ed. (S. B. Needleman, ed.), Vol. 2. Springer-Verlag, Berlin and New York, in press.
[6] R. A. Laursen, Ed., "Solid-Phase Methods in Protein Sequence Analysis." Pierce Chemical Co., Rockford, IL, 1975.
[7] M. J. Horn and R. A. Laursen, *FEBS Lett.* **36**, 285 (1973).
[8] R. A. Laursen, M. J. Horn, and A. G. Bonner, *FEBS Lett.* **21**, 67 (1972).
[9] E. Wachter, W. Machleidt, H. Hofner, and J. Otto, *FEBS Lett.* **35**, 97 (1973).
[10] J. Bridgen, *FEBS Lett.* **50**, 159 (1975).

thin-layer chromatography, gas–liquid chromatography, high-pressure liquid chromatography, and mass spectrometry (for a recent review, see Bridgen et al.[11]). Thin-layer chromatography of ^{35}S-labeled PTH's can be used for subnanomole quantities (see this volume [33]).

Coupling Procedures

NH₂-Terminal Blocking

In certain coupling procedures (e.g., carbonyldiimidazole) it is necessary to protect the peptide amino groups to prevent reaction with the coupling agent. Earlier we had felt that protection was necessary also to prevent reaction with the activated carboxyl. However, Shellenberger et al.[12] and others[13] have found that peptides can be coupled efficiently by the carbodiimide method without blocking the NH₂-terminal group. Recently we have found (unpublished) that amines, such as glycine methyl ester, react with carboxyl-activated peptides to an extent of only 1–2% even when the amine is present in 3-fold excess over the peptide. The reason for the slow reaction is simply that the reaction is bimolecular and the concentration of the reactants is very low (ca. 10^{-4} M). NH₂-terminal blocking is sometimes useful for improving the solubility of peptides in organic solvents.[13]

Procedure. Salt-free peptide samples are dissolved in a 2% triethylamine solution and evaporated to dryness under vacuum to remove traces of ammonia. To the residue is added 0.5 ml of DMF, 0.05 ml of N-methylmorpholine and 0.075 ml of t-butyloxycarbonyl (Boc) azide. The mixture is stirred by means of a small magnetic stirring bar until solution is complete. In cases where the peptide does not dissolve readily, up to 0.1 ml of water can be added. The tube is covered with Parafilm and heated at 50° for at least 5 hr, after which the solvent and excess reagents are removed by evaporation in a vacuum desiccator using a good vacuum pump. Evacuation of the desiccator should be cautious at first to prevent bumping. When the tube is dry, 0.5 ml more of DMF is added and evaporated. Drying under vacuum is continued for 10 hr or more at room temperature.

[11] J. Bridgen, A. P. Graffeo, B. L. Karger, and M. D. Waterfield, in "Instrumentation in Amino Acid Sequence Analysis" (R. N. Perham, ed.), p. 111. Academic Press, New York, 1975.

[12] A. Schellenberger, H. Graubaum, C. Meck, and G. Sternkop, Z. Chem. **12,** 62 (1972).

[13] E. Wachter, H. Hofner, and W. Machleidt, p. 31 of Laursen.[6]

Carbonyliimidazole Activation

Carbonyldiimidazole was first used for peptide activation in conjunction with solid-phase Edman degradation because of its reported[14] high yields and the absence of side reactions. Although this reagent does seem to couple peptides efficiently, it suffers from certain drawbacks. Since carbonyldiimidazole reacts rapidly with water and nucleophiles, such as amines, reactions must be carried out in scrupulously dry solvents, such as DMF, in which not all peptides are soluble. Also the peptide amino groups must be blocked, for example with Boc-azide. A more serious problem is that the side-chain carboxyls are also activated, with the result that glutamic acid becomes attached to the resin and aspartic acid undergoes cyclic imide formation, which prevents degradation past aspartic acid.[1]

Procedure. To a dry sample (up to 500 nmol) of Boc-peptide in a 13×100-mm test tube is added 0.2 ml of a solution carbonyldiimidazole in dry DMF (40 mg/ml). The tube is quickly flushed with dry nitrogen and carefully sealed with Parafilm. The mixture is stirred at room temperature for 30 min, and then 2 μl of water is added to destroy excess carbonyldiimidazole. After 10 min the mixture is immediately transferred to a second test tube containing 50 mg of TETA resin preswollen in 0.2 ml of DMF. The first tube is washed with 0.1 ml of DMF. The resin mixture is stirred at room temperature for at least 2 hr. Excess amino groups are then blocked as described below.

Carbodiimide Activation

Carbodiimides have been used in peptide chemistry for many years and have the advantage over carbonyldiimidazole that they are relatively insensitive to water. Thus coupling can be done in aqueous solution (usually at pH 5.0 with a water-soluble carbodiimide). These reagents share the common problem with carbonyldiimidazole that side chain carboxyls are also activated and become coupled to the resin; cyclic imide formation (except at Asp-Gly bonds) is not a problem, however. The major problem with carbodiimides is that coupling yields are extremely variable, and are often poor. The most likely explanation for this fact is rearrangement of the *O*-acylurea intermediate (Fig. 1) to the inactive *N*-acylurea. The degree of *N*-acylurea formation is probably related to the structure of the amino acid and to the solvent, but has not been studied in a systematic way in connection with solid-phase sequencing. To some extent one can control the side reaction (see Fig. 1).

---

[14] R. Paul and G. W. Anderson, *J. Am. Chem. Soc.* **82**, 4596 (1960).

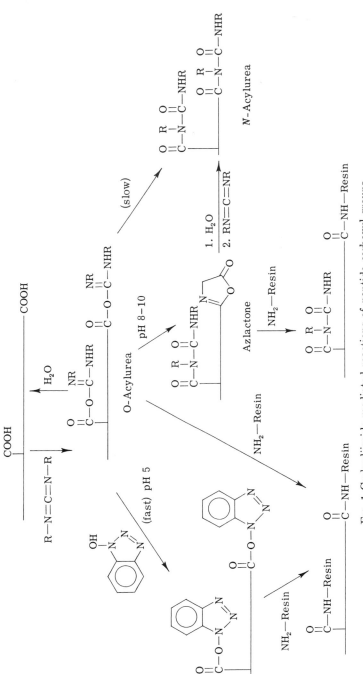

Fig. 1. Carbodiimide-mediated reactions of peptide carboxyl groups.

We have found recently (unpublished) that the addition of 1-hydroxybenzotriazole, a catalyst frequently used in peptide synthesis,[15] increases the rate of coupling to resins by a factor of 5–10, and sometimes also the yield, presumably by converting the peptide to an intermediate that cannot rearrange to an inactive form. On the other hand, N-acylurea formation can sometimes be made to work to advantage. If carbodiimide activation takes place at a high pH (8–10) in the absence of a nucleophile, the intermediate O-acylurea rearranges to the N-acylurea. However, at the COOH-terminal carboxyl, an azlactone can also form, and in this way one can effectively block the side chain and activate the COOH-terminal carboxyl in one step.[16] The observation that this procedure often occurs with low yields suggests that azlactone formation does not always compete favorably with rearrangement to the N-acylurea (Fig. 1).

Procedure A: Uncatalyzed (cf. reference 12). A 50-mg sample of TETA sequencing resin is washed by centrifugation with 1 ml of pyridine-HCl buffer (1.0 M, pH 5.0), then 8 ml of water and finally 2 ml of DMF. The excess reagents are pipetted off. The peptide sample (which may or may not be N-protected) is dissolved in 0.05 ml of pyridine-HCl buffer and 0.1 ml of DMF and added to the resin. The peptide sample tube is washed further with 0.1 ml of DMF, and this is also added to the resin. To the resin mixture is then added 0.1 ml of a freshly prepared solution containing 0.08 ml of DMF, 0.02 ml of H_2O, and 4 mg of N-dimethylaminopropyl-N'-ethyl carbodiimide. The reaction mixture is stirred at room temperature for about 4 hr and amino groups are blocked as usual with methyl isothiocyanate.

Procedure B: Catalysis with 1-Hydroxybenzotriazole. This procedure is identical to A above, except that the peptide sample is dissolved in 0.05 ml of pyridine-HCl buffer and 0.1 ml of DMF containing 1.0 mg of 1-hydroxybenzotriazole (Pierce).

Procedure C: Attachment with Simultaneous Side-Chain Blocking.[16] The Boc-peptide (up to about 250 nmol) is dissolved in 0.2 ml of DMF containing 1–2 μl of N-methylmorpholine. N-Ethyl-N'-dimethylaminopropyl carbodiimide hydrochloride (5 mg) is added and the reaction mixture is allowed to stand at 40°. After 90 min (or 30 min at 50°), 50 mg of TETA sequencing resin is added and this suspension is maintained at 40° for at least 3 hr. Excess resin amino groups are blocked as described below.

[15] G. A. Fletcher, M. A. Low, and G. T. Young, *J. Chem. Soc., Perkin Trans.* 1, 1162 (1973).
[16] A. Previero, J. Derancourt, M.-A. Coletti-Previero, and R. A. Laursen, *FEBS Lett.* 33, 135 (1973).

Attachment at Homoserine[7]

Cleavage of proteins at methionine with cyanogen bromide yields peptides which have homoserine at the COOH terminus. Treatment of the peptide with anhydrous trifluoroacetic acid results in a lactone that is sufficiently activated to undergo aminolysis by aliphatic amines[17] or sequencing resins such as TETA-polystyrene[7] or the aminoglass supports.[13] In this way peptides are selectively attached at the COOH terminus without the need for amino or carboxyl protection. Coupling yields are characteristically high (>80%), making this procedure the method of choice when homoserine peptides are encountered. As will be discussed under "strategies," another merit of the homoserine lactone method is that it can be used to selectively remove homoserine peptides from mixtures.

Procedure.[7] A solution (ammonia-free) of the COOH-terminal homoserine peptide (usually 20–200 nmol) is evaporated to dryness in a small vial or test tube, and the residue is redissolved in 1.0 ml of anhydrous trifluoroacetic acid. The solution is kept at room temperature for 1 hr, and the solvent is evaporated over KOH in a vacuum desiccator. The peptide sample is dissolved in 0.1 ml of DMF, and the solution is added to a 13 × 100-mm test tube containing 50 mg of TETA sequencing resin previously washed and swollen in 0.2 ml of DMF. The test tube is washed with 0.1 ml of DMF and 0.05 ml of triethylamine, which is added to the resin mixture, followed by 0.05 ml of DMF. (In cases where the peptide is poorly soluble in DMF, up to 20% of water can be added to promote solution.)

The resin mixture is then stirred at 45° for 2 hr. Excess amino groups on the resin are blocked as described below.

Diisothiocyanate Activation

Peptides that contain a lysine residue, such as those generated by cleavage of proteins with trypsin, can be coupled efficiently to supports using *p*-phenylene diisothiocyanate, which cross-links both the ε-amino groups of lysine and α-amino groups to the amino resin (aminopolystyrene gives the best results).[8] Since this compound is essentially a bifunctional Edman reagent, treatment of the bound peptide with trifluoroacetic acid results in cleavage of the first peptide bond, leaving the remainder of the peptide linked by the ε-amino group. If the lysine is COOH-terminal, the peptide can be degraded to the end; if it is internal, then the remainder of the peptide is liberated (and can some-

[17] M. J. Horn, *Anal. Biochem.* **69**, 583 (1975).

times be isolated) when degradation reaches lysine. In this process, the thiazolinones of both the NH_2-terminal residue and the lysines remain bound to the resin, and no phenylthiohydantoin is detected in these positions.

Proteins (and peptides) can be immobilized by reaction of the lysine amino groups with aminoglass supports which have first been activated by treatment with excess p-phenylene diisothiocyanate. Coupling yields by the diisothiocyanate method are generally good, and this procedure ranks with the homoserine lactone method in terms of reliability. The glass supports are usually better for peptides larger than about 30 residues, while small peptides seem to couple more efficiently to aminopolystyrene.

Diisothiocyanate coupling can be used for coupling other side chain amino groups, such as aminoethyl-cysteine or ornithine. Arginine in peptides can be deguanidinated to ornithine[8,18] by treatment with aqueous hydrazine, though this procedure is limited to smaller peptides because of side reactions (cleavage at asparagine, random peptide bond cleavage).[5]

p-Phenylene Diisothiocyanate Coupling of Peptides.[8] The peptide sample is evaporated to dryness in a 13 × 100-mm test tube and is redissolved in 0.5 ml of 5% triethylamine and evaporated again to remove traces of ammonia. The residue is then dissolved in 0.05 ml of N-methylmorpholine–water (1:1, adjusted to pH 9.5 with CF_3COOH). To this is added 1.0 mg (5 μmol) of p-phenylene diisothiocyanate (Pierce Sequanal grade; reagent from other suppliers should be recrystallized from acetic acid) in 0.1 ml of DMF. The solution is heated at 45° for 30 min and is transferred with a Pasteur pipette to a second (12 × 100-mm) test tube containing 35 mg of peptide resin (aminopolystyrene) previously swollen in 0.25 ml of DMF. The first tube is washed with 0.1 ml of DMF, which is also added to the resin. The resin mixture is then treated with methyl isothiocyanate to block resin amino groups as described below.

Hydrazinolysis of Arginine Peptides.[8] A peptide sample in a 13 × 100-mm test tube is heated in an oil bath at 75° for 15 min in 0.50 ml of 50% hydrazine. The solution is then evaporated on a rotary evaporator. Traces of hydrazine are removed by coevaporation twice with 0.5 ml of 5% aqueous N-methylmorpholine and finally by evacuation in a desiccator over P_2O_5 using an efficient vacuum pump. The residue is then attached to the peptide resin by the diisothiocyanate method described above. It is essential that all traces of hydrazine be removed, since it can react with the diisothiocyanate. Normally during coupling the re-

[18] H. R. Morris, R. J. Dickinson, and D. H. Williams, *Biochem. Biophys. Res. Commun.* **51**, 247 (1973).

action mixture becomes yellow; if all the hydrazine is not removed, the reaction remains colorless and yields are frequently very low.

Attachment of Proteins to Diisothiocyanate-Glass (cf. references 9, 10, and 19, and this volume [32]). The glass support (aminopropyl glass[9,19] or β-aminoethylaminopropyl glass) is activated by stirring with a 100-fold molar excess of *p*-phenylenediisothiocyanate over amino groups in DMF at room temperature for 90 min. Excess reagent is removed by filtration on a sintered-glass funnel and washing extensively with DMF. The peptide or protein is dissolved in *N*-methylmorpholine-water (1:1, v/v) adjusted to pH 9.0 with CF_3COOH, and this solution is added to the activated support. Approximately 0.5 mg of support is used per nanomole of peptide or protein amino group. The mixture is degassed for a few minutes in an evacuated flask or desiccator and is stirred at 45° for 30 min, after which 0.2 ml of ethanolamine is added to block excess NCS groups. The support is then washed several times with methanol by centrifugation and is dried under vacuum.

Combined Attachment Methods

In some cases the advantages of two of the procedures described above can be combined to overcome the disadvantages of both. Schiltz[20,21] has combined the diisothiocyanate and carbodiimide procedures to permit sequencing of nonlysine peptides. The peptide is first coupled by its NH_2 terminus to aminopolystyrene using *p*-phenylene diisothiocyanate. With the peptide bound to the resin the COOH-terminal is then coupled using a carbodiimide. This procedure is reported to be useful for larger peptides which do not couple efficiently with the carbodiimide alone.[20]

A second approach has been taken by Herbrink *et al.*,[22] who have combined the diisothiocyanate and homoserine lactone coupling methods. Peptides are treated with the bifunctional reagent *p*-isothiocyanato-benzoyl-DL-homoserine lactone (IBHL), which reacts with the amino groups of lysine peptides at the isothiocyanate end. The sequencing resin then reacts at the lactone end, giving a peptide cross-linked to the resin. An advantage of this process is that one need not add a large excess of cross-linking agent to avoid polymerization, as in the case of diisothiocyanate procedure (i.e., a 2-fold excess instead of a 50-fold excess).

[19] W. Machleidt, E. Wachter, M. Scheulen, and J. Otto, *FEBS Lett.* **37**, 217 (1973).
[20] E. Schiltz, p. 47 of Laursen.[6]
[21] E. Schiltz and J. Reinboldt, *Eur. J. Biochem.* **56**, 467 (1975).
[22] P. Herbrink, G. I. Tesser, and J. M. Lamberts, *FEBS Lett.* **60**, 313 (1975).

Diisothiocyanate-Carbodiimide Procedure.[20] The dry peptide is dissolved in 200 μl of DMF containing 2 μl of N-methylmorpholine. p-Phenylene diisothiocyanate (0.5 mg) in 50 μl of DMF is added, followed after 30 min at 50° by 30 mg of aminopolystyrene. After 60 min at 50°, the resin is washed by centrifugation with DMF, 1 M pyridine-HCl (pH 5), methanol and ether. To the dry resin are added 100 μl of DMF, 50 μl of 1 M pyridine-HCl (pH 5), 4 mg of N-dimethyl-aminopropyl-N'-ethylcarbodiimide in 80 μl of DMF and 20 μl of water. The suspension is stirred at 50° for 2 hr; amino groups are blocked as usual.

*p-Isothiocyanatobenzoyl-*DL-*homoserine Lactone (IBHL) Procedure.*[22] The peptide is dissolved in 100 μl of N-methylmorpholine–water (1:1; pH 10.7), and a 2-fold excess of IBHL[22] over amino groups in 100 μl of DMF is added. After 1 hr at room temperature, the mixture is evaporated to dryness under vacuum. The residue is dissolved in 1 ml of trifluoro-acetic acid, which is then evaporated under vacuum. The peptide is then coupled to 100 mg of aminoglass[9,10] using the conditions for homoserine lactone coupling (see above), except that a pH 9.7 buffer (N-methyl-morpholine–water, 1:1, acidified with trifluoroacetic acid to pH 9.7) is used.

Blocking of Excess Resin Amino Groups

After peptides have been attached to supports, the excess resin amino groups must be blocked to prevent consumption of phenyl isothiocyanate in the first cycle of the Edman degradation, which could lead to inefficient degradation yields. Isothiocyanates are ideal for this purpose, since, if they react with an unprotected NH_2-terminal, they give a derivative that is cleaved during Edman degradation. Although phenyl isothiocyanate would seem the logical reagent for this purpose, we have found that the resulting phenylthiourea resins give extraneous spots on thin-layer chromatography. Methyl isothiocyanate, on the other hand, seems to give less background, although UV absorbing material is found near PTH-threonine and the baseline. We have also used with some success (unpublished) p-dimethylaminophenyl isothiocyanate (Transworld Chemicals), the breakdown products of which are basic and are retained on the Dowex-50 columns[1] used to work up the phenylthio-hydantoins.

Procedure. To the peptide attachment mixture, containing resin (up to 50 mg), peptide, and activating agent, is added 0.1 ml of CH_3NCS–CH_3CN (1:1, v/v) and 0.1 ml of N-methylmorpholine–DMF (1:1, v/v). The mixture is stirred at room temperature for 75 min, and the resin is washed by centrifugation three times with 8-ml volumes of methanol, and finally with ether, the solvent being removed with a Pasteur pipette after each washing. The damp resin is then dried, by laying the tube on its side and allowing the ether to evaporate. At this point the resin is either stored in a refrigerator or mixed with glass beads and packed into a reaction column for sequencing.

Sequencing Strategies

Since a reliable, universal procedure for immobilizing peptides is not yet available, it is necessary to devise strategies that take maximal advantage of the procedures that are at hand. One of the more useful schemes makes use of the homoserine coupling method.

Large (more than 30 residues) cyanogen bromide peptides can generally be coupled to supports by either the homoserine or diisothiocyanate method and sequenced to give the first 20–30 residues at the NH_2 terminus. If in a second experiment a sample of the peptide is digested with trypsin (or chymotrypsin), and the resulting peptide mixture is subjected to the homoserine lactone coupling procedure, only the COOH-terminal homoserine peptide will be immobilized for sequencing. In this way it is relatively easy to obtain NH_2 and COOH-terminal sequences of cyanogen bromide peptides, which is of great help in their alignment with overlapping peptides, without having to isolate the fragments chromatographically. Longer COOH-terminal sequences can sometimes be obtained by blocking the lysines (or arginines) and then cleaving with trypsin.

Similarly, one can couple and selectively sequence COOH-terminal lysine peptides in a mixture after diisothiocyanate coupling. In this case all the other peptides also become attached by their α-amino groups, but the nonlysine peptides become detached after the first cycle of the Edman degradation. COOH-terminal lysine peptides containing methionine can be cleaved with cyanogen bromide and the mixture coupled by the homoserine lactone procedure, which will result in attachment of the NH_2-terminal fragment. The washings will contain the lysine fragment, which can be attached by the diisothiocyanate method. Alternatively, the intact peptide can be fixed at lysine and cleaved with cyanogen bromide on the support.

Future Developments

Still the greatest need in solid-phase sequencing is a completely general and reliable means of immobilizing peptides. A significant step in this direction could be made, ignoring for the moment the problem of side-chain carboxyl blocking, by finding a reagent that activates carboxyl groups without side reactions such as those noted with carbodiimides. The main problem with carbodiimides is the low attachment yield. Coupling of Asp and Glu side chains is really a problem only if both residues occur in the same peptide, since a gap in the sequence can usually be interpreted as an acidic residue. In preliminary studies of the carbodiimide reaction, F. Chin (unpublished) has found that if peptides are activated with a carbodiimide in the presence of a 2–3-fold excess of [^{14}C]glycine methyl ester before addition of the sequencing resin, about 2% of the carboxyls become labeled in a random manner. The remaining carboxyls remain free to couple to the resin. Thus during sequencing, the appearance of radioactivity in a cycle signals the presence of Asp or Glu in the sequence. Unfortunately, the overall coupling yields are unpredictable, as is the usual case in carbodiimide coupling.

Another approach may be the use of new types of resins. Horn[23] has used polymeric carbodiimides first described by Weinshenker and Shen.[24] In this case, activation is done under basic conditions which favor $O \rightarrow N$ acyl migration (cf. Fig. 1), and results in attachment of the peptides by an N-acylurea linkage. Presumably one could also pulse-label some of the side-chain carboxyls with a radioactive amine as described above. Another example is the use of the sulfonium resins described by Fontana et al.[25] Peptides react by attack of carboxylate on the benzylic sulfonium salt, resulting in linkage by an ester bond. Whether the ester bond would survive the conditions of the Edman degradation has not been determined, but there is a good chance it would, since benzyl esters are commonly used in solid-phase peptide synthesis.[26]

Besides a general attachment procedure, new selective methods would be of great benefit, since they can also be used to isolate particular peptides. A support specific for arginine would be very useful, in conjunction with the diisothiocyanate method, for sequencing tryptic peptides.

Acknowledgments

Portions of the work described here were supported by a grant from the National Science Foundation.

[23] M. J. Horn, p. 51 of Laursen.[6]
[24] N. M. Weinshenker and C.-M. Shen, *Tetrahedron Lett.*, 3281 (1972).
[25] A. Fontana, F. M. Veronese, and E. Boccu', *Z. Naturforsch.* **26b**, 314 (1971).
[26] R. B. Merrifield, *J. Am. Chem. Soc.* **85**, 2149 (1963).

[31] Alternative Reagents in Sequential Degradation on Solid-Phase Supports

By ALDO PREVIERO

Sequential analysis by solid-phase techniques[1] was initially proposed as a modification of the Edman degradation for automated sequencing of short peptides. The chemistry of solid-phase sequencing which has been developed essentially involves the preparation of suitable insoluble supports and the development of new or modified chemical reactions for the attachment of peptides to the support. The overall aim is the adjustment of Edman's conditions from a homogeneous to a heterogeneous reaction. For example, the reaction time and temperature have been optimized to complete the coupling reaction between the insolubilized peptide and phenylisothiocyanate and the same aqueous–nonaqueous buffers (i.e., pyridine/H_2O/N-alkylmorpholine trifluoroacetate are currently utilized as solvents).

The role of the solvent in the condensation step is not the same in the homogeneous as opposed to the heterogeneous reaction. The ideal solvent for the condensation reaction in a homogeneous phase can dissolve both the reagent and the peptide and is easily eliminated at the end of the reaction. Under heterogeneous conditions the solvent must dissolve only the reagent, must obviously assure a satisfactory reaction rate and eventually must achieve a good penetrability into the resin to which the peptide is linked. The first important consequence is that the condensation step in solid-phase procedures may be performed entirely in anhydrous media. This on the one hand eliminates the competitive nucleophilic attack of water, which is kinetically favored when compared to the desired reaction of the insolubilized terminal amino groups and, on the other hand, extends the possibility of finding suitable solvents to bring about the sequential degradation with reagents other than phenylisothiocyanate.[2] This should improve not only the automatic degradation of peptides, but also the automatic analysis of the amino acid residues released step by step.

Thioacetylating reagents as sequencing reagents are reported here as illustrative examples.

[1] R. A. Laursen, *Eur. J. Biochem.* **20**, 89 (1971).
[2] J. Rosmus and Z. Deyl, *Chromatogr. Rev.* **13**, 163 (1971).

Sequential Analysis by Thioacetylating Reagents

Peptides, linked to a solid support through their COOH terminus, are converted into their N-thioacetyl derivatives using dithioacetic acid esters as reagents [reaction (1)]. The thioacetyl peptides are subsequently split off in trifluoroacetic acid (TFA) yielding the 2-methyl-thiazol-5(4H)-one, which identifies the NH$_2$-terminal amino acid, and the resin-bound shortened peptide [reaction (2)].

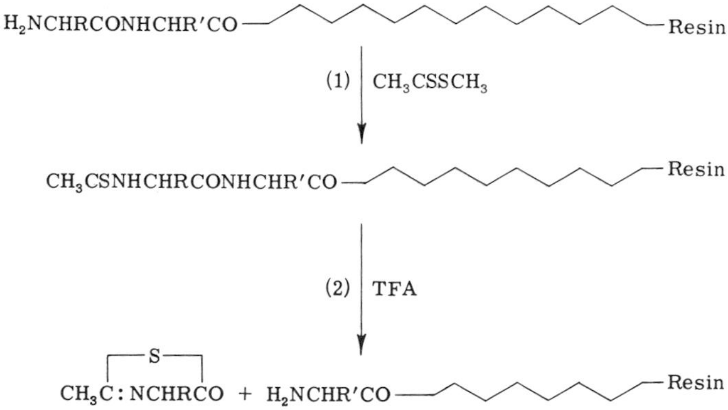

H$_2$NCHRCONHCHR'CO ——〰〰〰〰〰—— Resin

(1) | CH$_3$CSSCH$_3$

CH$_3$CSNHCHRCONHCHR'CO ——〰〰〰〰—— Resin

(2) | TFA

CH$_3$C : NCHRCO + H$_2$NCHR'CO ——〰〰〰—— Resin

(with S bridge over CH$_3$C : NCHRCO)

Methyldithioacetate

Preparation of the Reagent[3,4]

Sixty grams of N-acetylmorpholine (I), 60 g of phosphorus pentasul-

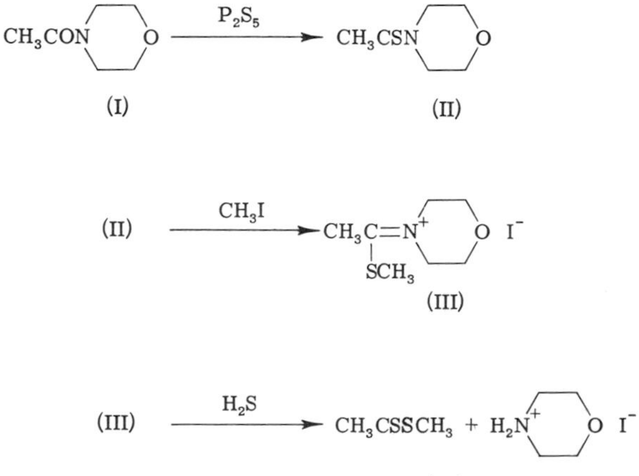

CH$_3$CON⟨O⟩ —P$_2$S$_5$→ CH$_3$CSN⟨O⟩

(I)　　　　　　　　　(II)

(II) —CH$_3$I→ CH$_3$C=N$^+$⟨O⟩ I$^-$ ， SCH$_3$

(III)

(III) —H$_2$S→ CH$_3$CSSCH$_3$ + H$_2$N$^+$⟨O⟩ I$^-$

(IV)

[3] D. A. Peak and F. Stansfield, *J. Chem. Soc.* 3, 4067 (1952).
[4] K. A. Jensen and C. Pedersen, *Acta. Chem. Scand.* 15, 1087 (1961).

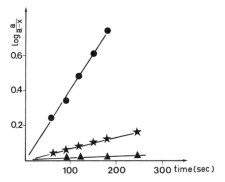

FIG. 1. Thioacetylation of LeuOMe in dimethylformamide by methyl dithio-acetate at 20°. [LeuOMe] = 50 mM = a; [CH₃CSSCH₃] = 1 M; ▲——▲, without catalyst; ✶——✶, 0.5 M triethylamine, ●——●, 0.5 M triethylammonium acetate.

fide, and 80 g of carbon disulfide are shaken for 2 hr at room temperature and then heated together under reflux for 2 hr; 100 ml of water are added, and the reaction mixture is allowed to stand overnight at room temperature. Thioacetylmorpholine (II) is extracted in chloroform, washed with 20% NaCl in water, and crystallized from ethanol–petroleum ether. The crude product (about 50 g) is utilized without further purification.

Compound (II) is dissolved in dry benzene (300 ml) and treated with methyl iodide (50 ml) at room temperature. After about 24 hr the separated methylthioethylidene morpholidium iodide (III) is collected by filtration and washed with ether. The crude product (about 70 g) is dissolved in methanol (300 ml), and the solution is maintained saturated with H₂S during about 24 hr at room temperature. After dilution with water (1500 ml) the methyldithioacetate (IV) is extracted in ethyl ether and washed three times with water. The organic layer is dried (Na₂SO₄) and the ether is removed under vacuum (12–15 mm) heating no higher than 20–25°. The crude methyl dithioacetate is purified by distillation under reduced pressure. Yield 15–18 g, b.p. 40–41° (15 mm).

Reactivity of Methyl Dithioacetate toward Amino Groups:
  Kinetic Data[5]

Kinetic studies in homogeneous solution show that the reaction of methyl dithioacetate with amino groups is strongly susceptible to base catalysis (Fig. 1) and to solvent effects (see the table).

[5] A. Previero, A. Gourdol, J. Derancourt, and M.-A. Coletti-Previero, *FEBS Lett.* **51**, 68 (1975).

EFFECT OF DIFFERENT SOLVENTS ON THE REACTION RATE BETWEEN
METHYLDITHIOACETATE AND LeuOMe[a]

Solvent	Pseudo first-order reaction constant $(k \times 10^4 \text{ sec}^{-1})$
Methanol	53.6
Dimethylformamide	15.6
n-Butanol	13.4
Hexamethyl phosphoric triamide	5.7
Dioxane	0.9

[a] $CH_3CSSCH_3 = 1 \ M$; LeuOMe $= 50 \ mM$; $N(C_2H_5)_3 = 0.2 \ M$; temperature $= 20°$. Formula utilized: $k = (2.303/t) \log [a/(a-x)]$.

Tertiary amines (curve ★—★, Fig. 1) or, better, acetate anion (curve ●—●, Fig. 1) are efficient catalysts, and dimethylformamide (DMF) and methanol are the best reaction solvents (see the table). The choice of the solvent to attain thioacetylation in solid-phase degradation is dependent on the nature of the solid support, as discussed below.

Automatic Sequencing

Peptides. The peptide is linked to an aminated polystyrene resin[1,6] by a general[1,7] or specific procedure[8,9] and introduced into the column of an automatic sequencer constructed as described by Laursen.[1] A typical sequencer program is reported in Fig. 2.

The sequencing operations differ from those of the solid-phase Edman degradation in that all the reactions take place entirely within organic solvents. DMF is chosen for the condensation step to swell the polystyrene solid support and triethylammonium acetate to speed up the thioacetylation reaction (Fig. 1). Being thioamides susceptible to oxidation, maintenance of reductive conditions is important until the cyclization step: these conditions are assured during the condensation step by liberation of methylmercaptan, which parallels the aminolysis of the reagent, while some mercaptoethanol is systematically added to the washing solvents. The cyclization reaction of thioacetyl peptides is comparable in rate to that of phenylthiocarbamyl peptides or perhaps more rapid. A rather prolonged acid treatment is performed in order to

[6] R. A. Laursen, in "Immobilized Enzymes, Antibodies, Antigens and Peptides" (H. H. Weetall, ed.), p. 567. Dekker, New York, 1975.

[7] A. Previero, J. Derancourt, M.-A. Coletti-Previero, and R. A. Laursen, FEBS Lett. 33, 135 (1973).

[8] M. J. Horn and R. A. Laursen, FEBS Lett. 36, 285 (1973).

[9] R. A. Laursen, M. J. Horn, and A. G. Bonner, FEBS Lett. 21, 67 (1972).

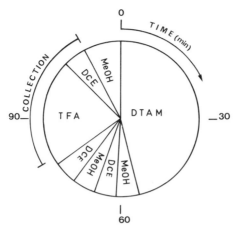

Fig. 2. Cycle of sequential degradation using methyldithioacetate at 45°. DTAM = a solution of methyl dithioacetate, 30% in dimethylformamide (flow rate = 1.2 ml/hr), and dimethylformamide containing acetic acid (0.45 M) and triethylamine (0.75 M) (flow rate = 2.4 ml/hr) simultaneously pumped through the column of the sequenator. MeOH = a 0.2% solution of mercaptoethanol in methanol (flow rate = 50 ml/hr). DCE = a 0.2% solution of mercaptoethanol in dichloroethane (flow rate = 50 ml/hr). TFA = anhydrous trifluoroacetic acid (flow rate = 2.4 ml/hr).

assure complete penetration of TFA within the resin, thus avoiding incomplete reaction at buried peptide chains.

Proteins. N-Aminoethyl polyacrylamide,[10] a rigid support possessing a high concentration of amino groups on its surface, has been recently prepared as follows:

$$(-CH-CH_2-)_n \xrightarrow[\text{2 hr reflux}]{\text{SOCl}_2 \text{ (pyridine 5\%)}} (-CH-CH_2-)_n$$
$$\quad | \qquad\qquad\qquad\qquad\qquad\qquad\qquad | $$
$$\text{COOH} \qquad\qquad\qquad\qquad\qquad\qquad \text{COCl}$$

$$(-CH-CH_2-)_n \xrightarrow[\text{2 hr 40}^\circ\text{ C}]{\text{H}_2\text{NCH}_2\text{CH}_2\text{NH}_2 \text{ (excess)}} (-CH-CH_2-)_n$$
$$\quad | \qquad\qquad\qquad\qquad\qquad\qquad\qquad\qquad | $$
$$\text{COCl} \qquad\qquad\qquad\qquad\qquad\qquad \text{CONHCH}_2\text{CH}_2\text{NH}_2$$

Lysine-containing proteins can be linked to this new resin in high yield by a modified phenylenediisothiocyanate technique.[9]

The protein (300–500 nmol) is dissolved in 50% aqueous pyridine (1 ml) and treated with phenylenediisothiocyanate (30–50-fold molar excess over protein amino groups) dissolved in DMF (1–2 ml).

The reaction mixture is allowed to stand 2 hr at 40° under a nitrogen

[10] J.-C. Cavadore, J. Derancourt, and A. Previero, *FEBS Lett.* **66**, 155 (1976); see also this volume [30].

atmosphere. N-Aminoethyl polyacrylamide (300–500 mg) is added, and the suspension is maintained for 2 hr at 40° under gentle shaking. The unreacted amino groups of the resin are blocked by adding methyliso-thiocyanate (200 mg). The insolubilized protein is introduced into the column of an automatic sequencer[1] and degraded with methyl dithio-acetate following the program reported in Fig. 2.

Methanol as well as DMF containing triethylamine or, better, tri-ethylammonium acetate can be utilized interchangeably as solvents in the condensation step, since the protein is linked at the surface of a resin which does not swell in any solvent. Twenty amino acid residues have been easily sequenced in egg white lysozyme and horse apomyo-globin with an average yield of 90% at each cycle.

Identification of Methylthiazolinones

Acid Hydrolysis. The thiazolinones released at each cycle of degrada-tion are usually identified through the regeneration of the free amino acid. The trifluoroacetic acid deriving from the cyclization step is re-moved under reduced pressure or in a nitrogen stream, the residue sub-mitted to acid hydrolysis (2 N HCl, 3 hr at 100°), and the free amino acid is identified and quantitated by automatic amino acid analysis. The yield on regeneration is about 80% with some exceptions: tryptophan is partially destroyed by acid hydrolysis and cannot be easily identified; glutamine and asparagine are completely converted into glutamic acid and aspartic acid; and β-hydroxy amino acids are recovered as β-thiol amino acids. The recommended procedure is to perform the acid hy-drolysis followed by performic acid oxidation; serine is identified as cysteic acid and threonine as 2-amino-3-sulfonylbutyric acid, which are detected by paper electrophoresis with reference compounds or by gas-liquid chromatography of their trimethylsilyl derivatives.[11]

Gas-Liquid Chromatography. Methylthiazolinones of most of the common amino acids can be identified by gas–liquid chromatography directly or, better, after 5-O-acetylation of their enolic form.[11]

In the first case, the trifluoroacetic acid solution of the thiazolinone is evaporated to dryness under vacuum and the residue is dried over KOH pellets. Ethyl acetate containing N-methylmorpholine (30%) is added to dissolve the thiazolinone as the free base and an aliquot is injected into the chromatograph (see Fig. 3).

[11] D. L. Simpson, J. Hranisewljevic, and E. A. Davidson, *Biochemistry* **11**, 1849 (1972).

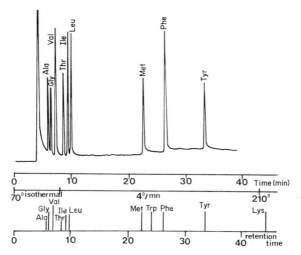

FIG. 3. Gas chromatographic analysis of amino acids as methylthiazolinones: retention times and illustrative examples. Glass capillary column 0.028 cm × 25 m coated with FFAP, 0.5% in (CH₂Cl)₂; nitrogen carrier.

In the second case, the trifluoroacetic acid solution of the thiazolinone is previously treated with acetyl chloride (100 μl/1 ml of TFA) and allowed to stand 20 min at 40°. After the elimination of the solvent under reduced pressure, the residue is dissolved in ethyl acetate:trimethylmorpholine (7:3) and analyzed (see Fig. 4).

The representative results of Figs. 3 and 4 have been obtained using reference thiazolinones or during the sequential degradation of model peptides by methyldithioacetate. These results are reported here to indicate their potential usefulness and to encourage a larger application, which should improve the experimental conditions for routine analysis.

Ammonolysis. The trifluoroacetic acid solution of methyl thiazolinones is evaporated to dryness, as previously described, and the residue is treated with concentrated aqueous ammonia for 1 hr at 40° in a sealed tube. This treatment converts the thiazolinones into the corresponding amino acid amides.[12]

$$\overset{\lceil S \rceil}{CH_3C:NCHRCO} \xrightarrow{NH_4OH} H_2NCHRCONH_2$$

The amide functions of asparagine and glutamine are perfectly stable under these conditions. Therefore this reaction is of particular interest to distinguish between Asp:Asn and Glu:Gln before regeneration by acid

[12] A. Previero and J.-C. Cavadore, *in* "Solid Phase Methods in Protein Sequence Analysis" (R. A. Laursen, ed.), p. 63. Pierce Chemical Co., Rockford, IL, 1975.

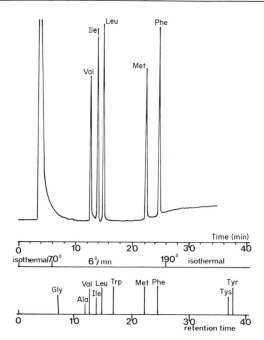

FIG. 4. Gas chromatographic analysis of methylthiazolinones after O-acetylation: retention times and illustrative examples. Glass capillary column 0.25 cm × 30 m coated with XE 60, 2% in (CH₂Cl)₂; nitrogen carrier.

hydrolysis of the free amino acid from the parent thiazolinone or the amide.

The positive result in gas–liquid chromatography as well as ammonolysis depends on the molecular integrity of the thiazolinone (or O-acylthiazolinone) molecules, which must be preserved during their isolation. Negative results or poor yields may be obtained when there is insufficient washing prior to the cyclization step and when degradative reactions of the peptide resin system cause the accumulation of secondary products possessing a nucleophilic character and therefore able to react with the released thiazolinones. The regeneration of free amino acids by acid hydrolysis, which is a rather unspecific reaction, however, gives reproducible results.

Thioacetylthioglycolic Acid[13]

The reagent is prepared from thioacetyl piperidine according to the method of Jensen and Pedersen.[4]

[13] G. A. Mross, Ph.D. Thesis, University of California, San Diego, 1971.

Sequential degradation using thioacetylthioglycolic acid has been performed essentially according to the procedure developed by Mross and Doolittle.[13,14] The peptide is linked to 2-aminoethylpolystyrene–1% divinyl benzene resin by ethyl-dimethylaminopropyl-carbodiimide[13,15] and introduced into the chambered vessel of a semiautomatic sequencer.[13]

The reaction vessel consists of a jacketed glass column of about 5–7 ml capacity, similar to the one used for solid-phase peptide synthesis, permitting the heating, shaking, and introduction of solvents and reagents and their draining or collection.

The sequencing operations are performed at 40° according to the following program:

1. Washing with 50% aqueous pyridine.

2. Thioacetylation with thioacetylthioglycolic acid, 0.3 M, in aqueous 60% pyridine–13% triethylamine adjusted to pH 9.5 ± 0.1 with trifluoroacetic acid. The reaction time is 1 hr at 40°.

3. Washing the resin with 50% aqueous pyridine–methanol and then drying it.

4. Cyclization with 50% trifluoroacetic acid in chlorobutane, 15 min at 40°; filtration and collection of trifluoroacetic acid.

5. Washing with chlorobutane, which is also collected.

6. Washing with methanol and start of the next degradative cycle.

The trifluoroacetic acid and chlorobutane fractions containing the methylthiazolinones can be analyzed as reported under the heading Methyldithioacetate.

The hydrolysis of thiazolinones with 50% hydriodic acid (110° for 16 hr) has been proposed as an alternative method for amino acid regeneration.[16] Serine is recovered in good yield with other products, while threonine is identified as α-aminobutyric acid.

[14] G. A. Mross and R. F. Doolittle, *Fed. Proc., Fed. Am. Soc. Exp. Biol.* **30**, 1241 (1971).

[15] A. Previero, J. Derancourt, and M.-A. Coletti-Previero, *in* "Recent Developments in the Chemical Study of Protein Structure" (A. Previero, J. F. Pechère, and M. A. Coletti-Previero, eds.), p. 29. INSERM, Paris, 1971.

[16] E. M. Prager, N. Arnheim, G. A. Mross, and A. C. Wilson, *J. Biol. Chem.* **247**, 2905 (1972).

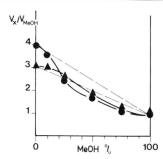

FIG. 5. Swelling of chloromethyl polystyrene–1% divinyl benzene suspended in solvent mixtures at 20°. ●——●, Dichloroethane–methanol; ▲——▲, dimethylformamide–methanol; V_x = volume of the resin in the mixture; V_{MeOH} = volume of the resin in methanol.

Correlations between Solid Support, Reaction Solvent, and Sequencing Reagent

Figure 5 shows the swelling of a polystyrene resin in different mixtures of DMF/methanol and dichloroethane/methanol. DMF and dichloroethane swell the resin, whereas the penetrability of methanol is practically negligible. The nonlinear decrease of the resin volume with progressive increase of methanol concentration in the suspending solvent suggests the possibility that preferential extraction may occur. Specifically, the swollen resin, together with the two solvents, is to be viewed as a system of three components with different reciprocal affinities. A consequence is that the solvent composition inside the resin is different from outside the resin. The reaction rate of a sequencing reagent (see the table) may thus be affected by solvent effects and cannot necessarily be predicted from kinetic experiments in homogeneous solution. The use of a single solvent able to swell the resin and to catalyze the condensation reaction eliminates this complication. Furthermore, factors such as differing physical properties arising from different preparations, or subsequent chemical treatments, and differing reaction temperatures which can be required using alternative degradating reagents, may considerably affect the composition and the amount of liquid absorbed by the resin from a solvent mixture, thus preventing reproducible results. By contrast, a large choice of solvent mixtures, even aqueous solvents, is possible when inert and nonswelling supports[10,17] with a suitable concentration of functional groups on their surface are utilized for the insolubilization of the peptides.

[17] E. Wachter, W. Machleidt, H. Hofner, and J. Otto, *FEBS Lett.* **35**, 97 (1973).

Acknowledgment

The help of the Fondation pour la Recherche Médicale Française is gratefully acknowledged.

[32] Solid-Phase Sequencing in Spinning-Cup Sequenators*

By DENNIS A. POWERS

Between 1950 and 1960 Edman published a series of papers that elaborated the chemistry and methodology of the isothiocyanate procedure for determining the amino acid sequence of peptides and proteins.[1-4] The Edman degradation rapidly became the method of choice and has remained so until the present day. Edman's procedure opened up the field of primary structural analysis, resulting in an explosion of studies that have unlocked fields that were previously unapproachable.

In 1967 Edman and Begg[5] introduced a machine that automatically performed the sequential degradation. This automatic sequenator stimulated an exponential growth in primary structural analysis. The heart of their machine consisted of a spinning cylindrical cup that immobilized protein in a thin film with reagents or extracting solvents being passed over the protein, allowing either reaction or extraction at the film interface. In general, the liquid-phase sequenator has worked remarkably well, but it has had its shortcomings. The sequenator is maximally effective on peptides and proteins between 50 and 150 residues. Sequencing becomes difficult for proteins much larger than 150 amino acid residues. Increasing bulk from large proteins and rising background in the detection system reduces the efficiency of the thin film approach. Furthermore, Edman and Begg pointed out that repetitive yields for small peptides are reduced because of losses to extracting solvents. Finally, poor yields are often obtained, even for peptides of optimal size, when they contain a large proportion of hydrophobic residues. This type of peptide is more soluble in extracting solvents and is therefore difficult to sequence in the liquid–phase sequenator. In the latter two cases modifications in programming and the use of different buffers have been helpful, but cannot be universally applicable to all peptides. Chemical modifications of some peptides to make them less soluble in extracting organic

* Contribution No. 906, Department of Biology, The Johns Hopkins University.

[1] P. Edman, *Acta Chem. Scand.* **4**, 283 (1950).
[2] P. Edman, *Acta Chem. Scand.* **10**, 761 (1956).
[3] P. Edman, *Proc. R. Aust. Chem. Inst.* **24**, 434 (1957).
[4] P. Edman, *Ann. N.Y. Acad. Sci.* **88**, 602 (1960).
[5] P. Edman and G. Begg, *Eur. J. Biochem.* **1**, 80 (1967).

solvents have improved sequencing procedures for those specialized groups of peptides. A number of other procedures have been introduced to further immobilize peptides, thereby reducing losses to extracting solvents. Unfortunately, no method has proved to be a panacea.

Laursen's introduction of the solid-phase sequenator[6] provided an alternative approach that was particularly useful when applied to small peptides. Improvements in the chemistry of peptide attachment and the introduction of new types of solid supports (see review in this volume) have greatly enhanced the usefulness of this method.

The purpose of the present article is to introduce the use of solid-phase sequencing in the liquid phase (i.e., Edman–Begg type) sequenator. There are two solid-phase approaches that are useful in the Edman–Begg machine. These approaches are (1) the use of solid supports directly in the spinning-cup, and (2) the temporary modification of a liquid-phase sequenator into a solid-phase machine. Both of these applications will be examined.

Reagents and Chemicals

Although most chemicals for sequential analyses of peptides can be purified from reagent grade material,[5-8] it is most convenient to purchase Sequanal grade reagents from Beckman Instruments (Palo Alto, California) and/or Pierce Chemical Co. (Rockford, Illinois). Many of the solid support resins and specialized reagents for derivatization are also available from the same commercial sources.

Attachment of Peptides to Supports

For solid-phase sequencing, peptides must be attached to a solid support. The development of new methodologies for peptide attachment and the evolution of solid-phase supports are two of the most critical research areas in solid-phase sequencing today. The current "state of the art" for each of these areas is reviewed in this volume by Laursen and by Machleidt and Wachter, respectively. Therefore, methods of peptide attachment will not be exhaustively discussed in the present article. We shall cover those methods that have proved to be successful in our laboratory, but this does not mean that a number of other pro-

---

[6] R. Laursen, this series Vol. 25, p. 344 (1972).

[7] P. Edman, in "Protein Sequence Determination" (S. B. Needleman, ed.), p. 211. Springer-Verlag, Berlin and New York, 1970.

[8] A. C. Cope and P. H. Towe, J. Am. Chem. Soc. 71, 3423 (1949).

cedures are not equally suitable or, perhaps, superior to those detailed here. There is considerable room for personal innovation and experimentation.

Attachment to Polystyrene Supports

Although polystyrene resins have been used in the liquid-phase sequenator, they have met with limited use because they tend to float in the extracting solvents. In addition, the resins cause backpressure problems in solid-phase modification of some liquid-phase sequenators. For these reasons and because the methodologies for peptide attachment are considered by Laursen in this volume, we shall not discuss them in detail.

Peptides can be attached to aminopolystyrene resins via their carboxyl groups according to the method of Laursen.[6] Peptides can be attached to lysine side chains via p-phenylene diisothiocyanate-modified aminopolystyrene resin using the procedure of Laursen, Horn, and Bonner.[9] Triethylenetetramine polystyrene is also useful as a solid support for peptides; in this case the coupling procedure of Bridgen is employed.[10] This resin is particularly useful for coupling homoserine lactone-activated peptides. Although these polystyrene resins can be synthesized, they are also available from commercial sources (Pierce Chemical Co., Rockford, Illinois).

Glass Supports

Aminopropyl-Glass (APG). Aminopropyl-glass (APG) can be purchased directly (Pierce Chemicals, Rockford, Illinois) or it can be made from Corning controlled-pore glass beads (Electro-Nucleonics Inc., Fairfield, New Jersey) as outlined by Robinson, Dunhill, and Lilly.[11]

Before derivatization, glass beads of 120–200, 200–400, or 325–400 mesh with 75 Å pore diameter should be pretreated in order to maximize eventual derivatization. The beads are mixed with 1 N HCl in a ratio of 1 g of beads per 5 ml of acid. The mixture is sonicated at maximum output in a sonifier but without allowing the sonicator tip to come into direct contact with the beads. Five volumes of 1 N HCl are added to the

---

[9] R. J. Laursen, H. Horn, and A. G. Bonner, *FEBS Lett.* **35,** 97 (1973).

[10] J. Bridgen, *in* "Solid-Phase Methods in Protein Sequence Analysis" (R. Laursen, ed.), p. 11. Pierce Chemical Co., Rockford, IL, 1975.

[11] P. J. Robinson, P. Dunhill, and M. D. Lilly, *Biochim. Biophys. Acta* **242,** 659 (1971).

sonicated bead-acid mixture in a graduated cylinder. The milky supernatant is decanted after the beads have settled to the bottom of the cylinder. The beads are then washed five more times with fresh 1 N HCl.

The glass beads are collected on a sintered-glass funnel and washed exhaustively with distilled deionized water. The beads are refluxed for 1 hr in 1 N HNO$_3$ according to the method of Scouten.[12] The beads are removed and washed exhaustively with distilled water on a sintered-glass funnel and dried under vacuum.

Clean, dry beads are just wetted or barely covered with a 2% solution of 3-aminopropyltriethoxysilane (Pierce Chemicals, Rockford, Illinois) in redistilled acetone. The solution is degassed and allowed to stand at 45° for 24 hr. Afterward, the glass beads are washed three times with five to ten volumes of redistilled acetone on a sintered-glass funnel. The glass beads are dried with a stream of nitrogen and stored at 4° under a nitrogen atmosphere.

The amino capacity of the aminopropyl-glass will vary between 50 and 200 nmol per milligram of beads. The capacity of any given batch can be determined by a number of procedures. Aminopropyl groups are continuously lost from glass beads in aqueous solutions. The process is pH-dependent with a maximum stability at pH 8–9. The aminopropyl groups are totally removed at low pH in aqueous acids. Therefore, derivatized beads can be incubated in an aqueous acid (pH 2–3) for 18 hr at 40° to remove the aminopropyl groups, and the free amino groups in solution can be quantified by any of the end-group reagents [e.g., trinitrobenzene sulfonic acid (TNBS), ninhydrin, dansyl chloride, fluorescamine].

The following procedure has proved to be useful. One gram of beads (smaller portions can be used if desired) is added to a test tube with 5 ml of 1 N HCl and incubated at 40° for 18 hr. One-half milliliter of the supernatant is added to each of four Pyrex (No. 9825) screw-cap test tubes and neutralized with 0.5 ml of 1 N NaOH. If desired, the base may be used for hydrolysis and the acid for neutralization. In either case, 0.5 ml of 0.2 M borate buffer (pH 8.0) and 0.5 ml of fresh 0.2% TNBS are added *to each test tube*. The tubes are covered with their screw caps and incubated *exactly* 90 min at 40° and then placed in an ice bath. Once cooled, 2 ml of 0.5 N HCl are added to each tube to hydrolyze excess TNBS, and the optical density is determined at 420 and 340 nm. The quantity of free amino groups is determined by employing a standard curve. Although aminopropanol may be used, a free amino acid (e.g., leucine) is adequate. The standard curve generally employs quantities between 50 and 500 nmol. If greater sensitivity is required, a

---

[12] W. H. Scouten, this series Vol. 34, p. 288 (1974).

fluorescamine assay can be employed, as in the method of Naoi and Lee.[13]

p-Phenylene Diisothiocyanate Treatment of Aminoderivatized Glass Beads. Ethanol-recrystallized *p*-phenylene diisothiocyanate (DITC) (Eastman Chemicals, Rochester, New York) is dissolved in dimethylformamide (DMF) and added in 25–100-fold excess of the free amino groups of the glass beads. If the amino capacity of the glass beads is not known, assume it to be 200 nmol per milligram of beads and add a 25-fold excess of the *p*-phenylene diisothiocyanate. Derivatized glass beads (APG) are slowly added to a degassed solution of DITC in DMF and buffered (pH 10) with triethylamine. The volume should be kept to a minimum, but there should be enough liquid to wet the beads. An atmosphere of nitrogen is maintained over the reaction mixture for 1 hr at 45°. Afterward the DITC-APG derivatized beads are washed on a sintered-glass funnel, initially with DMF and then with redistilled methanol. The operation is best executed in a portable tissue culture hood that is flushed with nitrogen. The DITC-APG beads are dried *in vacuo,* flushed with nitrogen, and stored in a sealed container under a nitrogen atmosphere at 4°. These storage conditions are critical. Derivatized beads in either high or low pH aqueous solutions are unstable. However, beads stored under the above conditions are stable for months.

Peptide Attachment to DITC-APG. Peptide, 200 nmol, is added to 0.5 ml of *N,N*-dimethyl-*N*-allylamine (DMAA) buffer, pH 9.5. The solution is stirred with 100 mg of DITC-APG derivatized glass beads. The mixture is degassed, flushed with nitrogen, and incubated under nitrogen at 45° for 1 hr. Afterward the beads are washed exhaustively with degassed, distilled methanol on a sintered-glass funnel inside a tissue culture hood that is flushed with nitrogen. The beads are dried under vacuum and stored under a nitrogen atmosphere below 0°C. Significant quantities of DITC couple to the amino groups of the APG beads; however, owing to cross-linking and size restrictions the availability of DITC-APG reactive groups for peptide coupling is often much lower. It is therefore advisable to assay the potential peptide coupling capacity of each batch of DITC-APG beads. Peptide coupling capacity is assayed via test peptides containing one or more lysine side chains. When total capacity is to be determined, the test peptide or protein should be in excess (e.g., 10–20-fold) of that stated above in order to assure saturation of the available DITC sites. The capacity is determined by hydrolyzing aliquots of peptide-coupled beads at 150°C in 6 N HCl for 18 hr and analyzing the acid-soluble portion in an amino acid analyzer. These

[13] M. Naoi and Y. C. Lee, *Anal. Biochem.* **57**, 640 (1974).

data provide one with the bead capacity in nanomoles of peptide coupled per milligram of DITC-APG beads. An alternative and/or supplemental method may be used that employs an amino acid analyzer to quantitate the uncoupled peptide that exists in the supernatant of the coupling mixture and successive buffer washes of the derivatized beads (see Fig. 1). The peptide in the pellet of the nth buffer wash is calculated from the peptide in the pellet of the previous wash $(n-1)$ minus the peptide in the supernatant of the nth buffer wash. This method is less accurate than the former, but it does not require the consumption of coupled peptides. An additional feature that is demonstrated in Fig. 1 is that coupled peptide is continuously lost from the derivatized beads in aqueous solutions. The rate of peptide loss is pH-dependent with a minimum at pH 8–9. Fortunately, the derivatives are fairly stable at acid pH under *anhydrous* conditions so that the use of anhydrous acid during the Edman degradation plays a minor role in uncoupling peptides from glass beads.

Once the potential capacity for a given batch of DITC-APG beads is known, the efficiency for coupling experimental peptides is simply:

$$\text{Coupling efficiency} = \frac{\text{nanomoles of peptide coupled}}{\text{DITC-APG peptide capacity in nanomoles}} \times 100$$

Some investigators report either capacities or coupling efficiencies based on PTH yields of the first stage of Edman degradation. These data are

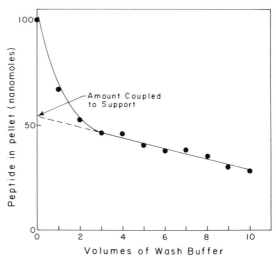

FIG. 1. Nanomoles of peptide in the pellet can be determined directly from the pellet or calculated from the total peptide available from previous wash minus peptide in the supernatant of the present wash. See the text for clarification.

not accurate coupling data, rather they should be considered as a guide or as a different parameter (i.e., yield efficiency). Yield efficiencies may be substantially different from the coupling efficiencies, especially when carboxyl coupling is employed. In a typical experiment one will find about 200 nmol of amino groups per milligram of AGP beads, but only about 30 nmol of DITC-APG groups available for peptide coupling. Furthermore, experimental peptide coupling efficiencies to the same batch of DITC-APG beads will vary between 20 and 100% depending on the particular peptide, its concentration, the pH of the coupling mixture, and the exclusion of oxygen.

Carboxyl-Terminal Attachment of Peptides to APG. There are a number of methods for the carboxyl-terminal attachment of peptides to solid supports (see review by Laursen, this volume [30]). The use of dicyclo-hexylcarbodiimide (DCCI) has been a particularly useful method. Although blocking of the amino terminus is no longer considered necessary in carboxyl attachment procedures, blocking of amino groups is useful when reactions are to be carried out in anhydrous reagents like DCCI.

Although S-Boc blocking of amino groups is easily done by a number of procedures, the method of Nagasawa et al.[14] is especially useful. In this method, 200 nmol of peptide are dissolved in 0.3 ml of 10% triethylamine in water; 0.2 ml of DMF containing 1.25 mg of S-Boc reagent is added to the peptide solution and incubated for 2 hr at room temperature. Afterward, 1.5 ml of ethyl acetate is added to the reaction mixture, the solution is vortexed and centrifuged in a table-top centrifuge, and the organic phase is removed and discarded. The extraction is repeated twice and then the aqueous phase is frozen and lyophilized. The presence or the absence of free amino groups can be detected qualitatively via the Kaiser ninhydrin method.[15]

The peptide is attached via its carboxyl terminus after activation with DCCI by the following procedure. The S-Boc blocked peptide is dissolved in 0.2 ml of anhydrous DMF. The dissolved peptide is added to dry APG (2 nmol of peptide per milligram of APG), 0.4 ml of 10% DCCI in anhydrous DMF is added, and the solution is degassed and incubated for 2 hr at 37°C under appropriate anhydrous conditions. After the peptide is coupled to the glass beads, the beads are washed on a sintered-glass funnel with several volumes of DMF and two volumes of methanol. The peptide-coupled glass beads are lyophilized and stored under nitrogen below 0°C. The above procedures were introduced to our laboratory by Dr. J. I. Ohms.[16]

[14] T. Nagasawa, *Bull. Chem. Soc. Jpn.* **46**, 1269 (1973).
[15] E. Kaiser, R. L. Colescott, C. D. Bossinger, and P. I. Cook, *Anal. Biochem.* **34**, 595 (1970).
[16] J. I. Ohms, personal communication.

Because the amino terminus is blocked, there is a slightly different procedure for sequencing this type of peptide than for the unblocked type (see later discussion on sequence procedures).

Attachment of CNBr peptides to APG via Homoserine. Horn and Laursen's method for attaching CNBr peptides to solid supports via their carboxyl-terminal homoserine is one of the most useful methods in solid-phase sequencing today.[17] The following modification of Horn and Laursen's polystyrene method appears to be useful for attaching peptides, between 12 and 45 amino acid residues, to APG beads. Peptides smaller than 12 residues should in general be attached to polystyrene (there are exceptions).

Dry peptide, 200 nmol, is dissolved in 1.0 ml of anhydrous trifluoroacetic acid (TFA) and incubated for 1 hr at room temperature. Afterward, the TFA is evaporated in a vacuum desiccator over pellets of KOH. Immediately after drying, the peptide is redissolved in 0.2 ml of anhydrous DMF and the solution, buffered with 0.05 ml of triethylamine, is added to 100 mg of APG beads (i.e., 2 nmol of peptide per milligram of APG beads). If the peptide is not soluble in the DMF, Horn and Laursen advise adding up to 20% water, but we have not found this to be necessary if the peptide is kept totally anhydrous after treatment with TFA. If the glass beads are not totally wetted, more DMF may be added. The mixture is incubated with occasional swirling, not stirring, at 45° for 2 hr under a nitrogen atmosphere. Excess amino groups are blocked with CH_3 CNS in CH_3CN (0.2 ml of a 1:1, w/v, solution) with 0.2 ml of DMAA-propanol, pH 8.6. Incubation time of the blocking mixture is 1 hr. After blocking is completed, the coupled glass beads are washed with 50 ml of distilled methanol, dried under nitrogen or by lyophylization, and stored under nitrogen below 0°C until needed.

Procedure for Solid-Phase Sequencing in the Spinning-Cup

Polystyrene Resins

The use of polystyrene solid supports in Edman–Begg sequenators has been attempted by a number of people. The results have generally been mixed. The method has not received wide acceptance owing to programming changes and the different solvents required and because the resin has a tendency to float in extracting solvents. Current research on high-density supports and the coupling of polystyrene resins to heavier supports may increase the usefulness of this technique.

The peptide-coupled resin in a methanol slurry is added to the cup

[17] M. J. Horn and R. A. Laursen, *FEBS Lett.* 36, 285 (1973).

with a plastic syringe fitted with a piece of Teflon tubing. The cup is rotated at low speed, and the resin is allowed to form a film not exceeding 15 mm in height. The sample is then dried using the same subroutine as that used for a liquid sample application.

The first step in the actual sequencing procedure is to swell the resin, in the cup, with $CHCl_3$. The amount of $CHCl_3$ that will be used depends on the amount of resin applied to the cup. Enough $CHCl_3$ must be added to swell the resin, but not to cause the polystyrene to rise above the 15 mm level of the cup. The solvent time should be adjusted specifically for each sample. Usually, between 10 and 100 mg of resin are added to the cup, and the ideal solvent time will vary between 14 and 30 sec. Both $CHCl_3$ (solvent 3) and CH_3OH (solvent 2) are placed in smaller vessels (i.e., 250 ml). The $CHCl_3$ is placed on the reagent shelf, but the CH_3OH remains on the solvent shelf.[16] These modifications are necessary for proper solvent delivery to the cup.

After the resin is swollen, it is dried and the peptide is coupled with PITC in DMAA buffer. The amount of buffer added must be adjusted to just cover the resin. Consequently, if the resin is at the 15-mm level, the buffer should be kept at 16 mm, even when additions are made later in the programming events.

Drying of the PITC-coupled peptide and cleavage with *n*-hepta-fluorobutyric acid (HFBA) is followed by an extraction with CH_3OH. After drying, the entire process is repeated for successive stages of Edman degradation. A model program (No. 032873/051073) for this procedure is available from Beckman Instruments (Palo Alto, California).

Following completion of the procedure, conversion of the sequenator back to the standard liquid-phase methodology should adhere to the following general approach. The scoop line should be flushed with methanol, then water, then acetone. Preferably, the waste/collector *solenoid* valve should be taken apart and cleaned to ensure that no resin has become lodged in the valve. The bottles containing the 250 ml of solvents 2 and 3 should be removed, and the lines flushed with ethyl acetate and chlorobutane, respectively. After repositioning the solvent 3 lines toward the solvent shelf, the proper larger vessels for solvents 2 and 3 should be reattached appropriately.

Peptides Coupled Directly to the Spinning Cup

The directions in this section refer to the Beckman instrument.

Since peptides can be attached to glass beads as outlined earlier, the same general procedures can be used to attach the peptides directly to the walls of the spinning cup when it is made of glass. This procedure has the advantage that the solid support cannot be lost in extracting solvents. However, there are two potential disadvantages: (1) the surface area is restricted, and (2) it is not known what long-term effect this application will have on the glass cup. These must be considered only as potential disadvantages since actual problems with the procedure have not been reported.

In this procedure it is critical that the glass cup be very clean. After the cup has been cleaned, it should be soaked in 1 N HCl for 15 min. The acid is then removed and the cup is filled again with 1 N HCl for 15 more minutes, then the acid is removed and the cup is washed exhaustively with distilled, deionized water. Afterward, the cup is filled with high-purity HNO_3, covered with a glass plate, and allowed to stand for 2 hr. The HNO_3 is removed and the cup is washed exhaustively with distilled deionized water and dried under vacuum.

Once the cup is prepared, it is rotated at low speed and a 2% solution of 3-aminopropyltriethoxysilane (Pierce Chemicals, Rockford, Illinois) in acetone is added until it reaches 16 mm from the bottom of the cup. This level of solution is maintained by additions from a stock bottle for 15 min. An alternative procedure that improves the coupling is to fill the nonrotating cup to 16 mm with 3-aminopropyltriethoxysilane, cover with a glass plate, and allow to stand for 24 hr. In either case, the cup is emptied of solution and washed with two exchanges of deionized distilled water, two exchanges of acetone, and one exchange of methanol. Afterward the cup is dried via the routine cup-drying procedure suggested by the sequenator manufacturer.

Carboxyl Attachment to Cup. Peptides can be attached to the aminopropyl groups on the glass cup via the carboxyl group by the following method. Peptide, 100–200 nmol, is blocked via the S-Boc reagent method described earlier or by one of the other amino-blocking groups. The unblocked peptide can be used directly if desired. Even though blocking is not always needed to prevent polymerization, it is often useful in increasing the solubility of peptides in anhydrous conditions.

The peptide is dissolved in 0.2 ml of anhydrous DMF and added to the cup as it rotates at low speed; 0.2 ml of a 20% solution of DCCI in anhydrous DMF is added to the spinning cup. The cup area is flushed with nitrogen for 15 min and allowed to rest for 3 hr, during which time the cup area is covered with a piece of black felt and the room lights are turned off. After 3 hr the cup is stopped, the mixture is removed, and 10% of the solution is dried, hydrolyzed, and analyzed in an amino acid analyzer to determine the coupling efficiency. An alternative procedure is to dry the peptide–DMF mixture in the cup so that losses are minimized. In either case, the cup is washed with two volumes of methanol and dried under vacuum. After drying, HFBA is added to the cup so that the S-Boc groups will be removed. The cleavage is followed by a chlorobutane wash, and the cup is dried. The first stage of Edman degradation will require a double or triple coupling step in order to ensure a minimum amount of overlapping. After the first stage a single coupling will suffice.

Homoserine Attachment to Cup. CNBr peptides can be attached via their carboxyl-terminal homoserine to the aminopropyl-derivatized glass cup by a modification of the polystyrene procedure of Horn and Laursen.[17] Dry peptide, 200 nmol, is dissolved in 1.0 ml of anhydrous TFA and kept for 1 hr at room temperature. The TFA is evaporated to dryness over pellets of KOH. Immediately after drying, the peptide is redissolved in 0.2 ml of anhydrous DMF buffered with 0.05 ml of triethylamine. The solution is added to the aminopropyl-derivatized glass cup as it spins at low speed. The peptide should not rise above 15 mm in the cup. The peptide is allowed to react with the cup surface, under a nitrogen atmosphere, for 2 hr at 55°C. Anhydrous DMF with buffer included may be added as needed to maintain the peptide at the 15 mm level. Afterward, the cup area is evacuated for 30 min, flushed with nitrogen, and the excess amino groups are blocked with a 1:1 (w/v) mixture of CH_3CNS in CH_3CN buffered with an equal volume of pH 8.6 DMAA-propanol (the blocking solution should be maintained at the 16-mm level of the cup). The reaction time of the blocking mixture is 1 hr. After blocking is completed, the cup is washed with methanol and dried under vacuum and the sequenator is started.

Lysine-DITC Attachment to Cup. Tryptic peptides that contain lysine are attached to the cup by *p*-phenylene diisothiocyanate (DITC). There are two different approaches. The aminopropyl-derivatized glass cup can be coupled with DITC and then the peptide can be attached or the peptide can be coupled to DITC and the DITC-peptide can be linked to the aminopropyl glass cup. Both methods have advantages and disadvantages.

First, let us consider the DITC derivation of the cup. Ethanol-recrystallized DITC, 20 μmol, is added to 0.5 ml of DMF (pH 10), and the mixture is added to the aminopropyl glass cup as it rotates at low speed. The cup area is flushed with nitrogen, and the reaction is allowed to take place at 55° for 1 hr. Afterward, the cup is washed three times with DMF and once with methanol. The cup is dried under vacuum and flushed with nitrogen. Peptide, 200 nmol, in 0.4 ml of DMAA-pyridine buffer, pH 9.5, is added to the spinning cup at low speed, the chamber is flushed with nitrogen, and the reaction is allowed to proceed at 55°C for 1 hr. After an hour an aliquot of the liquid is hydrolyzed and the hydrolyzate is analyzed in the amino acid analyzer to determine the coupling efficiency. The remainder of the liquid is dried in the cup by employing a drying subroutine. One-tenth milliliter of 50% methylisothiocyanate in acetonitrile is added to the cup with 0.4 ml of DMAA-propanol buffer, pH 9.5, and allowed to react under nitrogen for 30 min. Afterward, the sample is dried, and the sequenator can be started in the manner suggested by the manufacturer.

The second method of lysine attachment is to couple DITC with peptide in a 2:1 mixture in 0.3 ml of DMAA-propanol buffer, pH 9.5, at 55° for 1 hr. The mixture should be allowed to react under an atmosphere of nitrogen. After completion, the peptide is transferred to the aminopropyl-derivatized glass cup that is rotating at slow speed and allowed to react for 1 hr. Afterward, 0.1 ml of 50% methylisothiocyanate in acetonitrile as well as 0.1 ml of DMAA-propanol, pH 9.5, are added to the cup. The chamber is flushed with nitrogen and allowed to react for 30 min. The solution is then dried according to a normal vacuum subroutine. Once the sample is dry, the chamber is flushed with nitrogen and the sequenator is started.

Removal of Silanes from Cup. After each run the cup must be treated to remove the derivatives. High pH in aqueous solutions will hydrolyze and remove the derivatives from the glass. A 5% solution of NaOH is added to the cup until full and allowed to remain for 1 hr or more at room temperature. The NaOH is removed and the cup is washed exhaustively with water and cleaned according to the recommended procedure.

Procedure for Solid-Phase Sequencing of Derivatized Glass Beads in the Spinning-Cup[18]

The difference between liquid-phase sequencing and solid-phase sequencing of glass beads in the spinning cup is primarily one of sample application. Derivatized glass beads, 50–100 mg, are added, in a dry state, to the sequenator cup, then 0.5 ml of distilled methanol is added and the cup is rotated at low speed. The beads are distributed evenly along the cup walls but not higher than 16 mm. If necessary, a piece of Teflon tubing may be used to aid in the distribution of the glass beads. Once the beads are uniformly distributed, the cup area is closed, then flushed with nitrogen, and the methanol is evaporated by evacuation. After the glass beads are dried on the cup wall, the sequenator is started and allowed to proceed until the coupling and benzene extractions are

[18] D. A. Powers, *in* "Solid-Phase Methods in Protein Sequence Analysis" (R. Laursen, ed.), p. 99. Pierce Chemical Co., Rockford, IL, 1975.

completed; then the machine is restarted at step 1. This procedure will result in a double coupling of the first stage.

Treatment of S-Boc Blocked Peptides. The only departure from the above procedure is for peptides that have been S-Boc blocked to facilitate carboxyl-terminal attachment to glass beads. S-Boc-blocked peptides do not have free α-amino groups, but these can be generated by the removal of the S-Boc function with acid cleavage either in the sequenator (i.e., with HFBA) or in a test tube before addition to the sequenator. If the sequenator is used, the machine is started with the HFBA cleavage, allowed to run through the first coupling step, restarted at stage one for a recoupling and allowed to continue automatically.

Cleaning Procedure. During sequencing with glass beads, some of the beads are washed over into the groove of the cup and eventually into the extraction line. For this and other reasons, the cleaning procedure is important. After the sequenator run is completed, the cup is stopped, filled with water, and aspirated. The wash procedure is repeated several times until the glass beads have been removed. The cup is wiped clean to eliminate the last traces of beads, special attention being paid to the groove of the cup. The cup is filled with water and aspirated. Then the cup is turned manually and wiped clean. The waste line is flushed by opening the waste valve and injecting water through the line with a 50-ml syringe. The same procedure is repeated for the sample collector line. After this cleaning is completed, the cup is cleaned by the routine procedure recommended by the sequenator manufacturer.

Modification of the Liquid-Phase Sequenator into a Laursen-Type Solid-Phase Machine

Liquid-phase and solid-phase sequenators have many features in common. In fact, a solid-phase sequenator is essentially a liquid-phase machine with the spinning cup replaced by a column (see Figs. 2A and 2B). Since the liquid-phase sequenator is more expensive than its solid-phase counterpart, it would be counterproductive to permanently dedicate it to the solid-phase methodology. Temporary modification is, of course, useful for some selected peptides, but the ultimate purpose of these modifications is to develop within a single machine the capability of performing either solid-phase or liquid-phase function on a routine basis (see Fig. 2C). Within the next few years sequenators will probably be manufactured with this dual capability. Until that time, modifications must be considered temporary and experimental. Some sequenators use nitrogen pressure to drive reagents and solvents (see Fig. 2A). Conversion to a solid-phase machine can be facilitated in the Beckman 890C by reconnecting the delivery valve system according to the scheme in

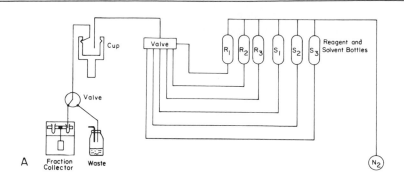

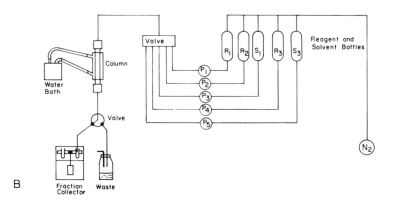

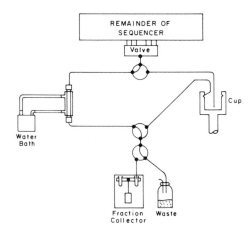

Fig. 2. Sequenator schematic. (A) Edman–Begg-type machine; (B) Laursen-type machine; (C) combined dual machine.

Fig. 3B. Unfortunately, the back pressures generated by many solid-phase columns exceed the operating pressures of the classical Edman–Begg machine. Modifications can be made to the nitrogen drive system to overcome these back pressures, but they must be extensive, including high-pressure bottles that allow operation in excess of 300 psi. In addition, the nitrogen pressures of each reagent and solvent system must be carefully calibrated for individual column systems.

An alternative approach is to introduce reagents and solvent with pumps like those employed by Laursen.[6] The Chromatronix valving system may be utilized,[6] or the Beckman delivery valve system can be modified according to the schematic in Figs. 3C and 3D. Pumps P_1, P_2, and P_3, in Fig. 3C, are Harvard portable, infusion-withdrawal pumps (No. 1100), while P_4 and P_5 are Milton-Roy "minipumps" (No. 196-31). Since these pumps are identical to those recommended for solid-phase machines,[6] Laursen's wiring modifications can be employed. Programming of the pump refill mode should be correlated with the delivery valve inlet mode and the pump infusion mode to operate when the delivery valve is open to the column. This adjustment is critical.

Modification of the JEOL sequenator is easier and less expensive than alteration of the Beckman machine.[18] This is primarily because the JEOL sequenator employs pumps to drive reagents and solvents rather than nitrogen pressure.

The JEOL sequenator can be modified according to the scheme in Fig. 4. A column like the one described by Laursen[6] can be employed, but it has been successfully substituted by a Teflon tube (2 mm × 400 mm) connected directly to the line from the valving system.[18] Glass wool is used in each end if the Teflon column, and the tubing is immersed in a 40-ml glass test tube that is filled with water and placed in the well created by removing the spinning cup. The well's heat source will maintain the temperature within 0.5°. However, for routine use, the Laursen-type glass column is superior. When the glass column is used, a Haake heating bath is employed to control column temperature. Although the JEOL sequenator is generally easier to modify than the Beckman 890C, it does require one wiring modification that should be done by factory-trained personnel. The JEOL Company is happy to provide this service.

Representative Results and Discussion

Polystyrene Resins in Spinning Cup

The use of polystyrene supports in the spinning cup is not yet a routine procedure. It is filled with difficulties, the majority of which re-

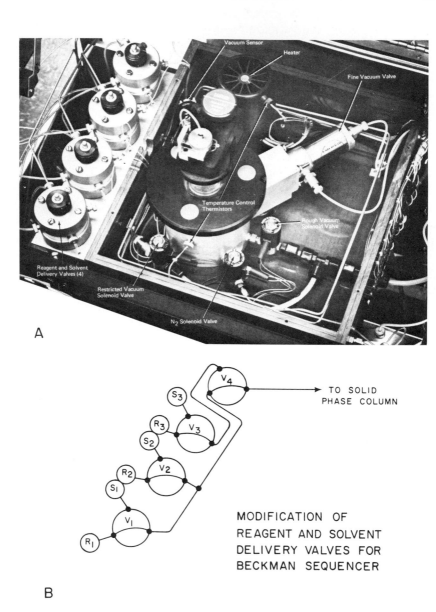

Fig. 3. Plumbing modification of solvent and reagent delivery valve system of Beckman machine. (A) Actual delivery valve system; (B) schematic of modification for nitrogen-pressure-driven conversion.

C

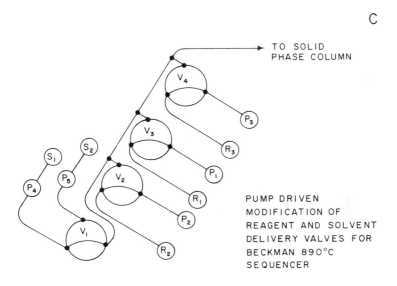

TO SOLID
PHASE COLUMN

PUMP DRIVEN
MODIFICATION OF
REAGENT AND SOLVENT
DELIVERY VALVES FOR
BECKMAN 890°C
SEQUENCER

D

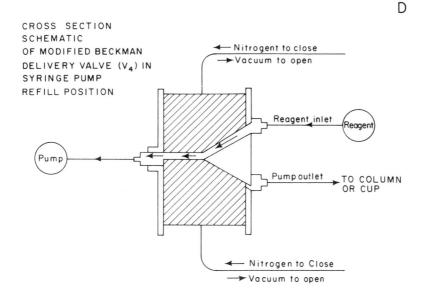

CROSS SECTION
SCHEMATIC
OF MODIFIED BECKMAN
DELIVERY VALVE (V$_4$) IN
SYRINGE PUMP
REFILL POSITION

← Nitrogent to close
→ Vacuum to open

Reagent inlet ← Reagent

Pump →

Pump outlet → TO COLUMN
OR CUP

← Nitrogen to Close
→ Vacuum to open

Fig. 3. (C) Schematic of modification to include pumps; (D) cross section of modified delivery valve with infusion pump.

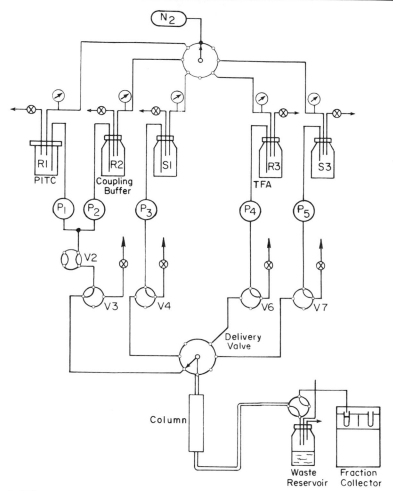

Fig. 4. Schematic of modification of JEOL sequenator into Laursen-type solid-phase machine.

volve about the problem of resin flotation in extracting solvents. Although many investigators have attempted this procedure, few have found the results satisfactory enough to warrant publication. Some laboratories use the method as a means to assess purity of synthetic peptides coupled to resins.[19] Once higher density polystyrene polymers are synthesized, this method should be more useful.

[19] H. D. Niall, G. W. Tregear, and J. W. Jacobs, in "Chemistry and Biology of Peptides" (J. Meienhofer, ed.), p. 696. Ann Arbor Sci. Publ., Ann Arbor, Michigan, 1972.

Sequencing of Peptides Coupled Directly to the Walls of the Glass Cup

Large peptides coupled to the walls of sequenator glass cups generally improve the relative yield by approximately 5% over uncoupled controls; however, the efficiency of coupling is quite variable. Although moderate results have been obtained employing homoserine lactone coupling of large peptides, DITC-coupled tryptic peptides often improve relative yields by as much as 30% over uncoupled controls. Results obtained on carboxyl-terminal coupled peptides are the most variable and even in the best experiments rarely improve relative yields significantly above uncoupled controls.

Solid-Phase Sequencing on Glass Beads in the Spinning-Cup

The use of glass beads in spinning-cup sequenators is clearly the best and most dependable solid support for this methodology. For large peptides, the use of the solid support is only slightly better than the uncoupled control (see Fig. 5.) However, smaller peptides demonstrate relative yield increases of 30% or more over uncoupled peptides (e.g., see Figs. 6 and 7). We have sequenced a number of peptides by this method, and the results have been consistently impressive.[18] Furthermore, different investigators have obtained similar successes on peptides isolated from other sources.[16]

A Comparison of the Two Solid-Phase Approaches

Peptides coupled to glass beads have been sequenced both in the spinning-cup and in a Laursen-type solid-phase machine. Figure 5 illustrates data obtained from Insulin B chain. Both solid-phase methods gave better results than the uncoupled liquid-phase technique, but no statistical difference between either of the solid-phase methods were observed.

Figure 6 illustrates the results obtained from an 18-amino acid tryptic peptide that was sequenced four different ways.[18,20] The methods were: (1) liquid-phase sequencing of an unmodified noncoupled peptide, (2) liquid-phase sequencing of a charged modified peptide,[21] (3) solid-phase sequencing in spinning cup, and (4) solid-phase sequencing in a column.

[20] The peptides αT7 and αT8 were isolated according to the procedure of: D. A. Powers and A. B. Edmundson, *J. Biol. Chem.* **247**, 6694 (1972).

[21] G. Braunitzer, B. Schrank, and A. Ruhfus, *Hoppe-Seyler's Z. Physiol. Chem.* **352**, 1730 (1971).

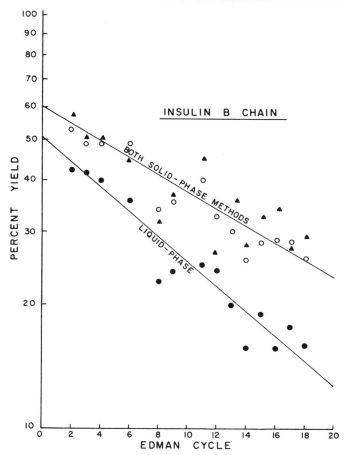

FIG. 5. Edman degradation of insulin B chain. ○, Results of solid phase sequencing in the spinning cup; ▲, results obtained from solid-phase sequencing in a column; ●, results of liquid-phase sequencing of the peptide without chemical modification.

The figure illustrates that both solid-phase methodologies are superior to chemical modification. Although both solid-phase methods were better than the controls, there was no statistical difference between the solid-phase techniques.

Figure 7 illustrates data obtained from a nine-residue peptide.[18,20] These results show essentially the same trend outlined for larger peptides.

We may conclude that glass beads may be used as a solid support in the spinning-cup of any liquid-phase sequenator and that the solid-phase spinning cup method yields results equivalent to solid-phase column

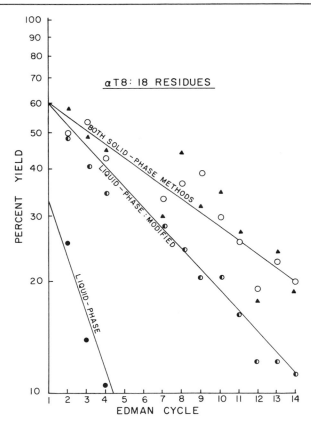

Fig. 6. Edman degradation of the αT8 peptide (Thr-Tyr-Phe-Ala-His-Trp-Ala-Asx-Leu-Ser-Pro-Gly-Ser-Gly-Pro-Val-Lys) [D. A. Powers and A. B. Edmundson, *J. Biol. Chem.* **247**, 6694 (1972)]. The symbols are the same as those of Fig. 5, with the addition of results obtained for the charged modified peptide ◐.

methodology. Although the spinning-cup technique is not yet routinely applicable to solid supports other than glass beads, we are optimistic that the technology will be forthcoming.

Modifications of the Edman–Begg type machine into a Laursen-type sequenator suggest that any commercial sequenator could have an alternate *secondary* use as a solid-phase machine. Although modifications are still being investigated, the prospect of success is promising and the potential of the dual capability appears real. However, if the spinning cup solid-phase methodology continues to equal the column solid-phase results, then sequenator modification would not be necessary. On the other hand, if the spinning-cup approach cannot be applied to other solid sup-

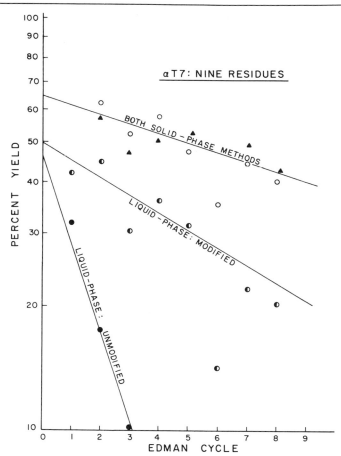

Fig. 7. Edman degradation of the αT7 peptide (Met-Leu-Thr-Val-Tyr-Pro-Gln-Thr-Lys) [D. A. Powers and A. B. Edmundson, *J. Biol. Chem.* **247**, 6694 (1972)]. The symbols are the same as in Figs. 5 and 6.

ports, then it will be advantageous to have the dual capability (i.e., liquid-phase and solid-phase).

Acknowledgments

This work was supported by a grant from The National Science Foundation (No. GB37548) to Dr. Powers and NIH Training Grants to The Department of Biology (Nos. GM57 and HD139). I would like to thank Dr. J. I. Ohms, whose help over the past seven years has been invaluable. Finally, I would like to acknowledge that numerous unnamed scientists have significantly influenced various aspects of the techniques described in the preceding pages.

[33] High-Sensitivity Sequence Determination of Immobilized Peptides and Proteins

By JOHN BRIDGEN

In solid-phase sequencing[1] the peptide or protein is first covalently attached[2-4] to an inert support.[1,3,5-7] The sample is then degraded on the support, using the conventional Edman reaction. Solid-phase methods are particularly suitable for micro sequencing for the following reasons: (1) Once the sample is attached to the resin there is no physical or mechanical loss of peptide or protein; this enables extensive solvent washes to be used between the various degradative steps; (2) the apparatus is readily adaptable for using small volumes of high specific-activity [^{35}S]-PITC[8] during the coupling reaction[9]; (3) the coupling reaction may take place in a wide variety of solvents and the prior removal of salts such as guanidine or SDS is unnecessary[9]; (4) the method is applicable to molecules varying from small peptides to large proteins.

In this article the preparation of supports, attachment of proteins and mechanical adaptations required for high-sensitivity operation as well as the specialized application of sequencing proteins eluted from SDS-polyacrylamide gels will be described.

Preparation of the Support

For molecules smaller than 15–20 amino acids, polystyrene-based supports[1,3] should be used. For larger molecules, a porous glass derivative[5,6] will generally give the best results. All the high-sensitivity work here has been done on porous glass supports of 200–400 mesh and with pore diameters ranging from 75 to 170 Å. Porous glass beads of this

[1] R. A. Laursen, *Eur. J. Biochem.* **20**, 89 (1971).

[2] R. A. Laursen, M. J. Horn, and A. G. Bonner, *FEBS Lett.* **21**, 67 (1972).

[3] M. J. Horn and R. A. Laursen, *FEBS Lett.* **36**, 285 (1973).

[4] A. Previero, J. Derancourt, M.-A. Coletti-Previero, and R. A. Laursen, *FEBS Lett.* **33**, 135 (1973).

[5] E. Wachter, W. Machleidt, H. Hofner, and J. Otto, *FEBS Lett.* **35**, 97 (1973).

[6] J. Bridgen, *FEBS Lett.* **50**, 159 (1975).

[7] E. Atherton, J. Bridgen, and R. C. Sheppard, *FEBS Lett.* **64**, 173 (1976).

[8] Abbreviations: DITC, *p*-phenylene diisothiocyanate; DMAA buffer, 0.4 *M* *N,N*'-dimethylallylamine trifluoroacetate, pH 9.5, in 60% aqueous pyridine; HFBA, heptafluorobutyric acid; PITC, phenyl isothiocyanate; PTH, phenylthiohydantoin; SDS, sodium dodecyl sulfate; TFA, trifluoroacetic acid.

[9] J. Bridgen, *Biochemistry* **15**, 3600 (1976).

type are available from Electronucleonics, Fairfield, New Jersey, and are used directly. Amination of these glasses and their activation with DITC has been described elsewhere.[5,6] Aminated and DITC-activated glasses are available from Pierce Chemical Co., Rockford, Ill. DITC is available from Pierce or Eastman, although batches from the latter supplier are sometimes orange or granular in appearance and require recrystallization from acetone. It is important to use good quality DITC which should be white or pale yellow and crystalline. Activated glasses (porous glass with isothiocyanate functional groups attached) are perfectly stable and may be stored at $-20°$ for at least 12 months.

Attachment of Polypeptides

The peptide or protein (50 pmol to 10 nmol) is dried down from 1% aqueous triethylamine (Pierce, Sequenal grade) in a small test tube (approximately 5×1 cm). The sample is redissolved in 200 μl of DMAA buffer[8] (Pierce, Sequenal grade), and the appropriate isothiocyanatoglass (100 mg) added. The mixture is gently shaken or stirred for 45 min under N_2 at room temperature, and 50 μl of ethanolamine is then added to block excess isothiocyanate groups. Agitation continues for a further 30 min, and the sample is then centrifuged at $\sim$2000 rpm for 5 min. The supernatant is removed and, since this will contain any unattached material, should be retained if the sample is particularly valuable. The protein-glass is resuspended twice in $\sim$5 ml of methanol(redistilled grade) and centrifuged; the supernatant is discarded. It is then washed with $\sim$2 ml dichloroethane and dried in a vacuum desiccator using a water-pump vacuum. Drying takes less than 1 hr.

The DMAA attachment buffer is the same as that used for liquid-phase sequencing of peptides. It has given consistently good results and is available ready-made (Pierce). When stored at $4°$ under N_2, it appears to be quite stable. In cases where the peptide or protein appears to be insoluble in this buffer or if the sample is not denatured, the buffer should be made 6 M in guanidine hydrochloride (Aristar grade, B. D. H., Poole, U. K., or equivalent) or 0.1% (w/v) in SDS (Sequenal grade, Pierce or equivalent). In these cases it may be necessary to check the pH before proceeding with the attachment. Other solvent systems, which on occasions have been successfully used for attachment, include 50% aqueous N-methylmorpholine-trifluoroacetate, pH 9.5, 0.1 M Tris-HCl, pH 9.0, and dimethylformamide containing 2% (v/v) triethylamine. In principle any solvent system within the pH range 9–10 will be usable provided the sample is soluble and that there are no functional groups present that will react with primary amine or isothiocyanate groups.

Before sequencing, the dried glass with protein attached is packed into a jacketed column (2 mm i.d. $\times$ 10 mm). The lower unjacketed part is filled with washed nonporous glass beads (Pierce), the derivatized support is added, and the column is filled to the top with packing glass. Altex-type columns have proved more durable than those of the Chromatronix design.

Sequencer

Commercially available solid-phase sequencers need slight modifications for high-sensitivity operation. The modifications are incorporated solely to increase the efficiency of incorporation of the [35S]PITC used for coupling. The instruments (Rank Hilger APS 24, Rank Hilger, margate, Kent, U.K.; LKB 4020, LKB-Biochrom Ltd., Science Park, Milton Road, Cambridge, U.K., and Sequemat Inc., Boston, Massachusetts, U.S.A.) have proved to be essentially identical under normal low-sensitivity operation. However, instruments with variable-speed pumping systems (LKB) are more easily adapted for high-sensitivity work.

The following modifications are necessary:

1. The reagent and solvent systems are changed according to Table I. The methanol and TFA systems remain as they are. In place of dichloroethane is buffer, in place of buffer is PITC and in place of PITC is [35S]PITC.

2. The [35S]PITC delivery system is modified so as to reduce the inherent volume as much as possible. The reservoir consists of a 5-ml Reacti-Vial (Pierce) with a low-pressure N_2 line passing just through the septum. This is the line normally used for flushing the conventional PITC reservoir. The Reacti-Vial is mounted close to the [35S]PITC de-

TABLE I
MODIFIED DELIVERY SYSTEMS FOR HIGH-SENSITIVITY SOLID-PHASE SEQUENCING

	Normal	High-sensitivity
Reagent 1	5% PITC	2% [35S]PITC
Reagent 2	Buffer	5% PITC
Reagent 3	TFA	TFA
Solvent 1	Methanol	Methanol
Solvent 2	Dichloroethane	Buffer

livery valve so that the volume of the Teflon tubing (0.3 mm i.d.) between the two is minimal. The vial is maintained under positive N_2
pressure.

3. The [35S]PITC delivery pump (normally the conventional PITC
pump) is connected to its valve by a short length of 0.3 mm i.d. Teflon
tubing. Where syringe pumps are used (Rank, Sequemat) the normal
5-ml syringe should be replaced by a 0.5-ml syringe (Hamilton Gas-
Tight).

4. In order that unlabeled PITC and buffer may flow concurrently
in the new configuration, it is necessary to modify the PITC delivery
valve. This may be achieved by incorporating an additional 4-way center
plus mixing-block as shown in Fig. 1 or by using a specially drilled valve-
slider.[1] The former arrangement is used by Rank to allow these
reagents to mix in conventional operation and the latter by LKB and
Sequemat. Incorporation of a second valve slider requires the operating
N_2 pressure for these valves to be raised to 95 psi. It may also be necessary to override electrical safety interlocks which would prohibit the
pumping of PITC and buffer in this new configuration. One manufacturer

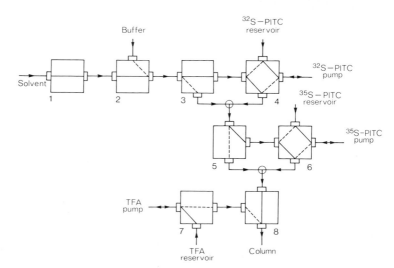

FIG. 1. Modified valve system for high-sensitivity sequencing. Valve 1 is used
for methanol and valve 2 for buffer. Valves 3 and 4 are controlled together by a
single actuator. In the off position, the PITC pump may refill and buffer alone may
be pumped through. In the on position either PITC alone or buffer and PITC
together may be pumped. A similar arrangement is used for valves 5 and 6 in the
delivery of [35S]PITC. In the off position the solid line marks the direction of flow
through the valve. The dashed line shows the flow direction after operation of
the valve.

(LKB) produces a solid-phase sequencer with these modifications already in place. Other manufacturers offer conversion kits using similar principles.

The Sequencing Program

A high-sensitivity sequencing program[9] is shown in Table II. This program may have to be modified to take account of the different physical characteristics of various instruments. The following points should be checked:

1. The initial buffer flow time is that required for the buffer to reach the end of the column. The 1-min deliveries of buffer are to ensure maximum utilization of [^{35}S]PITC in the delivery line.

2. The 1-min [^{35}S]PITC delivery time allows this reagent to come up to pressure before mixing with the faster-flowing buffer stream.

3. It is important to check that the combined flow rate of the buffer with both labeled and unlabeled PITC equals the sum of the individual flows.

4. The delay after the combined buffer and PITC delivery is to ensure that coupling is complete. A slower buffer flow rate may have the same effect.

TABLE II

PROGRAM USED FOR HIGH-SENSITIVITY SOLID-PHASE SEQUENCING[a]

Function	Program time (min)	Flow rate (ml/min)
Methanol	1–2	1.0
Buffer	3–9	0.07
Delay	10–14	—
[^{35}S]PITC	15	0.02
Buffer + [^{35}S]PITC	16–20	0.09
Delay	21–35	—
Buffer	36	0.07
Delay	37–40	—
Buffer	41	0.07
Delay	42–45	—
Buffer	46	0.07
Delay	47–50	—
Buffer + [^{35}S]PITC	51–58	0.12
Delay	59–73	—
Buffer	74–79	0.07
Methanol	80–94	1.0
TFA	95–97	0.15
Delay	98–99	—
Methanol	100	1.0

[a] All program times are inclusive. Thus, the initial methanol delivery lasts for a total of 2 min. Column temperature is 53°.

5. The TFA delivery should be timed so that the acid just reaches below the end of the column. After a 3-min delay, methanol delivers the thiazolinone to the fraction collector.

The program should be checked on a macro scale (50–100 nmol) before scaling down the amount of sample. Overlap should be no more than is encountered with the conventional program, and incorporation of radioactivity should be about 50% of theoretical. In this system excessive overlap generally arises from incomplete coupling rather than incomplete cleavage and in this case, the buffer and PITC flow rates, etc., should be carefully checked at the waste outlet with the column in position.

Chemicals

All the reagents used for the sequencing are obtained from Pierce (Sequenal grade). Any good quality glass-distilled grade of methanol should be suitable. Isotope is obtained from the Radiochemical Centre, Amersham, U.K., who will prepare 3 months supply, dispensed in sealed ampoules such that, if used evenly over this period, each ampoule will contain 2 mCi of ^{35}S at the time of use. The solvent for this material should be specified as acetonitrile-Sequenal grade (Pierce). Sulfur-35 has the advantages that it is relatively inexpensive, particularly if large quantities are purchased and that it may be detected by autoradiography. It has a half-life of approximately 87 days. Tritium has a longer half-life (12.3 years) and may be less expensive. However, it is much more difficult to detect on TLC plates. Presensitization of the X-ray film followed by exposure to the sample at $-70°$[10] has allowed 1000 cpm of [^{3}H]PTH-derivative to be detected on a polyamide layer in 3 days whereas the same number of cpm of ^{35}S could easily be detected in 24 hr. Carbon-14 is too expensive and not available with sufficiently high specific activity to be useful for this application.

Although it is true that the criteria for reagent and solvent purity are less demanding for solid-phase than for liquid-phase sequencing, it is worth investing in reagents prepared especially for sequencing since the small extra cost is adequately repaid by the gain in confidence, if not by better results.

Various solvents have been used for the postcoupling wash. These include dichloroethane, ethyl acetate, methanol–dichloroethane (1:1), and other mixtures of these solvents. The addition of 1% acetic acid and of dithiothreitol has also been tried. None of these combinations proved superior to methanol alone. The omission of dichloroethane made no difference to the repetitive yield using porous glass supports.

[10] R. A. Laskey and A. P. Mills, *Eur. J. Biochem.* **56**, 335 (1975).

Sequencer Operation

The reagent and solvent bottles are filled as follows:
1. Methanol, 1 liter
2. Buffer, 50 ml
3. PITC (5% in acetonitrile), 20 ml
4. PITC (2% in acetonitrile containing 2 mCi of [35S]PITC of approximately 300 mCi/mmol) 4 ml[11]
5. TFA, 100 ml

The buffer, TFA and PITC reservoirs are pressurized with nitrogen to 0.5 bar before use. The methanol vent is left open. The 35S vial is left under N_2 pressure, i.e., with the purge valve open. Marker mixture containing 1 nmol of the common PTH-derivatives is placed in each, preferably conical, fraction-collector tube. No other modifications are necessary.

Conversion to PTH-Derivatives

At the end of the run, the samples from the fraction collector are dried at 50° under a stream of N_2, 100 μl of 0.2 M HCl is added to each tube, and the fractions are then incubated at 80° for 10 min; 0.7 ml of ethyl acetate is then added. The two phases are mixed and centrifuged, and the supernatant organic phase is withdrawn into a conical glass tube (5 × 0.5 cm). The organic phases are dried under N_2, the sides of the tube are washed with 20 μl of ethyl acetate, and this is once again dried. The samples are then redissolved in 5μl of ethyl acetate prior to application to the TLC plate.

The use of HFBA[8] allows the PTH-derivatives of histidine and arginine to be extracted directly from the aqueous phase as their heptafluorobutyrate salts.[9] However, the amount of non-PTH material extracted from the aqueous phase is also increased, and since this may interfere with certain PTH identification systems, this system is no longer used.

Identification of [35S]PTH-Derivatives

Polyamide layers are cut with a guillotine into 5 × 5-cm squares. Layers from Cheng-Chin, Analtech (ML 1010) and Schleicher and Schuell (F1700) have all proved satisfactory. Plates from the latter source have consistently given the best results, and these are recommended. Plates from Analtech give very high contrast under UV light, but the softness of the layer makes spotting-out difficult.

[11] It may be necessary to increase this volume depending on the characteristics of the instrument being used.

An origin is marked on one side of each plate approximately 8 mm from two of the edges. The origin corner should be chosen so that solvent does not have to flow across a ragged or frayed edge. Half of each sample is applied to the origin with a 2-μl microcapillary tube such that the diameter of the origin spot is less than 2 mm. The plates are developed in square tanks (10×8 cm) with four plates to a tank. Each plate is placed into and removed from the tank using plastic forceps, and as far as possible the edge of the plate is kept parallel to the solvent layer.

Of the various solvent systems suggested there appears to be no important difference between two of these,[12,13] although the resolution on the 5×5-cm plates is better with these solvents than with the third type.[14]

The system recommended is:

Solvent 1: toluene-pentane-acetic acid (10:5:3.5, v/v)
Solvent 2: acetic acid–water (7:20, v/v)

The plates are developed, then the separated marker mixture is examined under UV light to ensure that the plates have been correctly developed. If not, the second half of the sample is applied to a fresh plate and this is then developed as above using fresh solvents. The plates are mounted in a cardboard folder using a strip of double-sided adhesive tape. A sheet of X-ray film (Kodak Auto-Process, or if this is unavailable, Kodak Blue Brand) is placed over the plates, and they are allowed to expose in the dark for 24–30 hr (slightly longer for Blue Brand).

After development of the film, the spots on the autoradiograph are matched to the internal marker spots to provide an unambiguous identification.

It should be noted that this method is, at best, only semiquantitative. In particular, overexposure of the films will give an apparently much higher background. It is possible to make the method quantitative by use of a separation technique other than TLC (see below) followed by scintillation counting. Cutting spots from the plate and counting these for radioactivity gives somewhat variable results.

Proteins Eluted from Polyacrylamide Gels

In selecting a gel system it should be remembered that anything that reacts with isothiocyanate—amino acids, urea, ampholytes, etc.—will

[12] M. R. Summers, G. W. Smythers, and S. Oroszlan, *Anal. Biochem.* **53**, 624 (1973).
[13] D. K. Kulbe, *Anal. Biochem.* **59**, 564 (1974).
[14] J. Silver and L. Hood, *Anal. Biochem.* **67**, 392 (1975).

have to be removed before attachment of the sample to the support and it is simpler if these molecules are not used initially. The SDS–sodium phosphate gel buffer system[15] can be recommended, but any system, with or without detergent, that satisfies the above criteria should be suitable. The percentage of acrylamide in the gel appears to make little difference. Unfortunately, commercial acrylamide, bis-acrylamide, and occasionally SDS, often contain impurities that may interfere with the running of the gel, the elution of protein bands from the gel, and the subsequent determination of their sequence. Clearly, these impurities must be removed, and the most efficient method is to presoak the gel in running buffer (see below). To date the method has been used with slab-gels but there is no practical reason why tube-gels should not be used. The method is as follows:

1. Prepare gel exactly as normal.

2. Remove polymerized gel from the gel plates and transfer to a vessel containing running buffer. Allow to soak for approximately 18 hr.

3. Replace gel on back plate. The gel swells during the dialysis so that 1–2 cm will have to be cut from the bottom and from one side with a razor blade.

4. Replace spacers and front plate and reseal edges with molten Vaseline petroleum jelly or equivalent.

5. Place gel in electrophoresis equipment and fill with running buffer. Check carefully for leaks.

6. Load sample. This may be as a single broad band across the gel or as two narrower bands side by side, one of which will be used for detection. A dye marker should be mixed with the sample to show the position of the elution front. It is advisable not to load samples close to the edges of the gel plates, since they may not run very evenly.

7. After the gel has run, the position of the dye marker is noted and the detection bands are then cut out as long strips with a razor blade or, if a single broad band is being run, guide strips are cut from both edges of the band. These are stained with Coomassie Blue and destained the normal way.

8. Bands with the same mobility relative to the marker as bands of interest detected on the stained strip are cut from the gel. These are further dissected into small cubes (approximately 1–2 mm square) and transferred to screw-top tubes (approximately 4×1 cm). DMAA buffer (0.5–1.0 ml depending on the size of the gel strip) containing 0.1% (w/v) SDS (Pierce, Sequenal grade) is added and the tubes shaken under N_2 at 37°C for 18–24 hr.

[15] K. Weber and M. Osborn, *J. Biol. Chem.* **244**, 4406 (1969).

9. The supernatants are removed with a 1-ml disposable hypodermic syringe, a 13-mm, 45 μm Millipore filter is fitted to the end of the syringe using a Swinney 13 adapter and the solution is passed through into a small test tube (5 $\times$ 1 cm). The procedure is repeated using 0.5 ml of DMAA buffer as a wash. The combined solutions are then concentrated under a stream of N_2 to approximately 0.5 ml.

10. Activated glass (100 mg) is added, and coupling and sequencing proceed as described above.

Alternative systems of elution have been investigated. These include: the use of buffers containing 6 M guanidine, increasing the SDS concentration to 1% (w/v), increasing the time and temperature of elution, the use of a variety of different buffer systems, and the use of a second extraction with the DMAA/SDS buffer. None of these significantly increased the elution yield when S-[^{14}C]carboxymethyl lysozyme was used as a test protein. Yields of protein recovered from bands that had been stained in Coomassie Blue or fixed in methanol:acetic acid:water (5:7:100, v/v) were considerably reduced. Attempts to solubilize stained bands by replacing the Coomassie Blue with trichloroacetic acid followed by solubilization in DMAA/SDS also were unsuccessful.

Addition of a dansylated protein to the sample can sometimes provide a useful guide to the straightness of the gel bands.

Examples

Some of the results obtained by this method are summarized in Table III. For peptides attached to polystyrene supports, it is necessary to include an additional dichloroethane wash into the program. In these cases a conventional program is used with a single coupling step using [^{35}S] PITC (4 mCi) in 20 ml of a 5% solution of unlabeled PITC in acetonitrile. The feasibility of attaching small quantities of peptide is demonstrated, however, and the use of a six-channel sequencer (see p. 323) should allow picomole quantities of peptides to be sequenced. An advantage of this technique is that amino acid contamination either from the purification media or from the atmosphere does not interfere.

From the results obtained so far, certain principles may be formulated: (1) The coupling medium has little effect on the subsequent sequencing efficiency. (2) The state of the cysteine residues appears to be unimportant. Thus, lysozyme may be attached and sequenced without prior reduction or oxidation of its disulfide bridges. It is, however, easier to identify cysteine residues after carboxymethylation. (3) Proteins eluted from SDS-gels give less clean sequences than those purified by other techniques. The reason for this is not yet clear. (4) Hydrophobic pep-

TABLE III

PROTEINS AND PEPTIDES SEQUENCED BY THE HIGH-SENSITIVITY SOLID-PHASE PROCEDURE[a]

Protein/peptide	Molecular weight	Quantity	Solvent	Sequence	Support[b]
Lysozyme[c]	14,300	1 µg (70 pmol)	Buffer (DMAA)	VFGRCELAAAM-RHGLDNY	APG-75
Lysozyme[c]	14,300	10 µg (700 pmol)	SDS (eluted from gel)	VFGR-ELAAAM-RHGLDNY	APG-75
Methionine tRNA synthetase[c]	65,000	50 µg (750 pmol)	6 M Guanidine HCl	E--TFYLTTPIYYP	AEAPG-125
Superoxide dismutase[c]	20,000	50 µg (2.5 nmol)	1% SDS	FELPALPYDYDALEPHID	APG-75
Trypanosome antigen I[c,d]	65,000	10 µg (140 pmol)	SDS (eluted from gel)	NHNGL-LQ-AEAIC	APG-125
Trypanosome antigen II[c,d]	65,000	50 µg (700 pmol)	6 M Guanidine HCl	EALEY-TWTNHCG	APG-75
Trypanosome antigen I[d] fragment	22,000	90 µg (4 nmol)	Buffer (DMAA)	NHNGL-LQ-AEAIC-M	APG-75
Retinal-binding peptide[e]	900	2 nmol	Dimethylformamide	See ref. 8	TETA
β_2-Microglobulin[f]	12,000	50 µg (4.2 nmol)	SDS (eluted from gel)	QRTP-IQVYSRHPAE	AEAPG-120
Histocompatibility antigen[f]	43,000	50 µg (1.1 nmol)	SDS (eluted from gel)	S-SMRY$^{\mathrm{FFTA}}_{\mathrm{YYSS}}V^{\mathrm{A}}_{\mathrm{S}}$RPG	AEAPG-120

[a] Except for the retinal-binding peptide, the sequences shown all begin at residue 2 since residue 1 remains attached to the resin and is not identified.

[b] APG, aminopropylglass [E. Wachter, W. Machleidt, H. Hofner, and J. Otto, FEBS Lett. **35**, 97 (1973)].; AEAPG, aminoethylaminopropylglass [J. Bridgen, FEBS Lett. **50**, 159 (1975)]. The number following a support refers to the pore diameter of the glass. TETA, triethylenetetraminepolystyrene [M. J. Horn and R. A. Laursen, FEBS Lett. **36**, 285 (1973)].

[c] J. Bridgen, Biochemistry **15**, 3600 (1976).

[d] P. J. Bridgen, G. A. M. Cross, and J. Bridgen, Nature (London) **263**, 613 (1976).

[e] J. Bridgen and I. D. Walker, Biochemistry **15**, 792 (1976).

[f] J. Bridgen, D. Snary, M. J. Crumpton, P. Goodfellow, and W. D. Bodmer, Nature (London) **261**, 200 (1976).

tides attach and sequence with better yields than hydrophilic peptides. This is also true for macro-scale solid-phase sequencing.

Problems

With this, as with any other high-sensitivity sequencing method, it is important to gain confidence and competence at normal sensitivities (10–100 nmol) first. Without this experience it will not be easy to set up the high-sensitivity program or to diagnose faults when these occur. Many of the problems that may be encountered with this high-sensitivity method are those that are also found when working at normal levels. Specific points are discussed below:

Low Incorporation of Radioactivity. Provided the isotope used is that described here, this is almost certainly due to an error in reagent delivery or programming. Check the flow rate of the [^{35}S]PITC and of the [^{35}S]-PITC/buffer combination. It may be necessary to increase the flow rate of the former or increase the delay time when [^{35}S]PITC pumps alone to allow this reagent to come up to pressure. It may also be found advantageous to increase the delay after the first coupling to allow completion of the reaction.

High Overlap. This is generally a coupling rather than a cleavage problem. If incorporation of radioactivity is satisfactory (check this by reversing the coupling order, see ref. 9), increase the flow rate of unlabeled PITC, and/or increase the delay-time after coupling. The delay must be timed to occur when both buffer and PITC are still on the column.

High Background. First, run some blank cycles (pack column with isothiocyanate glass which has been blocked with ethanolamine and work up the fractions as normal). If any background spots are observed, either the reagents or the glass is contaminated. If this test is negative, the background presumably arises from the protein sample. Possible remedies include decreasing the cleavage time and/or temperature. Since these changes may also result in an increase in overlap, it will be necessary to strike a balance between the two.

Background spots that do not correspond to PTH-amino acids often arise, providing the instrument is functioning correctly, from impurities in particular manufacturers' TFA. This reagent must be completely colorless and should not be overdried since this may result in anhydride formation.

Low Coupling Yield. It is not worth measuring coupling yields when working at high sensitivity. While it is true that whatever protein material remains in the supernatant after coupling is unattached, the converse does not hold. The protein may simply adsorb onto the glass surface without binding and will be washed from the column at the first TFA step. The only useful check on coupling yield at this level is to pack the presumptive sample into a column and attempt to sequence it.

Low coupling yields may be due to the poor quality of the support. Each fresh preparation of activated glass should be test-coupled with lysozyme, and the first ten or so residues of sequence determined. Activated glass should be stored at −20° when not in use. Another possibility is contamination of the sample with amines. Since the quantity of protein used is frequently in the picomole range, only a small quantity of, for example, an ammonium salt or amine contamination of the coupling medium, may interfere. Ammonia may be removed from protein samples by drying down from 1% aqueous triethylamine (Pierce, Sequenal grade). Urea solutions, at neutral or alkaline pH, will form appreciable quantities of cyanate and should be avoided wherever possible at all stages of the protein purification and coupling.

The protein may be aggregated or partly insoluble in the coupling buffer. If in doubt, make the coupling solution 6 M in highly purified guanidine. In somewhat limited experience, the state of the cysteine residues has not been found to be important. However, the safest course would be to oxidize these to cysteic acid. The oxidation may be performed directly in the sample tube and the performic acid freeze-dried *in situ* so that dilution or transfer steps are avoided. Carboxymethylation has the advantage that the carboxymethylcysteine residues formed are easier to identify, but this reaction requires the removal of excess reagents, an additional step which may be undesirable.

The choice of support may influence the final attachment yield, although this has not yet been fully investigated. It has been reported[16] that N-(2-aminoethyl)-3-aminopropyl glass gives better coupling of larger proteins. For small proteins (M_r less than 25,000) a pore diameter of 75 Å appears optimal, but larger molecules may not be able to penetrate these supports. Glasses with nominal pore diameters of 120 and 170 Å have been successfully used for proteins with M_r between 25,000 and 65,000, but the extent of penetration will depend on the shape as well as molecular weight. Hence, proteins coupled in solutions containing SDS may require a support with larger pore diameter than would be used for the same protein coupled in buffer. It should be remembered that in going from 75 to 120 Å pore diameter, the surface area of the glass is reduced by approximately 40%, and in going to 170 Å beads, the surface area is reduced by an additional 30–35%.

[16] E. Wachter, H. Hofner, and W. Machleidt, *in* "Solid-Phase Methods in Protein Sequence Analysis" (R. A. Laursen, ed.), p. 31. Pierce Chemical Co., Rockford, IL, 1975.

Identification Problems. Difficulties in resolving the PTH standards on the TLC plates often arise when fresh solvent 1 (p. 328) is not used. This solvent should be made up fresh daily. Variation of the concentration of acetic acid in this solvent may provide improved separation. Developing the plates in small ($10 \times 8 \times 8$ cm) chromatography tanks rather than beakers often gives better solvent fronts and eliminates edge affects. It is also faster since several plates may be run simultaneously in one tank.

Other Problems. The chromatographic systems used do not distinguish PTH-leucine from PTH-isoleucine. This is not a serious problem in microsequencing, and hopefully a suitable solvent system will soon be found for resolving this pair. The NH_2-terminal amino acid is not identified. This again is unimportant since the method only attempts to provide limited sequence information, and it is usually unimportant whether this includes residues 1–20 or 2–21. The method does not identify lysine residues. This is a more serious problem, whose solution lies in finding alternative reliable attachment procedures. It does appear to be possible[17] to identify small quantities of PTH-serine, a derivative that often undergoes dehydration and polymerization.

Future Developments

The major development in this method will not be higher sensitivity, nor will it be the production of longer sequences although both of these are clearly important. It will be the development of better coupling methods and the extension of the technique to give sequence information on small quantities of peptide. At present there are no good microsequence methods for peptides. This is mainly because the spinning-cup sequence methods are not applicable to small quantities of peptide. Solid-phase sequencing was originally developed for small peptides and works very well with these molecules. There is every prospect of fractionating CNBr fragments on polyacrylamide gels, eluting and sequencing the peptides. Methods exist for attaching CNBr fragments directly to glass supports[6] so that all the residues, including the NH_2-terminal and internal lysines, may be identified.

A further modification, currently under investigation, is the development of a quantitative microsequencing system. The sequencing is carried out exactly as described above, but each fraction is examined by

[17] J. Bridgen, D. Snary, M. J. Crumpton, P. Goodfellow, and W. F. Bodmer, *Nature* (*London*) **361**, 200 (1976).

high-pressure liquid chromatography. The eluent from the chromatograph is directed to a fraction collector that advances one step each time that a peak is detected. Thus, with a suitable time delay, each peak may be collected as a single fraction. Since all the samples contain internal PTH-derivative standards, a complete set of peaks will be found for each sequencer fraction. Scintillation counting of a known aliquot of the fractions then gives a quantitative value for the amount of each [^{35}S]PTH-derivative present in each peak. A similar system, using back-hydrolysis followed by amino acid analysis of each fraction has recently been described.[18]

[18] D. J. McKean, E. H. Peters, J. I. Waldby, and O. Smithies, *Biochemistry* **13**, 3048 (1974).

[34] Improved Manual Sequencing Methods

By GEORGE E. TARR

Manual sequencing methods that utilize the Edman chemistry[1] have provided most known amino acid sequences, but automation is assuming an increasingly important role and will provide most of the new information. The main reasons for the displacement of manual methods are the benefits generally associated with automatic operation: relief for the operator from the tedium of repetitive tasks, greater reproducibility, and less dependence upon human training and experience. The artistry of the manual mode of operation tends to inhibit the incorporation of new procedures into an adequately functioning system. For example, the dansyl-Edman approach[2] remains popular despite being intrinsically less efficient and informative than direct methods (because of repetitive removal of peptide and hydrolysis) and despite advances in identification of thiohydantoins,[3] which have eliminated most advantages of this formerly most sensitive technique. The current trend in fragmentation strategies is to produce a few large peptides[3,4] that are more appropriate to the extended capability of automatic equipment. This chapter will consider the properties and merits of the manual mode and its chemical and

[1] P. Edman and A. Henschen, *in* "Protein Sequence Determination" (S. B. Needleman, ed.), p. 232. Springer-Verlag, Berlin and New York, 1975.
[2] W. R. Gray, this series Vol. 25, p. 333 (1972).
[3] See elsewhere in this volume.
[4] R. Walter, *Biochim. Biophys. Acta* **422**, 138 (1976).

mechanical implementation. Information provided earlier[5] will not be duplicated.

Any sequencing strategy can be realized in either an automatic or manual mode. This option is well illustrated by the recommendation that the spinning-cup sequenator be operated manually in difficult cases.[6] The spinning-cup also has increasingly utilized the liquid-partition strategy often considered characteristic of manual sequencing.[7] Thus, the term "manual method" does not specify sensitivity, speed, ease of identification, repetitive efficiency, etc., for these are properties of a strategy rather than of a mode. In fact, manual methods have all the options and capabilities of automatic methods plus others not yet automated, such as multiple sample processing. Besides being flexible, manual methods are cheap, especially with regard to capital outlay. A comparison of the merits of each mode suggests the continued application of manual methods to: (1) limited sequencing projects with small budgets; (2) large numbers of smaller-sized peptides (to about 20 residues); (3) cases requiring special handling; (4) characterization of NH$_2$-terminal regions for identification; and (5) screening prior to attempts at complete, automatic degradation. The second has in fact been the typical application and will be the main concern of this review.

Abbreviations

The following abbreviations and proprietary names are used in this chapter: TMA, trimethylamine; PITC, phenyl isothiocyanate; PTC, phenylthiocarbamyl; PTH, phenylthiohydantoin; ATZ, anilinothiazolinone; HFIP, hexafluoroisopropanol; TFA, trifluoroacetic acid; HFBA, heptafluorobutyric acid; EtSH, ethanethiol; diox, dioxane; MeOH, methanol; t-BuOH, tert-butanol; hep, heptane; Pyr, pyridine; EA, ethyl acetate; PTFE, polytetrafluoroethylene (Teflon, Dupont Chemical Company); TLC, thin-layer chromatography; standard 3-letter IUB code is used for all amino acid residues.

The Edman Chemistry

Understanding of the Edman chemistry has advanced but little since Edman's original series of studies, while instrumentation has changed a great deal. However, successful sequencing depends more on fundamental

---

[5] See this series Vols. 11, 25, and 27.

[6] M. D. Waterfield and J. Bridgen, in "Instrumentation in Amino Acid Sequence Analysis" (R. N. Perham, ed.), p. 41. Academic Press, New York, 1975.

[7] For instance, see W. G. Crewther and A. S. Inglis, Anal. Biochem. 68, 572 (1975).

understanding than on hardware. The factors considered here are relevant to all approaches, but are most specifically related to liquid-partition with direct identification of the derivatized products. A detailed procedure for manual sequencing based on these considerations is given under Procedure and Fig. 4.

Coupling Reaction

Pyridine/water solvents are commonly used because of their excellent solvent power for peptides; a tertiary amine, usually volatile and often buffered, is added to deprotonate the α-amino group.[1] Automatic instruments and some manual methods[8] substitute an alcohol, usually n-propanol, for pyridine, apparently to eliminate products of side reactions often found with the latter solvent. Freshly distilled pyridine[9,10] poses no difficulties, whereas alcohols give poor discrimination of nonpolar peptides vs excess reagent in the subsequent wash step. TMA salts of peptides resist extraction ("wash out") more than triethylamine or dimethylbenzylamine salts, with dimethylallylamine being nearly as good. Buffering is not essential with these weakly basic amines and carboxylic or stronger acids encourage subsequent salt accumulation, but the weak acid HFIP ($pK_a = 9.3$) stabilizes the pH and has no ill effects.

PITC is often added neat, presumably to reduce the decomposition of stored reagent. However, small quantities are difficult to measure and a solution in good pyridine, especially with HFIP added, remains completely stable even if left for a week or more at 20° under N_2. The solution is good for months if kept in the freezer when not in use.[11]

The common reaction temperature for coupling is currently near 50°. Above 60°, excessive by-products and/or blocking reactions[12] are said to result, but there seems to be no advantage to temperatures below 50°. At 50° the reaction of a soluble peptide was reported to be complete within 20 min.[8,10] However, recent tests with the coupling medium de-

[8] J. D. Peterson, S. Nehrlich, P. E. Oyer, and D. F. Steiner, *J. Biol. Chem.* **247**, 4866 (1972).

[9] High-grade pyridine (e.g., spectro 99⁺% from Aldrich Chemical) should be distilled from NaOH pellets, ninhydrin, or p-toluenesulfonyl chloride and stored at 0° except for the small portion in current use. Other solvents from Burdick and Jackson Laboratories have been acceptable without redistillation, so their pyridine may be as well.

[10] G. E. Tarr, *Anal. Biochem.* **63**, 361 (1975).

[11] G. E. Tarr, unpublished data, 1976.

[12] R. A. Laursen, A. G. Bonner, and M. J. Horn, *in* "Instrumentation in Amino Acid Sequence Analysis" (R. N. Perham, ed.), p. 73. Academic Press, New York, 1975.

scribed here (Fig. 4) have shown first order half-lives of less than 9 sec, i.e., a reaction better than 99% complete at 1 min. Extensive use of a 3 min coupling time, resulting in a reduction of total cycle time to 14 min when one peptide was processed manually, has not revealed any significant incompleteness of reaction with any small peptide or readily soluble large one (including whole proteins).[11]

Washing

The purpose of washing the coupling mixture is to remove excess reagent and any by-products, ideally without removing any peptide or permitting further chemical reaction. In fact, only the most hydrophobic peptides are likely to be dissolved by the apolar solvents required, so the ideal case can very nearly be achieved. Three conditions must be met in the liquid-partition approach usually taken: (1) the extracting solvent must only be polar enough to remove most of the reagent in each wash; (2) the amines used in the coupling reaction must not form an extractable salt of the peptide (e.g., dimethylbenzylamine should be avoided); (3) emulsions must be prevented from forming or, failing that, be broken by centrifugation. These conditions appear to be met best with a graded series of washes that become more polar as amine is removed from the lower phase. The usual wash solvent is benzene, but benzene-based systems tend to emulsify. Hep/EA systems are superior, for they clear rapidly even after vigorous mixing and can be adjusted through the appropriate range of polarity (10:1 to 1:1); the long-chain hydrocarbon is the critical factor in clarification. A few peptides containing hydrophobic groups may still encourage emulsification to an unacceptable degree, so these will require a gentler mixing method or centrifugation between washes. More than 95% of the upper phase may be safely removed from individual vessels, but great care should be taken to avoid the interface, where most of the peptide may have precipitated. Chemical reactions are minimized by washing at room temperature rather than at 50°.

Adequate drying of the lower phase (usually reduced to a very small volume by the washing) is more important for the removal of residual amine than of water, as such amine at the cleavage step will lead to progressive salt accumulation, in turn limiting the number of cycles that can be performed with the liquid-partition strategy. As with washing, it seems best to dry at room temperature, although 50° has been used with apparent success.[8] Using a vacuum pump to dry and HCl cleavage, salt does not begin to interfere until cycle 15–20. An alternative or additional strategy to liquid-partition is the washing of dry residues with a moderately polar solvent. This may result in the loss of short or hydrophobic peptides because, despite the low solubility of the pure PTC-peptide, there are usually low levels of contaminants that enhance solution to an unacceptable degree.[10] However, with medum to large polar peptides, extraction of dry material leads to no loss of peptide, no salt accumulation, and an exceptionally clean analysis. The least polar solvent that adequately removed TMA-HCl was EA:acetone 1:1; the proportion of ethyl acetate may be increased for salts of TFA. Acetone particularly should be of the highest purity available[9] or redistilled to remove acidic and oxidizing contaminants.[1]

Cleavage

Virtually the only cleavage agents used currently are perfluorocarboxylic acids, particularly TFA. This acid has several advantages: great solvent power for peptides, anhydrous condition (which prevents hydrolysis of peptide bonds), high purity, and volatility. However, it also has drawbacks: tenacious association with peptides, enhancement of the solubility of peptides in organic solvents, and nearly complete dehydration of Ser and Thr derivatives. The popularity of TFA requires an elaboration of these points. The extraordinary solubility of peptides and proteins in organic solvents conferred by pretreatment with TFA (and presumably other perfluoroacids) is generally unappreciated. The solubilities of ATZ's are also enhanced, a phenomenon recognized in the spinning-cup literature[6]; indeed, the 1-chlorobutane extraction would be quite ineffective were it not for the residual acid. However, the author's impression is that the differential solubility of these two classes of TFA salts or complexes is less than that of the corresponding chlorides. The difficulty in removing residual TFA also leads to rapid salt accumulation when the wash solvent is nonpolar enough to extract only PITC and its by-products. After about 6 cycles, so much salt is present that a single-phase coupling medium cannot be readily obtained, drying times become very long, and efficiency is seriously decreased. If the washing is vigorous enough to remove these salts, then short or nonpolar peptides are necessarily removed as well.

The difficulty is recovering PTH-Ser and Thr generally has been attributed to the relative instability of such derivatives, which tend to dehydrate by β-elimination,[13] and was attributed to the anhydrous conditions of the usual cleavage reaction.[10] Dehydration has been considered unavoidable, and any attempt to quantify these derivatives as futile.[14]

---

[13] D. Ilse and P. Edman, *Aust. J. Chem.* **16**, 411 (1963).
[14] M. A. Hermodson, L. H. Ericsson, K. Titani, H. Neurath, and K. A. Walsh, *Biochemistry* **11**, 4493 (1972).

However, anhydrous HCl gives a good yield of these derivatives, and they may be stored in solution for years at room temperature in the dark without serious degradation. More probably, these losses arise because the hydroxyl groups of Ser and Thr become acylated during the cleavage reaction,[1] and the acid, rather than water, is subsequently eliminated. This scheme accounts for several disparate observations: (1) 90% formic acid gives good recovery of PTH-Ser and Thr as long as cleavage time is kept short (3 min at 50°), but long exposures (more than 12 min) lead to nearly complete losses[10]; a similar effect with lower recoveries has been noted with HFBA.[15] (2) Addition of thiols to the extraction solvent sometimes improves the recovery of these derivatives, presumably because the nucleophilic thiols attack any remaining ester, thus preventing acid elimination, rather than because of their suggested protection against oxidation[6,16] (see comments on air below). (3) Acids incapable of forming esters have much less tendency to decompose Ser and Thr derivatives. (4) Addition of water to TFA improves yields, but more than 30% is required to reach a half-way point for PTH-Ser recovery relative to that seen with conc. HCl. (5) Fairly good recovery of seryl and threonyl derivatives with intact side-chains can be achieved by reacting ATZs produced by TFA cleavage with alkylamines, which produce PTC amides[17] and would saponify any ester present. This implies that most elimination of TFA occurs during the acid conversion step and thus the PTHs may be obtained in good yield if the extract is dried and treated with amine or other base before conversion.

Avoiding cleavage with carboxylic acids entirely would seem to be a better solution. Edman's original cleavage agent was nitromethane saturated with HCl and 4 N HCl in dioxane (Pierce Chemical Co.) is similarly effective.[11] Neither of these is a good solvent for peptides, but solubility is not required for cleavage as long as the PTC-peptide is spread thinly enough for penetration. (These agents and the Lewis acid discussed below should be ideal in solid-phase applications.) Concentrated HCl (11.5 N) is a fairly good solvent, although not as good as TFA or formic acid, and allows nearly quantitative recovery of PTH-Ser and Thr.[10] The likelihood that water prevents other dehydration reactions, such as nitrile or imide formation, should be considered an additional virtue. Fortunately, the author was unaware of Edman's pronouncement that "cleavage in aqueous acids does not proceed smoothly,"[18] for several years of experi-

[15] A. W. Brauer, M. N. Margolies, and E. Haber, *Biochemistry* **14**, 3029 (1975).
[16] J. Thomsen, K. Kristiansen, and K. Brunfeldt, *FEBS Lett.* **21**, 315 (1972).
[17] J. K. Inman and E. Appella, *in* "Solid Phase Methods in Protein Sequence Analysis" (R. A. Laursen, ed.), p. 241. Pierce Chemical Co., Rockford, IL (1975).
[18] P. Edman, *Acta Chem. Scand.* **7**, 700 (1953).

ence has shown conc. HCl to be a completely reliable agent. There are two debits for conc. HCl: (1) a strong tendency to bump and foam during vacuum drying, which can be largely overcome by addition of *t*-BuOH but still requires care on the part of the operator to prevent physical loss of peptide, and (2) a catalysis of hydrolytic splitting of peptide bonds. While it could be demonstrated that this hydrolysis occurs much more rapidly than the corresponding acidolysis induced by TFA,[11] for small to medium-sized peptides neither reaction was quantitatively significant. Alcoholic HCl was also quickly abandoned by Edman,[18] which was correct, but for the wrong reason. This medium esterifies side-chain carboxyl groups, which leads to an efficient blocking reaction for both Asp and Glu residues (probably imide and pyroglutamic acid formation, respectively).[10] This property of the reagent may be useful under special circumstances—for instance, blocking one component of a binary mixture—but precludes its generality unless all carboxyl groups are first amidated.

A final alternative reagent, and one not previously reported, is the Lewis acid boron trifluoride. As a catalyst for the cleavage reaction, its diethyl ether complex was as potent as TFA or HCl, did not tend to bump, did not promote rapid salt accumulation, allowed excellent recoveries of PTH-Ser and Thr, and caused no detectable chain splitting or blocking of simple peptides.[11] Like the anhydrous HCl reagents, it is a poor solvent for peptides, so clumps of material must be avoided. The highly desirable properties of this reagent are somewhat mitigated by its sensitivity to moisture and its toxicity. The ether complex from Aldrich Chemical Co. was yellowish because of oxidation and should be cautiously redistilled as follows: add 2 ml of anhydrous and peroxide-free diethyl ether to 100 ml $BF_3 \cdot$etherate; distill from 0.5 g of CaH_2 in an all-glass still with PTFE joints under reduced pressure—e.g., 10 torr where the b.p. is 46°. The colorless distillate should be stored cold in portions of a few milliliters under N_2; screw-cap culture tubes (13 $\times$ 100 mm) with PTFE-coated liners in the caps are recommended for this as well as general storage purposes. The tube in current use should be fitted with a PTFE valve as described in section Reservoirs under Apparatus. HCl/diox, also peroxide-prone, should be stored in a similar fashion. Addition of a small amount of thiol (e.g., 10^{-3} M) might serve to scavenge peroxides and any other oxidants.

Since the isothiocyanate degradation of peptides was introduced in 1949, there have been virtually no kinetic data available on the cleavage reaction.[1] This is most unfortunate considering the relative harshness of the reaction conditions. Table I summarizes results with the cleavage acid as the variable, and Table II shows the effect of various residues in the

TABLE I
A Comparison of the Effectiveness of Acids in the Cleavage Reaction[a]

Cleavage acid	TFA	Conc. HCl	HCl/dioxane	Formic acid	BF₃·etherate
Normality	13.4	11.5	4.0	24.0	8.1
pK_a	0.3	<-1	<-1	3.8	—
Half-life of PTC-Gly-His-Gly (min)	7.0	2.7	5.6	36.0	2.9

[a] A mixture of several PTC-peptides was cleaved for 3 min at 19° with each acid. The solutions were frozen in Dry Ice/acetone and vacuum dried. Each sample was then analyzed for the amino acid or peptide released and compared with a fully cleaved control sample and 0 time. The reaction followed first-order kinetics in other tests, so this was assumed here for half-life calculation. Cleavage is essentially complete at 6–7 half-lives.

TABLE II
Rates of Cleavage as a Function of Residue[a]

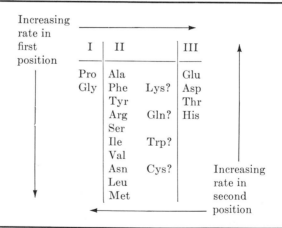

[a] Rates were determined for mixtures of PTC-peptides in TFA at 23°. A fairly complete series in the form of Gly dipeptides was available for testing the effect of the second residue, and a more limited set for testing first residue effects. These position effects were inversely related and so are combined in one listing. A factor of roughly 2 exists between rates for the first and last amino acids in each group, but these rates, and sometimes relative order, vary somewhat according to position and companion residue. Group I is distinctly set apart, and groups II and III grade into each other. An average half-life for group II pairs was about 10 sec.

first and second position.[11] Evidently the cleavage reaction varies directly with the strength of the acid, and both protic and Lewis acids are effective catalysts. It has been suggested that the cleavage reaction reaches a point of equilibrium, particularly with acidic residues, leaving a significant proportion of uncleaved material, hence the double cleavage often performed in the spinning-cup instrument[19] (see Scheme 1 below). If true, the phenomenon would be expected to be more prominent with weaker cleavage acids, and the relative strength of HCl over TFA to be an additional advantage. However, attempts to displace the postulated equilibrium with added alanine or aminoethanol (a better nucleophile) were completely unsuccessful, as might well be expected under conditions in which the amine would be always protonated. Also, recent experience with the spinning-cup instrument indicates single cleavages to be entirely adequate.[14,15]

Edman has surmised that the relative rates of cleavage of various residues are similar to those encountered in the better studied conversion reaction,[13] where the Gly derivative appeared to convert uniquely slowly. However, the situation is more complex.[11] While Gly and Pro derivatives are indeed slower to cleave than the others, in accord with observations made with alcoholic conversion media (see below), an unexpectedly strong influence by the residue following the peptide bond is also observed. There was an imperfect, but impressive, inverse correlation between the effects of the first and second residues, permitting both results to be roughly expressed by the same ordering (Table II). The important practical conclusions are: (1) cleavage rates vary a great deal—by an estimated three orders of magnitude between the slowest (Pro-His) and the fastest (His-Pro or Asp-Pro) linkages—and (2) the most difficult linkages to cleave are rare. Proline is by far the slowest to cleave in the first position, with PTC-Pro-Gly-Gly having a half-life of 406 sec at 23° in TFA, or nearly eight times that of the next slowest, PTC-Gly-Gly-Gly. The accelerations imparted by these residues in the second position are more nearly the same (half-lives of the PTCs of Gly-Gly, Gly-Pro and Gly-Ser are 40, 29 and 80 sec, respectively). While His and Asp in the first position cleave about equally rapidly, the inhibition by His in the second position is more striking. Thus linkages that tend to cleave incompletely contain Pro (sometimes Gly) in the first position and/or His (occasionally another residue from group III) in the second position. For example, one cleavage condition advocated (5 min at 20° in conc HCl)[10] would give about 75% cleavage of a PTC-Gly-His-X (see Table I). If such a linkage is suspected, appropriate extensions of cleavage

[19] P. Edman and G. Begg, *Eur. J. Biochem.* 1, 80 (1967).

time may avoid confusion by out-of-phase material. Conversely, if internal acid-sensitive bonds, such as X-Asn, Asp-Pro or modified sidechains, are present, a lower temperature or shorter time than usual should help. In any case, optimal cleavages may be scheduled in a repeated sequencing attempt. A considerable extension of the number of cycles usefully performed is possible with a marked reduction in aggregate cleavage time, and therefore in acidolysis and background from chain spitting.[1]

An extensive discussion of the theoretical implications of these studies is inappropriate here, but a few points bear mention. First, the great resistance to cleavage of PTC-Pro-X is undoubtedly due to steric interference of cyclization by the side chain. This is supported by the observation that any unconverted Pro is recovered as its thiazolinone, while unconverted Gly (slow but otherwise representative) is recovered in the PTC configuration. Although other investigators depict cleavage of the peptide bond as concurrent with cyclization, the relative resistance of terminal Gly residues seems to require the existence of an unstable intermediate, thus:

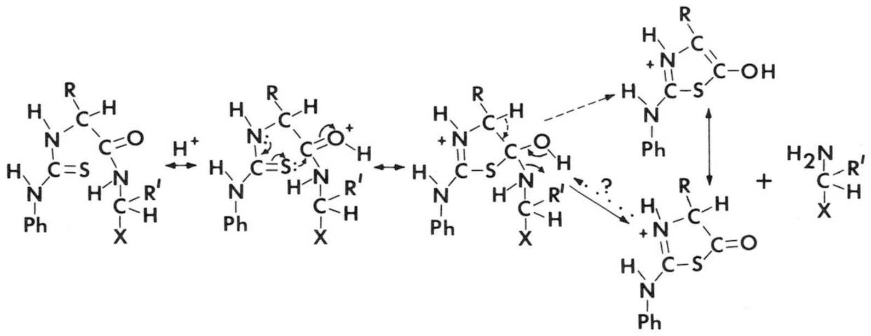

SCHEME 1

A detailed account has not been formulated, but this slow cleavage can be explained if removal of hydrogen from the α-carbon (which is tertiary in all amino acids except glycine) is important, leading presumably to the enol form of the thiazolinone. The inhibition of cleavage by His and others in the second position may be accounted for by a coordination between side chains and the peptide bond, perhaps increasing the stability of the cleavage intermediate. It is unclear whether the variable (but often complete) resistance to cleavage of the heme-linked cysteine residues in cytochrome c[20] is related to the above factors,

---

[20] G. E. Tarr and W. M. Fitch, *Biochem. J.* **159**, 193 (1976).

but the potential for anomalous cleavage by modified residues should be recognized.

While His in the second position slows cleavage, it has sometimes been observed to undergo "precleavage" in the first position. The testing of a variety of parameters, including pH and temperature, has not revealed the cause.[16,21] The present author observed precleavage, sometimes complete, with older procedures,[20] but nothing which obscured the correct sequence using the improved method described here; the reasons for this difference are not evident. The unhelpful conclusion is that the problem is completely avoidable and not due to any of the obvious factors.

Extraction

Separation of cleaved residues, generally ATZ's, from remaining peptides has been effected by a wash of the dry material and/or liquid partition. For reasons discussed in the previous section, a weakly polar solvent such as chlorobutane, dichloroethane, or diethyl ether[1,8] is appropriate for dry extraction after TFA cleavage, while the more polar solvent acetone is necessary after HCl cleavage.[10,11] To minimize any chemical reactions, the extraction is best performed at room temperature. The tendency to dissolve small or nonpolar peptides, particularly in the presence of contaminants, exists at this step as it does after coupling, so the more finely discriminating liquid-partition described below should be used exclusively for such peptides. Even with larger peptides, great care must be taken not to remove suspended material.

The usual liquid-partition has a strongly acidic aqueous phase, such as 0.1 N HCl or simply the residual acid after cleavage. It is naturally difficult to extract the two basic residues ATZ's or PTH-His and Arg from such a medium, and it has been claimed that the partition coefficients are too similar to separate nonpolar peptides from these derivatives by this technique.[19,22] However, a neutral aqueous phase permits, in theory and in practice, a better discrimination for several reasons: (1) ATZ's are weakly basic amines, and as such exist entirely as positively charged compounds at low pH; neutral conditions partially suppress this ionization. (2) The charge on the imidazole side chain is also suppressed near neutrality, and, while the guanidino group should remain ionized, Arg derivatives at least become more extractable. (3) Peptides would be encouraged to remain in the aqueous phase because ionization of all carboxyl groups would now add to the charge on the α-amino group. Derivatives of Asp and Glu become more hydrophilic for the same reason,

[21] W. A. Schroeder, this series. Vol. 25, p. 298 (1972).
[22] H. D. Niall, Agric. Food Chem. 19, 638 (1971).

but not seriously so. A roughly neutral medium may be achieved with 30% aqueous pyridine,[10] which is extracted twice with benzene:EA 1:2, as this is the least-polar mixture that removes Arg derivatives after conc. HCl cleavage. Vigorous mixing tends to make emulsions requiring centrifugation. As in the washing step (see above), heptane-based solvents clear much more readily, but a system containing sufficient heptane, yet polar enough to extract most of ATZ-Arg, is difficult to formulate. A 1:2 mixture with EA is adequate for all other derivatives and removes sufficient ATZ-Arg for the confident assignment of most sequences. Clearly, less extractive solvents are advantageous with peptides lacking highly polar residues and benzene alone or hep:EA 1:1 will give excellent results with residues less polar than the amides. In cases of incomplete extraction of derivatives, most of the remaining material will be recovered in subsequent cycles (an "echo"). If salt accumulates, extraction of Arg and His derivatives will become more difficult.

As in the washing step and in the extraction of dry residues, the polarity of the organic partitioning solvent mixture should be decreased when TFA is used instead of HCl. A benzene:EA ratio of 1:1 or 2:1 seems appropriate with 30% pyridine.

Protection against Oxidation

There has been considerable controversy over the importance of oxidative desulfurization for the Edman chemistry. PTC's and ATZ's are certainly sensitive to oxidation, even by dissolved oxygen,[13] so Edman and Begg argued strongly for completely anaerobic conditions,[19] and virtually all contemporary techniques attempt to provide a N_2 blanket for at least some stages of the cycle. However, the conversion reaction may be performed under air at 50° without apparent ill effect[10]; Niall[23] has suggested that the effect of air may be ascribed instead to atmospheric-borne oxidants other than O_2; and neither the removal of O_2 or addition of antioxidants improved the efficiency of a solid-phase sequencer.[12] Unpublished tests by the author detected no difference in efficiency when degradations were performed with various phases of the cycle under air instead of N_2. Two minor effects were observed: (1) Coupling under an atmosphere of air introduced several by-products that reduced the sensitivity of TLC to many nonpolar PTH's. (2) Unextracted ATZ-His was oxidized to a variable extent, and thus its PTH was not well recovered as an "echo" (Arg was not tested). Loss of Lys under aerobic conditions has been noted,[20,21] which may be related to oxidation

---

[23] H. D. Niall, this series Vol. 27, p. 942 (1973).

of its ϵ-PTC group. Coupling under an anaerobic atmosphere is easy to arrange, so there is no reason for not doing this routinely. As noted before, room temperature washing and extraction also minimize any oxidation that might occur at these stages. Whether there is an advantage to a continuous N_2 barrier for extended degradations or special cases remains an unresolved issue, but it seems likely that low levels of oxidation could be eliminated by antioxidants, such as a thiocarbamyl "carrier" modeled after a supposed protection in solid-phase sequencing[12] (e.g., react polyethylene imine with 4-sulfo-PITC) or 0.01–0.1% EtSH added to all solvents.

The purification procedures for solvents given by Edman[1] are geared largely to the removal of oxidants (even from ethyl acetate) that are far more potent than O_2. These should be used if commercial sources of high-quality material (Pierce Chemical, Burdick and Jackson) are found to be inadequate. The peroxide test based on the liberation of I_2 from 1% aqueous KI (mix with 2 volumes of solvent and evaporate to one phase) is particularly useful for screening batches of solvent, and it is made more sensitive by the addition of starch.

Conversion and Analysis

The traditional conversion medium is 1 N aqueous HCl at 80° for 10 min under N_2.[1] ATZ-Gly is incompletely converted to the PTH, so some is recovered as the PTC. Side chains of Asn and Gln are partially hydrolyzed under these conditions and some loss of Ser occurs. After conversion, the PTH's are extracted into an organic phase, with the exception of PTH-Arg and -His, which remain in the aqueous phase and must therefore be identified separately. These undesirable properties have prompted many alternative procedures, most involving a thermal conversion.[24,25] These give essentially quantitative recovery of PTH-Asn and -Gln and most other derivatives and leave all PTH's in the same sample (no partitioning), but any Ser and Thr derivatives that were originally present are completely dehydrated and Lys is often recovered as multiple spots of low yield on TLC.[25] The author's conversion medium—1 N methanolic HCl at 50° for 10 min under air[10]—avoids all the above difficulties. In this medium, ATZ's open to give PTC-methyl esters, which are about one order of magnitude more reactive in recyclization to the PTH than are the free acids, hence the milder conditions required. PTH-Ser and -Thr recovery is nearly quantitative, alcoholysis of

[24] R. L. Guyer and C. W. Todd, *Anal. Biochem.* **66**, 400 (1975).
[25] A. S. Inglis, P. W. Nicholls, and P. M. Strike, *J. Chromatogr.* **107**, 73 (1975).

amides is virtually nil, and the medium dries down readily, so that partitioning is not required. In contrast to other derivatives, the opening of the ring is a slow reaction for ATZ-Pro, so it is recovered at a level of about 10% in this form. A similar level of PTC-Gly-methyl ester remains after 10 min and the side chains of PTH-Asp and Glu are completely esterified, which is advantageous because their mobilities on silica are now that of neutral compounds. They elute conveniently on HPLC between PTH-Gly and Ala,[26] and they may be gas-chromatographed without further treatment.[11] If the migration of these methyl esters interferes with identification of PTH-Tyr, Trp, or Ala in the particular system used, the conversion may be carried out in 2-methoxyethanol instead, which yields esters chromatographing between PTH-Thr and Lys on silica. Carboxymethylcysteine behaves similarly. It is ironic that the principal developers of solid-phase sequencing carefully remove methanol from their eluent before doing the less effective aqueous conversion.[12]

Methods for analyzing PTH's are presented elsewhere in this volume,[27] but the following technique possesses an exceptional combination of advantages—low cost, multiple sample processing, sensitivity, one-dimensionality, rapidity and ability to discriminate all common PTH's in a single run. The original system[10] employed Adsorbosil-3, a fine-grained silica gel from Applied Science (State College, Pennsylvania), but recent lots are markedly inferior, and there seems to be little prospect for achieving the original resolving power. Adsorbosil-1 is nearly as good and is superior to all other silica gels so far tested. The aqueous suspension is spread on 25 or 50 × 76-mm microscope slides with a glass rod or tube taped at the edges so one thickness of laboratory tape (Time) forms a spacer between the slide and the tube. The gel must be poured within 1 min after mixing onto a queue of slides, which adhere to a glass or plastic board by capillary attraction, and the tube is pushed (not rotated) back and forth several times to make an even coating. The board is then tapped sharply to briefly shake the layer, removing any irregularities.

Slides are activated immediately prior to use on a hot plate at about 200° for at least 2 min. Samples are spotted with a very fine capillary (e.g., the pulled-out tip of a Pasteur pipette), with about 0.05–0.1 μl being drawn into the silica with each application. No more than 0.5 μl should be spotted normally. A standard mixture of about 10 PTH's should be

[26] R. L. Niece, in "Solid-Phase Methods in Protein Sequence Analysis" (R. A. Laursen, ed.), p. 233. Pierce Chemical Co., Rockford, IL, 1975.

[27] See also J. Bridgen, A. P. Graffeo, B. L. Karger, and M. D. Waterfield, in "Instrumentation in Protein Sequence Analysis" (R. N. Perham, ed.), p. 111. Academic Press, New York, 1975.

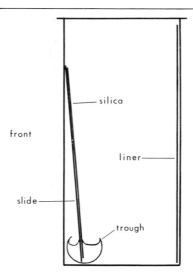

FIG. 1. Arrangement for gradient elution TLC of PTH's. A tank for 25×76-mm slides should have approximate inside dimensions 100 mm high $\times$ 47 mm wide $\times$ 31 mm deep. One for 50-mm slides should be twice as wide. Troughs 8 mm in diameter are cut from glass or polypropylene tubing. The atmosphere and liner are presaturated with the less polar solvent mixture; then the more polar mixture is placed in the trough.

adjacent to each sample, with a 25-mm slide accommodating two standards and four samples.

    The usual solvent system consists of two separate mixtures—hep:benzene:EA 4:3:1 and EA:acetone:HFIP:water:acetic acid 60:30:3:1:0.3. This is a gradient elution system[28] in which the first mixture saturates the liner and atmosphere of the tank and the second is placed in a small glass or polyethylene trough (cut from 8 mm o.d. tubing) just before the slide is inserted (see Fig. 1). A developing jar appropriate for 50-mm slides is available from Eastman and some museum jars are even better. After development at $22° \pm 2°$ (8 min or less) the slide should be dried briefly in a stream of air (blow on it for about 20 sec) so that background solvent is removed but spots of PTH still contain solvent. The slide is placed immediately in a chamber saturated with I_2 vapors. Sensitivity is about 0.02–0.1 nmol depending mostly on distance of migration, and the compact spots produced may permit even greater sensitivities with newer detection methods.[29] The separations achieved with this system and with a useful second system are diagrammed in Fig. 2.

[28] G. E. Tarr, *J. Chromatogr.* **52**, 357 (1970).
[29] A. S. Inglis and P. W. Nicholls, *J. Chromatogr.* **97**, 289 (1974).

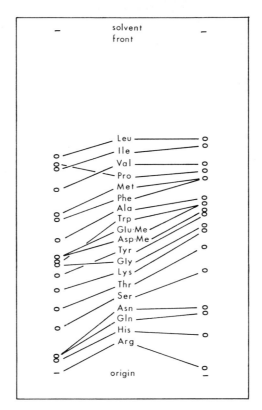

Fig. 2. Separation of PTH's on silica gel by gradient elution. Samples are dissolved in EA:MeOH 1:1 and about 0.4 nmol of each derivative is spotted on 76 mm-high glass slides coated with Adsorbosil-1. Silica layers are activated on a hot plate immediately before spotting and cannot be reused. System I, shown on the right, is hep:benzene:EA 4:3:1 saturating the tank and EA:HFIP:water:acetic acid, 60:30:3:1:0.3 in the trough. System II, on the left, is toluene:diethyl ether 9:1 and benzene:acetone 3:1. Solvents are stable for months if tightly closed. PTH's are indicated by 3-letter code for their constituent amino acid residues; Me, methyl ester. Positions of other derivatives and by-products in system I are: PTH-dehydrothreonine, PITC and diphenylthiourea between Val and Ile; unknown by-product and PTC-Gly-Me just above Ala; ATZ-Pro between Gly and Tyr; phenylthiourea between Lys and Gly; methoxyethyl esters of Glu and Asp separated between Thr and Lys; PTH-carboxymethylcysteine esters just below those of Asp; PTH-homoserine just below Ser; ϵ-acetyl-PTH-Lys between His and Gln.

Many alternative isothiocyanates —methyl, pentafluorophenyl (Pierce), p-dimethylaminobenzeneazobenzene[30] and acridine (Eastman) —produce thiohydantoins which can also be separated fairly well by these

[30] J. Y. Chang, E. H. Creaser, and K. W. Bentley, *Biochem. J.* **153**, 607 (1976).

systems. The last two are of particular interest as the sensitivity that can be achieved with them (about 0.1 pmol for the last by TLC[11]) is about the same as that reported for methods relying on radioactive reagents or substrates.[31] However, much experience will be needed with these more sensitive reagents before they can be recommended for routine use.

Regeneration of amino acids from PTH's permits accurate qualitative and quantitative analysis of residues. Yields have been highest and most consistent by hydrolysis at 150° for 3 hr with 57% HI containing about 5% ethanethiol.[11] Convenient individual tubes may be made by sealing the tapered region of disposable pipettes in a flame; tubes are labeled by etching; about 20 μl of HI is added to the dry sample in the cone, and the tubes are placed in Dry Ice. A constriction is made high on the tube, which is replaced in the Dry Ice. A set of such tubes on a manifold is brought through a vacuum/N_2 flush cycle three times, then each is sealed off under vacuum. The heating block (e.g., Hallikainen Instruments, Richmond, California) should contain silicone oil for good thermal contact. After hydrolysis, tubes are cooled, opened, and an equal volume of t-butanol is mixed in to convert the HI to t-butyl iodide and water. Vacuum drying is performed as described in the section Accessory Apparatus under Apparatus. Hydrolysis can also be performed in batch mode, with all of the HI placed outside the tubes in a reaction bomb.[27] It is difficult to clamp ordinary dessicators tightly enough to withstand the pressure at 150° without leaking, which leads to destructive oxidation. However, 20 or more 6 × 50 mm culture tubes can be accommodated within a 43 mm diameter cylindrical flask blown at the end of a 200-mm section of thick-walled 13 mm o.d. Pyrex tubing; the flask is then processed like an individual tube, except that the samples may be dried after hydrolysis in a dessicator without t-BuOH.

Apparatus

Reaction Vessels

The need for excluding air from the reaction cell and from reagents presumed by many manual operators, including the author, led to elaborate devices that provide a continuous nitrogen carrier.[10,32,33] However, results discussed above suggest that anaerobic conditions are advan-

[31] J. W. Jacobs, B. Kemper, H. D. Niall, J. F. Habener, and J. T. Potts, Jr., *Nature* (*London*) **249**, 155 (1974).
[32] R. B. Meagher, *Anal. Biochem.* **67**, 404 (1975).
[33] K. D. Lin and H. F. Deutsch, *Anal. Biochem.* **56**, 155 (1973).

tageous only at the coupling step for small and medium-sized peptides and a continuous barrier against O_2 may be unnecessary for all applications. So, while more complicated devices may be helpful in some cases, none will be detailed here.

One common system,[8] both simple and convenient, utilizes small (1–3 ml) glass-stoppered centrifuge tubes that are flushed with nitrogen, at least for the coupling step. The seal can be made more secure with a vessel assembled as shown in Fig. 3A. This chamber is a Reacti-Vial fitted with a PTFE valve (Pierce Chemical Co. and others). The silicone plug supplied with the valve is removed and the bore of the valve slightly enlarged and aligned exactly with the hole in the slider by reaming out the open valve with a 1–1.2 mm drill; a larger hole weakens the slider. The valve with its attached vial is reversibly connected to a 3 mm o.d. nitrogen line inserted part way into the hole for the silicone plug (simi-

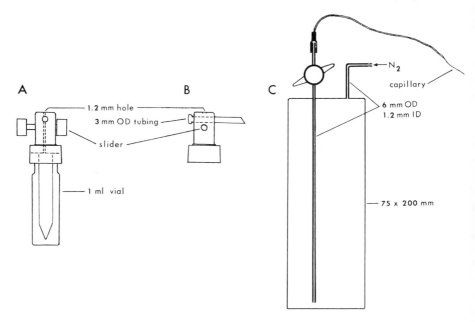

FIG. 3. Reaction vessels and reservoirs. (A) Reaction vessel: the silicone plug supplied with the PTFE valve is removed; the valve bore is slightly enlarged by drilling. (B) Modified valve for use with small reservoirs: 3-mm o.d. or slightly larger PTFE or polyethylene tubing is force-fitted into the hole above the slider and heat sealed at one end. The bore is then drilled as in (A). Larger sized valves may be similarly modified. (C) Large glass reservoir: inert tubing about 1.2-mm o.d. may be forced into the bore of the outlet line, attached via a Chromatronix fitting, or sealed against the glass with heat-shrinkable PTFE tubing as shown here. The stopcock has a 1.2-mm bore (Interflon). The container may be filled by placing a vacuum on the inlet line and should be flushed after filling by reversing the N_2 flow.

lar to Fig. 3B) and a completely anaerobic atmosphere is established by briefly inserting a capillary connected to another line. The inert atmosphere will now persist during the addition of reagents and after the valve is closed and detached from the line.

Reacti-Vials are usually adequately cleaned with aqueous pyridine and/or formic acid. The best vigorous treatment is annealing, but a chromate/H_2SO_4 cleaning solution is acceptable if any oxidant remaining after rinsing is destroyed with thiol.

Reservoirs

Reservoirs for coupling reagents, and preferably for wash solvents as well, should permit the removal of portions under nitrogen barrier. This prevents oxidation of reagents and the possible generation of potent oxidizing materials. The PTFE valve for Reacti-Vials is easily modified for small reservoirs (Fig. 3B). PTFE or polyethylene tubing (3 mm o.d. or slightly larger) is force-fitted into the hole that bore the silicone plug. One end is sealed (e.g., by pinching with a hot hemostat) and the other can be connected to a nitrogen line. The entire assembly is drilled out as indicated above. The initial anaerobic atmosphere is maintained by having the nitrogen source turned on whenever the valve is open. The PITC reagent is best disconnected and stored in ice and in the dark when not in immediate use or at $-20°$ for prolonged periods. The body of the reservoir may be any tube or bottle that may be fitted with a 1 ml or 5 ml size of PTFE valve. The 13×100 mm-screw cap culture tubes used for general storage are appropriate with the 1 ml size, while many of the sequencing reagents from Pierce are supplied in bottles compatible with the larger valves. Small volumes of fluid may be transferred in a piece of PTFE or polyethylene tubing attached to a 1-ml syringe or to a "pipet suction apparatus" (Scientific Products). The capillary end of the tubing may be any inert material or simply the heat stretched tubing itself (see Fig. 3C). A simple way of keeping the tubing clean is to pick up a small volume of solvent, followed by an air or N_2 spacer, before drawing up the reagent or upper phase. For wash solvents, large reservoirs made of glass and having a PTFE stopcock on the outlet line may be fabricated inexpensively (Fig. 3C; Norman D. Erway, Glass Blowing, Oregon, Wisconsin). Inlet and outlet lines may be attached through Chromatronix-type fittings or with heat-shrinkable PTFE or similar tubing. In the absence of a stopcock, the outlet line will require a means of stopping flow; this is conveniently provided by a hemostat (prevented by tape from closing completely at the tip) acting on a short section of silicone tubing. Many designs of reservoirs and delivery line controls are acceptable, but rubbers and other materials not completely inert to nonpolar solvents and weak bases should be avoided.

Accessory Apparatus

Other equipment useful for manual sequencing includes a tabletop centrifuge with swinging-bucket rotor (angle heads are decidedly inferior), a heating block with 14-mm or larger holes (water improves thermal contact), small vacuum pump with traps (cold Dewar flask or a series of H_2SO_4 and NaOH traps; either design is adequate as volumes to be collected are small, but change the H_2SO_4 when it gets dark), and a desiccator fitted with rubber O-ring and retainer.

Mixing. Mixing within reaction vessels is rapid and efficient with a Vortex mixer and, if the washing chemistry described here is followed, few difficulties with emulsions will be encountered. An alternative is magnetic stirring with paddles bearing PTFE-coated magnets (Pierce). One stirring device will drive many paddles. This is much gentler and so requires longer extraction times (about 20 sec), but does not encourage emulsions.[11] Also, mixing is usefully continued during the coupling reaction with partially soluble material and also during drying to improve efficiency and reduce the tendency to bump. The latter advantage is offset by the difficulty in processing a few microliters of lower phase in the presence of this obstacle. In the absence of extensive experience, magnetic stirring is recommended only for special cases and for resin-bound peptides.

Drying. Drying of samples may be accomplished in a variety of ways: (1) The simplest is to place all vials or tubes in a 125-ml Erlenmeyer vacuum flask, stopper, and attach to a vacuum line. With the flask on its side or nearly so, the tubes are at roughly a 20° angle to the horizontal and so expose a large surface area for evaporation. It is unsafe to pull a full vacuum immediately, so the stopcock on the vacuum line should be turned initially only enough to begin evacuation of the chamber. The rapidity with which the vacuum may be allowed to develop depends on the fluid; those with low surface tension such as TFA, HCl/dioxane, and HCl/methanol may be brought to full vacuum quickly, but conc. HCl and other aqueous media will require more care. (2) Desiccators have larger volumes and limit inspection of contents, but are otherwise similar to flasks. (3) Aqueous media such as lower phases and conc. HCl may be dried more quickly and safely by vigorous mixing while applying vacuum. Some foaming is acceptable and even helpful. A trick that usually helps with conc. HCl is to mix in a drop or two of *t*-butanol before drying. A simple manifold for the vials can be constructed from

PTFE-valves with sliders removed and expanded bores, short pieces of plastic or rubber tubing, and plastic Y-connectors. Small sets (up to 4) of vials can be hand-held and mixed at one time on a Vortex while drying, or larger sets accommodated with magnetic stirring.

For most purposes, it is advantageous and convenient to treat each sample individually, but batch processing of sequencing products is sometimes necessary. Manifold devices such as those available from general suppliers will improve the efficiency of such processing.

Procedure

Figure 4 is a flow chart for manual degradation. Quantities of 5–50 nmol are most appropriate, but the lower limit is basically set by the sensitivity of analytical methods. Alternative processing is offered for different classes of peptide and different cleavage agents. Note that a given peptide may not remain in the same class throughout its degradation. Residual salt and small peptide fragments can usually be removed from a large peptide before sequencing by precipitating it from aqueous solution with acetone, then washing with ethanol. It is in some cases beneficial, and rarely detrimental, to "precycle" a sample beginning at the cleavage step. Identification should be a routine procedure lagging degradation by only a few cycles, so that processing may be modified or interrupted as needed. Cycle time may be as short as 14 min with one peptide, with 20–30 min being more typical with several samples. Repetitive efficiency achieved with this procedure averaged about 92%,[10] which is entirely adequate for small peptides; it was higher for longer and more polar peptides.

Discussion

The information on the Edman reaction presented here extends previous knowledge, but is still far from adequate. Optimal utilization of isothiocyanates in protein sequencing, by whatever strategy and device, will require extensive testing of coupling conditions as well as a completion of studies on the cleavage step. It is hoped that all users of the Edman chemistry, not just manual operators, will note the points made here.

The procedure described is most appropriate for peptides of up to 20 residues, which form the bulk of samples appropriate for manual investigation. With good reagents, solvents, and care to avoid mechanical loss of peptide, there is no particular difficulty as long as the peptide begins with a free NH_2 terminus. If the procedure is to be used for longer deg-

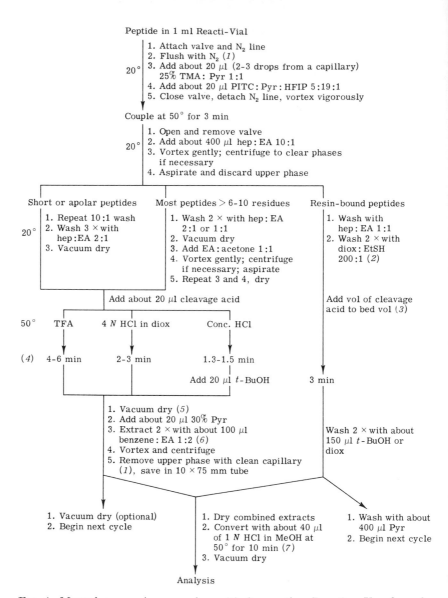

Fig. 4. Manual sequencing procedure. (1) See section Reaction Vessels under Apparatus. (2) Peroxides destroy PTC-groups, so the dioxane must be redistilled and stored under N_2 in the dark. The thiol should remove any last traces of oxidant. (3) If the peptide is not covalently attached, use anhydrous HCl or incorporate a drying step before extraction. (4) Vials are best recapped with valves; atmosphere of air or N_2. Timing begins when vial is inserted into heating block and may be adjusted beyond these limits in accordance with information in Table II. (5) Be especially careful with conc. HCl to avoid bumping (see section Accessory Apparatus). (6) In the absence of residues more polar than Ser, weaker extraction solvents, such as hep:EA 1:1, are appropriate (see section Extraction under the Edman Chemistry). (7) Alternative formulations are: $SOCl_2$:MeOH 1:25; acetyl chloride:MeOH 1:14 (mix cold); 4 N HCl in diox:MeOH 1:3. Store all of these at or below 0° when not in use to minimize reaction of HCl and methanol. Conversion under air is fine, but evaporation should be retarded, e.g., by placing a glass sphere on top of the tube.

radations, which are perfectly possible,[10] it may be necessary to give more attention to providing anaerobic conditions and to processing insoluble material. Washing and drying of proteins sometimes creates refractory lumps, so it is advantageous to anchor such large peptides in a dispersed condition. Solid-phase methods of covalent attachment to porous glass beads[12] or adsorption to such beads[34], to ion-exchange resins such as AG 1 or 50 $\times$ 2 (Bio-Rad),[11] or to paper[21,35] are useful. The advantages of adsorption are its simplicity and the recoverability of all residues, but high-yield reactions at the COOH terminus, such as attachment of cyanogen bromide fragments, are similarly useful. Any of these techniques may be used with the apparatus described here.

[34] Durrum Instruments, unpublished, 1975.
[35] J. Jentsch, *in* "Solid-Phase Methods in Protein Sequence Analysis" (R. A. Laursen, ed.), p. 193. Pierce Chemical Co., Rockford, IL, 1975.

Acknowledgments

The author wishes to express his gratitude to the following people for support of the research forming the basis of this article: Dr. Walter M. Fitch (University of Wisconsin, Madison), Dr. George D. Cain (University of Iowa, Iowa City), and Dr. Emanuel Margoliash (Northwestern University).

[35] Solid-Phase COOH-Terminal Sequential Degradation

By ANDRÉ DARBRE

Despite some recent progress in the use of carboxypeptidases (see other chapters in this volume), there is a real need for a chemical method for protein sequencing from the COOH terminus. Only the COOH-terminal amino acid was identified by hydrazinolysis[1] and by selective ^{3}H labeling.[2] A new method for determination of the COOH-terminal amino acid in high yield was successfully applied to 20 different peptides.[3] The peptide COOH was activated with 1-ethyl-3-(3-dimethylaminopropyl) carbodiimide and this was reacted with *O*-pivaloylhydroxylamine to form the hydroxamate, pep-CO·NH·CHR·CO·NH·O·CO·C(CH$_3$)$_3$. At pH 8.5, a Lossen rearrangement occurred at 50° over a period of 20 hr, leading to the formation of a mixture of ureas, pep-CO·NH·CHR·NH·CO·NH$_2$ and (pep-CO·NH·CHR·NH)$_2$CO. These were hydrolyzed at

[1] S. Akabori, K. Ohno, and K. Narita, *Bull. Chem. Soc. Jpn.* **25**, 214 (1952).
[2] T. Baba, H. Sugujama, and S. Seto, *J. Biochem. (Tokyo)* **72**, 1571 (1972).
[3] M. J. Miller and G. M. Loudon, *J. Am. Chem. Soc.* **97**, 5295 (1975).

pH 1–2 at 50° for 2 hr to generate 1 mol of ammonia, the peptide-amide-1 residue (pep-CO·NH$_2$), and the aldehyde corresponding to the terminal amino acid, R·CHO. Alternatively, the peptide-ureas were hydrolyzed with 6 M HCl at 110° for 22 hr for subtractive amino acid analysis. Studies are in progress to develop a sequential COOH-terminal degradation method.[3]

A method of peptide sequencing from the COOH terminus, using a chemical method, was introduced recently.[4] One procedure described (see Scheme 1) was to couple the peptide with S-n-butyl

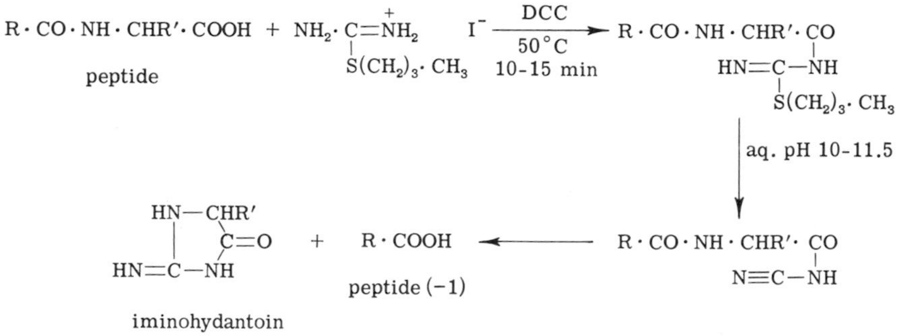

$$R \cdot CO \cdot NH \cdot CHR' \cdot COOH \; + \; \overset{+}{NH_2 \cdot} \underset{\underset{S(CH_2)_3 \cdot CH_3}{|}}{C} {=} NH_2 \quad I^- \; \xrightarrow[\substack{50°C \\ 10\text{-}15\ min}]{DCC} \; R \cdot CO \cdot NH \cdot CHR' \cdot \underset{\underset{S(CH_2)_3 \cdot CH_3}{\underset{|}{HN{=}C{-}NH}}}{CO}$$

peptide

$$\Bigg\downarrow \text{aq. pH 10-11.5}$$

$$\underset{\text{iminohydantoin}}{\overset{HN-CHR'}{\underset{HN{=}C-NH}{\big| \quad \overset{\diagdown}{\underset{\diagup}{C}}{=}O}}} \; + \; R \cdot COOH \; \longleftarrow \; R \cdot CO \cdot NH \cdot CHR' \cdot \underset{N{\equiv}C-NH}{CO}$$

peptide (−1)

SCHEME 1. After G. E. Tarr, in "Solid-Phase Methods in Protein Sequence Analysis" (R. A. Laursen, ed.), p. 139. Pierce Chemical Co., Rockford IL, 1975.

thiuronium iodide with dicyclohexylcarbodiimide by heating at 50° for 10–15 min. Cleavage was effected with aqueous base at pH 10–11.5. First the n-thiobutanol was released and then the peptide acyl cynamide cyclized, to be followed by the subsequent cleavage of the iminohydantoin from the peptide-1 residue. The iminohydantoin was detected by thin-layer chromatography (TLC) with nitroprusside/ferricyanide or as the trimethylsilylated derivative by gas–liquid chromatography (GLC) on a column with 1.5% OV-101. A method of recovering the parent amino acid from the iminohydantoin and sequencing up to five residues on small peptides was achieved, but the method is still at an exploratory stage (private communication, G. E. Tarr).

One other chemical sequencing method was reported in detail by Stark.[5] Protein and peptidyl thiohydantoins were prepared and sequentially cleaved with the release of the amino acid thiohydantoin from the carboxyl terminus. Three cycles of degradation were usually satisfactory. Using a similar procedure,[6] some dipeptides and the four COOH-terminal

[4] G. E. Tarr, in "Solid-Phase Methods in Protein Sequence Analysis" (R. A. Laursen, ed.), p. 139. Pierce Chemical Co., Rockford, IL, 1975.

[5] G. R. Stark, this series Vol. 25, p. 369 (1972).

[6] S. Yamashita, $Biochim. Biophys. Acta$ **229**, 301 (1971).

amino acids of ribonuclease were analyzed with TLC identification (s = 100 nmol) of the cleaved amino acid thiohydantoins. The yield dropped to approximately 10% for aspartic acid at the fourth cycle.[6] A further report indicated that 10 COOH-terminal amino acids in ribonuclease and 14 in papain could be identified. About 10 COOH-terminal amino acids from polypeptides were analyzed, starting with 0.01–0.1 μmol of material, but no details were given.[7]

The Stark method in free solution was studied with the objective of improving on the yields previously reported, but with little success.[8] At the fourth cycle of degradation the yield dropped dramatically to about 10%. Two Sephadex filtrations and lyophilization stages were required, and for each degradation cycle 3.5 days (in our hands) were required. The use of solid-phase sequencing, in conjunction with the Stark method[8,9] offered advantages for removing both excess reagents and the amino acid thiohydantoin from the residual attached peptide, and this method is described here.

Principle

The peptide was coupled by its amino terminus to the COOH group of a modified polystyrene polymer, using dicyclohexylcarbodiimide coupling reagent. Alternatively, the peptide was attached to N-hydroxysuccinimide-activated glass beads (Pierce).

The free COOH group of the peptide was allowed to react with ammonium thiocyanate in acetic anhydride—acetic acid solution to prepare the peptidyl thiohydantoin.[5] Cleavage with 12 M HCl enabled the terminal amino acid thiohydantoin to be released, and this was washed from the solid support for identification directly by GLC, GLC-mass spectrometry (MS), and TLC, or indirectly after hydrolysis, to its parent amino acid by GLC.

Procedure

Preparation of Reagents and Reference Thiohydantoins. Recrystallize ammonium thiocyanate from absolute ethanol.[5] Acetic anhydride is redistilled through a Vigreux column.[5] Acetic acid is refluxed with chromic acid and redistilled through a Vigreux column. Hexafluoroacetone trihydrate is prepared by distillation from the sesquihydrate.[5] The meth-

[7] S. Yamashita and N. Ishikawa, *Proc. Hoshi Coll. Pharm.* **13**, 136 (1971).
[8] M. Rangarajan and A. Darbre, *Biochem. J.* **157**, 307 (1976).
[9] M. J. Williams and B. Kassel, *FEBS Lett.* **54**, 353 (1975).

TABLE I
STAGES OF PREPARATION OF MODIFIED POLYSTYRENE-p-DIVINYLBENZENE
POLYMER (1% CROSS-LINKED)[a]

Polymer	Substituent groups
I	$-CO \cdot N(C_6H_5)_2$
II	$-COOH$
III	$-COOH$
	$-NO_2$
IV	$-COOH$
	$-NH_2$
V	$-COOH$
	$-NH-CS-NH-CH_3$
VI	$-CO-NH(CH_2)_3COOCH_3$
	$-NH-CS-NH-CH_3$
VII	$-CO-NH(CH_2)_3COOH$
	$-NH-CS-NH-CH_3$

[a] M. Rangarajan and A. Darbre, *Biochem. J.* **157**, 307 (1976).

ods for preparing many crystalline thiohydantoins were published.[5,6,10]

Preparation of Methylthiocarbamoyl Polymer with γ-Amino-n-butyric Acid Side Chain. The polystyrene-p-divinylbenzene polymer (1% cross-linked) was purchased from Bio-Rad Laboratories, Richmond, California. It was modified via the stages shown in Table I. Diphenyl-carbamyl chloride (7.4 g, 0.032 mol) in 24.0 ml of nitrobenzene was added over a period of 15 min to a well-stirred mixture of 16.64 g (0.016 mol) of polymer and 10 g of anhydrous aluminum chloride (0.075 mol) in 150 ml of dry nitrobenzene.[11] The mixture was heated at 80° for 2.5 hr, during which time it turned dark brown. The mixture was cooled and treated with 200 ml of water. The polymer was filtered off and washed successively with 2 M HCl, methanol and ether and dried. The carboxamido polymer [20 g, (I) in Table I] was hydrolyzed by heating with 360 ml of a mixture of glacial acetic acid, conc. sulfuric acid and water (2:1·5:1, by volume) at 135° for 40 hr. The mixture was cooled and filtered. The pale green polymer (II) was washed with water until free of acid, then with methanol and ether and dried.

The carboxyl content was determined by boiling 500 mg of polymer

[10] T. Suzuki, K.-D. Song, Y. Itagaki, and K. Tuzimura, *Org. Mass Spectrom.* **11**, 557 (1976).

[11] R. L. Letsinger, M. J. Kornet, V. Mahadevan, and D. M. Jerina, *J. Am. Chem. Soc.* **86**, 5163 (1964).

in a solution consisting of 25 ml of 95% (v/v) aqueous ethanol and 25 ml of 0.1 M NaOH for 3 min, cooling and back-titrating excess NaOH with 0.1 M HCl to phenolphthalein end point.[11] Values of approximately 1.25 mmol of COOH per gram were obtained.

The polymer (II) (15 g) was mixed intimately with 30 g of KNO$_3$ with a pestle and mortar and added slowly with stirring to 35 ml of ice-cold conc. sulfuric acid. The temperature was kept below 8°. After addition was complete, the mixture was heated to 90° for 30 min. The reaction mixture was cooled in ice, and 500 ml of ice-cold water were added. The polymer was filtered off, washed with water until the washings were pH 3–4 (pH paper), then with methanol and ether, and dried. There was a weight gain of about 40%, and the polymer (III) was pale yellowish brown. The nitro groups were reduced by adding 30 g of granulated tin slowly to a well-stirred suspension of 15 g of polymer in 165 ml of conc. HCl. The reaction was kept under control by cooling when necessary. The mixture was finally heated at 80° for 2 hr until all the tin had dissolved. The dark-brown polymer (IV) was filtered off, washed first with 1 M HCl to remove tin salts, then with water, methanol and ether, and dried. The amino groups were blocked by stirring at 40° for 4 hr 10 g of polymer in 35 ml of pyridine, N-methylmorpholine trifluoroacetate buffer at pH 9.5 with the addition of 10 ml of methylisothiocyanate.[12] The methylthiocarbamoyl polymer (V) was filtered off, washed alternately with dimethylformamide (DMF) and methanol until the washings were colorless, and finally with ether, and dried.

γ-Amino-n-butyric acid methyl ester hydrochloride (1.5 mmol) was dissolved in 3.0 ml of dry DMF and 200 μl of triethylamine was added slowly to liberate the free base. The methylthiocarbamoyl polymer (V) (1.0 g) was suspended in 7.0 ml of dry DMF and allowed to swell for about 20 min, then 50 μl of triethylamine was added, followed by the solution of γ-amino-n-butyric acid methyl ester (1.5 mmol) containing dicyclohexylcarbodiimide (1.6 mmol).[13] The reaction mixture was stirred at about 20° for 18 hr in a stoppered vessel. The light-brown polymer with attached γ-amino-n-butyric acid methyl ester (VI) was filtered off and washed successively with DMF, methanol, and ether, and dried.

The methyl ester groups were hydrolyzed by stirring 1 g of the polymer with 10 ml of 0.8 M NaOH in methanol/acetone solution (1:1, v/v) at room temperature for 4 hr. The solution was adjusted to about pH 3.0 with 1 M HCl and the polymer (VII) was filtered off, washed with water, methanol and ether, and dried.

[12] R. A. Laursen, *Eur. J. Biochem.* **20**, 89 (1971).
[13] J. C. Sheehan and G. P. Hess, *J. Am. Chem. Soc.* **77**, 1067 (1955).

Attachment of Peptide to Methylthiocarbamoyl γ-Amino-n-butyric Acid Polymer. The peptide methyl ester hydrochloride was dissolved in 0.5 ml of DMF and 30 μl of triethylamine added to liberate the free base. About 800 mg of polymer was allowed to swell in 4.0 ml of DMF for about 20 min when the peptide methyl ester solution was added, followed by a solution of dicyclohexylcarbodiimide (0.2 mmol) in DMF. The mixture was stirred at room temperature for 18 hr. The polymer was filtered off, washed extensively with DMF and methanol, and dried.

Attachment of Protein to Glass Beads. The N-hydroxysuccinimide-controlled-pore glass beads were used as purchased. Buffers (pH 5-9) were prepared from 0.1 M sodium acetate and 0.1 M acetic acid or 0.1 M NaH_2PO_4 and 0.1 M Na_2HPO_4. The protein dissolved in the buffer solution (5.7 mg/ml) usually at pH 6 to 7.5 was cooled to 4° and, for every 5 ml of solution, between 0.5 g and 1.0 g of the active glass beads was added. The vessel was degassed in the cold and gently rocked at 0–4° for 4 hr. Magnetic stirring caused fracturing of the beads. The beads were washed with ice-cold buffer on a sintered-glass funnel and any remaining active ester groups were allowed to react with 1.0 M ethanolamine in the same buffer solution at room temperature for 2 hr. The beads were filtered off and washed with buffer, then deionized water, and finally 1,4-dioxane and methanol. They were dried with a stream of nitrogen gas.

Monitoring Attachment to Solid-Phase. The amount of polymer-attached γ-amino-n-butyric acid and of peptide or protein coupling to polymer and controlled-pore glass beads was determined by hydrolyzing a weighed sample of the solid phase *in vacuo* in a Rotaflo stopcock[14] with aqueous 6 M HCl containing 10^{-3} M β-mercaptoethanol and 0.1% phenol (w/v) at 110° for 24, 48, and 72 hr. The dried amino acid residue was esterified with 4 M HCl in methanol for 90 min at 70° and trifluoro-acetylated with 0.1 ml of trifluoroacetic anhydride for 10–20 min at 20° for determination of the amino acids by GLC, using norleucine as internal standard.[15]

Sequential Degradation of Peptides Attached to Polymer. A solution of ammonium thiocyanate (2.0 mmol) in acetic anhydride (26 mmol), and glacial acetic acid (6.0 mmol) was added to 100 mg of polymer (about 0.4 μmol of attached peptide) in a small stoppered vessel and stirred at 80° for 1.5 hr. Excess acetic anhydride was hydrolyzed by adding 3.0 ml of water, and the polymer with the attached peptidyl thiohydantoin was filtered off through a sintered-glass funnel, washed with water, then with acetone, and dried.

[14] A. Darbre, *Lab. Pract.* **20**, 726 (1971).
[15] A. Darbre and A. Islam, *Biochem. J.* **106**, 923 (1968).

The peptidyl thiohydantoin attached to the polymer was cleaved with aqueous 12 M HCl at 30° for 1.5 hr. The HCl solution was removed through a sintered-glass funnel and the polymer was washed with a few drops of water to remove excess HCl. Any remaining HCl was removed with an acetone wash to prevent any methyl ester formation. Because not all thiohydantoins were soluble in acetone, the polymer was then stirred with 3.0 ml of methanol at room temperature for 30 min. This was sucked off and the polymer was further washed with methanol and dried. The aqueous filtrate was taken to dryness and then the acetone and methanol filtrates were added and taken to dryness.

Sequential Degradation of Proteins Attached to Glass Beads. About 500–750 mg of glass beads containing 1 μmol of attached protein was treated with 1 ml of hexafluoroacetone trihydrate followed by a freshly prepared solution of ammonium thiocyanate (2.0 mmol) in a mixture of acetic anhydride (54 mmol) and glacial acetic acid (11.0 mmol). The stoppered flask was gently rocked in a shaking water-bath at 50° for 2 hr. A further 2.0 mmol of solid ammonium thiocyanate was added and shaking continued at 50° for 18 hr. The beads were filtered off through a sintered-glass funnel and washed exhaustively with 50% aqueous acetic acid (1:1, v/v) and then with water until the washings were colorless. The beads were finally washed with 1,4-dioxane and transferred to a dry conical flask. Cleavage was carried out by shaking for 1 hr at room temperature with 2 ml of 12 M HCl. The glass beads were then taken to dryness with a stream of nitrogen gas. Acetone (3.0 ml) was added to extract any remaining HCl. The extract was dried down with nitrogen gas. The beads were extracted three times with 1.0-ml portions of methanol and all the extracts were combined for drying and subsequent analysis.

Identification of Amino Acid Thiohydantoins

Directly by Thin-Layer Chromatography.[16] Polygram polyamide-6/UV$_{254}$ or polygram polyamide-6 precoated (0.1 mm) plastic sheets (20 cm × 20 cm) (Camlab, Cambridge, U.K.) were prewashed with chloroform or a solution of chloroform–95% (v/v) ethanol–acetic acid (20:10:3, by volume). Ten-centimeter squares were cut, and the plates were sectioned as shown in Fig. 1 by removing 1-mm strips of coating at right angles to each other 8.5 cm from the starting edges to stop the solvent flow. The origins for the unknown sample and the standard amino acid thiohydantoins were 1 cm from the edges of their respective plates.

[16] M. Rangarajan and A. Darbre, *Biochem. J.* **147,** 435 (1975).

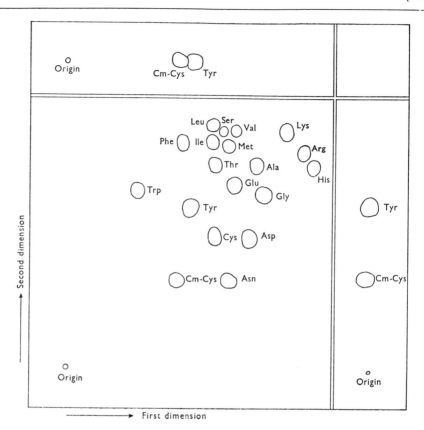

Fig. 1. Schematic representation of two-dimensional thin-layer chromatography of 19 amino acid thiohydantoins. From M. Rangarajan and A. Darbre, *Biochem. J.* **147**, 435 (1975).

Carboxymethylcysteine and tyrosine thiohydantoins were usually used as markers in the outer zones. Solvent 1 was acetic acid–water (7:13, v/v) and solvent 2 was chloroform–95% (v/v) ethanol–acetic acid (20:10:3, by vol.) containing 0.025% (w/v) of BBOT [2,5–bis(5-*t*-butylbenzoxazol-2-yl)thiophen] (Koch-Light, Colnbrook, Bucks., U.K.). Under UV at 254 nm the amino acid thiohydantoins showed up as dark spots on a pale blue fluorescent background. The BBOT was not soluble in solvent 1, but the plates could be prewashed with chloroform solvent containing BBOT and the plate could then be viewed after running in the first dimension. Polygram polyamide-6/UV$_{254}$ plates without indicator or with the added indicator butyl PBD did not give such satisfactory contrast. The solvents kept well for at least 1 week. The chromatography tank was used without equilibration of the atmosphere. The schematic

TABLE II

R_f ($\times$ 100) Values for Amino Acid Thiohydantoins[a,b]

Amino acid thiohydantoin	Solvent system		Amino acid thiohydantoin	Solvent system	
	1[c]	2[d]		1[c]	2[d]
Ala	72	76	Leu	56	90
Arg	90	80	Lys	83	88
Asn	63	33	Met	61	84
Asp	70	49	Phe	45	84
Cys	57	50	Ser	58	87
Cm-Cys	43	33	Thr	57	76
Glu	64	68	Trp	27	67
Gly	75	65	Tyr	45	60
His	93	75	Val	63	88
Ile	56	85			

[a] M. Rangarajan and A. Darbre, *Biochem. J.* **147**, 435 (1975).

[b] Polygram polyamide-6 precoated (0.1 mm) plastic sheets with 7.5 cm run (see text).

[c] Solvent 1: acetic acid–water (7:13, v/v).

[d] Solvent 2: chloroform–95% (v/v) ethanol–acetic acid (2:1:1, by volume) containing 0.025% (w/v) of BBOT.

representation of the separation of 19 amino acid thiohydantoins is shown in Fig. 1. The R_f values are given in Table II. The limit of detection was approximately 0.5 nmol for alanine thiohydantoin. Glutamine thiohydantoin was probably hydrolyzed to glutamic acid thiohydantoin, and no distinction could be made between these.

Directly by Gas–Liquid Chromatography. Chromatography is unsatisfactory without prior trimethylsilylation.[17] To the dried residue of amino acid thiohydantoin add 40 μl of a mixture of pyridine-*N,O*-bis (trimethylsilyl) trifluoroacetamide (1:1, v/v) containing 40 μg of bibenzyl and 40 μg of pyrene and heat at 50° for 10 min in a small closed reaction vial. Inject 1 μl onto the GLC column. Identification of the silylated derivative was made by its retention time relative to that of the standard. Absolute retention times cannot be recommended for identification. Some of the derivatives gave two peaks, which were used for identification. The relative retention times are given in Table III.

Directly by Gas-Liquid Chromatography–Mass Spectrometry. The mass spectra of trimethylsilyl derivatives of alanine, asparagine, glutamic acid, glycine (2), isoleucine, leucine, methionine, phenylalanine, dehydrothreonine (2), and valine were reported.[17]

[17] M. Rangarajan, R. E. Ardrey, and A. Darbre, *J. Chromatogr.* **87**, 499 (1973).

TABLE III

RETENTION TIMES OF TRIMETHYLSILYLATED AMINO ACID THIOHYDANTOINS
RELATIVE TO BIBENZYL OR PYRENE TAKEN AS 1.0[a,b]

Compound	Dexsil 300 GC 1% w/w	OV-17 5% w/w
	120°	160°
Alanine	1.95	1.26
N-Acetylcysteine	2.87	1.71
Glycine	2.49	1.57
Glycine	7.16	2.54
Isoleucine	4.35	2.07
Isoleucine	5.03	2.24
Leucine	4.54	2.15
Proline	1.04	1.05
Serine	1.94	1.25
Serine	2.45	1.75
Dehydrothreonine	4.34	2.53
Dehydrothreonine	5.11	3.54
Valine	3.07	1.58
Bibenzyl	(10.50)[c]	(11.22)[c]
	190°	190°
Asparagine	0.72	0.40
Aspartic acid	0.56	0.40
Glutamic acid	0.75	0.55
Methionine	0.64	0.50
Phenylalanine	0.94	0.81
Pyrene	(9.00)[c]	(58.00)[c]
	210°	260°
Histidine	2.26	1.43
N-Acetyllysine	3.04	1.28
Tyrosine	2.16	1.28
Tryptophan	—	4.48
Pyrene	(6.10)[c]	(6.40)[c]
	240°	
Tryptophan	5.37	
Pyrene	(3.00)[c]	

[a] M. Rangarajan, R. E. Ardrey, and A. Darbre, *J. Chromatogr.* **87**, 499 (1973).
[b] Glass GLC columns 3.25m × 2.5 mm internal diameter, packed with HP Chromosorb W 80–100 mesh coated with stationary phase.
[c] Retention times in minutes.

TABLE IV
MASS SPECTRA OF AMINO ACID THIOHYDANTOINS[a]

Thio-hydantoin	M+ [m/e (%)]	Base peak (m/e)	Major fragment peaks [m/e (%)]						
Ala	130 (100)	130	60	(24)	43	(33)		
Arg[b]	215 (3)	60	155	(95)	153	(30)	59	(46)
Asn	173 (100)	173	156	(48)	128	(93)	59	(54)
Asp	174 (100)	174	128	(73)	60	(50)		
Glu	188 (100)	188	170	(53)	128	(92)	60	(50)
Gly	116 (100)	116	60	(25)				
His	196 (75)	81	82	(63)				
Hyp[b]	172 (100)	172	56	(55)	69	(41)	128	(25)
Ile	172 (66)	116						
Leu	172 (77)	116	129	(36)	43	(66)		
Met	190 (100)	190	129	(63)	116	(76)	75	(22)
Phe	206 (39)	91						
Pro[b]	156 (3)	70	112	(65)	113	(52)	68	(48)
Tyr	222 (3)	107	77	(30)				
Trp	245 (2)	130	77	(25)				
Val	158 (78)	116						
Lys(Ac)	229 (100)	229	187	(49)	170	(52)	129	(49)
Met(O₂)[b]	222 (88)	142	129	(42)	55	(40)	60	(36)
Dehydroser[b]	128 (34)	43	41	(36)	59	(26)		
Ser(Ac)[b]	188 (4)	59	43	(99)	128	(46)	44	(45)
Dehydrothr	142 (100)	142	55	(57)	54	(46)		
Cys(Ac)[b]	204 (22)	43	130	(54)	128	(78)	59	(60)
Cm-Cys[b]	220 (11)	218	60	(70)	85	(68)	59	(40)
Cys[c]	162 (8)	41	128	(86)	60	(29)	59	(24)
CySO₃H[c]	210 (0)	64	128	(23)	59	(21)	48	(99)
Gln[c]	187 (43)	170	128	(65)	60	(36)	59	(86)
Ser[c]	146 (15)	41	128	(56)	60	(58)	59	(76)

[a] M. Rangarajan, R. E., Ardrey, and A. Darbre, *J. Chromatogr.* **87**, 499 (1973).
[b] M. Rangarajan, University of London, Ph.D. Thesis, 1975.
[c] T. Suzuki, K.-D. Song, Y. Itagaki, and K. Tuzimura, *Org. Mass Spectrom.* **11**, 557 (1976).

Directly by Mass Spectrometry. Mass spectra of amino acid thiohydantoins were obtained with an ionizing beam[10,17,18] and chemical ionization[10] by direct insertion probe. The molecular ions and major fragment peaks with m/e greater than 20% are given in Table IV. The mass spectra for the same derivatives are remarkably similar,[10,17,18] but comparison showed some differences, particularly in the mass spectra of arginine, 5-carboxymethylcysteine, glutamic acid, and tyrosine.[10,17] Also,

[18] T. Suzuki, S. Matsui, and K. Tuzimura, *Agric. Biol. Chem.* **36**, 1061 (1972).

Suzuki *et al.*[10] did not detect a molecular ion for S-carboxymethylcysteine, but they did record a spectrum for serine. We were unable to prepare glutamine thiohydantoin.[16] Three μg of amino acid thiohydantoin was usually required for mass spectrometric analysis.[10]

Indirectly by Gas-Liquid Chromatography after Acid Hydrolysis. This was carried out as described on p. 362 (Monitoring Attachment to Solid-Phase) except that hydrolysis was at 135° for 23 hr. The recoveries are given in Table V.

Discussion

Sequencing was originally carried out with a polystyrene-*p*-divinyl-benzene (1% cross-linked) polymer with a methyloxycarbonyl chloride polymer. Some di- and tripeptides were attached in good yield.[8] Longer peptides were attached with yields of less than 10%, probably because of the low swelling ratios of the polymer in the solvents used and also because there was no spacer molecule. The polymer described here had suitable swelling ratios (see Laursen[12]) ranging from DMF (3.2) to acetic acid (1.4). Some simple peptides were attached to the extent of about 4.4 μmol per gram of polymer, e.g., leucylalanylvalylglycyl-phenylalanylglycine, with yields of about 90%. These were degraded sequentially, and the amino acid thiohydantoins were identified successfully. With GLC it was possible to deduce that the yield at each cycle was nearly quantitative, because the graph made by the pen recorder did not show increased "noise" with each successive cycle. However, it must be stressed that only a limited number of amino acids was considered. Larger peptides were not attached because the use of activated glass beads appeared to be an improvement.

Bovine ribonuclease A, lysozyme, and lysozyme reduced and carboxymethylated, were attached to glass beads, with amounts of 19–29 mg of protein per gram. Coupling was carried out in 0.1 M acetate or phosphate

TABLE V
RECOVERY OF AMINO ACIDS FROM THEIR THIOHYDANTOIN
DERIVATIVES AFTER ACID HYDROLYSIS[a]

Amino acid thiohydantoin	Recovery (%)	Amino acid thiohydantoin	Recovery (%)	Amino acid thiohydantoin	Recovery (%)
Ala	86	Gly	102	Phe	90
Arg	73	His	70	Ser	29
Asn	65	Ile	106	Thr	74
Asp	81	Leu	81	Trp	25
Cys	33	Lys	69	Tyr	92
Cm-Cys	93	Met	66	Val	100
Glu	80				

[a] M. Rangarajan and A. Darbre, *Biochem. J.* **147,** 435 (1975).

buffer at pH 6.2 to pH 7.9. In two experiments with ribonuclease, 6 M urea was included in the buffer, but this did not improve the coupling. Ribonuclease was taken through six cycles to confirm the known COOH terminus of -His-Phe-Asp-Ala-Ser-Val. No difficulty was experienced with aspartic acid or the preceding amino acid, alanine.[19] Williams and Kassel[9] attached eight different peptides possessing between two and five amino acids to similar glass beads by coupling at pH 8.3, with average attachment yields of 49% to 71%. They followed the cycles of degradation by subtractive analysis.

The sequencing reaction was formerly carried out in a 5-ml separating funnel with a porous disk at the bottom. This was made from a sintered-glass filter funnel with a 1-cm diameter disk, with a female ground-glass joint above and a tap below.

Because thiocyanate is unstable in acid solution, and this becomes deep yellow, an improved method has been developed using a small Omnifit water-jacketed chromatography column (10 cm $\times$ 0.7 cm internal diameter) with tap fittings possessing Tefzel keys. Stainless steel keys were attacked by the reagents, and this was shown by the dark red color of ferric thiocyanate in the system.

Reagents are forced through narrow-bore Teflon tubing through the column containing the solid-phase (100–200 mg) with nitrogen pressure at 4–5 psi. The ammonium thiocyanate solution remains almost colorless when standing in an ice-bath. The Omnifit precision miniature fittings were purchased from Biolab Limited, 51 Norfolk Street, Cambridge, CB1 2LE, U.K.

Conclusion

Solid-phase attachment of the peptide for sequencing from the COOH terminus is a big improvement on the method in free solution described by Stark.[5] Further work is required to extend both the variety of peptides attached to polymers or glass beads, and also the amino acids to be sequenced. Detailed work will be required to consider the effects of varying the conditions for sequencing. At present about 5–6 hr are required for one cycle of degradation.

[19] G. R. Stark, *Biochemistry* **7**, 1796 (1968).

[36] Regeneration of Amino Acids from Anilinothiazolinones

By C. Y. LAI

In practice, direct identification of amino acid phenylthiohydantoins (PTH) is often beset by difficulties arising from contaminating reagent impurities, buffer salts, by-products of the degradation reaction and/or from the unexpected instability of the PTH itself. One way to circumvent this problem is to regenerate the amino acid from its derivative and subject it to amino acid analysis, the precision and reliability of which is time-proved. Regeneration of amino acids may be accomplished by hydrolysis of their PTH or anilinothiazolinone derivatives, either in acid or in alkali.[1-3] A procedure that appears to be simple and most generally applicable is presented here and discussed in reference to other hydrolytic conditions. With the availability today of automated amino acid analyzers capable of performing multisample analyses at subnanomole levels, the procedure should be of practical value in the routine analysis of anilinothiazolinones (ATZ) obtained in the Edman degradation of peptides or proteins.

Reagents

HCl, 5.7 N, constant boiling: Concentrated HCl is diluted 1:1 with water and distilled. The fraction with the constant boiling point of 108.5° at 760 mm Hg is collected for use.

$SnCl_2$, stock solution: Several grains of crystalline $SnCl_2 \cdot H_2O$ are stirred in acetone in a 250-ml beaker for a few minutes. When the white insoluble matter loosens and is shaken off the surface of the crystals, these are picked up with forceps onto filter paper, dried briefly, and weighed. The stock solution, 10% w/v in 5.7 N HCl, is made from this material, stored in a refrigerator and used within 2 weeks.

Procedure

The ATZ, extracted into an organic solvent after the cleavage reaction in the Edman procedure, is washed once with a small amount of water[4] and dried at the bottom of a glass hydrolysis tube under a nitrogen stream. Freshly made 5.7 N HCl containing 0.1% $SnCl_2 \cdot 2 H_2O$ (1 part stock solution of $SnCl_2$ mixed with 100 parts of 5.7 N HCl), 0.5–1 ml, is

[1] H. O. Van Orden and F. H. Carpenter, *Biochem. Biophys. Res. Commun.* **14**, 399 (1964).
[2] O. Smithies, D. M. Gibson, E. M. Fanning, R. M. Goodfliesch, J. G. Gilman, and D. L. Ballantyne, *Biochemistry* **10**, 4912 (1971).
[3] E. Mendez and C. Y. Lai, *Anal. Biochem.* **68**, 47 (1975).
[4] This is essential, especially when peptides are undergoing Edman degradation. Contamination by even a minute quantity of peptides in the extract can confuse the results of analysis. The ATZ solution in organic solvent (4–5 ml) is Vortex-mixed with 0.2–0.3 ml of H_2O for several seconds and centrifuged.

added, and the tube is evacuated and sealed. Hydrolysis is carried out at 150° for 4 hr.[5]

Discussion

In the cleavage step of the Edman degradation procedure, the NH_2-terminal residue is removed from the peptide chain as the ATZ derivative and extracted into an organic solvent.[6] Amino acid ATZ's are relatively unstable and readily isomerize to PTH's on heating in acid,[7] which are then characterized by various chromatographic methods.[8] In the procedure described above, the ATZ derivatives of amino acids are directly subjected to acid hydrolysis in the presence of 0.1% $SnCl_2 \cdot 2\ H_2O$. All but a few amino acids may be identified in this procedure. Threonine and serine ATZ are converted to α-aminobutyric acid (90%) and alanine (68%), respectively, and the tryptophan derivative, to glycine (70%) and alanine (27%) (Table I). As in other hydrolysis procedures, the presence or the absence of cysteine, and of amide groups on aspartic or glutamic acid cannot be determined.

The percentage yield of amino acids from their ATZ under various hydrolytic conditions are shown in Table I. In the absence of 0.1% $SnCl_2 \cdot 2\ H_2O$, hydrolysis in 5.7 N HCl resulted in the loss of most threonine and serine with poor recoveries in either α-aminobutyric acid or alanine. With HI, little or no methionine is recovered from its ATZ derivative.[2,3] In 4 M CH_3SO_3H, the recoveries of serine and threonine are better than in HCl or HI, but still are less than 10% after 4 hr at 150°, with small conversion to α-aminobutyric acid (11%) and alanine (40%), respectively. In general, the yields on regeneration of amino acids with CH_3SO_3H approached those with HCl containing $SnCl_2$.[3] Hydrolysis of ATZs in dilute alkali[2,3] resulted in lower recoveries of most amino acids than that in acids with no obvious advantage (see the table).

When PTH-derivatives (rather than ATZ's) are hydrolyzed under the conditions described here, the recoveries of amino acids have been somewhat lower but close to the values shown in the table.[9]

Tests of the hydrolytic regeneration method in the actual Edman

---

[5] A small hot plate oven, e.g., Fisher 11-494-5, connected to a time switch is quite adequate.

[6] P. Edman, *Acta Chem. Scand.* **4**, 283 (1950).

[7] P. Edman and G. Begg, *Eur. J. Biochem.* **1**, 80 (1967).

[8] See, for example: J. J. Pisano, T. J. Bronzert, and H. B. Brewer, Jr., *Anal. Biochem.* **45**, 43 (1972); D. K. Kulbe, *Anal. Biochem.* **59**, 564 (1974); J. Jeppson and J. Sjoquist, *Anal. Biochem.* **18**, 264 (1967); and C. L. Zimmerman and J. J. Pisano, *Biochem. Biophys. Res. Commun.* **55**, 1220 (1973).

[9] Based on the experiments carried out by Dr. J. Moshera in the author's laboratory.

TABLE I
Recovery of Amino Acids from Their Anilinothiazolinones[a]

Amino acid	5.7 N HCl[b] 150°C, 4 hr 0.1% SnCl₂	—	47% HI[c] 110°C 24 hr	47% HI[c] 150°C 4 hr	4 M CH₃SO₂H[c] 110°C 24 hr	4 M CH₃SO₂H[c] 150°C 4 hr	0.01 M NaOH[c] 110°C 24 hr
Lys	92	98	7	100	14	84	60
His	99	94	10	100	18	93	24
Arg	99	84	6	81	15	85	—
Asp	97	99	94	100	78	100	64
Thr	0	5	10	1	15	7	5
Aab[d]	90	5	ND[e]	89	ND	11	ND
Ser	0 (68)[d]	4	13	1	20	9	7
Glu	97	99	85	100	74	100	50
Pro	94	93	40	78	20	91	72
Gly	99	90	21	95	26	93	107[f]
Ala	99	84	57[g]	214[g]	45[g]	136[g]	60[g]
Val	61	83	19	72	33	69	40
Met	92	83	3	0	25	75	24
Ile[h]	60	82	13	65	30	68	34
Leu	82	73	14	82	30	81	28
Tyr	67	79	ND	80	5	70	36
Phe	70	75	ND	72	5	72	34
Trp	0 (97)[i]	0 (40)[i]	ND	ND	ND	ND	ND

[a] Amino acids (20–30 nmol each) were converted to ATZ's [E. Mendez and C. Y. Lai, *Anal. Biochem.* **68**, 47 (1975)] in hydrolysis tubes (16 × 125 mm, Corning No. 9860), dried under a stream of nitrogen, and subjected directly to hydrolysis. No extractions were made. Amino acid analysis of an aliquot similarly treated, but not hydrolyzed, showed that the conversion was quantitative. Recoveries were calculated in percentage from the amount of amino acids in the hydrolyzate and those in the untreated aliquot. The values represent the average of triplicate experiments except for those from HI hydrolysis, which are of a single determination.

[b] Values for Thr, Ser, Gly, Ala, and Trp were obtained by experiments with the individual amino acids. For other amino acids, the data obtained with a mixture of amino acids are presented.

[c] The data obtained with a standard mixture of amino acids (Hamilton Co., Whittier, CA).

[d] Aab, α-aminobutyric acid, is formed from the ATZ of Thr, and Ala, from that of Ser, in acid hydrolysis.

[e] ND, not determined.

[f] With alkali, glycine is formed from the derivatives of Thr, Ser, and Gly.

[g] Values represent those from Ala- and Ser-ATZ's.

[h] Sum of Ile and aIle recovered from Ile-ATZ.

[i] Trp-ATZ yielded Gly (70%) and Ala (27%) in 5.7 N HCl containing 0.1% SnCl₂, and Gly (40%) in HCl without added SnCl₂.

TABLE II
SEQUENCE ANALYSIS OF APOMYOGLOBIN[a]

Steps	Amino acid found	% yield[b] Exp. 1	% yield[b] Exp. 2	% yield, calculated values[c]
1	Val	79.5	100	92
2	Leu	73.0	78.5	85
4	Glu	71.7	78	74
6	Glu	64.8	66.5	66
8	Glu	51.3	—	58
9	Leu	—	43.3	54
10	Val	46.2	—	51
12	His	27.5	25.9	46

[a] Sperm whale apomyoglobin (5 mg, 290 nmol; Beckman No. 339182) was sequentially degraded in a Model 890C Spinning-cup Automatic Sequencer (Beckman Instruments, Palo Alto, CA), using a high yield program (Beckman Peptide Program No. 102974). The anilinothiazolinones obtained in the indicated steps were identified after regeneration of amino acids as described below.

[b] An aliquot containing 50 nmole (Exp. 1) or 100 nmol (Exp. 2) of PTH-norleucine was added to the anilinothiazolinone obtained in each step. About one-fourth of each sample was then hydrolyzed with 0.4 ml 5.7 NHCl containing 0.1% $SnCl_2 \cdot H_2O$ at 150° for 4 hr in a sealed, evacuated tube. After the regeneration, amino acids were analyzed with a Jeol 6 AH Automatic Amino Acid Analyzer according to Spackman et al.[11] Values were expressed in mole % of the starting material

[c] Values were taken from Fig. 3, In Sequence No. 7 (1975), Beckman Instruments Inc., Palo Alto, CA. In the experiment the repetitive yield of 93.8% was achieved with this protein through the use of Program No. 102974.

degradation of peptides have shown its general applicability; the recoveries of amino acids from the ATZ's are in good agreement with those expected from the data presented in the table. In general, recoveries of amino acids, including conversion of threonine to α-aminobutyric acid and tryptophan to glycine and alanine, are near quantitative. The yield of conversion of Ser to Ala, of regeneration of Val, Ile, Leu, Tyr, and Phe range from 60 to 80%, using the procedure described here. The cleavage yield at each step of Edman degradation of peptides can thus be estimated from these values. An example of application of this procedure in the automated sequence analysis of sperm whale apomyoglobin[10] is presented in Table II. It serves to illustrate the usefulness of the regeneration procedure.

[10] The author is indebted to Mr. Rusty Kutny for performing the Sequencer runs.
[11] D. H. Spackman, W. H. Stein and S. Moore, Anal. Chem. 30, 1190 (1958).
   and were corrected against that of PTH-norleucine, the hydrolytic yields of which were 90% and 93% in Exp. 1 and Exp. 2, respectively.

[37] Identification of Anilinothiazolinones after Rapid Conversion to N^α-Phenylthiocarbamyl-Amino Acid Methylamides

By JOHN K. INMAN and ETTORE APPELLA

When cleavage steps of an Edman degradation are carried out with a strong, anhydrous acid, the terminal amino acids are split off as 2-anilinothiazolinones. These reactive compounds are usually converted to the more stable phenylthiohydantoins (PTH)[1] by heating with aqueous acid; the hydantoins are subsequently extracted into an organic solvent. In the process some derivatives are obtained in low yield. In the case of PTH-serine, often little will remain of the original side-chain structure. Ambiguities in identification commonly result which make it necessary to use additional methods of analysis. Further, time-consuming operations are required to recover PTH-histidine and arginine, which are not extracted into solvent.

Recently, we proposed an alternative means for derivatizing thiazolinones that offers some promise of avoiding problems encountered with PTH methods.[2] Thiazolinones (see second structure in pathway 1 of Fig. 1) are cyclic thiol esters and, as such, can be effective acylating agents. Barrett et al.,[3,4] recognizing this fact, carried out the aminolysis of thiazolinones with nucleophiles such as aniline to yield derivatives useful in sequence analysis. Elevated temperatures (111°) were required to complete the reaction with aniline. Under such conditions a serine thiazolinone would be quickly destroyed through side reactions. In contrast with aniline, we found that certain other nucleophiles, especially small primary alkylamines (but not ammonia), react very rapidly with thiazolinones at room temperature. Anilinothiazolinones gave high yields of N^α-phenylthiocarbamyl (PTC)-amino acid alkylamides by the reaction exemplified in pathway 1 of Fig. 1. Hydrazine and dialkylamines readily yielded analogous products.

PTC-amino acid alkylamides can be formed in several minutes, following removal of cyclizing acid and solvent, by simply adding a large

---

[1] Abbreviations used: PTH, phenylthiohydantoin; PTC, phenylthiocarbamyl; PTMA, N^α-phenylthiocarbamyl methylamide; TFA, trifluoroacetic acid; Boc, t-butyloxycarbonyl; Z, benzyloxycarbonyl; TLC, thin-layer chromatography; APAM, N-(3-aminopropyl)aminomethylated; THF, tetrahydrofuran; DMF, N,N-dimethylformamide.

[2] J. K. Inman and E. Appella, in "Solid-Phase Methods in Protein Sequence Analysis" (R. A. Laursen, ed.), p. 241. Pierce Chemical Company, Rockford, IL 1975.

[3] G. C. Barrett, J. Chem. Soc., Chem. Commun. 1967, 487 (1967).

[4] G. C. Barrett and A. R. Khokhar, J. Chromatogr. 39, 47 (1969).

excess of an alkylamine and removing the latter with a stream of nitrogen. Small alkylamines are either gases or highly volatile liquids and thus can be removed quickly. The residual material is ready for chromatographic identification without further steps. The chromatographic properties of PTC alkylamide derivatives on silica thin layers or on columns are very good, and these derivatives have high UV absorptivity, which makes sensitive detection easy. PTC-serine and threonine methylamides have been recovered in 57 and 75% yields, respectively, without apparent alteration of side-chain structure. Seventeen of the other 18 amino acids have been recovered from PTC-amino acid polystyrene resin-amides as clean PTC-amino acid methylamides (PTMA-amino acids) by successive treatment with trifluoroacetic acid (TFA) and methylamine. Derivatives of cysteine have not yet been tried.

The major limitation of the above approach is the requirement that thiazolinones be chemically intact and free from extraneous substances which would interfere with subsequent identification. Nearly all the cyclizing acid must be removed, as it forms a troublesome salt with the amine. The latter also reacts with halocarbons such as 1-chlorobutane. Thiazolinones are quickly hydrolyzed by moisture and react rapidly enough with primary and secondary alcohols that the latter must be absent prior to addition of the amine. However, since aminolysis is quite rapid, alkylamines may be added as solutions in various alcohols.

Attempts to remove impurities from PTC-amino acid alkylamides by extraction into solvents have so far not been successful. The more effective strongly acid systems cause loss of the derivatives probably through cyclization back to the thiazolinone and subsequent hydrolysis. Besides, an extraction step violates the simplicity of the method. It therefore seems best to adopt a degradation procedure which yields clean, unaltered thiazolinones. Solid-phase sequencing methods appear to offer the greatest chance of achieving this goal.

Currently, we are endeavoring to adapt the new identification method to automated solid- and liquid-phase sequencing. Manual solid-phase sequencing has been successfully carried out with identification of the resulting PTC methylamide derivatives on thin-layer chromatograms and with a high-pressure liquid chromatographic system. Below are described some procedures which have been used in studies on the application of the alkylamine conversion reaction in Edman-type degradations.

Conversion Reaction

Methylamine has been tentatively selected as the best alkylamine for the conversion reaction on the basis that the resulting methylamides possess the least shared structure. Consequently, they should be more readily separable by virtue of differences in the amino acid side chains. Anhydrous methylamine (b.p. −6°) is stored and sold in metal cylinders and is conveniently drawn off as a gas. Anilinothiazolinones can be quickly converted to PTMA-amino acids by treatment with a nonaqueous solution of methylamine or by direct exposure to the gaseous amine. This operation is best carried out shortly after each cyclization step has been completed because serine and threonine thiazolinones are not stable in a strongly acidic environment. At 23°, in TFA-solvent mixtures, the former disappears with a half-life of approximately 2.5 hr; similarly, threonine thiazolinone is half destroyed in about 13 hr.

Conversion in Solution. Cyclizing acid and/or solvent[5] are removed with a stream of N_2. Contamination with moisture must be avoided by using a pure, dry form of N_2 (such as the prepurified grade) and by applying gentle heat in order to avoid cooling the system below the dew point. In some cases moist room air can be drawn into the sample tube by the N_2 stream. Heat should be applied by means of a heating block or glycerol bath (set at 40°–50°) rather than a water bath. A small amount (0.2–0.4 ml) of 2.4 M methylamine in 2-propanol is added; the tube is swirled and allowed to stand 1–2 min at room temperature. Solvent and methylamine are then removed with N_2 under gentle heating, and the remaining PTMA-amino acid is dissolved in starting solvent for chromatographic identification.

A 2.4 M solution of methylamine in 2-propanol may be prepared as follows: A flask containing 80 ml of 2-propanol is tared on a top-loading balance. Methylamine gas is passed into the alcohol through a disposable pipette until a weight gain of 7.75 g (0.24 mol) is achieved. The solution is finally made up to 100 ml with 2-propanol and appears to be usable for many months when stored under refrigeration.

Conversion with Gaseous Amine. The thiazolinone is dried down with N_2 as described above. A gentle stream of methylamine gas is led into the tube (near the bottom) for about 20 sec (room temperature). The tube is covered and allowed to stand for 2 min. The amine is then expelled by directing a stream of N_2 near the bottom of the tube for several minutes. The PTMA-amino acid remains and is ready for identification.

The solution and gas conversions appear to give equally satisfactory results. The former avoids the bother of opening a storage cylinder, the

---

[5] The cyclizing acid, if present in significant amount, may be anhydrous trifluoroacetic acid, but not heptafluorobutyric acid, which cannot be removed with N_2. The presence of methanol, 2-propanol, other primary or secondary alcohols, or aqueous acid at this stage is not compatible with the method.

latter allows much easier and faster removal of the excess amine. The last-mentioned advantage can be especially important for automated conversion systems. We are currently designing and evaluating such a system to be used with an automated solid-phase sequenator[6] where N_2, $CH_3NH_2(g)$, and N_2 are sequentially metered through a lead that is directed into the collection tube by an elevator mechanism. When the conversion is completed, the jet is lifted out of the tube before the sample changer is actuated. The cyclizing acid (TFA) is removed very slowly in order to avoid condensation of moisture, since no heat is applied. TFA and methylamine vapors are carried away by aspiration through an inverted funnel (also attached to the elevator) and led into a water trap.

Identification and Quantitation

PTMA-amino acids are readily separated by liquid chromatography; satisfactory conditions for their separation on unactivated thin layers of silica have been found. These derivatives, synthesized by several routes, have been employed as identification markers and as standards for investigating separation methods. The three routes which have been used are shown in Fig. 1. So far, all PTC-amino acid alkylamides which have been prepared by several of these pathways chromatograph identically in various thin-layer systems, in support of the reaction scheme as shown. PTMA-isoleucine, prepared from the anilinothiazolinone (pathway 1), usually chromatographs as a double spot; presumably this circumstance is due to racemization of the α-carbon configuration leading to diastereomers. PTMA-L-isoleucine prepared by pathway 2 is not racemized and cochromatographs with the higher R_f component. PTC-threonine n-propylamide (via pathway 1) chromatographs as a double spot in some solvent mixtures, but the nature of the alternate product has not yet been established.

PTMA-amino acids in nonaqueous solutions appear to be usable for many months as markers when stored in a refrigerator. The solid derivatives last only a week or two in the laboratory environment. Desulfurization of the PTC group by atmospheric oxygen may be one process responsible for their decomposition.

PTMA-Amino Acids from Resin Standards. To 25 mg of PTC-amino acid APAM-polystyrene resin-amide (see below under Syntheses) in a centrifuge tube are added 0.9 ml of 1-chlorobutane and 0.3 ml of TFA.

---

[6] R. A. Laursen, this series Vol. 25, p. 344; and *Eur. J. Biochem.* **20**, 89 (1971). The solid-phase sequenator and part of the automated converter was engineered and built by Sequemat, Inc., Watertown, MA 02172.

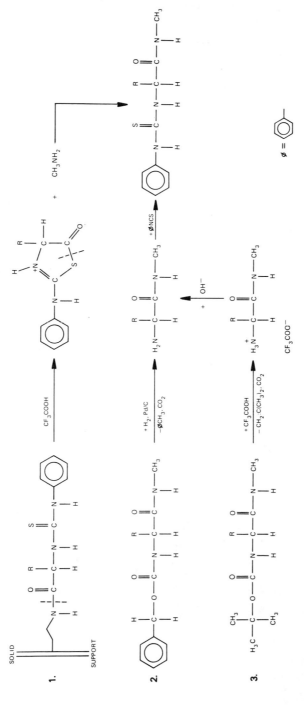

FIG. 1. Reaction pathways leading to PTC-amino acid methylamides (PTMA-amino acids). Pathway **1** shows the route through 2-anilinothiazolinones occurring in solid-phase sequencing or in synthesis from the PTC-amino acid APAM-polystyrene resin-amides. Pathways **2** and **3** show alternative routes, yielding optically pure PTMA-amino acids, which may be used for producing standards from stable precursors.

The mixture is left at room temperature and occasionally agitated with a thin glass rod. After a certain period of time (25 min for Ser and Thr, 1 hr for Trp, and 2–3 hr for other derivatives), the mixture is centrifuged briefly and its supernatant withdrawn. The resin is washed with 0.6 ml of 1-chlorobutane, and the two supernatants are combined, blown dry with N_2, and converted to the PTMA amino acid as described above under Conversion Reaction, *Conversion in Solution*. The residue is dissolved in 3.0 ml of a 2:1 v/v 1-chlorobutane-2-propanol mixture. The concentration of the solution is approximately 5 μmol/ml.

PTMA Derivatives from Boc- and Z-Amino Acid Methylamides.[1] Fifty micromoles (12.2 mg) of Boc-L-leucine methylamide (see below under Syntheses) are dissolved in 0.30 ml of TFA (sequenator grade) in a 2-ml glass-stoppered tube. The solution is allowed to stand for 30 min at room temperature, and the TFA is removed with N_2 for 15 min (using a 25° covered water bath). The oily residue (which contains some bound TFA) is dissolved in 0.50 ml of acetonitrile (spectrophotometric grade), and 20 μl each of N,N-dimethylbenzylamine and phenylisothiocyanate are added. The latter two reagents should be highly pure (sequenator grade). The tube is swirled, flushed with N_2, and heated at 50° for 30 min. The bulk of the acetonitrile is removed with N_2; the remaining liquid is diluted with 0.8 ml of chromatography solvent and passed through a column with a bed of silica gel (chromatographic grade[7]) using a mixture of 1,2-dichloroethane and 2-propanol, 95:5 v/v.

Benzyloxycarbonyl(Z)-L-amino acid methylamides can be converted to PTMA derivatives by pathway 2 (Fig. 1). Fifty micromoles of the Z-amino acid methylamide (see Syntheses) are dissolved in a suspension of 3.5 mg of 10% palladium-on-charcoal catalyst in 0.50 ml of methanol. The mixture is equilibrated with hydrogen at 1 atm for 1 hr and centrifuged; the supernatant is dried down with N_2. The residue is dissolved in 0.50 ml of acetonitrile plus 5 μl of TFA and treated in the same way as the acetonitrile solution in the preceding paragraph. The conditions for chromatographic purification will be different for each derivative. In most instances a different ratio of 1,2-dichloroethane to 2-propanol will be all that is required. This final step is currently being investigated.

Thin-Layer Chromatography (TLC). Silica thin-layer chromatograms can be visualized directly under shortwave UV light if the layer contains a fluorescent indicator (available in commercially prepared TLC plates). The limit of detection is around 0.5 nmol. Greater sensitivity may be achieved by exposing a developed, dried plate to dilute

---

[7] The authors have used J. T. Baker No. 3405, silica gel, 60–200-mesh powder. Extra fine particles were removed by gravity sedimentation in water. The material was dried in an oven and equilibrated with room air.

chlorine gas for about 30 sec, airing it (at the front of a fume hood) for 30 min, and spraying the layer with a fine mist of 1% KI in 1% soluble starch solution. With this technique, detection of as little as 50–100 pmol is possible.

On 0.25-mm layers of E. Merck's silica gel 60, F-254, 19 different PTMA derivatives can be separated by the use of two solvent mixtures as shown in Fig. 2. The 13 least polar derivatives are separated by solvent mixture A (see Fig. 2 legend). The separation is improved for the lower R_f components by a second development with the same solvent. The 6 most polar derivatives are separated by solvent mixture B; they remain near the origin in solvent A. As discussed above, PTMA-isoleucine appears as a double spot, which is a useful circumstance for identification. Only cases of mixed sequences or high backgrounds

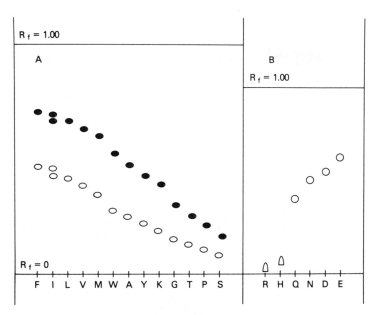

FIG. 2. Thin-layer chromatograms showing the separation of 19 PTMA-amino acids using two solvent mixtures. The derivatives are identified by the standard single-letter code for the parent amino acids; K refers to N^α,N^ϵ-di-PTC-lysine methylamide. Glass plates coated with 0.25-mm layers of silica 60, F254 (by E. Merck, Darmstadt, Germany; sold through EM Laboratories) were used. Solvent A has the composition: ethyl formate (60 ml), bromochloromethane (15 ml), cyclohexane (25 ml), stored with water (2 ml) plus $KHCO_3$ (4 g). The mixture is shaken well and filtered through Whatman No. 1 filter paper on the day of use. Solvent B has the composition: 1,2-dichloroethane (80 ml), 2-propanol (20 ml), and glacial acetic acid (8 ml). Open spots are single developments; filled-in spots represent a double development.

yielding comparable amounts of Phe and Leu at a given step would be ambiguous.

Standards may be comprised of two mixtures of PTMA-amino acids consisting of alternate members in the R_f series as illustrated in Fig. 6 of Inman and Appella,[2] where Quantum Industries, Q6F silica plates were used and solvent A had the following composition (by volume): ethyl acetate 10, benzene 8, 1,2-dichloroethane 2, acetonitrile 1. Solvent B is the same for both sources of plates.

Quantitation. PTMA-amino acids absorb UV light generally to the same extent as do PTH-derivatives. PTMA-leucine exhibits 3 maxima in 2-propanol (Fig. 3) with peak molar absorptivities of 19,100 (208 nm), 13,600 (245 nm), and 11,200 (268 nm). At 254 nm, where highly sensitive and stable UV detectors operate, the molar absorptivity is 12,100. Quantitative determination of PTMA-amino acids is carried out reliably with a high-pressure liquid chromatograph. We have used the Du Pont Model 830 instrument with Zorbax-ODS-packed columns (for

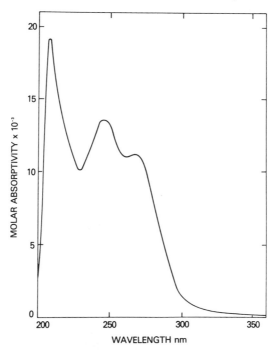

Fig. 3. Ultraviolet absorption spectrum of PTC-L-leucine methylamide (PTMA-leucine) in 2-propanol. A Beckman Model 25, double-beam, scanning spectrophotometer was used. The sample of PTMA-leucine was prepared from Boc-L-leucine methylamide (see pathway 3, Fig. 1), separated on a thin-layer plate, and quantitatively eluted immediately after development.

reverse-phase systems) with gradient mixtures of 10 mM (aq.) sodium acetate and acetonitrile[8] for quantitative assessment of yields in cyclizations. Sharp peaks and very low backgrounds are obtained, and detection limits may be as low as 20 pmol. At present, we are investigating high-pressure liquid chromatographic systems for identifying all principal PTMA-amino acids by this means.

Syntheses

Materials. The following sources of chemicals were found convenient and satisfactory: (Mono) methylamine (lecture bottle) from Matheson Gas Products (E. Rutherford, NJ 07073); most Boc-amino acids from Pierce Chemical Co. (Rockford, IL 61105), and Beckman Instruments, Inc. (Palo Alto, CA 94304); specifically, Boc-histidine and arginine (free side chains) from Bachem, Inc. (Marina Del Rey, CA 90291); Boc-aspartic acid-β-t-butyl ester, dicyclohexylammonium salt and Boc-glutamic acid-γ-t-butyl ester from Sigma Chemical Co. (St. Louis, MO 63178); and N^α,N^ε-bis-Boc-L-lysine from Vega-Fox Biochemicals (Tucson, AZ 85719); Z-L-Tyr hydrazide from Sigma; *tert*-butyl nitrite from ICN Pharmaceuticals, K&K Labs (Plainview, NY 11803); N,N-diethylethanolamine (puriss) from Aldrich Chemical Co. (Milwaukee, WI 53233). The following reagents were Sequenal Grade of Pierce Chemical Co.: TFA, N-methylmorpholine, N-ethylmorpholine, phenylisothiocyanate, triethylamine, 1-chlorobutane, and 4 N HCl-in-dioxane.

1-Hydroxybenzotriazole monohydrate (Aldrich Chemical Co.) is recrystallized as follows: Forty grams of the material are dissolved in 210 ml of ethanol plus 50 ml of water by heating to boiling. Activated charcoal (2.5 g) is added, and the hot solution is stirred for 30 min and filtered. The filtrate is mixed with 750 ml of 85° water, and the mixture is cooled slowly to +5°. Crystals are collected and dried under vacuum.

Boc-aspartic acid-β-t-butyl ester, dicyclohexylammonium salt is converted to the free acid by equilibrating 500 mg with 25 ml of CH_2Cl_2 plus 10 ml of 15% w/v citric acid (aq.) and subsequently shaking the lower layer twice with 8 ml of water. The final lower layer is dried over anhydrous $MgSO_4$, filtered and evaporated. About 1.0 mmol of the free acid is obtained.

Dioxane is tested for the presence of peroxides by mixing equal volumes (2 ml) with 4% w/v KI. If a yellow-brown color results within 5 min, then more than the required amount of dioxane is stirred with activated alumina (ICN Pharmaceuticals, Inc., Woelm basic 200, activity I), and filtered.

[8] C. L. Zimmerman, E. Appella, and J. J. Pisano, *Anal. Biochem.* **75**, 77 (1976).

N-(3-Aminopropyl)aminomethylated   (APAM)-Polystyrene   Resin.
Twenty-two grams of chloromethylated polystyrene beads[9] are suspended
in 200 ml of benzene for 30 min, and 100 ml of 1,3-propanediamine is
then added. The mixture is stirred and heated to 80° for 1 hr. The resin
is allowed to cool, then is washed on a glass Büchner funnel with generous amounts (at least 500 ml) of benzene, tetrahydrofuran (THF),
80% v/v THF (20% H_2O), and 70% THF. The resin is stirred overnight in 70% THF and washed on the funnel again with 70% THF and
finally methanol. The APAM resin is dried in air (overnight) and then
under vacuum.

PTC-Amino Acid APAM-Polystyrene Resin-Amides. One gram of the
above APAM resin is suspended in a solution of 2.00 mmol of Boc-
amino acid plus 2.50 mmol (383 mg) of 1-hydroxybenzotriazole monohydrate in 15 ml of *N,N*-dimethylformamide (DMF) (spectrophotometric grade) in a 30-ml siliconized beaker. The mixture is stirred for
0.5 hr and allowed to stand overnight at room temperature. 1-Ethyl-3-
(3-dimethylaminopropyl)carbodiimide hydrochloride (Sigma Chemical
Co.) (1.50 mmol, 288 mg) is added, and the mixture is stirred at room
temperature for 5 hr. The resin is then washed on a siliconized glass
Büchner funnel (15-ml, coarse porosity) with DMF, DMF-CH_3OH 1:1,
and CH_2Cl_2, and then suspended in CH_2Cl_2-TFA 1:1 v/v for 30 min.
The latter operation may be carried out in the washing funnel by
plugging the stem with a rubber bulb. The bulb is removed, and the resin
is washed with CH_2Cl_2, CH_2Cl_2-dioxane 1:1, and DMF-*N*-ethylmorpho-
line 15:2 v/v[10] and suspended in the latter mixture to a volume of 18
ml. The suspension is heated to 50°, and 0.80 ml of phenylisothiocyanate
(sequenator grade) is added. Stirring with heating is continued for 30
min. The resin is washed thoroughly (in the siliconized funnel) with
DMF, DMF-CH_3OH 1:1, CH_3OH, and ethyl ether, then dried in air
(briefly) and under vacuum. The resulting PTC-amino acid resin amide
should be stored under N_2 (with desiccant) in a freezer.

The N^α-Boc derivatives of Ser, Thr, Pro, Ala, Asn, Gln, Leu, Ile, Val,
Met, Gly, Phe, Tyr, Trp, His, and Arg are used without side-chain protecting groups because of the protective effect (rapid active ester formation) of the coupling catalyst, 1-hydroxybenzotriazole.[11,12] Lysine is em-

---

[9] Bio-Beads, S-X1, 200–400 mesh, chloromethylated to 0.75 meq/g are available from
  Bio-Rad Laboratories, Richmond, CA.
[10] A. Previero and J.-F. Pechère, *Biochem. Biophys. Res. Commun.* **40**, 549 (1970).
[11] W. König and R. Geiger, *Chem. Ber.* **103**, 788 (1970).
[12] V. J. Hruby, F. Muscio, C. M. Groginsky, P. M. Gitu, D. Saba, and W. Y. Chan,
  J. Med. Chem. **16**, 624 (1973).

ployed as its N^α,N^ε-bis-Boc derivative, and the resulting resin-amide is the corresponding di-PTC form. Asp and Glu are introduced as their N^α-Boc, ω-t-butyl esters. These protecting groups are removed by the CH_2Cl_2-TFA treatment.

Boc-L-Leucine Methylamide. Boc-L-leucine monohydrate (2.22 g, 8.92 mmol) is dissolved in 45 ml of CH_2Cl_2 and stirred with 1.07 g of anhydrous $MgSO_4$ for 20 min. N-Methylmorpholine (1.00 ml, 8.92 mmol) is added, and the mixture is cooled to $-20°$ ($MgSO_4$ need not be removed). Isobutyl chloroformate (1.10 ml, 8.37 mmol) is added, and the mixture is stirred at $-20°$ for 25 min (to give a mixed anhydride solution).

Methylamine hydrochloride (Eastman Organic Chemicals) is ground in a mortar and 0.68 g (10.1 mmol) is suspended in 20 ml of CH_2Cl_2 plus 1.40 ml (10.1 mmol) of triethylamine. The suspension is cooled to $-20°$, and the mixed anhydride solution is added. The mixture is stirred and allowed to warm slowly (30 min) to $0°$, then stirred 1 hr at $0°$ and overnight (17 hr) at $+5°$.

The mixture is filtered, and the precipitate is rinsed with 30 ml of CH_2Cl_2; the combined filtrates are shaken in a separatory funnel twice with 30 ml of 0.20 N HCl and twice with 30 ml of 0.50 M $KHCO_3$. The lower layers are returned to the funnel, and upper layers are discarded. The final lower layer is stirred with 7 g of anhydrous Na_2SO_4, filtered and evaporated to remove solvent.

The white, solid product is stirred at reflux with 55 ml of hexane for about 0.5 hr to dissolve all material. The solution is allowed to cool in a covered beaker to room temperature and stand for several hours (crystallization should begin). The beaker is placed in a refrigerator (tightly covered) overnight, and the crystalline mass is collected on a glass Büchner funnel, washed with 6 ml of hexane and dried in air and under vacuum. Expected yield is about 1.63 g (80% of theory); the m.p. is 122–123°. The material should be pure by TLC (silica) in $CHCl_3$-CH_3OH, 98:2 v/v, with $R_f = 0.37$. Benzyloxycarbonyl(Z)-L-leucine methylamide may be prepared in a similar way and crystallized from ethyl acetate–hexane 1:3 v/v.

Benzyloxycarbonyl(Z)L-Tyrosine Methylamide. The following procedure is an example of an alternative method of synthesis of methylamides, which may be employed where mixed anhydride couplings cannot be used with unprotected amino acid side chains: Z-L-tyrosine hydrazide (4.20 g, 12.5 mmol) is dissolved with warming in 160 ml of DMF. The solution is cooled to $-30°$ and 10.0 ml of 4 N HCl dioxane and 1.76 ml of *tert*-butyl nitrite are added. The mixture is stirred at $-25°$ for 30 min; then 13.8 ml of 0.3 M sulfamic acid in DMF is added, and stirring ($-25°$) is continued for 10 min longer. The temperature is lowered to −50°, and 4.2 ml of N,N-diethylethanolamine (Aldrich), 5.0 ml of N-ethylmorpholine, and 7.0 ml of 2.4 M methylamine in 2-propanol (see under Conversion Reaction, *Conversion in Solution*) are added. The mixture is stirred for 24 hr at 0° and for 48–72 hr at +5°. The solvent is evaporated with a mechanical pump, and the residue is shaken with a mixture of 150 ml of ethyl acetate, 50 ml of 1-butanol, and 60 ml of 0.5 N HCl. The upper layer is shaken successively with 60-ml volumes of 0.5 N HCl, 0.5 M KHCO$_3$ (twice), and 0.8 M NaCl, then dried over 17 g of anhydrous Na$_2$SO$_4$, filtered and evaporated to dryness.

The crude product is dissolved in 60 ml of boiling ethyl acetate, refluxed with 0.6 g of active charcoal and filtered. The filtrate is heated to 65° and 50 ml of hexane (60°) is added. The mixture is allowed to stand 2 hr at room temperature. The white precipitate is collected on a suction filter, dried, and recrystallized from ethanol (40 ml) plus water (80 ml), and finally from ethyl acetate (15 ml) to +5°. Expected yield is 1.58 g (38.5% of theory) and m.p. is 146–147°. The material is essentially pure by TLC (silica) in CHCl$_3$—CH$_3$OH, 9:1 v/v, with R_f = 0.57.

[38] Automated Conversion of Anilinothiazolinones into PTH-Derivatives

By JOHN BRIDGEN

In the Edman degradation amino acids are released sequentially from the NH$_2$ terminus of the protein as 2-anilino-5-thiazolinone derivatives. For every 1–2 hr of operation, an automatic sequencer will produce a fraction containing one of these derivatives. This must then be manually converted into the more stable phenylthiohydantoin (PTH) by heating in aqueous acid at 80° for 10 min, followed by an extraction of the PTH-derivative into ethyl acetate before identification. Automation of this process would have two advantages. First, the amount of labor and time required for sample workup would be reduced. Second, yields of the less stable thiazolinones may be improved by their immediate conversion into phenylthiohydantoins.

Since the design and mode of operation of the solid-phase sequencer differs considerably from the spinning-cup instruments, the design of automatic converters will also differ for these two instruments. The principle of operation will, however, remain very similar.

Principle of Operation

Solid-Phase Sequencer. The thiazolinone is removed from the reaction column in a stream of trifluoracetic acid (TFA). This is mixed with water

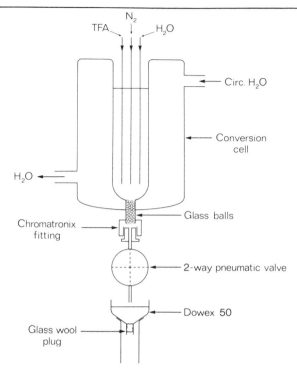

FIG. 1. Schematic diagram of an automatic converter for use with a solid-phase sequencer. For dimensions, see text.

to form an approximately 20% solution, and the conversion takes place either in a jacketed reaction vessel (see below) or by passing the mixture through a heated reaction coil.[1] The PTH-derivative is then delivered to the fraction collector.

Liquid-Phase Sequencer. Here, the thiazolinone is washed from the spinning-cup in organic solvent which passes into a reaction vessel. Excess solvent is removed and 20% trifluoroacetic acid added. Conversion takes place at 55° and the PTH-derivative is then transferred to the fraction collector in a stream of dichloroethane:methanol (7:3).[2]

Converters for Solid-Phase Sequencing

The converter to be described here is of the reaction cell type. It can be incorporated into almost any solid-phase sequencer and requires no additional programming capacity for operation. It is shown diagrammatically in Fig. 1.

[1] C. Birr and R. Frank, *FEBS Lett.* **55**, 61 (1975).
[2] B. Wittman-Liebold, H. Graffunder, and H. Kohls, *Anal. Biochem.* in press.

The following modifications are required to the instrument:
1. A jacketed glass reaction vessel (approx. 4 cm long, 1.5 cm diameter) is mounted in series between the water-jacket of the reaction column and the circulating water bath and is maintained at 50°.
2. The outlet of the effluent line from the reaction column to the fraction collector is redirected into the top of the conversion cell.
3. A Teflon line is connected via a three-way valve (e.g., Bio-Lab, Cambridge, U.K.) to the low-pressure N_2 delivery system and is led into the top of the converter.
4. A peristaltic pump is electrically connected between the terminals of the waste-collect valve solenoid so that when the effluent from the reaction column is directed to the fraction collector the pump is activated. The inlet to the pump is connected to a water reservoir. The outlet leads into the converter. The pump flow rate should be four times that of the trifluoracetic acid used for cleavage. In this case the TFA flow rate was 0.15 ml/min and the water flow rate was 0.6 ml/min.
5. A two-way pneumatically operated valve (e.g., Chromatronix, Altex, Durrum) is mounted directly beneath the converter. The outlet from the bottom of the conversion cell, a 3-mm internal diameter tube filled with glass beads to reduce the volume, should have a fitting compatible with this valve (i.e., Chromatronix, Altex or Luer). The valve is activated by N_2 pressure from the same supply as the other pneumatic valves in the instruments. The supply is controlled by a solenoid valve (Schrader 460SA) operated either from a spare channel in the programming unit or from a channel which controls an indicator rather than an operational function.
6. A small glass cup (2 × 1.5 cm diameter) containing a plug of silanized glass wool is placed on top of each fraction collector tube. Dowex 50 (H^+ form), previously washed with 1 N HCl and methanol, is added to each cup as a suspension in methanol, so as to form a layer about 0.5 cm deep. The two-way valve must be positioned to give adequate clearance between its outlet and the top of the resin cups.

The position of the various components in the fraction collector compartment is shown in Fig. 2.

Operation

The conventional sequencing program is used except that the final methanol wash is directed to waste, and only the first 1 ml of TFA, which contains virtually all the thiazolinone, is collected in the cell. Thus, during cleavage of each amino acid derivative from the peptide, the effluent passes into the reaction cell rather than directly into the fraction

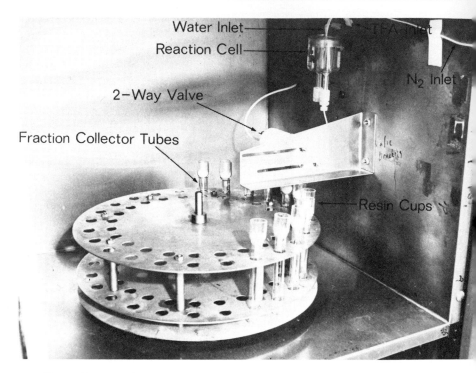

Fig. 2. Layout of components in the fraction collector compartment of a solid-phase sequencer (Sequemat) fitted with an automatic convertor. The conversion cell shown is suitable for two-column operation although it is currently being used on one column only. For convenience of operation a smaller fraction collector turntable has been fitted to the instrument.

collector. There it is mixed with water from the ancillary pump to form a 20% aqueous solution. A slow stream of N_2 bubbles continuously through the cell to facilitate the mixing. After incubation at 50° for 30 min, the pneumatic valve opens according to an instruction on the program tape and the contents of the cell pass under gravity through the resin cup into the fraction collector. Each fraction contains approximately 5 ml of liquid which is removed by rotary evaporation prior to identification.

Results

Using the B chain of insulin, attached via phenylenediisothiocyanate to aminopolystyrene, as a test peptide, results using the automatic converter appeared to be very similar to those obtained by manual conversion (200 μl of 1 M HCl at 80° for 10 min). However, a significantly improved yield of PTH-serine was seen at position 9 and the extent of deamidation of the asparagine and glutamine residues was reduced. There was also a significant saving in the time and labor normally needed for the workup of the fractions. Although the efficiency of conversion was as good if not better than by manual methods, there was a partial loss of some of the PTH-derivatives on the resin, and PTH-histidine and PTH-arginine were completely retained by the Dowex. Better results were obtained by omitting the resin cups, drying the contents of the fraction collector tubes, redissolving in a small volume of methanol and passing this solution through a Pasteur pipette filled with the Dowex resin.[3] This, of course, requires additional time and labor. Unfortunately, simply redissolving the converted fractions in methanol or ethyl acetate and applying a suitable aliquot to the TLC plate was unsatisfactory since contaminating spots running close to PTH-threonine, PTH-glycine, and PTH-leucine tended to obscure the identification. These contaminating spots may not interfere with the identification when other methods such as high pressure liquid chromatography are used.

An alternative version of this form of converter has been described by Birr and Frank.[1] In this system the TFA and water are mixed through a T-piece in the effluent line and conversion takes place in a heated reaction coil similar to those used in amino acid analyzers. This method appears to work well but has the disadvantage that the conversion time becomes part of the overall cycle time.

Converters for Liquid-Phase Sequencing

To date, results have been presented from only one laboratory using an automatic converter in conjunction with a spinning-cup sequencer.[4] This converter[2] consists of a jacketed pear-shaped glass reaction vessel 80–100 ml in volume (Fig. 3). The thiazolinone, dissolved in chlorobutane (S_3), is transferred from the cup into the conversion flask and the solvent is then evaporated under a stream of N_2. Aqueous trifluoroacetic acid (20%, v/v containing 0.02% dithioerythritol, R_4) is added and conversion takes place at 55° for 20–30 min. The acid is removed by a combination of N_2 blowing and vacuum and the PTH-derivative transferred to the fraction collector in a stream of dichloroethane:methanol (7:3, R_5) ready for identification.

The complete conversion program may be readily incorporated into that used for the sequence determination.

Results using this converter are impressive, but it should be remembered that considerable modifications were also made to the sequencer

---

[3] R. A. Laursen, *Eur. J. Biochem.* **20**, 89 (1971).
[4] B. Wittman-Liebold, *Hoppe-Seyler's Z. Physiol. Chem.* **354**, 1415 (1973).

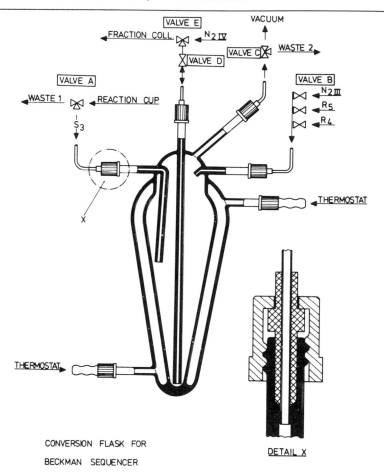

Fig. 3. Automatic conversion apparatus for use with a spinning-cup sequencer. B. Wittman-Liebold, H. Graffunder, and H. Kohls, *Anal. Biochem,* in press.

and how many of these are necessary for efficient operation of the converter is not yet clear.

Alternative Methods of Conversion

There are at least three alternatives to using aqueous acid for the conversion step, both of which would lend themselves quite readily to automation.

One of these is thermal conversion[5,6] where the sample is dried under a stream of N_2 and is then incubated at 80° for 30 min. Conversion

[5] P. Edman and G. Begg, *Eur. J. Biochem.* **1,** 80 (1967).
[6] R. L. Guyer and C. W. Todd, *Anal. Biochem.* **66,** 400 (1975).

yields compare well[6] with those obtained by the conventional procedure, and PTH-asparagine and PTH-glutamine are obtained in better than normal yield (L. Hood, personal communication). In addition, there is no separation of the derivatives of histidine, arginine, or cysteic acid. The second technique[7,8] is to treat the dried thiazolinone fraction with methylamine gas to form a PTC-amide although this method has the disadvantage that separate chromatographic systems have to be developed for the identification of these derivatives. Conversion in 1 N methanolic HCl[9] has been reported to be faster than aqueous HCl and this system has clear advantages for automation.

Conclusions

It is clearly easier to install an automatic converter on a solid-phase sequencer than on its spinning-cup counterpart. The solid-phase converters described here may be installed on almost any commercial instrument currently available.

It may be felt, however, that developments which improve the performance and applicability of automatic sequencing should take precedence over those which are designed primarily to increase the convenience. Whether one installs on automatic converter may therefore depend on one's reliance on additional automation and the availability of technical assistance.

Acknowledgment

I am indebted to Dr. B. Wittman-Liebold for a preprint of her paper describing the automatic converter used on the spinning-cup sequencer.

[7] J. K. Inman and E. Apella *in* "Solid-Phase Methods in Protein Sequence Analysis" (R. A. Laursen, ed.), 241. Pierce Chemical Co., Rockford, IL, 1975.
[8] J. K. Inman, and E. Appella, this volume [37].
[9] G. E. Tarr, *Analyt. Biochem.* **63**, 361 (1975).

[39] Analysis of Dipeptides by Gas Chromatography–Mass Spectrometry and Application to Sequencing with Dipeptidyl Aminopeptidases

By HENRY C. KRUTZSCH and JOHN J. PISANO

Dipeptides formed by dipeptidyl aminopeptidase I (DAP I) digestion of polypeptides[1,2] have been determined by ion-exchange and paper

[1] J. K. McDonald, B. B. Zeitman, T. J. Reilly, and S. Ellis, *J. Biol. Chem.* **244**, 2693 (1969).
[2] J. K. McDonald, P. X. Callahan, and S. Ellis, this series Vol. 25, p. 272 (1972).

chromatography,[3] but the procedures are cumbersome, time-consuming, and limit the rate of sequencing. Dipeptides also have been analyzed by gas chromatography–mass spectrometry (GC–MS) after the preparation of volatile derivatives, such as N-perfluoropropionylamide methyl esters,[4,5] N-ethoxycarbonylpropenyl methyl esters,[6] and, following reduction with $LiAlD_4$, trimethylsilylaminoalcohols.[7] We have investigated the trimethylsilyl (Me_3Si) derivatives of dipeptides and have not experienced reported difficulties with gas chromatographic[8] and mass spectrometric analysis.[9] The Me_3Si derivatives may be prepared and identified at the nanomolar level in less than 1 hr, as most have excellent gas chromatographic properties, and exhibit simple fragmentation patterns in the mass spectrometer.[10] Polypeptide sequencing with DAP[11,12] is greatly facilitated by the rapid and sensitive determination of dipeptides.

Preparation of Dipeptidyl Aminopeptidases

DAP I. The enzyme is prepared by a modification of the original procedure of Metrione *et al.*[13] A 2-kg quantity of frozen bovine spleen previously frozen and thawed twice is partially thawed, and the outer membrane is removed. The semifrozen spleens are then cut into pieces and homogenized batchwise for 3 min in a Waring Blendor using 1–2 liters of water. The homogenate is adjusted to pH 3.5 with 6 N H_2SO_4, stirred 1–2 hr at room temperature, the pH is readjusted if necessary, and then it is stirred overnight at 37°. The suspension is centrifuged at 10,000 g for 30 min in a refrigerated unit. The clear supernatant solution is taken to the cold room and ammonium sulfate is added to 40% of saturation.

[3] P. X. Callahan, J. A. Shepard, T. J. Reilly, J. K. McDonald, and S. Ellis, *Anal. Biochem.* 38, 330 (1970).
[4] Y. A. Ovchinikov and A. A. Kirushkin, *FEBS Lett.* 21, 300 (1972).
[5] R. M. Caprioli and W. E. Seifert, Jr., *Biochem. Biophys. Res. Commun.* 64, 295 (1975).
[6] G. M. Schier, P. D. Bolton, and B. Halpern, *Biomed. Mass Spec.* 3, 32 (1976).
[7] J. A. Kelley, H. Nau, H. J. Forster, and K. Biemann, *Biomed. Mass Spec.* 2, 313 (1975).
[8] K. Rühlmann, H. Simon, and M. Becker, *Chem. Ber.* 99, 780 (1966).
[9] K. M. Baker, M. A. Shaw, and D. H. Williams, *J. Chem. Soc., Chem. Commun.* 1108 (1969).
[10] H. C. Krutzsch and J. J. Pisano, *in* "Peptides: Chemistry, Structure and Biology," Proceedings of the Fourth American Peptide Symposium (R. Walter and J. Meienhofer, eds.), p. 985. Ann Arbor Sci. Publ., Ann Arbor, Michigan, 1975.
[11] P. X. Callahan, J. K. McDonald, and S. Ellis, this series Vol. 25, p. 282 (1972).
[12] P. X. Callahan, J. K. McDonald, and S. Ellis, *Fed. Proc., Fed. Am. Soc. Exp. Biol.* 28, 661 (1969); 30, 1045 (1971).
[13] R. M. Metrione, A. G. Neves, and J. S. Fruton, *Biochemistry* 5, 1597 (1966).

After standing for 2–4 hr, the mixture is again centrifuged at 10,000 g for 30 min; the supernatant is brought to 80% of saturation, and allowed to stand overnight. The precipitate is recovered by centrifugation, taken up in 50–100 ml of water, and dialyzed for 3 days against 1% NaCl, which is changed after 3 hr and each day. The dialyzed solution is adjusted to pH 5.0, dispensed into 13-ml conical glass centrifuge tubes, and the tubes are placed in a 65° bath for 40 min. The tubes are centrifuged and the supernatants are pooled and concentrated to 5–10 ml using a 50-ml Amicon stirred cell fitted with a PM-30 membrane.

The concentrate is filtered in the cold room through a 3.2 × 94-cm Bio-Gel A-0.5m column with 0.1 M NaOAc–0.1 M NaCl buffer, pH 5.0 (flow, 6 ml/hr; fractions, 3.1 ml). Enzyme assays are performed with 0.1 M N-methylmorpholine[14] HCl buffer, pH 7.0, containing 1% NaCl and 0.5 ml mercaptoethanol[15] per 100 ml. A small sample (usually ≤ 20 μl of appropriately diluted fraction aliquot) is added to 1.5 ml of buffer, and the reaction is started by the timed addition of 0.1 ml of 0.5 mM Gly-Phe-β-naphthylamide. The samples are incubated at 37° for 15 min, and, in the order that substrate is added, liberated β-naphthylamine is determined fluorometrically; excitation 335 nm, emission 410 nm. Active fractions are then tested for mercaptan-activated monoaminopeptidase activity using the DAP I buffer and Leu-β-naphthylamide in place of Gly-Phe-β-naphthylamide. DAP I fractions which contain aminopeptidase are rejected. Remaining DAP I fractions are heated 2 hr at 65° to destroy residual dipeptidase and catheptic carboxypeptidase C. Then the pooled fractions are adjusted to pH 6.0 with NaOH, clarified by centrifugation, and concentrated to approximately 5 ml in a 50-ml Amicon stirred cell fitted with a PM-30 membrane. Buffer is exchanged with water by adding approximately 30 ml of water to the concentrate and again reducing the volume to approximately 5 ml. This procedure is performed three times. The desalted sample is transferred to a 13-ml conical centrifuge tube and further concentrated at room temperature with a nitrogen stream. The viscous residue deposited on the wall of the tube is dried under vacuum at room temperature. Approximately 0.1 mg is required to determine the number of units by the procedure previously described.[1]

---

[14] N-Methylmorpholine used throughout this work was Sequanal grade, Pierce Chemical Co., or was redistilled from phthalic anhydride.

[15] Mercaptoethanol used throughout this work is purified by vacuum distillation into a receiver cooled in Dry Ice and stored in the cold under nitrogen. This treatment helps to eliminate contaminants (most notably a large peak due to the trimethylsilylated dimer which appears early in the GC run) which may interfere in the GC-MS analysis. Recently, we have replaced mecaptoethanol with 2-methoxyethylmercaptan ($CH_3OCH_2CH_2SH$), which does not cause any interference.

DAP IV. This enzyme, while not possessing the broad specificity of DAP I, has the unique property of hydrolyzing peptide bonds at the carboxyl group of proline. It is purified by the published procedure[16] with some modifications.

After DEAE-cellulose chromatography,[16] active fractions are combined and concentrated in the cold room to 2–5 ml using a 50-ml Amicon stirred cell and a PM-30 membrane. The concentrate is filtered through a 1.5 × 90-cm column of Bio-Gel A-0.5 m with 50 mM N-methylmorpholine·HCl–0.5 M NaCl buffer, pH 7.0 (flow, 6 ml/hr; fractions, 1.7 ml). Portions are assayed by the procedure previously described.[16] Active fractions are combined and concentrated to about 5 ml in an Amicon stirred cell; the buffer is exchanged with water as previously described for DAP I. The concentrate is freeze dried. Unlike DAP I, DAP IV is stable to freeze-drying but the preparation contains dipeptidases that must be inhibited with 1,10-phenanthroline (see below).

Polypeptide Digestion

Polypeptides are degraded in Reacti-Vials (Pierce Chemical Co.) cleaned in a solution of HNO_3–H_2SO_4, 1:4. With <25 nmol of peptide a 0.4-ml vial and 0.05 ml of DAP solution are used; with >25 nmol, a 1.0-ml vial and 0.2 ml DAP solution are used. To the dry sample is added the appropriate volume of DAP I prepared by dissolving 3–5 units of DAP I in 0.1 M N-methylmorpholine·CH_3COOH, pH 7.0, containing 1% NaCl and 5 μl of mercaptoethanol per 0.2 ml of buffer. When proline is present, 1 unit of DAP IV and 0.3 mg of 1,10-phenanthroline also are included. Vials are sealed with Teflon-lined caps, and the samples are digested for 4 hr at 37°. The reaction is stopped by freeze-drying.

Another aliquot of the polypeptide unknown in a Reacti-Vial is shortened by a single Edman degradation to obtain the "overlapping" dipeptides.[17] The shortened polypeptide is dissolved in 75 μl of a solution consisting of 50 μl of pyridine and 25 μl of water by stirring on a Vortex mixer for about 10 min. Then 100 μl of water is added and the solution concentrated to about 10 μl with a nitrogen stream. DAP digestion of

[16] V. K. Hopsu-Havu, P. Rintola, and G. G. Glenner, *Acta Chem. Scand.* **22,** 299 (1968).

[17] After the Edman cleavage, the solution of shortened peptide is evaporated to dryness with a N_2 stream and extracted with ethyl acetate. Excessive background is observed when benzene is used instead of ethyl acetate. Extraction losses of peptides are minimized when dry peptides are treated with ethyl acetate.

the shortened polypeptide is carried out as described above (ε-phenyl-thiocarbamyl lysine does not interfere).

Freeze-dried digests are stirred with 0.2–0.3 ml of benzene for 5–10 min. The suspension is centrifuged and the benzene removed. The extraction is repeated and the last traces of benzene are removed first with a gentle nitrogen stream, then under vacuum. Benzene extraction removes interfering 1,10-phenanthroline and by-products of mercaptoethanol. To the dry sample is added 0.1 ml of N,O-bis(trimethylsilyl)trifluoroacetamide (BSTFA) (Supelco) and 0.1 ml of dry acetonitrile (Burdick and Jackson). If <25 nmol of peptide is digested, use 25 μl of each reagent. Vials are tightly capped with Teflon-lined screw caps and heated for 10 min at 140°. A 10-μl aliquot of this solution is taken for GC-MS analysis. Trimethylsilylation occurs with the following functional groups: amino, carboxyl, hydroxyl, imidazoyl NH, indole NH, primary amide, and guanidino. One proton on the α-amino and primary amido group is exchanged; both protons are replaced on the ε-amino group of lysine and the α-amino group of glycine, and three of the four protons of the guanidino group of arginine. Protons of the ring amides of diketopiperazines also react, but protons of the amido group of dipeptides usually do not.

Gas Chromatography of Trimethylsilylated Dipeptides

Gas chromatography of trimethylsilylated dipeptides is performed with short columns, low loading of stationary phase and rigorous masking of active sites. The procedures for cleaning the column, solid support, and glass wool, masking active sites and coating support with stationary phase are adaptations of the procedure described in Volume 25 of this series.[18] The support, usually 25 g of 100–120 mesh Chromsorb W (Supelco), is soaked overnight in HNO_3–HCl (3:1) and then washed (by suspension and decantation) with concentrated HCl until the washes are colorless. Acid is removed by washing the support to neutrality in deionized water. It is then allowed to stand for 10 min in 5% Na_2CO_3 and again washed with water to neutrality. Any material not settled in 5 min is decanted. The support is then dried at 110°C and treated overnight at room temperature by allowing it to stand in a 20% (v/v) solution of dichlorodimethylsilane (Supelco) in toluene (Burdick and Jackson). To ensure reagent penetration, the support is degassed with an aspirator after the reagent is added. The silylated support is rinsed several times with toluene. Care should be taken to keep

---

[18] J. J. Pisano, this series Vol. 25, p. 27 (1972).

the support covered with toluene to prevent reaction of any remaining -Si-Cl groups with atmospheric moisture. Replacement of the latter with methoxy groups is effected by allowing the support to stand for 10 min in anhydrous methanol (Burdick and Jackson). The deactivated support is finally washed several times with toluene and dried by heating 2–3 hr at 110°. Glass GC-columns and glass wool plugs are treated with 1:4 HNO_3–H_2SO_4 and also are washed in HNO_3–HCl and treated with dichlorodimethylsilane and anhydrous methanol. The filtration technique is used to coat the support with stationary phase.[18] An OV-1 (Supelco) column (nominally 1%) is prepared by adding 1 g of support to 20 ml of chloroform containing 0.2 g of OV-1. The mixture is degassed by gently swirling under reduced pressure (aspirator) and filtered on a sintered-glass funnel for about 15 min (house vacuum). It is then carefully spread in a dish, thoroughly dried by heating for 5 min at 110°. The section of siliconized column used in the LKB GC-MS which fits into the column oven is 0.2 cm × 60 cm; that which fits into the flash heater, 0.3 × 15 cm. The column is filled up to the 0.3 cm section using suction and gentle tapping. Columns 0.2 cm throughout presumably are equally satisfactory. Column packing is held in place at both ends with siliconized glass wool. Silicone rubber septums are soaked overnight in chloroform to remove low molecular weight dimethylsiloxanes and then air-dried. Columns are conditioned by heating for 48 hr at 290° with a 30 ml/min helium flow rate. Periodically, 10 μl of BSTFA are injected to silylate any active sites that may have been missed or were exposed during filling or conditioning. New or reused columns require 1 or 2 injections of sample before reliable results are obtained. Samples (usually 2–10 μl, 0.2–100 nmol) are analyzed with the flash heater set at 280°, the helium flow 30 ml/min, and the column temperature initially at 100°. After injection of the sample, the column temperature is raised to 280° at the rate of 10°/min. A gas chromatogram of a DAP digest of a dodecapeptide is given in Fig. 1. Notice that Ser-Gln is the smallest peak. When columns begin to deteriorate after about a month of steady use, peptides containing Asn, Gln, or His give no peaks; presumably, they adsorb or decompose on the column.

Of the possible 400 dipeptides, it appears by extrapolation from the 180 we have examined, that at least 300 will chromatograph. Arginine-containing dipeptides are not sufficiently volatile, and those containing ε-phenylthiocarbamyl lysine apparently decomposed. Also, His-Gln does not elute and presumably such other (as yet untested) dipeptides as His-Asn, Gln-His, and Gln-Trp. The nonvolatile (or unstable) trimethylsilylated dipeptides are determined by direct introduction into the mass spectrometer. A 10- to 40-μl aliquot is placed in the probe cup, concen-

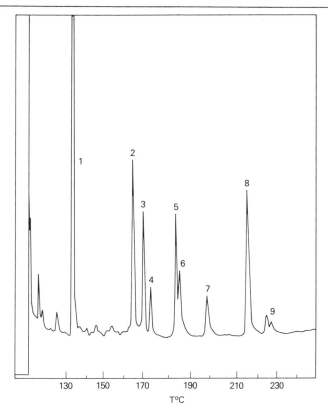

Fig. 1. Gas chromatogram of the Me₃Si-dipeptides obtained from Tyr-Thr-Met-Ser-Ser-Gln-Leu-Thr-Leu-Pro-Thr-Val digested with mixture of DAP I and IV. Peaks 1, 4, 5 come from mercaptoethanol; 3, Leu-Pro and Thr-Val; 5, Leu-Thr; 6, Met-Ser; 7, Ser-Gln; 8, Tyr-Thr; 9, contaminants.

trated to about 10 μl with a stream of nitrogen, capped with a siliconized glass wool plug, and introduced into the ion source with the external heater at room temperature. Then the temperature is raised approximately 10° per minute to 150°. Virtually all volatile derivatives are evaporated before the source temperature reaches 80°, when the less-volatile derivatives of interest begin to appear.

Replacing 100–120 mesh Chromosorb W with 80–100 mesh Chromosorb 750 has allowed the chromatography of His-Gln, His-Lys, His-Phe, His-Trp, His-Tyr, Asn-Trp, Gln-Trp, Tyr-Gln, and Trp-Gln. However, the responses at the 5 nmol level are slightly less and below 1 nmol significantly less than the responses of nonpolar volatile dipeptides. Presumably, all Me₃Si-dipeptides except those containing arginine now can be analyzed by GC-MS.

TABLE I
IMPORTANT MASSES FOR THE IDENTIFICATION OF TRIMETHYLSILYL DIPEPTIDES

Amino acid 1	Mass of fragment ion (a) when amino acid is amino-terminal 2	Masses used to calculate [M-15]$^+$ when amino acid is carboxyl-terminal[a] 3
Gly	174[b]	159
Ala	116	173
Pro	142[c]	199
Val	144	201
Leu	158[d]	215
Ile	158[d]	215
Met	176	233
Phe	192[e]	249
Ser	204	261
Thr	218	275
Asn	231	288
Asp	None[f]	289
Gln	156[g]	302
Glu	246	303
His	254[h]	311
Tyr	280[i,j]	337
Trp	303[i,k]	360
Lys	317[l]	374
Arg	417[l,m]	474
CYS(CH$_2$CO$_2$H)	278	335
CYS(CH$_2$CONH$_2$)	277	334
CYS[(CH$_2$)$_2$NH$_2$]	345	None

[a] These values equal the mass of the carboxyl-terminal fragment formed by scission of the central CH—CO bond minus 15. These fragments usually are not seen in spectra. When added to the mass of ion (a) they give the mass of ion (b).

[b] Since amino-terminal Gly usually forms the Bis-Me$_3$Si derivative, fragment ion 174 usually is observed instead of the monosubstituted 102 ion (a). During silylation, diketopiperazines form with the Me$_3$Si derivatives of Gly-Gly (M-15 = 243), Gly-Ala (M-15 = 257), Gly-Gln (M-15 = 401), Gly-Glu (M-15 = 402), Gly-Met (M-15 = 317), and Gly-Lys (M-15 = 458). In these cases, ion (a) is not observed.

[c] Instead of silylation, Pro-Pro undergoes cyclization to the diketopiperazine, and the molecular ion m/e 194 is one of the most abundant ions in the spectrum.

[d] Dipeptides containing Leu and Ile are distinguished by noting the intensities of the 41 and 43 amu fragment ions. With Leu 43 > 41; with Ile 43 ≤ 41.

[e] Dipeptides containing Phe (in either position) also give a fragment ion at m/e 91.

[f] Ion (a) is not seen with amino-terminal aspartyl dipeptides owing to the formation of the Me$_3$Si cyclic imide. The [M-15]$^+$ peak of the cyclic dipeptide is determined by adding 142 to the number in column 3.

[g] Samples introduced directly into the spectrometer via the probe usually give ion (a) m/e 245, but with GC-MS, cyclization usually occurs (especially with older columns) and gives ion m/e 156 corresponding to Me$_3$Si-pyrrolidonyl ion, (Me$_3$Si-N = CHCH$_2$CH$_2$C = 0)$^+$. In the latter case add 156 to the number in column 3 to obtain the M-15 peak for the cyclized derivative.

[h] Dipeptides containing His (in either position) also give a fragment ion at m/e 154.

[i] Instead of ion (a), dipeptides containing Tyr or Trp in either position sometimes give an ion 72 amu less than the ion (a) due to the loss of Me$_3$Si from the amino group and its replacement by H during gas chromatography. When this occurs, add the lower mass to column 3 to obtain the respective [M-15]$^+$ peak.

[j] Dipeptides containing Tyr (in either position) also give a fragment ion at m/e 179.

[k] Dipeptides containing Trp (in either position) also give a fragment ion at m/e 202.

[l] Dipeptides containing Arg, Lys, or Trp also give a molecular ion [M]$^+$.

[m] Ion (a) m/e 417 is seen in low abundance because it loses the (Me$_3$Si)$_3$-guanido group yielding fragment m/e 142 in high abundance.

Mass Spectral Identification of Trimethylsilyl Dipeptides

An LKB model 9000 mass spectrometer with coupled gas chromatograph has been used. The total ion plot is determined at 20 eV ionizing voltage, spectra are scanned at 70 eV, ionizing current, 60 μamp, Dipeptides usually can be identified by the appearance of two significant ions (Fig. 2). Ion (a) is derived from the amino-terminal amino acid by β-cleavage of the central CH-CO bond typical of peptides[19] and is usually the most abundant ion in the spectrum. Ion (b) is the molecular ion minus a methyl group (M-15), typical of trimethylsilylated compounds. In practice (M-15) is not always easily identified because it may not be the highest mass ion in the spectrum owing to the presence of traces of other peptides or contaminants. Peptides are identified from the data given in Table I, which contains the masses of ion (a) for each amino acid when it is amino-terminal and calculated masses when each is carboxyl-terminal. Carboxyl-terminal masses are calculated by subtracting ion (a) from (M-15). The sum of masses from the two columns provides (M-15) the highest mass expected from a dipeptide. When pure peptides are analyzed (M-15) is directly discerned, as it is the highest mass ion (except in the cases cited in Table I). In this situation, one subtracts ion (a) from (M-15) to obtain the carboxyl-terminal amino acid. A model spectrum is shown in Fig. 3. Exceptions to the typical fragmentation are given in the footnotes to Table I.

Dipeptide Alignment

The amino acid composition of the original dipeptide is usually required because it provides the only means of determining when more than one equivalent of a dipeptide is present in a digest. Separate lists are prepared of the dipeptides generated from the original polypeptide and the peptide shortened by a single Edman degradation.[20] The amino-

---

[19] E. Stenhangen, Z. Anal. Chem. 181, 462 (1961).

[20] Overlapping dipeptides also may be generated by adding an amino acid to the original polypeptide.[1] In preliminary experiments we have found that coupling the t-Boc-N-hydroxyphthalimide ester of α-aminobutyric acid in N,N-dimethylacetamide-pyridine proceeds smoothly at the 20-nmol level. Solvent is evaporated with a N_2 stream and excess ester and by-products are extracted with benzene. DAP's readily form α-aminobutyryl dipeptides which are determined by the usual GC-MS method. Since the α-aminobutyryl dipeptide contains the amino-terminal amino acid, it is not necessary to set up the Edman degradation procedure or methods to identify PTH's. α-Aminobutyric acid will also couple to the ε-amino group of lysine. It is presumed that the DAP's will hydrolyze such polypeptides. However, the digest will contain lysine tripeptides instead of lysine dipeptides, since α-aminobutyric acid is coupled to the ε-amino group of lysine. These tripeptides should be amenable to direct (probe) analysis by mass spectrometry if not by GC-MS.

$$
\begin{array}{cc}
\text{R} & \text{R}' \\
| & | \\
\text{Me}_3\text{Si}-\text{NHCH}-\text{CONHCHCO}_2-\text{Me}_3\text{Si}
\end{array}
\begin{array}{l}
\xrightarrow{e^-} [\text{Me}_3\text{Si}-\text{NH}{=}\text{CHR}]^{\oplus}\text{ (a)} \\
\xrightarrow{e^-} [\text{M-15}]^{\oplus}\text{ (b)}
\end{array}
$$

Fig. 2. Fragments from Me₃Si-dipeptides used in their identification.

TABLE II
Peptides Digested with DAP

$$
\begin{array}{ccc}
\text{HO}_2\text{CCH}_2\text{S} & \text{SCH}_2\text{CO}_2\text{H} & \text{SCH}_2\text{CO}_2\text{H} \\
| & | & |
\end{array}
$$

1. Gly-Ile-Val-Glu-Gln-Cy-Cy-Ala-Ser-Val-Cy-Ser-Leu-Tyr-Gln-Leu-Glu-Asn-
$$
\begin{array}{l}
\text{SCH}_2\text{CO}_2\text{H} \\
|
\end{array}
$$
   Tyr-Cy-Asn (CMC insulin A chain)[a]

2. Asp-Arg-Val-Tyr-Ile-His-Pro-Phe (angiotensin II)[b]
            5   7
$$
\begin{array}{l}
\text{SO}_3\text{H} \\
|
\end{array}
$$

3. Phe-Val-Asn-Gln-His-Leu-Cy-Gly-Ser-His-Leu-Val-Glu-Ala-Leu-Tyr-Leu-Val-
$$
\begin{array}{l}
\text{SO}_3\text{H} \\
|
\end{array}
$$
   Cy-Gly-Glu-Arg-Gly-Phe-Phe-Tyr-Thr-Pro-Lys-Ala (oxidized insulin B chain)[c]
        22   26   28

4. Ser-Tyr-Ser-Met-Glu-His-Phe-Arg-Trp-Gly-Lys-Pro-Val-Gly-Lys-Lys-Arg-Arg-
                                11 12   17
           39
   Pro-Val----Phe (ACTH)[d]

5. Asp-Ile-Asn-Val-Lys

6. Glu-Ala-Thr-His-Lys
        3   5

7. Thr-Ser-Thr-Ser-Pro-Ile-Val-Lys[e]

8. Ile-Asp-Gly-Ser-Glu-Arg

9. Asp-Glu-Tyr-Glu-Arg

10. Ser-Phe-Asn-Arg

11. Lys-Phe-Ile-Gly-Leu-Met
         5   7

12. Phe-Leu-Asn-Asn-Phe-Tyr-Pro-Lys[f]

13. His-Asn-Ser-Tyr-Thr-Cys(CH₂CH₂NH₂)

14. Lys-Lys-Lys-Lys-Lys

---

[a] Completely digested with DAP I before and after single Edman degradation.

[b] Digestion with DAP I or DAP I + DAP IV stops at Ile⁵ because peptide bonds at the imino group of proline are not hydrolyzed.

[c] Native peptide is completely digested by mixture of DAP I and IV. After a single Edman degradation, the digestion stops at Tyr²⁶.

[d] Digestion with DAP I stops at Lys¹¹; with DAP I + IV, digestion stops at Arg¹⁷.

[e] Digestion with DAP I + IV stops at Thr³. After single Edman degradation, digestion is complete.

[f] Digestion with DAP I stops at Phe⁵. After single Edman degradation, it would go to completion with DAP I + IV.

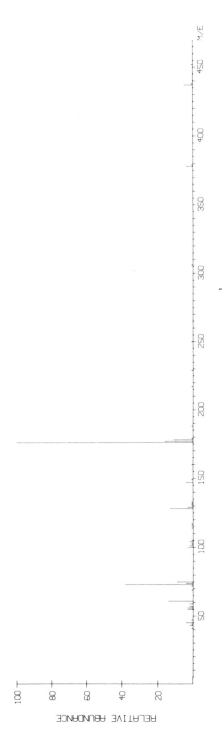

FIG. 3. Mass spectrum of Me₃Si-Met-Ser. $m/e = 176$ is ion (a) [Me₃Si-NH = CHCH₂SCH₃]; $m/e = 437$ is (M-15)⁺.

TABLE III
POLYPEPTIDE UNKNOWNS SEQUENCED

1. Asp-Ser-Thr-Tyr-Ser-Met-Ser-Ser-Thr-Leu-Thr-Leu-Thr-Lys
2. Phe-Ser-Gly-Ser-Gly-Ser-Gly-Thr-Asp-Phe-Thr-Leu-Gln-Ile-Ser-Arg[a]
3. Phe-Ser-Gly-Ser-Gly-Ser-Gly-Lys
4. Thr-Phe-Gly-Gly-Gly-Thr-Lys
5. Val-Lys-Ala-Gly-Asp-Val-Gly-Val-Tyr-Cys(CH$_2$CH$_2$NH$_2$)[b]
6. Ala-Ser-Gly-Val-Ser-Asn-Arg[a]
7. Ser-Gly-Ala-Gly-Ala-Gly
8. Arg-Leu-Leu-Gln-Gly-Leu-Val[a]
9. His-Ser-Gln-Gly-Thr-Phe
10. Trp-Met-Asp-Phe
11. Phe-Asp-Ala-Ser-Val
12. Phe-Val-Gln-Trp-Leu-Met-Asn-Thr
13. Tyr-Thr-Met-Ser-Ser-Gln-Leu-Thr-Leu-Pro-Thr-Val[c]

[a] Ser-Arg, Asn-Arg, and Arg-Leu where identified by direct (probe) analysis with the mass spectrometer.
[b] After a single Edman degradation, the ε-phenylthiocarbamyl lysine peptide was completely digested by DAP I.
[c] Digested with DAP I and IV. All dipeptides from a digest of 1 nmol of peptide were identified by GC-MS when 20% of the derivatized sample (0.2 nmol) was injected.

terminal dipeptide is usually found first because the amino-terminal amino acid is known. The carboxyl-terminal dipeptide is also easily identified when a tryptic peptide is degraded because Lys or Arg will be carboxyl-terminal. One proceeds to determine the correct sequence by selecting dipeptides alternately from each list until the dipeptides are accounted for. When more than one sequence is possible, the correct sequence can be determined only by a time study in which one looks for the rate of appearance of the dipeptides. This may be illustrated with polypeptide 1 in Table III, Asp-Ser-Thr-Tyr-Ser-Met-Ser-Ser-Thr-Leu-Thr-Leu-Thr-Lys.

The original and shortened peptide (single Edman) gave the following sets of dipeptides, respectively:

1	2
Asp-Ser	Ser-Thr
Thr-Tyr	Tyr-Ser
Ser-Met	Met-Ser
Ser-Ser	Ser-Thr
Thr-Leu	Leu-Thr
Thr-Leu	Leu-Thr
Thr-Lys	Lys

Starting with column *1* and knowing that Asp is amino-terminal, 7 sequences can be initiated; 3 accommodate all the dipeptides.

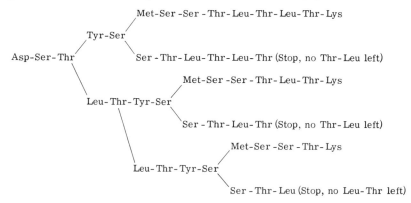

Of the 3 possible sequences, the correct sequence (top) was determined by the time study in which Thr³-Tyr⁴ appeared before Thr³-Leu⁴, another possibility. By alternating between columns *1* and *2* and checking off the dipeptides as they are used, the correct sequence of most polypeptides may be found without a time study, including polypeptides such as No. 2 of Table III, which has the repeating -Ser-Gly-Ser-Gly-Ser-Gly- sequence.

Comments

The scope of the DAP method for sequencing polypeptides depends upon the extent to which the polypeptide can be degraded and the facility of dipeptide identification. Contrary to previous reports, we have found that DAP I preparations can hydrolyze polypeptides containing amino-terminal Lys or Arg when the incubations are performed at pH 7.0 instead of pH 5.0. Dipeptide polymerization[13,21] is apparently insignificant. This observation extends the utility of DAP I preparations, but the inability to completely degrade proline-containing peptides continues to limit the method. This problem has been partly overcome by the use of DAP IV, which cleaves bonds at the carboxyl group of proline (or hydroxyproline). Bonds at the imino group of proline are not hydrolyzed. However, when only one proline residue is present, degradation can be completed by performing one or a combination of the follow-

²¹ J. K. McDonald, P. X. Callahan, S. Ellis, and R. E. Smith, *in* "Tissue Proteinases" (A. J. Barret and T. J. Dingle, eds.), p. 69. North-Holland Publ., Amsterdam, 1971.

ing steps: Edman degradation, LAP digestion,[22] or a time study of dipeptide release. With large polypeptides, all three procedures may be required. When two or more proline residues are present, these operations can be cumbersome. Recently, an endopeptidase has been described which cleaves bonds at the carboxyl group of proline giving peptides with carboxyl-terminal proline.[23] Such peptides would be completely degraded by DAP I. This enzyme should be useful if all proline bonds are cleaved and if interfering proteases can be removed.

Preliminary data have been obtained in our laboratory for a new dipeptidyl aminopeptidase, DAP V, which hydrolyzes bonds at the imino group of proline. It is hoped that it will be possible to degrade all polypeptides by the combined use of DAP's I, IV, and V.

Acknowledgments

We gratefully acknowledge the helpful suggestions of Dr. Henry Fales and the generous gifts of peptides from Dr. Ettore Appella.

[22] K. D. Vosbieck, B. D. Greenburg, and W. M. Awad, Jr., *J. Biol. Chem.* **250,** 3981 (1975).

[23] R. Walter, *Biochim. Biophys. Acta* **422,** 138 (1976).

Section IX

Chemical Modification

[40] Novel Sulfhydryl Reagents

By GEORGE L. KENYON and THOMAS W. BRUICE

Sulfhydryl groups of cysteinyl residues of peptides and proteins are generally the most reactive of all amino acid side-chain functionalities under normal physiological conditions. They may be readily alkylated, acylated, arylated, and oxidized, and will form complexes with many heavy-metal ions.

Several reviews on sulfhydryl reagents (Fontana and Scoffone,[1,1a] Friedman,[2] Torchinskii,[3] Jocelyn[4]) have recently appeared. And, since the publication of "Chemical Modification of Proteins" by Means and Feeney in 1971,[5] in which approximately forty types of sulfhydryl reagents were described and extensively referenced, there has been burgeoning interest in the development of new such reagents.

The purpose of this review, completed in June 1976, is to present sulfhydryl reagents developed since 1971. These reagents deliver groups that have been categorized as follows: Blocking and Labeling Groups, Reporter Groups, Cross-linking Groups, and Affinity Labeling Groups. First, each of these classes will be defined and some examples from the literature prior to 1971 (most of which are to be found in Means and Feeney) will be presented as illustrations.

Most of these earlier sulfhydryl reagents deliver groups that fall in the category of Blocking and Labeling Groups, which may be either *reversible* or *irreversible*. In general, these groups are designed to be structurally and chemically relatively innocuous so that when covalently bound they act merely to block the activity of, or to titrate or label (sometimes isotopically) any number of sulfhydryl groups. Of the *irreversible* reagents in this class, N-ethylmaleimide, organomercurials (p-chloromercuribenzoate), iodoacetate and iodoacetamide have shown

---

[1] A. Fontana and E. Scoffone, this series Vol. 25, p. 482 (1972).

[1a] A. Fontana and E. Scoffone, in "Mechanisms of Reactions of Sulfur Compounds" (N. Kharasch, ed.), Vol. IV, p. 15. Intra-Science Res. Found., Santa Monica, California, 1969.

[2] M. Friedman, "The Chemistry and Biochemistry of the Sulfhydryl Group in Amino Acids, Peptides and Proteins." Pergamon, New York, 1973.

[3] Y. M. Torchinskii, "Sulfhydryl and Disulfide Groups of Proteins." Consultants Bureau, Plenum, New York, 1974.

[4] P. C. Jocelyn, "Biochemistry of the SH Group." Academic Press, New York, 1972.

[5] G. E. Means and R. E. Feeney, "Chemical Modification of Proteins." Holden-Day, San Francisco, California, 1971.

most utility. Among the most useful, easily *reversible*[5a] reagents are aryl halides (dinitrofluorobenzene), sulfite, sodium tetrathionate, 4,4'-dithiodipyridine, and 5,5'-dithiobis-(2-nitrobenzoic acid); the latter two produce highly light-absorbing chromophores that allow for quantitative spectrophotometric determination of sulfhydryls.

A second class of sulfhydryl reagents deliver Cross-linking Groups. They possess one functionality that reacts initially and selectively with sulfhydryls and then another functionality that reacts (often under altered conditions) with another nearby group, which may or may not be another sulfhydryl. This second reaction is sometimes, but not always, selective. Older reagents in this class have almost exclusively been structurally symmetrical and used to generate cross-linking bridges between two nearby sulfhydryl groups. Among such reagents are mercuric ion, dibromoacetone,[6] bismaleimides and bisiodoacetamides.

For the other two major categories mentioned, Reporter Groups and Affinity Labeling Groups, there is a scarcity of older reagents that may be so classified, which react satisfactorily and are of any general practical utility. It will be seen in later discussions that a major effort has been expended in the early 1970's toward the design of suitable sulfhydryl-specific reagents for these purposes.

Reporter Groups generally serve to provide chemical and physical information about the environment of the sulfhydryl group that is labeled (and sometimes to quantitate such groups), with or without a wide variety of post-labeling perturbations. They exhibit easily detectable physical or chemical changes that are exquisitely sensitive to variable aspects of their microenvironment, such as polarity and hydrophobicity. These groups most often fall into one of three subclasses: *Chromophoric Groups* (e.g., 2-chloromercuri-4-nitrophenol[7,8]), *Fluorescent Probes* [e.g., fluorescein mercuric acetate,[9,10] 7-chloro-4-nitrobenzo-2-oxa-1,3-diazole (NBD-chloride),[11]   *N*-(*p*-(2-benzimidazolyl)phenyl)   maleimide

[5a] The reactions of many sulfhydryl blocking groups may be reversed. However, in many instances the reversals are extremely slow processes and/or require harsh conditions destructive to protein structure. By easily *reversible,* we mean removable in a reasonably short period of time under conditions that preserve the native structure of a polypeptide or protein; any reagent that cannot be thus removed will be defined as *irreversible.*

[6] S. S. Husain and G. Lowe, *Biochem. J.* 117, 341 (1970).

[7] F. A. Quiocho and J. W. Thomson, *Proc. Natl. Acad. Sci. U.S.A.* 70, 2858 (1973).

[8] F. A. Quiocho and J. S. Olson, *J. Biol. Chem.* 249, 5885 (1974).

[9] E. Karusch, N. R. Klinman, and R. Marks, *Anal. Biochem.* 9, 100 (1964).

[10] J. R. Heitz and B. M. Anderson, *Arch. Biochem. Biophys.* 127, 637 (1968).

[11] D. Birkett, N. Price, G. Radda, and A. Salmon, *FEBS Lett.* 6, 346 (1970).

(BIPM)[12]] and *Stable Nitroxide Radical Spin Labels* [e.g., *N*-(1-oxyl-2,2,5,5-tetramethyl-3-pyrrolidinyl) iodoacetamide, *N*-(1-oxyl-2,2,6,6-tetramethyl-4-piperidinyl) iodoacetamide[13]].

Sulfhydryl-specific Affinity Labeling Groups are designed to be structurally analogous to some ligand that normally shows specificity in association to a protein-binding site containing at least one sulfhydryl. Once the affinity labeling group is bound noncovalently, a reactive moiety is designed to be so positioned in approximation to the binding site sulfhydryl that it can react with it to anchor the affinity label covalently. Thus selectivity in modification of only binding site sulfhydryls is achieved. These reagents may be designed to accomplish further purposes once they are thus secured, but their outstanding features are those just described. Two examples of such reportedly sulfhydryl-specific affinity labeling reagents are the antibiotics penicillin[14] and fosfomycin (previously known as phosphonomycin).[15]

Sulfhydryl reagents may be placed into yet another general class, that of Simple Oxidizing Agents. Prior to 1971 a number of such reagents had been used and seemed to fulfill their task satisfactorily, and so very little new interest in designing new reagents in this class has been generated. Among the well-known older sulfhydryl oxidizing agents are iodosobenzoic acid, performic acid, tetranitromethane, and *N*-bromosuccinimide.

In general, each of the novel sulfhydryl reagents will be classified in one of the above categories. In addition, one further recently developed, chemically distinct class of sulfhydryl reagents of great potential, the Alkyl Alkanethiolsulfonates, will be discussed separately and in some detail. Because of space limitations, as a rule, syntheses of the novel sulfhydryl reagents will not be given, but suitable references will be cited. Optimal conditions for their use will, however, be briefly presented (to be given in parentheses after naming of the reagent). Usually, we will not comment on the relative merits of a given reagent unless it appears to possess some conspicuous positive or negative attribute(s).

Blocking and Labeling Groups

Irreversible. Three new simple alkylating agents have been developed which introduce charged groups at cysteinyl residues of proteins. Two

[12] Y. Kanaoka, M. Machida, M. Machida, and T. Sekine, *Biochim. Biophys. Acta* **207**, 269 (1973).
[13] H. M. McConnell, W. Deal, and R. T. Ogata, *Biochemistry* **8**, 2580 (1969).
[14] P. G. Lawrence and J. C. Strominger, *J. Biol. Chem.* **245**, 3653 (1970).
[15] P. J. Cassidy and F. M. Kahan, *Biochemistry* **12**, 1364 (1973).

of these, sodium 2-bromoethanesulfonate[16] [(I) ; Tris buffer, 0.2% EDTA, 7 M guanidinium chloride, pH 8.6, 12–48 hr , 20–100-molar excess reagent/SH] and 1,3-propane sultone[17] [(II); 0.5 M NaHCO$_3$:propanol (1:1), pH 8.3, 16 hr, 2-molar excess reagent/-SH] generate $^-$O$_3$S-(CH$_2$)$_n$-S-protein, where n equals 2 and 3, respectively. The third, (2-bromoethyl)-trimethylammonium (BETA) bromide[18] [(III) ; 0.2 M KHCO$_3$-8 M urea, pH 8.5, 24–48 hr, 7–17-molar excess reagent/-SH] generates $(CH_3)_3\overset{+}{N}$-CH$_2$-CH$_2$-S-protein. All three sulfhydryl-blocked forms are potentially useful in electrophoretic studies, maintaining their respective charges over a broad pH range. However, for other purposes, a probable serious drawback for these reagents is the prolonged reaction time apparently required for complete modification.

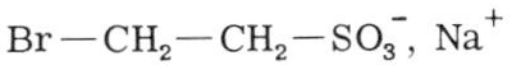

$$Br-CH_2-CH_2-SO_3^-, Na^+$$

(I)

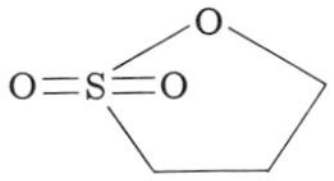

(II)

$$CH_3-\overset{\underset{\displaystyle CH_3}{|}}{\overset{Br^- \quad CH_3}{\overset{+}{N}}}-CH_2-CH_2-Br$$

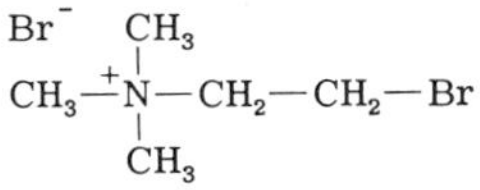

(III)

As alternatives to the trityl blocking group, often used in peptide chemistry, 5-dibenzosuberol[19] [(IV); trifluoroacetic acid, 15 min, 25°, 1:1 molar ratio reagent/-SH] and 2-picolylchloride 1-oxide[20] [(VI) ; 0.4 N NaOH, 5 hr, 0–5°, 1-molar equivalent reagent/-SH] place the 5-dibenzosuberyl and 1-oxido-2-picolyl groups on sulfhydryls to generate (V) and (VII), respectively. The 5-dibenzosuberyl blocking group can be removed upon catalytic hydrogenation.

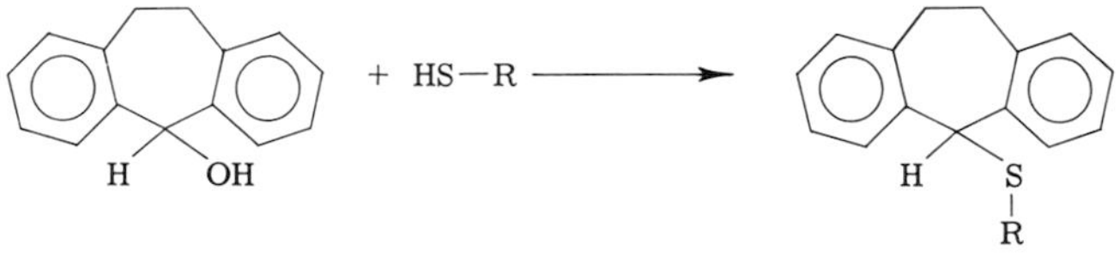

+ HS—R ⟶

(IV)   (V)

+ R—SH ⟶

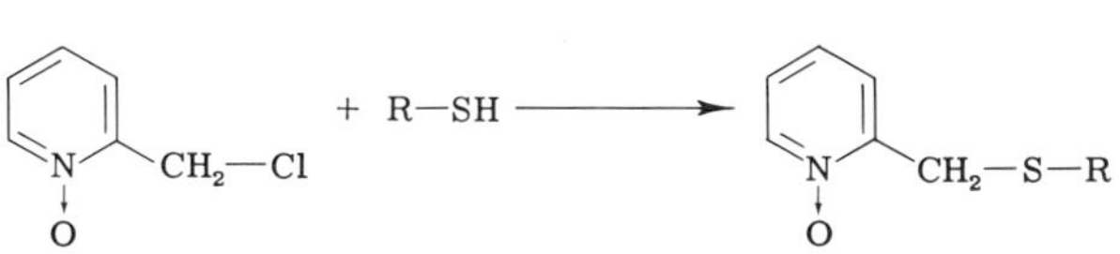

(VI)   (VII)

[16] V. Niketic, J. Thomsen, and K. Kristiansen, *Eur. J. Biochem.* **46**, 547 (1974).
[17] U. T. Rüegg and J. Rudinger, *Int. J. Peptide Protein Res.* **6**, 447 (1974).
[18] H. A. Itano and E. A. Robinson, *J. Biol. Chem.* **247**, 4819 (1972).
[19] A. G. Sandoz, *Helv. Chim. Acta* **59**, 499 (1976).
[20] Y. Mizuno and K. Ikeda, *Chem. Pharm. Bull.* **22**, 2889 (1974).

2-Nitro-5-thiocyanobenzoic acid (NTCB)[21] [(VIII); 6 M guanidine-HCl or 8 M urea, 0.2 M Tris-acetate, pH 8.0–8.5, 15 min, room temperature, 5–10 molar excess NTCB/-SH][22] has been developed for the direct cyanylation of thiol groups under mild conditions to generate protein-SCN. It has a distinct advantage of releasing the chromophoric thionitrobenzoate anion (IX) upon cyanylation. It may be prepared as shown, either from Ellman's reagent (X) or from (IX). This direct cyanylating reagent may be used in place of the older, indirect method

of Vanaman and Stark,[23] which utilizes treatment with (X) followed by addition of cyanide. At high pH values, excess thiolate anion will slowly reverse the cyanylation.[24] S-Cyanylated peptides may be chemically cleaved at the N-peptide bond of the modified cysteinyl residue,[22,24] providing a new tool for peptide sequencing, as discussed in another section of this volume. It should be pointed out, however, that (VIII) is a close structural analog of cyanogen bromide (Br—C≡N) and may therefore also modify methionine residues in some cases.

1-Cyano-4-dimethylaminopyridinum salts[25] [(XI); 7 M urea, 0.1 M acetate, pH 2–7, 11 min, 25°, 3-molar excess reagent/-SH] also react with thiols in neutral or acidic media to generate protein-SCN in reactions that again may be followed spectrophotometrically. As for (VIII), (XI) may also be expected to modify methionine residues.

[21] Y. Degani and A. Patchornik, *J. Org. Chem.* **36**, 2727 (1971).
[22] G. R. Jacobson, M. H. Schaffer, G. R. Stark, and T. C. Vanaman, *J. Biol. Chem.* **248**, 6583 (1973).
[23] T. C. Vanaman and G. R. Stark, *J. Biol. Chem.* **245**, 3565 (1970).
[24] Y. Degani and A. Patchornik, *Biochemistry* **13**, 1 (1974).
[25] M. Wakselman, E. Guibé-Jampel, A. Raoult, and W. D. Busse, *J. Chem. Soc., Chem. Commun.* **1**, 21 (1976).

$$
\begin{array}{c}
H_3C \diagdown_{N} \diagup CH_3 \\
\end{array}
$$

(structure with pyridinium ring, N-CN, X⁻)

$$X^- = Br^-$$
$$BF_4^-$$
$$ClO_4^-$$

(XI)

Reversible. Since mixed disulfides are easily cleaved upon addition of excess exogenous thiols, in recent years interest in developing new reversible sulfhydryl reagents has focused primarily on those that generate blocked-sulfhydryl mixed disulfides. One such reagent is thiocyanogen[26] (XII), which rapidly reacts with sulfhydryl groups to form sulfenyl thiocyanates (XIII). The reagent may also be employed in low

$$(SCN)_2 + R\text{—}SH \rightarrow R\text{—}S\text{—}SCN + SCN^- + H^+$$
(XII)   (XIII)

molar ratios to affect net oxidation of sulfhydryls. If the initial modification product (XIII) is hydrolyzed, R-SOH (sulfenic acid) is formed, regenerating SCN⁻, which in the presence of a suitable oxidant can reform (XII) to reenter the cycle, such that one SCN⁻ ion can effectively participate in oxidation of many sulfhydryls to the sulfenic acid.

Another reagent capable of generating mixed disulfides is carboxyethyldisulfide monosulfoxide[27] [(XIV); 0.2 N triethanolamine-HCl, pH 7.0, 23–25°, 90 min, 3-molar excess reagent/-SH optimally]. In experiments with (XIV), the authors assumed, but did not demonstrate directly, formation of a mixed disulfide (XV) in modified bovine liver crotonase. Besides its advantage of placing a readily removable group on sulfhydryls, (XIV) has been shown to be useful in modifying selec-

$$
\begin{array}{c}
O \\
\parallel \\
HOOC\text{—}CH_2\text{—}CH_2\text{—}S\text{—}S\text{—}CH_2\text{—}CH_2\text{—}COOH + R\text{—}SH \longrightarrow HOOC\text{—}CH_2\text{—}CH_2\text{—}S\text{—}S\text{—}R
\end{array}
$$

(XIV)   (XV)

tively only one hyperactive sulfhydryl out of five in this enzyme. Significantly, this modification did not alter the enzyme's catalytic activity.

Another chemically distinct class of sulfhydryl reagents which also rapidly forms a variety of reversible mixed disulfides are the Alkyl Alkanethiolsulfonates, which are to be discussed separately.

[26] T. M. Aune and E. L. Thomas, *Fed. Proc., Fed. Am. Soc. Exp. Biol.* **35,** 1630 (1976).
[27] H. M. Steinman and R. L. Hill, *J. Biol. Chem.* **248,** 892 (1973).

Reporter Groups

Reagents That Elicit Chromophoric Changes upon Reaction. The reagent 4,4'-bisdimethylaminodiphenylcarbinol (Michler's hydrol)[28-30] [(XVI); 0.1 M sodium acetate, pH 4–6, 30 min, 25°, 23-molar excess reagent/-SH] rapidly dissociates in aqueous solution to give the intensely blue resonance-stabilized carbonium ion (XVII).

(XVI)   (XVII)

Upon introduction of a thiol to this solution, the blue color (λ_{max} 606 nm) dissipates as S-(4,4'-bisdimethylaminodiphenylmethyl) derivatives form.

A related reagent is the disulfide of thio-Michler's ketone[31] [(XVIII); 0.1 M acetate, pH 5.1, 30 min, 0°). Similar to (XVI), this reagent is also intensely blue in solution, but differs in that the product of its reaction with sulfhydryls (XIX) retains color at an analytically convenient, shifted λ_{max}. Thus, further manipulations of (XIX) may be monitored by its stable color. An obvious disadvantage of both (XVI) and (XVIII) is their steric bulk, which allows them to react only with the most accessible cysteinyl residues in proteins.

Another reagent which delivers a chromophore to sulfhydryl residues is *p*-nitrobenzenediazonium fluoroborate[32] [(XX); 0.07 M phosphate, pH 6.8, 60 min, 4°, 3-molar excess reagent/-SH], which has been used to modify a single sulfhydryl of rabbit muscle myosin. Although the identity of the coupling product has not been determined, it absorbs strongly with λ_{max} in the range of 305 to 315 nm.

[28] B. A. Humphries, M. S. Rohrbach, and J. H. Harrison, *Biochem. Biophys. Res. Commun.* **50**, 493 (1973).
[29] M. S. Rohrbach, B. A. Humphries, F. J. Yost, Jr., W. G. Rhodes, S. Boatman, R. G. Hiskey, and J. H. Harrison, *Anal. Biochem.* **52**, 127 (1973).
[30] B. A. Humphries, Ph.D. Thesis, University of North Carolina, 1973.
[31] L. Jirousek and M. Soodak, *Biochem. Biophys. Res. Commun.* **59**, 927 (1974).
[32] S. Kobayashi, I. Kabasawa, M. Kimura, and T. Sekine, *J. Biochem.* **78**, 287 (1975).

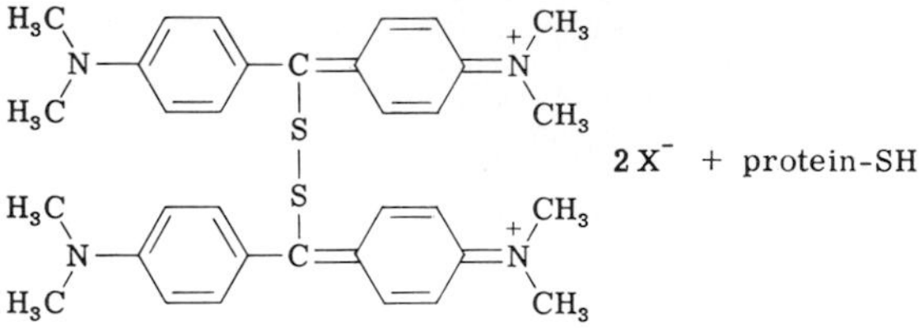

ultramarine
λ_{max} 630, 590, 530 nm

(XVIII)

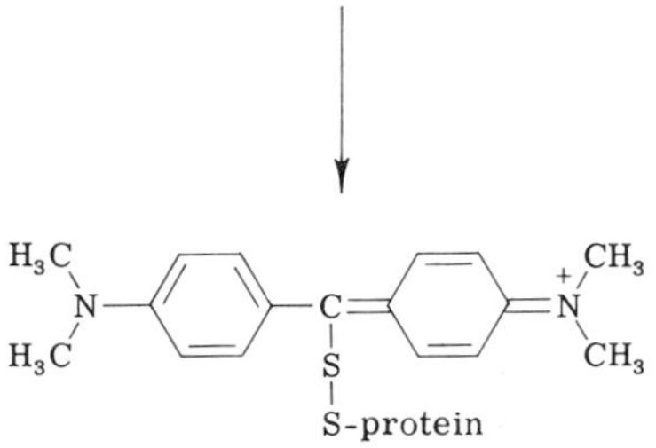

$2\,X^-$ + protein-SH

turquoise-green
λ_{max} 630, 590, 450 nm

(XIX)

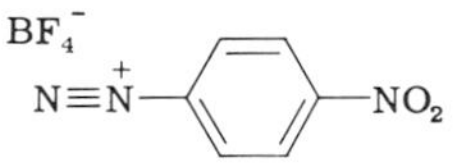

(XX)

Fluorescent Groups. As mentioned in the introductory comments, considerable advances in the development of sulfhydryl-specific fluorescent
probes have recently been realized. In general, such reagents are compos-
ites of older, chemically well-defined functionalities: they consist of a
fluorescent moiety attached to a sulfhydryl-reactive moiety. The latter
are often the old standbys, iodoacetamides, N-substituted maleimides, or
disulfides, and the former are usually derivatives of one of the following:
salicylic acid, fluorescein, pyrene, or dansyl. As the sulfhydryl reactive
moieties of these reagents are by no means novel, conditions for modification for each reagent need not be discussed, but may be readily found in
the literature, as they will be approximately the same as for the parent
sulfhydryl reactive compound. In the following discussion, it will be most
useful to classify the novel reagents according to the type of fluorescent
probe delivered. In addition, structures of each will be shown, adjacent
to which will be indicated the λ_{max} of absorption excitation (λ_e, nm) and
the λ_{max} of fluorescent emission (λ_f, nm), where available.

An attractive reagent because of its relatively small size is 5-iodo-acetamidosalicylic acid[33] (XXI). The intensity of fluorescence emission is lower than for many larger molecules, however.

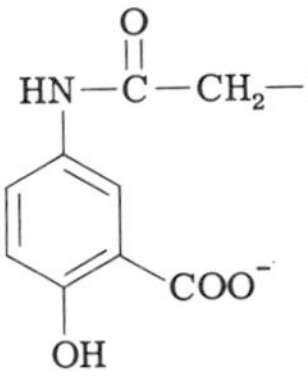

$$\lambda_e\ 323$$
$$\lambda_f\ 405$$

(XXI)

Two new derivatives of fluorescein are difluorescein isothiocarbamido-cystamine[33] (XXII) and "monomercurated dibromofluorescein acetate"[34] [(XXIII); λ_e 490, λ_f 550]; the latter may represent any of several possible isomers (the exact structure is not specified in reference 34). The outstanding feature of these reagents is the high intensity of fluorescence

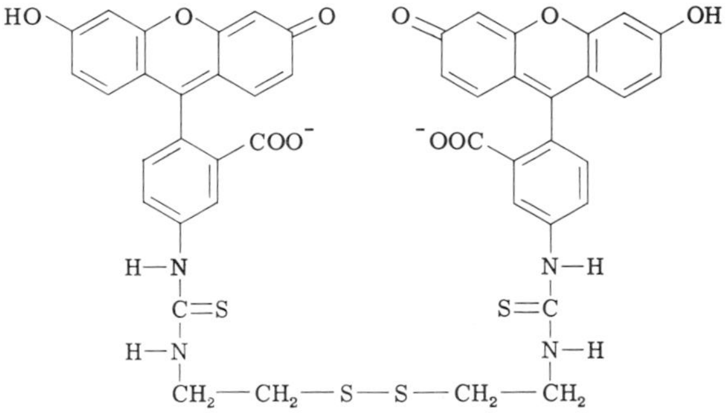

$$\lambda_e\ 495$$
$$\lambda_f\ 518$$

(XXII)

emission of the fluorescein moiety. In general, fluorescein labels such as these do have an inherent limitation for practical use which is that they exhibit changes in spectra in the neutral pH region.[35] And, although (XXII) does react rapidly, the nature of the reaction necessitates the use of large excesses of reagent.

Another type of fluorescent probe that may be introduced onto

[33] C.-W. Wu and L. Stryer, *Proc. Natl. Acad. Sci. U.S.A.* **69**, 1104 (1972).
[34] J. K. Weltman, A. R. Frackelton, Jr., R. P. Szaro, and R. M. Dowben, *Biophys. Soc. Abstr.* **12**, 277a (1972).
[35] P. C. Leavis and S. S. Lehrer, *Biochemistry* **13**, 3042 (1974).

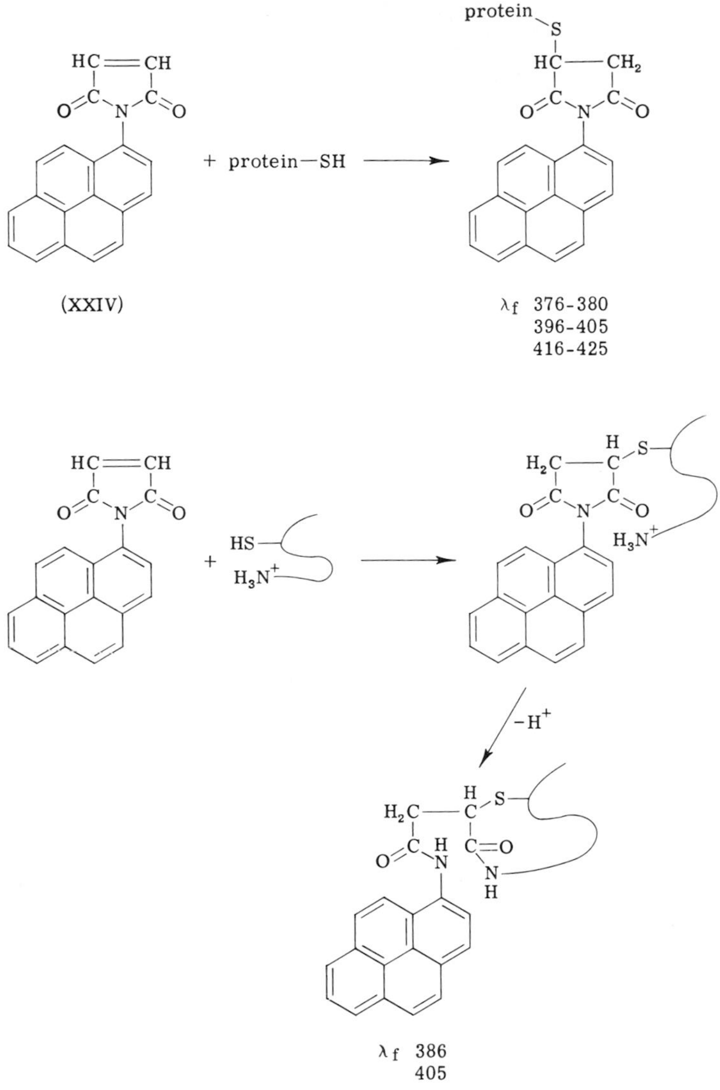

λ_f 376-380
396-405
416-425

λ_f 386
405

Scheme 1

SCHEME 1. Intramolecular aminolysis of N-(1- or N-(3-pyrene)maleimide-modified bovine serum albumin.

sulfhydryls of proteins is pyrene. N-(1-Pyrene)maleimide [or N-(3-pyrene)maleimide][36-38] (PM) (XXIV) is an example of this kind of

[36] J. K. Weltman, R. P. Szaro, A. R. Frackelton, Jr., J. R. Bunting, and R. E. Cathou, *J. Biol. Chem.* **248**, 3173 (1973)

reagent. It is nonfluorescent in aqueous solution, but forms a strongly fluorescent adduct with sulfhydryl groups of either simple organic compounds or proteins. The adducts formed with a variety of proteins exhibit two distinct fluorescent lifetimes, the longer component having a value on the order of 100 nsec. This long lifetime is useful for fluorescence polarization studies of high molecular weight, reactive sulfhydryl-containing proteins. This reagent also shows potential for fulfilling a bifunctional role. The PM-bovine serum albumin adduct exhibits a characteristic red shift of the fluorescence emission spectrum, which apparently occurs concomitant with intramolecular aminolysis of the succinimido ring (Scheme 1). Thus, (XXIV) "can serve as a fluorescent cross-linking reagent which provides information about the spatial proximity of SH and NH$_2$ groups."[37] Despite these attractive features, general use of this reagent may be somewhat restricted owing to its limited solubility in aqueous solutions.[35]

The final and largest class of fluorescent probe-delivering sulfhydryl reagents are the dansyl compounds. They have also generally seen the most use in protein studies. These naphthylaminesulfonic acid derivatives most often possess the following spectral properties: "(a) invariance over a wide pH range, (b) stability in most solvents, (c) sensitivity to environment, (d) distinctness from spectral properties of protein chromophores, and (e) suitability of lifetimes for the determination of a useful range of rotational relaxation times."[39]

In recent years, N-(iodoacetylaminoethyl)-5-naphthylamine-1-sulfonic acid (1,5-I-AEDANS) (XXV) and the 1,8 isomer (1,8-I-AEDANS) (XXVI), initially synthesized and characterized by Hudson and Weber,[39] and whose power as practical tools was first demonstrated by Wu and Stryer[33] in elegant studies on proximity relationships in rhodopsin, have become well popularized. Both reagents are water-soluble, stable for long

λ$_e$ 350
λ$_f$ 495
(XXV)

λ$_e$ 350
λ$_f$ 495
(XXVI)

[37] C.-W. Wu, L. R. Yarbrough and F. Y.-H. Wu, *Biochemistry* **15**, 2863 (1976).
[38] R. P. Haugland, Ph.D. Thesis, Stanford University, 1970.
[39] E. N. Hudson and G. Weber, *Biochemistry* **12**, 4154 (1973).

periods of time in crystalline form, and readily react with sulfhydryl groups in proteins to yield photostable covalent conjugates. Since their introduction, however, the following difficulties in their use have sometimes been encountered: the requirements of long reaction times and/or excess reagent to achieve complete SH modification, low stability of the reagents in solution (due to photocatalyzed degradation),[39] and fluorescence heterogeneity of the protein adducts formed.

N-(1-Anilinonaphthyl-4) maleimide[40] (XXVII), which is practically nonfluorescent, reacts selectively with protein thiols to give stable addition products (XXVIII) which are intensely fluorescent. Although (XXVII) cannot literally be classified as a naphthylaminesulfonic acid derivative, it is very similar in structure, function, and fluorescence properties to the older, less stable 2-aminonaphthalene-6-sulfonate derivatives (ANS) (XXIX) of Seliskar and Brand,[41] after which it is modeled. Like (XXIX), the outstanding feature of (XXVII) is that it

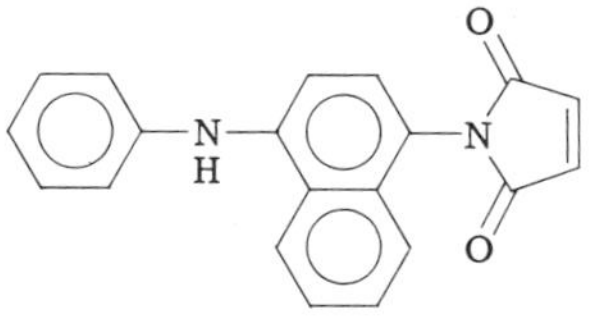

(XXVII)

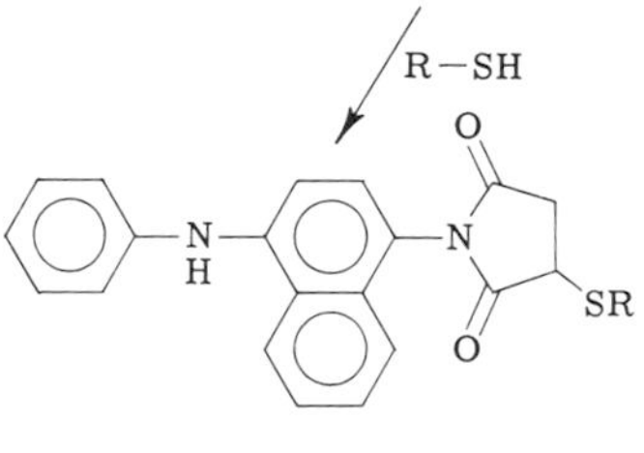

R—SH

(XXVIII) λ_e 355 $\big\}$ in ethanol   (XXIX)
                  λ_f 488

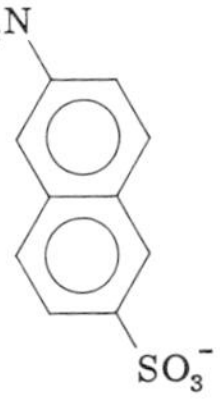

introduces a hydrophobic fluorescent probe whose fluorescence is extremely solvent-dependent, and thus it is capable of providing information about the polarity of its microenvironment. Serious drawbacks for its use are that it shows limited solubility in aqueous solutions and that excess reagent and extremely long reaction times are necessary.

N-Dansylaziridine[42] (XXX) is a fluorescent reagent that shows high

[40] Y. Kanaoka, M. Machida, M. Machida, and T. Sekine, Biochim. Biophys. Acta 317, 563 (1973).
[41] C. J. Seliskar and L. Brand, J. Am. Chem. Soc. 93, 5414 (1971).
[42] W. H. Scouten, R. Lubcher, and W. Baughman, Biochim. Biophys. Acta 336, 421 (1974).

selectivity for protein thiols, forming S-(2-dansylaminoethyl)-protein adducts (XXXI) with altered emission spectra. These adducts are extremely stable and will survive mild acid hydrolysis (105°, 4 hr, 6.6 M

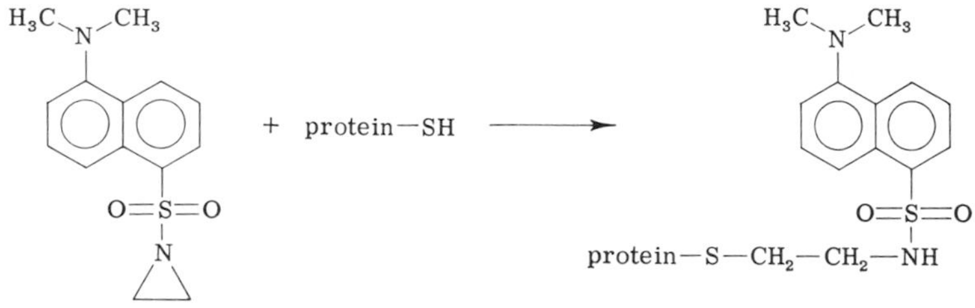

λ_e 345
λ_f 535

(XXX)

λ_e 345
λ_f 485

(XXXI)

HCl). Although (XXX) may be used in near stoichiometric molar ratios with SH groups and at neutral or slightly alkaline pH, it shows very sparse solubility in aqueous solutions and requires moderately long reaction times. Also, to date, an integral reactant in the preparation of (XXX) has been ethylenimine, which has been found to be such a potent carcinogen that its use, even for basic research, is restricted by Federal agencies.

Examination of the reactivities, fluorescence properties and limitations as a function of structure of a rapidly growing number of fluorescent sulfhydryl reagents,[42a] has recently enabled the development of a new, more versatile and less restricted such reagent, S-mercuric N-dansyl-cysteine[35] (XXXII).

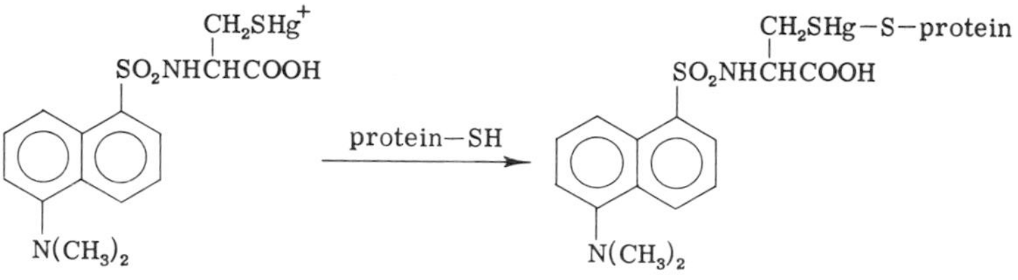

λ_e 330
λ_f 535

(XXXII)

λ_f 540-555

(XXXIII)

The reagent (XXXII) can be used both to estimate sulfhydryl groups (by fluorometric titration) and to label proteins with a fluorescent probe.

[42a] It has recently been brought to our attention that a large number of fluorescent, sulfhydryl (and other group) specific reagents have been developed for marketing by R. P. Haugland and colleagues of "Molecular Probes," Roseville, MN.

Mercury-bridged mercaptides (XXXIII) with extremely low dissociation constants (stable even upon sodium dodecyl sulfate-polyacrylamide gel electrophoresis) are formed with proteins. Concomitant are fluorescent enhancements of from 2.8- to 20-fold and spectral shifts of from 5 to 20 nm (blue shifted) upon binding of the label to various muscle proteins, for example.[35,43] Although (XXXII) has perhaps not been tested with enough different systems to allow final judgment, it apparently possesses many advantageous characteristics, which together are not matched by any one of the previously discussed reagents of this class.[35] This reagent is easily prepared, water-soluble, and stable. It reacts with sulfhydryls rapidly and over a wide pH range, stoichiometrically to form the high-affinity mercury-bridged mercaptide complexes, which may be easily dissociated upon addition of excess, exogenous thiols. It is obvious from the indicated structure of the reagent, however, that its size could be expected to impose serious limitations on its effectiveness in certain systems. Also, upon treatment of some proteins in the absence of certain precautionary measures, nonspecific labeling has been observed.[35]

Cross-Linking Groups

The only recently developed *nonphotoactivated* thiol-specific cross-linking reagent is N-(1-pyrene)maleimide (XXIV), discussed above, which has been shown to attach to an amino group of bovine serum albumin.[37,38]

Several new cross-linking reagents have been developed which have an inherent initial specificity for sulfhydryls (in the dark), and which possess a *photochemically labile* moiety that generates a highly reactive, nonspecific function (e.g., nitrene) upon photolysis. A general review of "Photogenerated Reagents for Biological Receptor-Site Labeling"[44] has been published. One specific example is p-azidophenacyl bromide[45] [(XXXIV); 0.05 M Tris, pH 7.0, about 1 hr, 0°, 100-molar excess reagent/-SH], commercially available from Pierce Chemical Company. Even though (XXXIV) binds covalently to sulfhydryls, in its reaction with rabbit muscle glyceraldehyde-3-phosphate dehydrogenase, non-stoichiometric binding was observed. This has been attributed both to incomplete alkylation of the active sulfhydryl groups and to slow attack of other nucleophilic groups of the protein. In these experiments the sites of nitrene insertion have not been elucidated. This specific example illustrates the common difficulty in using these types of reagents: random

[43] P. C. Leavis, *Fed. Proc., Fed. Am. Soc. Exp. Biol.* **35**, 1746 (1976).
[44] J. R. Knowles, *Acc. Chem. Res.* **5**, 155 (1972).
[45] S. H. Hixson and S. S. Hixson, *Biochemistry* **14**, 4251 (1975).

labeling. In light of this conclusion, the use of noncovalently bound aryl azides should probably not be further attempted, and new bifunctional photochemical reagents which may be bound covalently at a known locus of the protein prior to photolysis need to be developed.[45]

Two other compounds closely related to (XXXIV) which show potential as sulfhydryl-directed photochemical cross-linking reagents are 4-(β-bromoaminoethyl)-3-nitrophenyl azide[46] (XXXV) and 4-fluoro-3-nitrophenyl azide[47-50] (XXXVI). These should not be expected to show any particular advantages over (XXXIV), however, in terms of selectivity.

A promising but relatively untried new candidate in this class is 2-hydroxy-4-maleimidobenzoylazide[51] (XXXVII) (reported in a prelim-

[46] G. Rudnick, H. R. Kaback, and R. Weil, *J. Biol. Chem.* **250**, 1371 (1975).
[47] M. B. Perry and L. L. W. Heung, *Can. J. Biochem.* **50**, 510 (1972).
[48] G. W. J. Fleet, R. R. Porter, and J. R. Knowles, *Nature (London)* **224**, 511 (1969).
[49] Anonymous, *Chem. Eng. News* **48**, 28 (1970).
[50] J. R. Knowles, *Yale Sci. Mag.* **45**, 12 (1971).
[51] W. E. Trommer, H. Kolkenbrock, and G. Pfleider, *Hoppe-Seyler's Z. Physiol. Chem.* **356**, 1455 (1975).

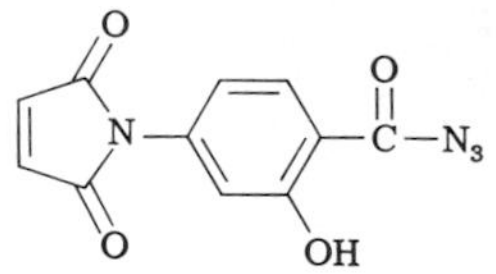

$$\text{(XXXVII)}$$

(XXXVII)

inary communication). Its azide moiety may be activated presumably either photochemically or by raising the pH.[51]

One last example of a reagent incorporating the aryl azide moiety is di-N-(2-nitro-4-azidophenyl)cystamine S,S-dioxide[52] (XXXVIII), which is an alkyl alkanethiolsulfonate. The characteristics of this class of sulfhydryl reagents will be discussed later.

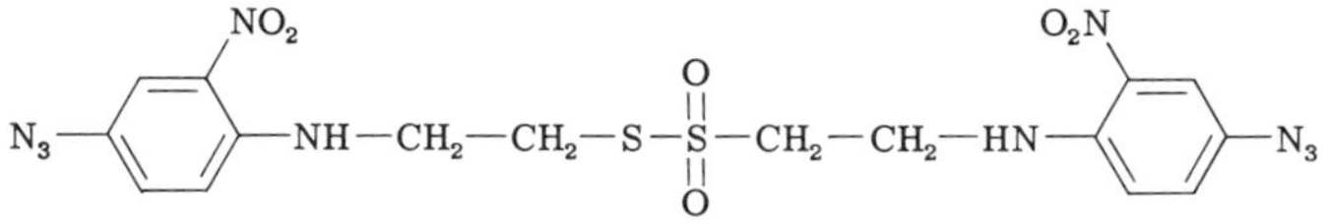

(XXXVIII)

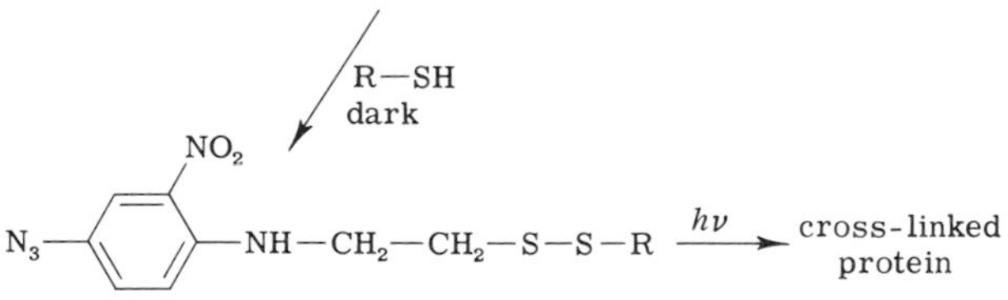

R—SH
dark

hv ⟶ cross-linked protein

Affinity Labeling Groups

The reagents in this class have been designed to combine specifically with either a single protein or class of proteins. As such, widespread utility of any one of these reagents cannot be expected. Consideration of the structures of these existing reagents, however, may serve as a guide for the design of suitable new affinity reagents for other proteins.

The first of several bromoacetyl-based reagents is β-bromoacetamido-*trans*-cinnamoyl-CoA[27] [(XXXIX); 0.20 triethanolamine, pH 7.0, 2 hr, 1.4–18-molar excess reagent/subunit], a probe for CoA binding sites, which has been used to label bovine liver crotonase. It has been shown that under conditions where (XXXIX) labels only one reactive sulfhydryl in the substrate binding site of the enzyme, activity is completely abolished, presumably owing to steric blockage of substrate binding.

[52] C.-K. Huang and F. M. Richards, *Fed. Proc., Fed. Am. Soc. Exp. Biol.* 35, 1378 (1976).

$$Br-CH_2-\overset{\overset{\displaystyle O}{\|}}{C}-HN-\overset{}{\underset{}{\bigcirc}}-\overset{\overset{\displaystyle H}{|}}{C}=\overset{\overset{}{\underset{\underset{\displaystyle H}{|}}{}}}{C}-\overset{\overset{}{\underset{\underset{\displaystyle O}{\|}}{}}}{C}-SCoA$$

(XXXIX)

The second reagent is N-bromoacetylethanolamine phosphate[53] (XL), a structural analog of 2,3-bisphosphoglycerate, and an effective competitive inhibitor of phosphoglycerate mutase on which it labels an -SH group at or near the active site.

$$
\begin{array}{ll}
CH_2-O-PO_3^{2-} & \qquad\quad \overset{-O}{\diagdown}\overset{}{\underset{}{C}}\overset{O}{\diagup} \\
\quad| \qquad\quad H & \qquad\quad CH-O-PO_3^{2-} \\
CH_2-N-\underset{\underset{\displaystyle O}{\|}}{C}-CH_2Br & \qquad\quad CH_2-O-\underset{\underset{\displaystyle O}{\|}}{\overset{\overset{\displaystyle -O}{|}}{P}}-O^{-}
\end{array}
$$

(XL)   2,3-Bisphospho-glycerate

Two further bromoacetyl-containing derivatives that are both analogs of NAD^+ are [3-(3-bromoacetylpyridinio)propyl]adenosine pyrophosphate[54,55] (XLI) and the 4-bromoacetylpyridinio isomer [(XLII); 0.2 M phosphate, pH 6.5, 20–30 min, moderate excess reagent/-SH]. Both analogs reacted with cysteine-43 of yeast alcohol dehydrogenase, incorporating one molecule of affinity label per protein subunit, and resulting in the enzyme's complete inactivation.

$$AMP-O-\overset{\overset{\displaystyle O}{\|}}{\underset{\underset{\displaystyle O^-}{|}}{P}}-O-(CH_2)_3-\overset{+}{N}\overset{}{\underset{}{\bigcirc}}-X$$

(XLI) X = —H

$$Y = -C\overset{O}{\diagup}CH_2-Br$$

(XLII) X = $-C\underset{\diagdown O}{-}CH_2-Br$

Y = —H

Another NAD^+ analog, possessing an organomercuriothioinosine as the sulfhydryl reactive moiety, has been designed to label a variety of NAD^+-requiring enzymes. It is nicotinamide-(S-methylmercurythioinosine) di-

[53] F. C. Hartman and I. L. Norton, *Fed. Proc., Fed. Am. Soc. Exp. Biol.* **35**, 1747 (1976).

[54] C. Woenckhaus, R. Jeck, G. Dietz, and G. Jentsch, *FEBS Lett.* **34**, 175 (1973).

[55] H. Jörnvall, C. Woenckhaus, E. Schättle, and R. Jeck, *FEBS Lett.* **54**, 297 (1975).

nucleotide[56] [(XLIII) ; 0.2 M phosphate, pH 6.5, 20 hr, 25°, large excess reagent], which exhibits coenzyme activity with both lactate dehydrogenase and liver alcohol dehydrogenase, but inactivates both yeast alcohol dehydrogenase and glyceraldehyde-3-phosphate dehydrogenase, presumably by methylmercuration of specific sulfhydryl groups, with concomitant release of nicotinamide-6-thiopurine dinucleotide (XLIV).

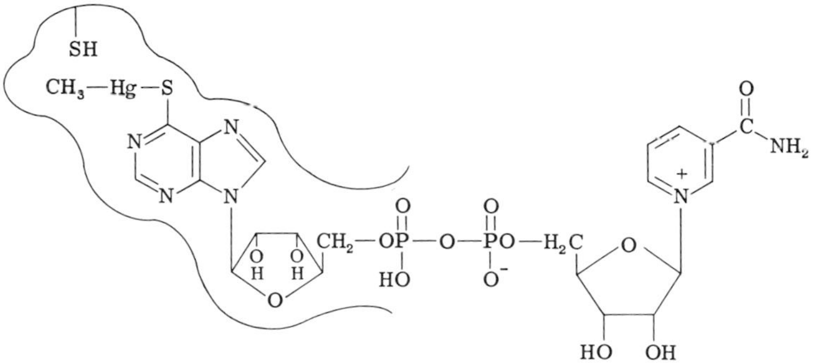

(XLIII)

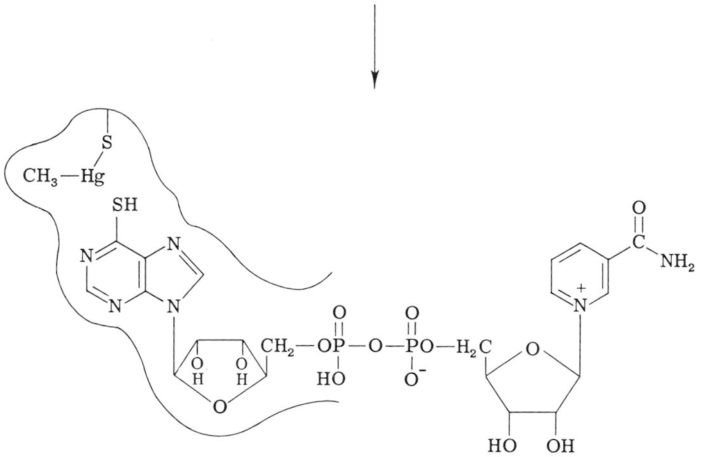

(XLIV)

Still another chemically totally unrelated affinity label has also been used to inactivate yeast alcohol dehydrogenase. The reagent is styrene oxide[57] [(XLV) ; 40 mM pyrophosphate, 140 mM glycine, 5 mM KCl, pH 8.5, approximately 3 hr, 25°, excess reagent], which closely structurally mimics both a known substrate, benzyl alcohol (XLVI) and the

[56] C. Woenckhaus and H. Duchmann, Z. Naturforsch. 30 c, 562 (1975).
[57] J. P. Klinman, Biochemistry 14, 2568 (1975).

simple organomercurial sulfhydryl blocking group, p-chloromercuriben-
zoate (pCMB) (XLVII). Both the epoxide and pCMB in this case

(XLV)   (XLVI)   (XLVII)

appear to alkylate approximately two sulfhydryl groups per subunit of
enzyme. With the epoxide, the inactivation has been shown to be a single
exponential process up to 93% inactivation. The two cysteines which are
modified may be ligands for the active site zinc atom.[57]

Another clever use (preliminary report) of an epoxide-based affinity
label is the reaction of glycidol phosphate[58] (XLVIII) with glyceralde-
hyde-3-phosphate dehydrogenase. The reaction is 10^4 times more rapid
than the reaction of (XLVIII) with cysteine alone, and presumably
produces a stable analog (XLIX) of the widely postulated thiolhemi-
acetal intermediate (L), formed upon reaction with the natural substrate,
glyceraldehyde-3-phosphate (LI).

(LI)   (L)   Postulated thiohemiacetal
                                       intermediate

(XLVIII)   (XLIX)   Presumed structure
                                       of blocked enzyme

Alkyl Alkanethiolsulfonates

A class of mixed disulfide-forming sulfhydryl reagents are the alkyl
alkanethiolsulfonates (LII). Their fundamental reaction with protein

[58] L. D. Byers, *Fed. Proc., Fed. Am. Soc. Exp. Biol.* **35**, 1498 (1976).

sulfhydryls is shown below:

$$\text{protein—SH} + \text{R—S—}\underset{\underset{O}{||}}{\overset{\overset{O}{||}}{S}}\text{—R'} \longrightarrow \text{protein—S—S—R} + \text{R'—}\underset{\underset{O}{||}}{\overset{\overset{O}{||}}{S}}\text{—H}$$

+ thiol

(LII)   (LIII)

As far as has been investigated it appears to proceed: (1) independent of the nature of the R and R' groups of the alkyl alkanethiolsulfonate; (2) under mild conditions, nondestructive to proteins; (3) extremely rapidly; (4) with high selectivity for cysteinyl sulfhydryls of proteins; (5) quantitatively to complete conversion to the disulfide (LIII) without large excesses of reagent; and (6) as illustrated, it is generally completely and rapidly reversible upon addition of thiols such as β-mercaptoethanol or dithiothreitol. These characteristics have prompted us to consider these reagents separately as a chemical class, rather than grouping specific alkyl alkanethiolsulfonates according to function as has been done with other sulfhydryl reagents.

Alkyl alkanethiolsulfonates had been used previously for the chemical synthesis of mixed disulfides[59–62] of simple organic compounds by Boldyrev et al.[63–65] and Dunbar and Rogers.[66] A closely related aryl arenethiolsulfonate had been used by Field and Giles[67] to block a sulfhydryl group of creatine kinase, but the conclusion was drawn that this reagent "shows no advantages over Ellman's reagent."[67]

The simplest and thus far most widely used reagent in this class is methyl methanethiolsulfonate[61,68] [(LIV); 0.01 M glycine-HCl, pH 6–8, 0°, virtually instantaneous, 1/1 molar ratio reagent/-SH], which delivers the small, uncharged, non-hydrogen-bonding CH$_3$S- group. In addition, since this group is a portion of the side-chain moiety of methionine, perturbations of a protein as a result of its introduction are generally as

[59] Mixed disulfides may also be conveniently synthesized using alkoxycarbonylalkyl disulfides,[60,61] or by using N-(alkanethio) phthalimides.[62]

[60] S. J. Brois, J. F. Pilot, and H. H. Barnham, J. Am. Chem. Soc. 92, 7629 (1970).

[61] D. J. Smith, E. T. Maggio, and G. L. Kenyon, Biochemistry 14, 766 (1975).

[32] D. N. Harpp and T. G. Back, J. Org. Chem. 36, 3828 (1971).

[63] B. G. Boldyrev, S. A. Gorelova, and A. T. Dovarko, J. Gen. Chem. USSR 31, 2238 (1961).

[64] B. G. Boldyrev, L. P. Slesarchuck, E. E. Gataza, T. A. Trotimova, and E. N. Vasenko, J. Org. Chem. USSR 2, 91 (1966).

[65] B. G. Boldyrev and A. T. Zakharchuk, Dokl. Akad. Nauk SSSR 94, 877 (1954).

[66] J. E. Dunbar and J. H. Rogers, J. Org. Chem. 31, 2842 (1966).

[67] L. Field and M. Giles, Jr., J. Org. Chem. 36, 309 (1971).

[68] D. J. Smith and G. L. Kenyon, J. Biol. Chem. 249, 3317 (1974).

$$CH_3—S—\overset{\overset{O}{\|}}{\underset{\underset{O}{\|}}{S}}—CH_3$$

(LIV)

slight as those observed upon modification with any other sulfhydryl reagent.

Methyl methanethiolsulfonate (MMTS) has been used to titrate two enzymes, papain and glyceraldehyde-3-phosphate dehydrogenase, that had been previously shown, using a variety of methods, to contain active site sulfhydryls essential for their catalytic activity.[61] Complete and rapid inhibition was observed and 1.0 mol of CH_3S- group was incorporated per mole of papain. The clean stoichiometry of CH_3S- incorporation paralleling loss of sulfhydryl-dependent activity confirms the absolute selectivity of MMTS for active and/or accessible sulfhydryl groups in preference to all other reactive protein groups. When rabbit muscle glyceraldehyde-3-phosphate dehydrogenase was similarly titrated with MMTS, two thiol groups per subunit were found to be modified, one much more rapidly than the other. Complete inactivation concurrent with blocking of the more reactive thiol was observed.

In contrast to these expected results, the behavior of creatine kinase with MMTS was anomalous.[61,68] It had been previously suggested, after titrations with reagents such as iodoacetamide of what appeared to be one unique, active-site sulfhydryl per subunit of this enzyme, that this thiol was required for activity. When 1.0 equivalent of CH_3S- was incorporated per subunit, however, approximately 20% residual activity remained. The presence of this CH_3S- group afforded complete protection against further inactivation by normally inhibitory iodoacetamide.

These results are consistent with the conclusion that the iodoacetamide-sensitive active-site sulfhydryl group is not essential for enzymic activity. Preliminary evidence based on both EPR and kinetic studies indicates that blocking of this sulfhydryl residue with CH_3S- interferes with nucleotide binding, not with binding of the guanidino substrates.[69]

Further evidence for relatively small protein structure perturbations upon treatment with MMTS was provided in a study of the blocking of the sulfhydryl groups of *Escherichia coli* succinic thiokinase.[70] At a point where approximately 4 mol of CH_3S- were incorporated per mole of enzyme, 80% of thiokinase activity was lost, but no loss of antigenicity

---

[69] G. D. Markham, G. H. Reed, D. J. Smith, E. T. Maggio, and G. L. Kenyon, *Fed. Proc., Fed. Am. Soc. Exp. Biol.* **34**, 545 (1975).

[70] J. S. Nishimura, G. L. Kenyon, and D. J. Smith, *Arch. Biochem. Biophys.* **170**, 461 (1975).

(as measured by microcomplement fixation) was observed. In contrast, losses of both thiokinase activity and antigenicity were observed upon similar treatment with 5,5'-dithiobis-(2-nitrobenzoic acid), ethylmercurithiosalicylate, and p-hydroxymercuribenzoate.

The behavior of MMTS as a sulfhydryl titrant of yeast alcohol dehydrogenase is also unusual.[71] The kinetics of inactivation by excess reagent are biphasic; 35% of activity is lost in a burst, corresponding to incorporation of one CH_3S- group per subunit, followed by a slower first-order loss of greater than 98% of the activity concurrent with incorporation of a second equivalent of CH_3S-. No such biphasic kinetic behavior is observed with similar titrations of the affinity label, styrene oxide (XLV), discussed in the previous section.[57]

MMTS has also been used for the potentiometric determination of ionizations at the active site of papain,[72] to block apparent intramolecular disulfide interchange in this same enzyme,[73] to titrate active-site sulfhydryls of beef heart isocitrate dehydrogenase,[74] and to synthesize the mixed methyldisulfide of CoA.[75,76] This CoA derivative is a potent inhibitor of choline acetyltransferase (squid ganglia).[75,76]

MMTS, as well as a wide variety of other alkyl methanethiolsulfonates of use for many different biochemical purposes, may be prepared from the common precursor, sodium methanethiolsulfonate (LV). For this reason, the synthesis of (LV) is given below.[77]

Preparative Scale Synthesis of Sodium Methanethiolsulfonate (LV). *Note of caution:* Sodium sulfide (LVI) and methanesulfonyl chloride (mesyl chloride) (LVII) are both potent lachrymators, and all work should be carried out in a well-ventilated fume-hood.

All materials used should be analytical reagent grade. Sodium sulfide ($Na_2S \cdot 9H_2O$, 72.1 g, 0.30 mol) is dissolved in 75 ml of water with careful heating to $\geq 80°$ ($Na_2S \cdot 9H_2O$ has a high positive heat of solution) in a 250-ml round-bottom flask, resulting in a clear solution. A large magnetic stirring bar is added, and the flask is fitted with both a reflux condenser and a dropping funnel. Freshly distilled mesyl chloride (34.5 g, 23.3 ml, 0.30 mol) is added slowly, dropwise, maintaining vigorous stirring, over a 10–15-min period. The ensuing reaction is highly exothermic, liberating a dense, white, noxious gas. The reaction mixture is heated at reflux for

[71] J. P. Klinman, *Fed. Proc., Fed. Am. Soc. Exp. Biol.* **34**, 600 (1975).
[72] S. D. Lewis, F. A. Johnson, and J. A. Shafer, *Biochemistry* **15**, 5009 (1976).
[73] A. Fink and K. Angelides, unpublished observations, 1976.
[74] R. S. Levy and J. J. Villafranca, submitted for publication (1977).
[75] S. F. Currier and H. G. Mautner, *Biochem. Biophys. Res. Commun.* **69**, 431 (1976).
[76] S. F. Currier and H. G. Mautner, *Fed. Proc., Fed. Am. Soc. Exp. Biol.* **35**, 1498 (1976).
[77] T. W. Bruice, E. T. Maggio, and G. L. Kenyon, unpublished observations, 1976.

about 12 hr, at the end of which it is a slightly opaque, dark orange-brown color and may contain a small amount of yellow tar.

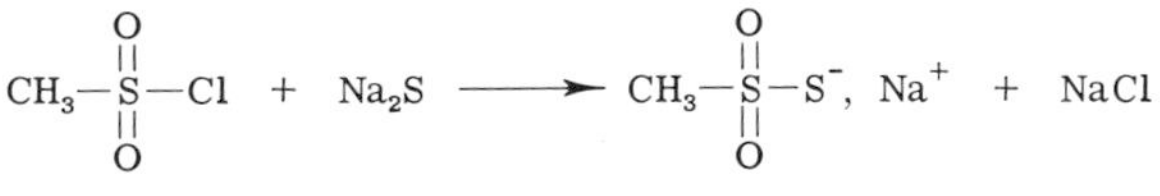

$$CH_3-\overset{\overset{O}{\|}}{\underset{\underset{O}{\|}}{S}}-Cl \; + \; Na_2S \longrightarrow CH_3-\overset{\overset{O}{\|}}{\underset{\underset{O}{\|}}{S}}-S^-,\; Na^+ \; + \; NaCl$$

(LVII)   (LVI)   (LV)

This mixture is then dried using a rotary evaporator under high vacuum with moderate heat. The resulting hard yellow cake is ground finely in a mortar and pestle and further dried overnight in a high-vacuum oven at 40° (it is very important for the success of the following step that all water be completely removed). The yellow powder is filter-extracted with 50–100-ml portions (350 ml total) of absolute ethanol (at room temperature), the mother liquor is slowly cooled, and the crystals that form are collected by filtration and washed with <50 ml of ice-cold absolute ethanol. The mother liquor from this filtration is evaporated using a rotary evaporator until more shiny, flaky crystals of product begin to fall out, then heated to boiling to resolubilize all solids, and cooled slowly. The crystals that form are collected by filtration as before. This concentration/filtration procedure is repeated (about 3 times total) until either the yield or purity decrease so as to preclude further efforts. All collected crystalline solid is combined, dissolved in a minimum amount of absolute ethanol at 25°, fine-filtered to remove remaining NaCl powder, taken to dryness, and then recrystallized from a minimum amount of boiling, absolute ethanol and dried under vacuum, without heating, for 12 hr. Product (LV) is obtained as large shiny, slightly off-white, flaky crystals (hygroscopic). Melting point: 272–273.5°. Yield: >60%.

Examples of useful alkyl methanethiolsulfonates which have to date been obtained by simple S_N2 displacement reactions using (LV) are shown below[61,68,77,78]:

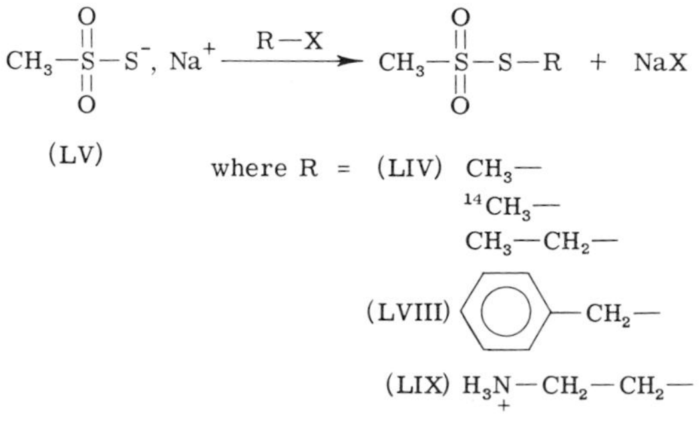

$$CH_3-\overset{\overset{O}{\|}}{\underset{\underset{O}{\|}}{S}}-S^-,\; Na^+ \xrightarrow{\;R-X\;} CH_3-\overset{\overset{O}{\|}}{\underset{\underset{O}{\|}}{S}}-S-R \; + \; NaX$$

(LV)

where R = (LIV) CH_3-
$^{14}CH_3-$
CH_3-CH_2-

(LVIII) ⟨○⟩$-CH_2-$

(LIX) $H_3\overset{+}{N}-CH_2-CH_2-$

[78] T. W. Bruice, E. T. Maggio, and G. L. Kenyon, *Fed. Proc., Fed. Am. Soc. Exp. Biol.* **35**, 1475 (1976).

That alkyl alkanethiolsulfonates with R groups that are larger and more complex than methyl still react with sulfhydryls with fulfillment of the six characteristics listed in the introduction to this section, and thus can be used to deliver RS- groups for varied biochemical purposes, has been demonstrated with studies of the ethyl ester (LX) and p-nitroanilide (LXI) of S-(β-aminoethanethiol)-N-acetyl-L-cysteine, and of the ethyl ester (LXII) and p-nitroanilide (LXIII) of S-(benzylthiol)-N-acetyl-L-cysteine.[77,78] These mixed disulfide model compounds, the analogs of the N-acetyl ethyl esters and p-nitroanilides of lysine and phenylalanine, respectively, were easily made by treatment of the N-acetyl ethyl esters and p-nitroanilides of cysteine with (LVIII) and (LIX). They have also been shown to be substrates for either trypsin or chymotrypsin, respectively.

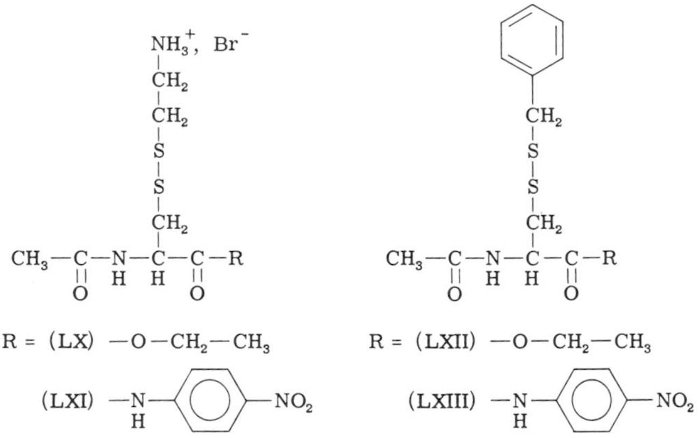

It is clear that a wide variety of new alkyl alkanethiolsulfonates of biochemical interest should be easily synthetically accessible. For example, it is likely that most of the sulfhydryl-specific reagents discussed in this review (even the complex ones), the majority of which are irreversible S-alkylating reagents, may be chemically transformed to their correspondingly alkyl methanethiolsulfonate in a one-step S-alkylation using sodium methanethiolsulfonate, in procedures analogous to reactions shown above. This should improve both the rapidity and selectivity of reaction of the resulting reagent with protein sulfhydryls and remove the usual requirement of excesses of reagent needed to afford complete modification. The resulting blocking groups could then also be readily removed, if desired.

[41] Modification of Histidyl Residues in Proteins by Diethylpyrocarbonate

By EDITH WILSON MILES

Diethylpyrocarbonate[1] reacts with histidyl residues in model systems[2-4] and in proteins[4,5] according to Eq. (1) to yield an N-carbethoxyhistidyl derivative. This reaction may be conveniently followed spectrophotometrically by the increase in absorbance, which has a maximum between 230 and 250 nm.[4,5] The number of modified histidyl residues is calculated from the molar absorption difference for N-carbethoxyhistidine at 240 nm: $\Delta\epsilon_{240 \text{ nm}} = 3200$ M^{-1} cm^{-1}.[4] Excess diethylpyrocarbonate is hydrolyzed by Eq. (2). Hydroxylamine removes the carbethoxy group from N-carbethoxyhistidyl residues.[6] Inactivation of an enzyme by diethylpyrocarbonate may be correlated with the modification of a histidyl residue if hydroxylamine reactivates the enzyme. Although diethylpyrocarbonate has been used successfully for the quantitative determination of total histidyl residues in five denatured proteins,[4] Morris and McKinley-McKee reported that calculation of the extent of modification of liver alcohol dehydrogenase by a 200-fold excess of diethylpyrocarbonate gave a number of modified histidyl residues greater than the known histidine content of the enzyme.[7] This result led to the study of model reactions of histidyl derivatives with excess diethylpyrocarbonate[7,8] and the recent conclusion that excess diethylpyrocarbonate can react to form a disubstituted histidyl derivative [(II), Eq. (3)]. Treatment of (II) with hydroxylamine or NaOH results in ring cleavage and the production of compound (III) (Eq. 4).[8] Since the dicarbethoxyhistidyl derivative (II) has a higher absorbance at 242 nm than the N-carbethoxyhistidyl derivative (I), the formation of (II) could lead to overestimation of the number of modified histidyl residues and to modification which is not reversed by hydroxylamine.[8] Treatment of ribonuclease with a large excess of diethylpyrocarbonate results in a 50%

[1] Diethylpyrocarbonate, $(C_2H_5OCO)_2O$, has also been termed ethoxyformic anhydride, diethyloxydiformate, or diethyldicarbonate, and the trade name Baycovin.

[2] L. Ehrenberg, I. Fedorcsák, and F. Solymosy, *Prog. Nucl. Acid Res. Mol. Biol.* **16,** 189 (1976).

[3] A. Mühlrád, G. Hegyi, and G. Tóth, *Biochim. Biophys. Acta* **2,** 19 (1967).

[4] J. Ovádi, S. Libor, and P. Elödi, *Biochim. Biophys. Acta* **2,** 455 (1967).

[5] J. Ovádi and T. Keleti, *Biochim. Biophys. Acta* **4,** 365 (1969).

[6] W. B. Melchior, Jr., and D. Fahrney, *Biochemistry* **9,** 251 (1970).

[7] D. L. Morris and J. S. McKinley-McKee, *Eur. J. Biochem.* **29,** 515 (1972).

[8] S. M. Avaeva and V. I. Krasnova, *Bioorg. Khim.* **1,** 1600 (1975).

loss in histidyl residues determined after acid hydrolysis.[8] Therefore, the use of a large excess of diethylcarbonate in modification of proteins should be avoided, and caution should be used in interpreting results. Treatment of adenosine with excess diethylpyrocarbonate has been shown to result in a ring-opening reaction of the imidazole ring of adenosine as well as reaction of the amino group of adenosine [Eq. (5)].[9]

$$\text{I } (230 - 250\text{nm}) \tag{1}$$

$$(C_2H_5OCO)_2O + H_2O \longrightarrow 2CO_2 + 2C_2H_5OH \tag{2}$$

$$\text{II } (220 - 240\text{nm}) \tag{3}$$

$$\text{III } (230 - 250\text{nm}) \tag{4}$$

$$(5)$$

Diethylpyrocarbonate has been shown to react specifically or stoichiometrically with a single histidyl residue in certain proteins (see the table, Class A). In other cases (see the table, Class B), the modification of activity has been correlated with the modification of one or more histidyl residues despite the possible modification of other residues; this correlation is facilitated by the fact that hydroxylamine removes the carbethoxy group from modified histidyl residues and tyrosyl residues, but not that of modified lysyl or sulfhydryl residues.[6] Several enzymes (see the table, Class C) have been shown to be inactivated by modification of a residue other than a histidyl residue.

Thus, although diethylpyrocarbonate does not always react specifically with histidyl residues in proteins, it is more selective than other acylating agents, and can give useful information about the role of histidyl

[9] N. J. Leonard, J. J. McDonald, R. E. L. Henderson, and M. E. Reichmann, *Biochemistry* **10**, 3335 (1971).

Selected Examples of Modification by Diethylpyrocarbonate

Protein or polypeptide	pH	Buffer	Concentration of diethyl-pyrocarbonate (mM)	No. of histidines modified	Other residues modified	References
Class A: Modification of 1 histidyl residue and no other residues						
Apamin	7.5	Phosphate	1.7	1	0	a
Yeast alcohol dehydrogenase	6.0	Phosphate	0.01	1	0	b
Lactate dehydrogenase	6.0	Phosphate	0.1	1	0	c
Glutamate dehydrogenase	6.0	Phosphate	0.016	1	0	d
Class B: Modification of activity correlated with modification of histidyl residue(s)						
Ribonuclease	4.0	Acetate	40	3	4 NH₂ groups	e
Thermolysin	7.2	HEPES[i]	3	2	2 Tyrosyl, 8 lysyl	f
Tyrosine phenol-lyase	6.0	Phosphate	1	2	—	g
Class C: Proteins with residues other than histidine modified exclusively						
Pepsin	4.0	Acetate	10	0	1 α-Amino group	e
α-Chymotrypsin	4.0	Acetate	0.1	0	1 Active site serine	e
Phospholipase A₂	6.0	TES[j]	0.03	0	1 ε-Amino group of lysine	h

[a] J.-P. Vincent, H. Schweitz, and M. Lazdunski, *Biochemistry* **14**, 2521 (1975).
[b] V. Leskovac and D. Pavkov-Pericin, *Biochem. J.* **145**, 581 (1975).
[c] J. J. Holbrook and V. A. Ingram, *Biochem. J.* **131**, 729 (1973).
[d] R. B. Wallis and J. J. Holbrook, *Biochem. J.* **133**, 183 (1973).
[e] W. B. Melchior, Jr., and D. Fahrney, *Biochemistry* **9**, 251 (1970).
[f] Y. Burstein, K. A. Walsh, and H. Neurath, *Biochemistry* **13**, 205 (1974).
[g] H. Kumagai, T. Utagawa, and H. Yamada, *J. Biol. Chem.* **250**, 1661 (1975).
[h] M. A. Wells, *Biochemistry* **12**, 1086 (1973).
[i] HEPES (*N*-2-hydroxyethylpiperazine-*N*-2-ethanesulfonic acid).
[j] Tes (*N*-tris(hydroxymethyl)methyl-2-aminoethanesulfonic acid).

residues in many proteins. The section below will describe how optimal conditions for reaction should be determined and how possible side reaction should be examined.

Methods

Reagent

Commercially available diethylpyrocarbonate[1] is used (Aldrich, Eastman, Bayer AG, ICN). This liquid (sp. gr. 1.12; concentration if pure =

6.9 M) is stored at 4° and diluted with anhydrous ethanol or anhydrous acetonitrile before use using a manual pipetting device and care not to come into skin contact or to breathe the vapors, which can cause irritation of the eyes and mucous membranes.[2] Since diethylpyrocarbonate is subject to hydrolysis by Eq. (2), care should be taken to store the container in a desiccator at 4° and to warm it to room temperature before opening it quickly.[2] A stock solution in anhydrous acetonitrile is reported to be stable for several months at 4°.[10] Since the purity of the commercial reagent may be variable owing to hydrolysis [Eq. (2)], the concentration of the dilution should be determined quantitatively. An aliquot of the dilution is added to 3 ml of 10 mM imidazole at pH 7.5 in a cuvette having a 1-cm light path; the increase in absorption at 230 nm due to N-carbethoxyimidazole ($\Delta\epsilon = 3.0 \times 10^3$ cm^{-1} M^{-1}) is determined; the increase is rapid and quantitative.[6] An alternative assay for diethylpyrocarbonate based on measuring the rate of reaction of diethylpyrocarbonate with colored 5-thio-2-nitrobenzoate to form a colorless product has recently been reported.[11] The syntheses of [^{14}C]diethylpyrocarbonate[6,12,13] and [^{3}H]diethylpyrocarbonate[14] have been reported.

Procedure for Carbethoxylation

A typical detailed procedure is illustrated by the following example[15] and by Fig. 1. Possible variations in pH, buffer, temperature, protein concentration, and reagent concentration are discussed below. Two cuvettes containing 1.0-ml aliquots of protein (1.0 mg/ml in 0.1 M potassium phosphate buffer, pH 6.5) are placed in the sample and reference compartments of a double-beam spectrophotometer at 4°. Difference spectra are recorded between 320 nm and 237 nm (Fig. 1A) before and at various time intervals after the addition of 5 µl or 25 µl of 10 mM diethylpyrocarbonate in ethanol to the sample cuvette and an equal volume of ethanol to the reference cuvette to give a final concentration of 0.05 mM (dashed curves) or 0.25 mM (solid curves) diethylpyrocarbonate. Spectra are also recorded of the enzyme solutions before (— - —) and after (— - - —) treatment for 60 min with 0.25 mM diethylpyrocarbonate using a buffer blank. An aliquot of the treated enzyme is removed immediately after each difference spectrum is recorded and is assayed. The relation between activity and the number of histidyl residues modified calculated

[10] B. Setlow and T. E. Mansour, *J. Biol. Chem.* **245**, 5524 (1970).
[11] S. L. Berger, *Anal. Biochem.* **67**, 428 (1975).
[12] E. Fisher and R. Schelenz, *J. Labelled Compd.* **5**, 333 (1969).
[13] B. Duhm, W. Haul, H. Medenwald, K. Patzschke, and L. A. Wegner, *Z. Lebensm.-Unters. Forsch.* **132**, 200 (1966).
[14] B. Öberg, *Eur. J. Biochem.* **19**, 496 (1971).
[15] E. W. Miles and H. Kumagai, *J. Biol. Chem.* **249**, 2843 (1974).

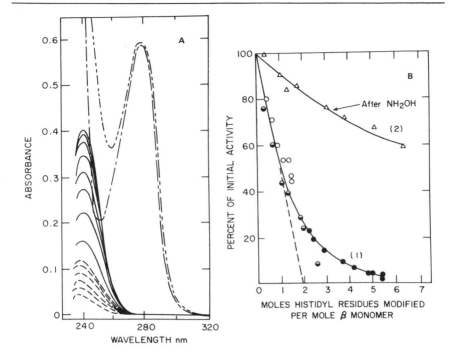

Fig. 1. Effects of diethylpyrocarbonate on the spectrum (A) and on the enzymic activities and the histidine content on the apo β_2 subunit of tryptophan synthetase (B). See text for experimental details. (A) Difference spectra with 25 mM diethylpyrocarbonate were recorded at 5, 10, 15, 20, 30, 40, and 60 min (——); difference spectra with 0.05 mM diethylpyrocarbonate were recorded at 3, 5, 10, 20, 30, and 45 min (-----).

from the absorbance at 240 nm ($\Delta\epsilon_{240}$ = 3200 cm^{-1} M^{-1})[4] is shown by Fig. 1B, curve 1.

Procedure for Decarbethoxylation

Aliquots of reaction mixtures and untreated controls are incubated with 0.3 ml of 1 M NH$_2$OH·HCl (adjusted to pH 7.0 with KOH) for 20 hr at 22° and dialyzed for 6 hr against 4 changes of 0.1 M potassium phosphate buffer, pH 7.8, before assay (Fig. 1B, curve 2; the extent of modification before hydroxylamine treatment is shown on the abscissa).

Comments on Procedure and Conditions

pH and Buffer

Since nucleophiles are reactive in their unprotonated forms, Ovádi et al.[4] proposed that diethylpyrocarbonate should be selective for histidyl residues in proteins at pH 6 and showed that this was the case for several proteins. Subsequently, Melchior and Fahrney[6] demonstrated that diethylpyrocarbonate can react with other groups in proteins at pH 4, including the active-site serine of chymotrypsin and amino groups of ribonuclease and pepsin at pH values as low as 4. Thus, the reactivity of nucleophilic groups in some proteins may not be the same as in model systems owing to different environments of the residues in these proteins. Holbrook and Ingram[16] have studied the pH dependence of the rate of reaction of diethylpyrocarbonate with imidazole and with the essential histidyl residue of lactate dehydrogenase; the results show pK's of 6.95 and 6.8, respectively, and indicate that only the unprotonated imidazole or histidyl residue is reactive. Studies at pH 6 are advantageous because the N-carbethoxyhistidyl residues are most stable at this pH.[6] However, in one case, an essential histidyl residue has been found to be reactive at pH 7.5, but not at pH 6.[17] It is therefore advisable to try more than one pH value and to consider the effect of pH on the stability and conformation of the enzyme being studied. The most frequently used buffer is potassium phosphate in the pH range 6.0–8.0 (see the table). Although acetate and succinate buffers have been used in the pH range 4.6–6.0, recent findings that diethylpyrocarbonate may react with sulfhydryl groups in carboxylate buffers to form thiol esters that absorb at 242 nm indicate that carboxylate buffers should be avoided.[18] Some effects of buffers on the decomposition of diethylpyrocarbonate have recently been reported.[11] Diethylpyrocarbonate has a half-life at 25° of 24 min in 60 mM sodium phosphate buffer, pH 6.0, and 9 min in 37 mM sodium phosphate buffer, pH 7.0.[11] Tris buffer accelerates the decomposition of diethylpyrocarbonate and presumably reacts with it as a nucleophile; the half-life of diethylpyrocarbonate in Tris-chloride at 25° is 1.25 min at pH 7.5 and 0.37 min at pH 8.2. Tris buffer should thus be avoided if possible.

Reagent Concentration and Temperature

Various investigators have used diethylpyrocarbonate over a wide range of concentrations from 0.01 mM to 40 mM (the maximum solubility in water) (see the table). This probably reflects variation in the reactivity or the exposure of histidyl residues in different proteins as well as the pH dependence of the rate of reaction.[16] However, recent

[16] J. J. Holbrook and V. A. Ingram, *Biochem. J.* **131**, 729 (1973).
[17] N. Tudball, R. Bailey-Wood, and P. Thomas, *Biochem. J.* **129**, 419 (1972).
[18] C. K. Garrison and R. H. Himes, *Biochem. Biophys. Res. Commun.* **67**, 1251 (1975).

studies showing that excess diethylpyrocarbonate can undergo a second reaction with histidyl residues [Eq. (3)] which is not reversed by hydroxylamine[8] indicate that use of excess diethylpyrocarbonate may lead to inaccurate quantitation and irreversible modification. Therefore, the investigator should attempt to find the lowest concentration of reagent necessary for modification and should be especially careful in interpreting his results if higher concentrations are required. One way to reduce the excess of diethylpyrocarbonate necessary is to increase the concentration of protein treated. Although the concentration of protein used is usually about 1 mg/ml or 20 μM if the reaction is followed spectrophotometrically as described above, Holbrook and Ingram[16] found stoichiometric inactivation of 0.7 mM lactate dehydrogenase. Modification can be carried out on small volumes of concentrated enzyme followed by dilution for determination of the extent of reaction and inactivation. No systematic study of the effect of temperature has been made and reactions are usually carried out at 4° or 25°. The latter is more convenient, especially for spectrophotometric studies, if the enzyme is stable at this temperature.

Determination of the Extent of Reaction

The extent of modification of histidyl residues by diethylpyrocarbonate is conveniently determined by measuring the increase in absorbance at 242 nm or by determining the absorbance maximum in difference spectra as described above and demonstrated in Fig. 1. The wavelength of this difference peak has been found to vary from about 238 nm to 246 nm. The most frequently used difference extinction coefficient, $\Delta\epsilon_{240} = 3200$ cm^{-1} M^{-1}, is derived from model experiments by Ovádi et al.[4] Holbrook and Ingram[16] have determined a slightly higher difference extinction coefficient, $\Delta\epsilon_{240} = 3600$ cm^{-1} M^{-1}, for N-α-acetyl-N'-carbethoxyhistidine and have found it to be constant within 10% between pH 5 and 9; Tudball et al.[17] have used a difference extinction coefficient of $\Delta\epsilon_{242} = 2900$ cm^{-1} M^{-1} at pH 7.5. Although it is possible that the extinction coefficient of modified histidyl residues might be different in proteins than in model systems, consistent stoichiometry has been obtained using the spectrophotometric method in several cases.[4,10,19] The number of [14C]carbethoxy groups incorporated into phosphofructokinase[10] based on specific activity coincided very well with the number of N-carbethoxyhistidyl residues determined spectrophotometrically at pH 6 up to 4 mol per 10^5 g of enzyme using $\Delta\epsilon_{242} = 3200$ cm^{-1} M^{-1}. Application of the

[19] J.-P. Vincent, H. Schweitz, and M. Lazdunski, Biochemistry 14, 2521 (1975).

spectrophotometric method at pH 7.5 (using $\Delta\epsilon_{242} = 2900$ cm^{-1} M^{-1} [17]) to apamin,[19] a polypeptide of 18 amino acids, gave the known value of one histidine. However, quantitation may be inaccurate when an excess of diethylpyrocarbonate is used as discussed above.[7,8] Quantitation of the modification of histidyl residues by the increase of absorption at 240 nm may be complicated if O-carbethoxylation of tyrosyl residues occurs simultaneously.[20] O-Carbethoxylation of N-acetyl-L-tyrosine ethyl ester results in a difference spectrum which shows a minimum at 278 nm ($\Delta\epsilon_{278} = 1310$ cm^{-1} M^{-1}) and a major decrease at wavelengths below 240 nm.[20] Thus, simultaneous modification of histidyl and tyrosyl residues results in an anomalously small change in the difference absorbance at 240 nm due to N-carbethoxyhistidyl residues.[20] Burstein et al. found that spectral changes during reactivation of thermolysin by mild hydroxylamine treatment were easier to interpret since O-carbethoxytyrosyl residues were not hydrolyzed by this treatment (see below and[20]).

The extent of reaction of histidyl residues can also be determined by measuring the incorporation of label from [^{14}C]diethylpyrocarbonate[6,10] or [^{3}H]diethylpyrocarbonate[14] before and after treatment with hydroxylamine; the possible occurrence of Eqs. (3) and (4) should be considered.[8]

Decarbethoxylation by Hydroxylamine

Hydroxylamine has been shown to remove the carbethoxy groups from modified histidyl and tyrosyl residues and from the active-site serine in native chymotrypsin.[6] The reaction is usually carried out at pH 7 and at either room temperature or 4°. The concentration of hydroxylamine and the time for reactivation seem to vary for each protein which has been studied and range from 20 mM for 30 min at 25° and pH 7.2[20] to 0.75 M for 20 hr at 22° and pH 7.0.[15] O-Carbethoxy-N-acetyltyrosine ethyl ester is 4 times less reactive with hydroxylamine than N-carbethoxyimidazole.[6] Burstein et al. found that 20 mM NH$_2$OH selectively decarbethoxylated the essential histidyl residue under conditions which did not decarbethoxylate 2 O-carbethoxytyrosyl residues.[20] The failure of hydroxylamine to fully reactivate a modified enzyme (see Fig. 1B and Miles and Kumagai[15]) has usually been ascribed to the modification of other residues, such as lysyl residues which have some secondary effects on activity or conformation of the enzyme. However, Avaeva and Krasnova[8] have suggested that this result may be due to the occurrence of the irreversible Eqs. (3) and (4) in the presence of excess diethylpyrocarbonate.

[20] Y. Burstein, K. A. Walsh, and H. Neurath, *Biochemistry* **13**, 205 (1974).

Agents necessary for stabilization of the enzyme during decarbethoxylation by hydroxylamine should be included. Removal of NH_2OH by dialysis or by gel filtration on Sephadex G-25 may be necessary before assay of the enzymic activity, particularly if the enzyme has a pyridoxal 5'-phosphate cofactor.[15]

Reaction of Diethylpyrocarbonate with Other Residues

Mühlrád *et al.*[3] demonstrated in model reactions that diethylpyrocarbonate can react with a variety of nucleophilic residues which occur in proteins, including histidyl, sulfhydryl, arginyl, and tyrosyl residues as well as with α- and ε-amino groups. O-Carbethoxylation of tyrosyl residues is readily detected by a decrease in the difference absorbance at 278 nm.[20] Although this side reaction has only been rarely reported,[20] its possible occurrence should be kept in mind, particularly if reactivation by hydroxylamine is used as a criterion of inactivation of essential histidyl residues; hydroxylamine also reverses O-carbethoxylation of tyrosyl residues at a lower rate.[6] Diethylpyrocarbonate has also been shown to modify the active site serine of chymotrypsin; this reaction is also reversed by hydroxylamine treatment of the native, but not of the denatured, enzyme.[6] The possible modification of sulfhydryl residues by diethylpyrocarbonate should be tested by titrating sulfhydryl residues in untreated and modified proteins with Ellman's reagent.[21] It is surprising that the modification of protein sulfhydryl residues by diethylpyrocarbonate has not been reported, since diethylpyrocarbonate reacts with sulfhydryl residues in model systems[11,22] at a 5-fold lower rate than primary amines with corresponding dissociation constants.[22] Although primary amines react about 50 times more slowly with diethylpyrocarbonate than imidazole at neutral pH in model systems,[22] Melchior and Fahrney reported that several amino groups in ribonuclease are as reactive as the histidyl residues at pH 4 and that the α-amino group is the only group in pepsin to react at pH 4.[6] These studies were facilitated by the use of [14C]diethylpyrocarbonate; the retention of label after treatment with hydroxylamine indicates that an amino group or a sulfhydryl group has been modified. Similar studies on heart phosphofructokinase[10] showed that the first 4 moles of [14C]diethylpyrocarbonate incorporated into 10^5 g enzyme were hydroxylamine-labile; at higher concentration of [14C]diethylpyrocarbonate, carbethoxy groups stable to hydroxylamine treatment were incorporated. Incomplete restoration of

[21] A. F. S. A. Habeeb, this series Vol. 25, p. 457 (1972).
[22] S. Osterman-Golkar, L. Ehrenberg, and F. Solymosy, *Acta Chem. Scand.* **B28,** 215 (1974).

activity by hydroxylamine treatment (see Fig. 1B) also suggests that some carbethoxy groups have been introduced that are stable to hydroxylamine and have some effects on activity. Estimation of the number of amino groups that have reacted with diethylpyrocarbonate can be accomplished by quantitative estimation of amino groups before and after modification with [^{14}C]acetic anhydride or fluorescamine.[19] Application of these determinations to apamin[19] showed that no amino group had been modified.

Although the tryptophan nucleus does not react with diethylpyrocarbonate in model studies in aqueous solution,[3] N-carbethoxylation of the indole nucleus has been demonstrated in ethanol[23]; this model reaction results in a decrease in the fluorescence of tryptophan. Since a decrease in the tryptophan fluorescence of bovine serum albumin has been found to accompany modification by diethylpyrocarbonate, Rosén and coworkers have postulated that diethylpyrocarbonate might be able to react with tryptophanyl residues in the hydrophobic interior of some proteins where diethylpyrocarbonate would be hydrolyzed less rapidly.[23] Although these observations have not been extended, investigators should be aware of the possibility that tryptophanyl residues might be modified and consider checking the fluorescent properties of diethylpyrocarbonate-modified proteins.

Other Side Reactions of Proteins with Diethylpyrocarbonate

Reichman and associates have reported that treatment of ribonuclease with very high concentrations of diethylpyrocarbonate results in inactivation and some polymerization.[24] They have proposed that diethylpyrocarbonate forms amide bonds between ε-amino groups of lysine and carboxylic groups of aspartic or glutamic, both intermolecularly and intramolecularly. Such possible side reactions are probably not the cause of inactivation of ribonuclease, however, since the activity of modified ribonuclease is restored by treatment with hydroxylamine.[2,6] These results suggest that investigators should be careful in interpreting their results with diethylpyrocarbonate, particularly when high concentrations of reagent are used, since all the possible reactions of diethylpyrocarbonate have not been thoroughly established and since chemical groups in different protein environments may have different reactivities. Physicochemical studies of modified proteins would be advisable to rule out possible polymerization or conformational change.

[23] C. G. Rosén, T. Gejvall, and L.-O. Andersson, *Biochim. Biophys. Acta* **221**, 207 (1970).
[24] B. Wolf, J. A. Lesnaw, and M. E. Reichmann, *Eur. J. Biochem.* **13**, 519 (1970).

Use of Substrates or Inhibitors during Studies of Reactions with Diethylpyrocarbonate

The finding that inactivation of an enzyme by diethylpyrocarbonate is prevented by the presence of a substrate or inhibitor may be used as evidence that a histidyl residue is a catalytic group or is located at the substrate binding site.

Several examples of the successful use of substrates or inhibitors in this way will be mentioned: 2'-adenylate prevents the inactivation of ribonuclease U_2[25]; NADH and sodium oxamate prevent the inactivation of lactate dehydrogenase[16]; NADH prevents the inactivation of glutamate dehydrogenase by a stoichiometric amount of diethylpyrocarbonate[26]; and benzyloxy-carbonyl-L-phenylalanine prevents the incorporation of a single carbethoxy group into an essential histidyl residue of thermolysin but does not prevent the incorporation of about 12 additional moles of reagent per mole of enzyme.[20] However, investigators should consider the possibility that the substrate or inhibitor used might react with diethylpyrocarbonate and produce apparent protection by lowering the effective concentration of diethylprocarbonate. For example, a high concentration of an amino acid might react with the diethylpyrocarbonate. This could be tested under the experimental conditions used by assaying the concentration of diethylpyrocarbonate with imidazole (see above) at time intervals after addition to the various reaction components minus the enzyme.

Stability of Carbethoxyhistidyl Residues and Isolation of Modified Peptides

The stability of N-carbethoxyimidazole at 25° in water has been shown to be about 2 hr at pH 2, 55 hr at pH 7, and 18 min at pH 10.[6] Although the instability of N-carbethoxyhistidyl residue severely limits the isolation of modified peptides, the isolation of a modified peptide has been accomplished in one case using procedures at neutral pH, including tryptic digestion, gel filtration, and electrophoresis before and after decarbethoxylation.[27]

The Use of Diethylpyrocarbonate in Nucleic Acid Research

Diethylpyrocarbonate has been used as a nuclease inhibitor and protein denaturant instead of the traditional phenol in the extraction of many

[25] S. Sato and T. Uchida, *J. Biochem.* (*Tokyo*) **77**, 795 (1975).
[26] R. B. Wallis and J. J. Holbrook, *Biochem. J.* **133**, 183 (1973).
[27] G. Hegyi, G. Premecz, B. Sain, and A. Mühlrád, *Eur. J. Biochem.* **44**, 7 (1974).

types of nucleic acids from many sources (for a recent review, see Ehrenberg et al.[2]). It has also been used to sterilize glassware and solutions to be used in nucleic acid research.[2] Although diethylpyrocarbonate has also been used for the preparation of subcellular systems, such as mitochondria and ribosomes, these uses can be complicated by the fact that diethylpyrocarbonate reacts with the protein components.[2] The usefulness of diethylpyrocarbonate in isolation of nucleic acids has also been compromised by the finding that it reacts under certain conditions, including high reagent concentration, with nucleic acids, in particular with single-stranded RNA or single-stranded regions.[2,14] Öberg showed that [3H] diethylpyrocarbonate reacts with poly(A), poly(C), and poly(G), but not with poly(U), and with single-stranded poliovirus RNA, single-stranded regions of transfer RNA, and with denatured yeast tRNA^{Ala}.[14] The mechanism of the reaction of diethylpyrocarbonate with adenosine has been studied by Leonard et al.[9] who found that it opens the imidazole ring of adenosine by Eq. 5. Although diethylpyrocarbonate treatment does not appear to degrade RNA, it does affect its solubility.[28] Modification of bases in single-stranded RNA clearly affects infectivity[14] and should also interfere with sequence studies of RNA. Therefore, investigators should use caution in the use of diethylpyrocarbonate in isolation of nucleic acids and try to find conditions that produce inhibition of nucleases without modification of nucleic acids where modification is likely and undesirable.[2]

[28] I. Fedorcsák, L. Ehrenberg, and F. Solymosy, *Biochem. Biophys. Res. Commun.* **65**, 490 (1975).

[42] Modification of Tryptophan to Oxindolylalanine by Dimethyl Sulfoxide–Hydrochloric Acid

By WALTER E. SAVIGE[1] and ANGELO FONTANA

The high reactivity of the indole nucleus of tryptophan toward electrophilic reagents has evoked much interest in finding techniques for the selective chemical modification of tryptophan residues in proteins. However, in spite of the variety of reagents introduced over the years for this purpose, only a few are specific for tryptophan residues and yield a single well-defined product.[2-4]

[1] This article is based on work carried out during a stay of W. E. S. at the Institute of Organic Chemistry, University of Padova, Padova, Italy.
[2] B. Witkop, *Adv. Protein Chem.* **16**, 221 (1961).
[3] T. F. Spande, B. Witkop, Y. Degani, and A. Patchornik, *Adv. Protein Chem.* **24**, 97 (1970).
[4] A. Fontana and C. Toniolo, *Fortschr. Chem. Org. Naturst.* **33**, 309 (1976).

Conversion of tryptophan to oxindolylalanine (OIA)[5] (2-hydroxy-tryptophan)[6] (I) residues can be effected by oxidation using halogenating

(I)

agents, of which N-bromosuccinimide (NBS)[7-10] is the most widely used example. This is a powerful oxidizing agent, which has been reported also to modify other amino acid residues in proteins.[11,12] BNPS-skatole[12,13] or 2,4,6-tribromo-4-methylcyclohexadienone[14] has latterly been found to give more specific modification than NBS. However, the use of any of the above three reagents can also give rise to cleavage of the tryptophanyl peptide bond.[4]

A new method for oxidation of 3-substituted indoles to the corresponding oxindoles has been recently proposed.[15-17] The method involves the use of a mixture of DMSO and concentrated aqueous HCl.

The interactions between organic sulfoxides and haloacids have been reviewed recently.[18,19] The currently accepted view is that a series of equilibrium reactions is established, viz.,

[5] Abbreviations used: DMSO, dimethyl sulfoxide; OIA, oxindolylalanine; DIA, dioxindolylalanine; NBS, N-bromosuccinimide; BNPS-skatole, 2-(2-nitrophenyl-sulfenyl)-3-methyl-3-bromoindolenine; IPA, indolyl-3-propionic acid; OIA-lysozyme, lysozyme derivative with six tryptophan residues selectively oxidized to oxindolylalanine. All amino acids and amino acid derivatives have the L configuration.

[6] The amino acid has been variously referred to as 2-hydroxytryptophan, α-hydroxy-tryptophan, hydroxytryptophan, β-oxindole-3-alanine, and β-3-oxindolylalanine. The oxindolylalanine nomenclature has been preferred over the hydroxytryptophan nomenclature, since the former more correctly reflects the structure of the amino acid.

[7] A. Patchornik, W. B. Lawson, and B. Witkop, $J.$ $Am.$ $Chem.$ $Soc.$ **80,** 4747 (1958).

[8] A. Patchornik, W. B. Lawson, E. Gross, and B. Witkop, $J.$ $Am.$ $Chem.$ $Soc.$ **82,** 5923 (1960).

[9] N. M. Green and B. Witkop, $Trans.$ $N.$ $Y.$ $Acad.$ $Sci.$ $Ser.$ II **26,** 659 (1964).

[10] T. F. Spande, and B. Witkop, this series Vol. 11, pp. 498, 506, and 522.

[11] M. J. Kronman, and F. M. Robbins, in "Fine Structure of Proteins and Nucleic Acids (G. D. Fasman, and S. N. Timasheff, eds.), p. 271. Dekker, New York, 1970.

[12] G. S. Omenn, A. Fontana, and C. B. Anfinsen, $J.$ $Biol.$ $Chem.$ **245,** 1895 (1970).

[13] A. Fontana, this series Vol. 25, p. 419.

[14] Y. Burstein and A. Patchornik, $Biochemistry$ **11,** 4641 (1972).

[15] W. E. Savige and A. Fontana, "Peptides 1976" (A. Loffet, ed.), p. 135. Presse Universitaire, Brussels.

[16] W. E. Savige and A. Fontana, $J.$ $Chem.$ $Soc.,$ $Chem.$ $Commun.$ 599 (1976).

[17] W. E. Savige and A. Fontana, in preparation (1977).

[18] G. Modena, $Int.$ $J.$ $Sulfur$ $Chem.,$ $Part$ C **7,** 95 (1972), and references cited therein.

[19] G. Scorrano, $Acc.$ $Chem.$ $Res.$ **6,** 132 (1973), and references cited therein.

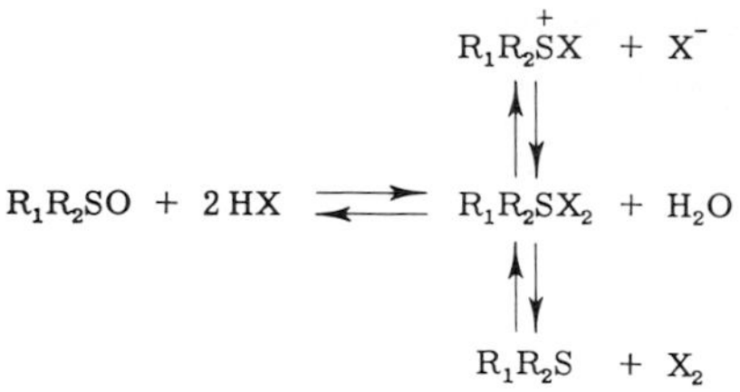

$$R_1R_2\overset{+}{S}X \ + \ X^-$$

$$R_1R_2SO \ + \ 2\,HX \ \rightleftharpoons \ R_1R_2SX_2 \ + \ H_2O$$

$$R_1R_2S \ + \ X_2$$

Any of the three species, sulfide dihalide, halosulfonium ion, or free halogen, can act as halogenating agent. In fact, mixtures of DMSO and haloacids have proved to be useful halogenating systems for aromatic substrates.[20-23]

Halogenation of 3-substituted indoles in a solvent containing small amounts of water leads to the corresponding oxindole, together with other products according to the nature of the halogenating agent and/or indole.[24-26] A likely mechanism for oxidation of 3-alkylindoles by DMSO/HCl is shown below.[27]

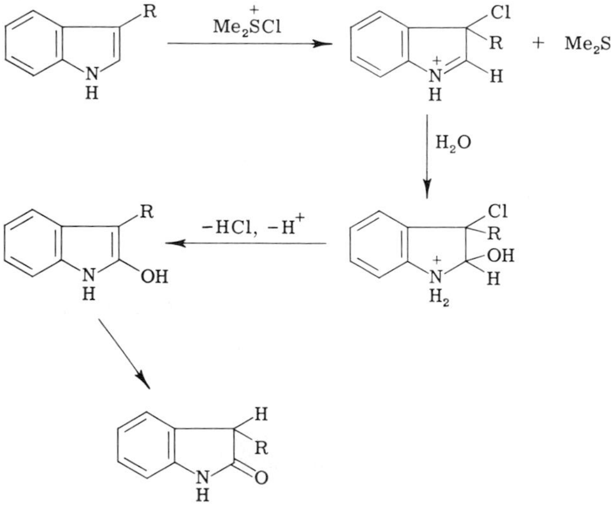

[20] H. Gilman and J. Eish, *J. Am. Chem. Soc.* **77**, 3862 (1955).

[21] T. L. Fletcher, and H. L. Pan, *J. Chem. Soc. (London)* **1965**, 4588 (1965).

[22] T. L. Fletcher, *Quart. Rep. Sulfur Chem.* **3**, 107 (1968).

[23] D. Martin, and H. G. Hauthal, "Dimethyl Sulfoxide," p. 269. Van Nostrand-Reinhold, Princeton, New Jersey. 1976.

[24] W. B. Lawson, A. Patchornik, and B. Witkop, *J. Am. Chem. Soc.* **82**, 5918 (1960).

[25] R. L. Hinman and C. P. Bauman, *J. Org. Chem.* **29**, 1206 (1964).

[26] J. C. Powers, *J. Org. Chem.* **31**, 2627 (1966).

[27] See R. J. Sundberg, "The Chemistry of Indoles," p. 14. Academic Press, New York, 1970, for a discussion on the oxidation of substituted indoles by halogenation.

The oxidation of tryptophan derivatives can be effected by treatment for 10–30 min at room temperature with a mixture of glacial acetic acid containing DMSO (5–10 equiv) and concentrated aqueous HCl. The oxidation is evidenced by the evolution of dimethyl sulfide (odor) and by a concomitant shift in the wavelength of the ultraviolet absorption maximum of the reaction mixture from 280 to 250 nm, due to the transformation of indole to the oxindole chromophore.[10]

It is worthy of mention that oxidation does not occur when any of the acids, formic, trifluoroacetic, sulfuric, or p-toluenesulfonic, is substituted for hydrochloric in the above-described treatment. Other sulfoxides can be used in combination with concentrated HCl, and in particular treatment of a 1:1 mixture of methionine sulfoxide and tryptophan gives high yields of oxindolylalanine and methionine.[16,17]

The described procedure allows for various permutations, since different haloacids and sulfoxides can be used in combination. Effective bromination and iodination systems can be produced by using HBr or HI with a sulfoxide. However, whereas treatment of a protein with DMSO/HCl under the above conditions brings about negligible overoxidation to the dioxindole level or fission of the tryptophanyl bond, with DMSO/HBr both of the secondary reactions occur to a considerable extent.[17,28]

Selectivity. The selectivity of the procedure employed was studied by exposure to the DMSO/HCl system (100 μl of acetic acid, 50 μl of 12 N HCl and 10 μl of DMSO) of a standard mixture of amino acids (0.25 μmol each), as is used for calibration purposes in automated amino acid analyzer systems, i.e., without tryptophan and cysteine. Methionine alone was modified, being quantitatively oxidized to the sulfoxide, provided that an excess of DMSO (20 equiv) is used. Cysteine was tested separately and found to be oxidized to cystine.

The results on the reactivity of the free amino acids were also checked with ribonuclease, a protein containing all common amino acids except tryptophan and cysteine. In this case, only the four methionine residues of the protein molecule were modified, all being quantitatively oxidized to the sulfoxide.

Analysis of the Course of Reaction

The oxidation of tryptophan to oxindolylalanine residues in peptides and proteins by DMSO/HCl can be followed by determining the ultraviolet absorption changes that accompany oxidation of the indole nucleus (ϵ_{280nm} 5600) to the oxindole chromophore (ϵ_{250nm} 7200).[10] To this end,

[28] W. E. Savige and A. Fontana, this volume [44].

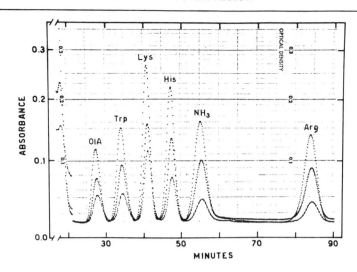

FIG. 1. Elution pattern of oxindolylalanine (OIA) (0.2 μmol), tryptophan (0.1 μmol) and the basic amino acids (0.1 μmol each) from the short column of the amino acid analyzer. The column (0.9 $\times$ 12 cm) was eluted with citrate buffer, pH 5.23, according to D. H. Spackman, W. H. Stein, and S. Moore, *Anal. Chem.* **30**, 1190 (1958).

aliquots from the reaction mixture can be taken and diluted with an aqueous buffer and the spectra recorded. With tyrosine-containing proteins the contribution of the phenolic chromophore (ϵ_{275nm} 1200) should also be considered.

Oxindolylalanine has been found stable to acid hydrolysis with 3 N p-toluenesulfonic acid containing 0.2% 3-(2-aminoethyl)indole, following the procedure described by Liu and Chang[29,30] for the determination of the tryptophan content in proteins. Hydrolysis of the DMSO/HCl-treated peptide or protein allows simultaneous determination of both tryptophan and oxindolylalanine. Quantitative determination of these amino acids can be obtained by the standard short-column procedure of Spackman *et al.*[31] using a 12-cm column (Fig. 1). The peak of oxindolylalanine precedes that of tryptophan. The color yield of oxindolylalanine is much lower than that of tryptophan. It is convenient to relate the color factor to that of tryptophan; oxindolylalanine gives a color yield which is 30% that of tryptophan under our conditions, but the color factor should be checked for each analyzer. A pure sample of oxindolylala-

[29] T.-Y. Liu and Y. H. Chang, *J. Biol. Chem.* **246**, 2842 (1971).
[30] T.-Y. Liu, this series Vol. 25, p. 44.
[31] D. H. Spackman, W. H. Stein, and S. Moore, *Anal. Chem.* **30**, 1190 (1958).

nine can be prepared following described procedures[32-35] or using the DMSO/HCl treatment of tryptophan herein reported.

Oxindolylalanine. L-Tryptophan (20.4 mg; 0.1 mmol) is dissolved in a mixture of 0.5 ml of glacial acetic acid, 0.5 ml of 12 N HCl, and 20 μl of DMSO. The solution is kept at room temperature for 15 min and then diluted with water (1 ml). The reaction mixture is then applied to a Sephadex LH-20 column (2.2 × 72 cm) equilibrated and eluted with water.[36] The effluent of the column is monitored at 255 nm. The major peak of oxindolylalanine (over 90% yield) is preceded by trace amounts of dioxindolylalanine[37] and followed by unchanged tryptophan. The fractions containing oxindolylalanine are concentrated under vacuum to a low volume and kept at 0°. The crystalline precipitate is separated by filtration and dried under vacuum over P_2O_5 (12 mg); m.p. 246–250°, R_f 0.65 (ninhydrin) in thin-layer chromatography (SiO_2; butanol:pyridine:acetic acid:water, 10:15:3:12), λ_{max} (H_2O) 250 nm (ϵ 7200) and shoulder at 280 nm (ϵ 1500).

Oxindolylalanine is sensitive to atmospheric oxygen in aqueous solution in alkaline media and is oxidized to dioxindolylalanine ($\lambda_{max}(H_2O)$ 253 nm (ϵ 4070) and 287 (ϵ 1220), λ_{min} 230 and 277 nm).[37,38] Dioxindolylalanine is eluted from the long column of the amino acid analyzer between phenylalanine and tyrosine. However, oxindolylalanine residues in proteins have been found to be less unstable than the free amino acid.[17]

Modification of Tryptophan-Containing Peptides

Several tryptophan-containing peptides were treated with DMSO/HCl in the presence of glacial acetic acid for 15–30 min at room temperature.[17]

[32] J. W. Cornforth, R. H. Cornforth, C. E. Dalgliesh, and A. Neuberger, *Biochem. J.* **48**, 591 (1951).

[33] T. Wieland, O. Weiberg, W. Dilger, and E. Fischer, *Justus Liebigs Ann. Chem.* **592**, 69 (1955).

[34] F. M. Veronese, A. Fontana, E. Boccù, and C. A. Benassi, *Z. Naturforsch. Teil B* **23**, 1319 (1968).

[35] M. Ohno, T. F. Spande, and B. Witkop, *J. Org. Chem.* **39**, 2635 (1970).

[36] In the course of these studies, it has been found that Sephadex LH-20 columns equilibrated and eluted with water are extremely efficient tools in separating on a preparative scale water-soluble tryptophan derivatives, including peptides, from their OIA derivatives. The nature of the interaction of tryptophan and derivatives with the LH-20 gel matrix, which is a hydroxypropyl derivative of Sephadex G-25, is hydrophobic in character, and an OIA-peptide is eluted from one of these columns considerably faster than the corresponding Trp-peptide (and clearly slower than DIA-peptides when present).

[37] P. L. Julian, E. E. Dailey, H. C. Printy, H. L. Cohen, and S. Hamashige, *J. Am. Chem. Soc.* **78**, 3503 (1956).

[38] W. E. Savige, *Aust. J. Chem.* **28**, 2275 (1975).

It was found that 5–10 equiv of DMSO were sufficient to effect complete oxidation of the tryptophan residues. The rate of reaction showed a marked dependence on the concentration of HCl and amount of glacial acetic acid used as diluent in the reaction mixture. The reaction can be terminated simply by dilution with water.[39]

Phe-Val-Gln-Trp-Leu. The pentapeptide (5 mg), which corresponds to the 22–26 sequence of glucagon, was dissolved in a mixture of 100 μl of glacial acetic acid and 50 μl of 12 N HCl, and then 5 μl of DMSO were added. The reaction mixture was kept at room temperature in the dark for 15 min. After dilution with 0.5 ml of water, the solution was poured on a column of Sephadex G-25 (3 $\times$ 54 cm) equilibrated and eluted with 10% acetic acid. The effluent was monitored at 255 nm and 280 nm. The elution profile illustrated in Fig. 2 indicates that modification of the peptide occurred in over 90% yield. The main peak of OIA-peptide (λ_{max} 250 nm) was preceded by traces of unidentified material and followed by some unchanged pentapeptide (λ_{max} 280 nm). Amino acid analysis of an acid hydrolyzate with 3 N p-toluenesulfonic acid[29,30] of the OIA-peptide gave the following values: Glu (1.0), Val (0.9), Leu (1.0), Phe (0.9), and OIA (0.9).

Modification of Hen Egg White Lysozyme

Lysozyme (10 mg) (Worthington, LY 8 HB) was dissolved in a mixture of glacial acetic acid (200 μl), 12 N HCl (100 μl) and DMSO (10 μl), and the solution was kept at room temperature in the dark for 30 min. After dilution with water (1 ml), the reaction mixture was passed through a column of Sephadex G-25 (1.8 $\times$ 35 cm) equilibrated and eluted with 10% acetic acid. The protein was located in the effluent by absorption measurements at 255 nm. The fractions of the protein peak were combined and the oxidized protein recovered by lyophilization (8 mg).

As shown in the table, amino acid analysis of an acid hydrolyzate

---

[39] The system DMSO/12 N HCl oxidizes tryptophan, methionine, and cysteine. The rate of oxidation of tryptophan is appreciably slower with DMSO/10 N HCl (Analar S. G. 1.16, 31.5% or 10 N). It has also been found [S. H. Lipton and C. E. Bodwell, *J. Agric. Food Chem.* **24**, 26 (1976)] that with DMSO/6 N HCl (30 min, 50°) methionine is completely oxidized to the sulfoxide, while tryptophan is scarcely affected. On the other hand, DMSO/0.5 N HCl oxidizes cysteine (to cystine), but not methionine or tryptophan. The retarding effect of water in the oxidation of thiols to disulfides by the DMSO/HCl system has also been documented by O. G. Lowe [*J. Org. Chem.* **40**, 2096 (1975)]. On the basis of these results, it appears that it could be possible to achieve preferential modification of tryptophan, methionine or cysteine by using HCl of different concentrations.

AMINO ACID COMPOSITION OF LYSOZYME AND ITS MODIFICATION
PRODUCTS OBTAINED BY TREATMENT WITH DMSO/HCl[a]

Amino acid	Native	Treated with DMSO/HCl	Treated with DMSO/HCl and then with IPA[b]
Cysteic acid	0.3 (0)	0.2 (0)	0.3 (0)
Aspartic acid	21.9 (21)	21.2 (21)	21.6 (21)
Threonine	6.8 (7)	7.0 (7)	7.3 (7)
Serine	9.2 (10)	9.6 (10)	10.0 (10)
Glutamic acid	5.1 (5)	5.2 (5)	5.7 (5)
Proline	2.2 (2)	2.3 (2)	2.2 (2)
Glycine	12.5 (12)	11.9 (12)	13.4 (12)
Alanine	11.9 (12)	11.9 (12)	12.6 (12)
Hemicystine	6.5 (8)	6.8 (8)	7.2 (8)
Valine	4.6 (6)	4.7 (6)	5.1 (6)
Methionine	1.8 (2)	Traces (0)	1.9 (2)
Methionine sulfoxide	0.2 (0)	1.9 (2)	Traces (0)
Isoleucine	5.1 (6)	4.7 (6)	5.2 (6)
Leucine	7.9 (8)	7.7 (8)	7.9 (8)
Tyrosine	3.0 (3)	2.8 (3)	3.1 (3)
Phenylalanine	3.1 (3)	2.7 (3)	3.0 (3)
Lysine	5.8 (6)	5.8 (6)	5.9 (6)
Histidine	0.9 (1)	0.9 (1)	1.0 (1)
Arginine	10.2 (11)	9.8 (11)	10.6 (11)
Tryptophan	5.7 (6)	0 (0)	0 (0)
Oxindolylalanine	0 (0)	6.2 (6)	5.7 (6)

[a] Lysozyme was modified with DMSO/HCl in glacial acetic acid and then the oxidized protein was isolated by gel filtration on a Sephadex G-25 column.
[b] The reaction was performed as for a, and then indolyl-3-propionic acid (IPA) was added to the reaction mixture to reduce the methionine sulfoxide residues.

with 3 N p-toluenesulfonic acid[29,30] of the product indicated that the 6 tryptophan residues of the protein were completely oxidized. Methionine was simultaneously oxidized to its sulfoxide, while all other amino acids, including cystine, tyrosine and histidine, were found unchanged.

In a separate experiment, the modification of lysozyme with DMSO/ HCl was followed, after 30 min at room temperature, by addition of an excess (40 mg; 100 equiv) of indolyl-3-propionic acid (IPA) to the reaction mixture in order to reduce the methionine sulfoxide residues of the oxidized protein. After an additional 15 min at room temperature, the solution was diluted with water and extracted several times with ethyl acetate to remove IPA and its oxidation products. The solution was then concentrated under vacuum at 37° and then the modified protein isolated by gel filtration.

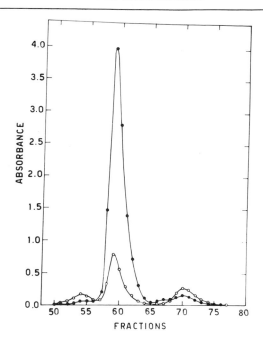

Fɪɢ. 2. Chromatography of the pentapeptide Phe-Val-Gln-Trp-Leu (4.8 mg) treated with DMSO/HCl on a Sephadex G-25 column (3 × 54 cm) equilibrated and eluted with 10% acetic acid at a flow rate of 15 ml/hr. Fractions of 4 ml were collected. ●——●, Absorbance at 255 nm; ○——○, absorbance at 280 nm.

Amino acid analysis of the oxidized protein showed that using the subsequent treatment with IPA the methionine residues had been effectively reduced back to methionine, the DMSO-oxidation being then specific for the 6 tryptophan residues of lysozyme (see the table).

The modified protein (OIA-lysozyme) was eluted as a single peak from a Sephadex G-25 column and no other peaks of peptide material were detected. In order to show more accurately that the oxidized protein was a single molecular species, the sample of OIA-lysozyme was subjected to ion-exchange chromatography on a column of carboxymethyl cellulose, equilibrated with ammonium acetate buffer at pH 6.5 and eluted with a linear gradient of the same buffer. As shown in Fig. 3, OIA-lysozyme is eluted as a single peak at about 0.35 M concentration, whereas native lysozyme is eluted at 0.5 M concentration. The shape of the peak shows some tailing, which could indicate a small degree of heterogeneity.

The ultraviolet spectrum of OIA-lysozyme (λ_{max} 250 nm and shoulder near 280 nm) was consistent with conversion of the indole to the oxin-

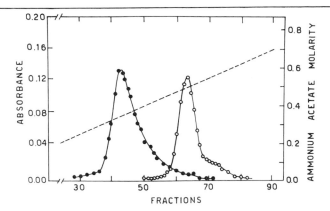

Fig. 3. Elution pattern of lysozyme (○——○) (3.5 mg) and of OIA-lysozyme (●——●) (3.1 mg) adsorbed on a carboxymethylcellulose column (1.6 × 12 cm) equilibrated with 0.1 M ammonium acetate buffer, pH 6.6, and eluted with a linear gradient (- - -) of the same buffer at a flow rate of 40 ml/hr. Fractions of 4 ml were collected and lysozyme was located in the effluent by absorption measurements at 280 nm and OIA-lysozyme at 255 nm.

dole chromophore. In addition, OIA-lysozyme does not show any tryptophan fluorescence upon excitation at 280 nm, as would be expected for a complete modification of the indole fluorophores of the protein.[12] Only a little tyrosine fluorescence is evident, since the oxidized enzyme shows an emission maximum near 304 nm. As expected from the essential role of tryptophan residues in the biological function of lysozyme,[44] OIA-lysozyme was devoid of lytic activity toward *Micrococcus lysodeikticus* cell walls.

Comments

The oxidation with DMSO/HCl provides a useful procedure for the modification of tryptophan to oxindolylalanine in peptides and proteins. A disadvantage of the method lies in the fact that the strongly acid conditions employed seemingly limit its applicability to those tryptophan peptides or proteins that can undergo reversible denaturation in acid solutions. However, the method has several particularly valuable features that deserve comment. In the first place, the high selectivity for tryptophan in peptides and proteins lacking cysteine residues is remarkable. Procedures that allow reversible blocking of the SH-groups, if present, are available and in particular their transformation to mixed disulfides appear appropriate.[40] Methionine is the only other amino acid affected

[40] G. L. Kenyon and T. W. Bruice this volume [40].

by the DMSO/HCl treatment, being oxidized to the sulfoxide. However, it has been clearly shown in the case of the modification of lysozyme that subsequent treatment with an excess of a 3-substituted indole such as IPA brings about quantitative reconversion of the sulfoxide to the thioether.

Incidentally, the oxidation of tryptophan to oxindolylalanine by methionine sulfoxide in concentrated HCl would indicate that this reaction represents a pathway of degradation of tryptophan during acid hydrolysis with HCl. It is known that the presence of oxygen in the hydrolysis mixture leads to partial oxidation of methionine to the sulfoxide.[41] In addition, since oxidation of tryptophan by DMSO does not occur after substitution of HCl by sulfuric or p-toluenesulfonic acid, a sound basis is given for the better recovery of tryptophan when proteins are hydrolyzed with p-toluenesulfonic,[29,30] methanesulfonic,[42] or mercaptoethanesulfonic acid.[43]

The "clean" oxidation reaction of tryptophan derivatives by DMSO/HCl contrasts with the usual complex mixture of products obtained when indoles are reacted with other electrophilic reagents or, in particular, other halogenating agents.[27] In fact, many of the existing methods for tryptophan modification are often accompanied by side reactions leading to the formation of multiple, poorly characterized products.[4]

It is pertinent to recall that NBS, although extensively used,[2-4] is an extremely reactive reagent which can bring about modification of other amino acid residues besides tryptophan, and the "ultimate" modification product of the indole nucleus has not as yet been clearly defined. Although the dual effect of NBS in bringing about pyrrole ring oxidation and benzene ring bromination (in the 5-position) was clearly demonstrated in early studies on the reaction of NBS with tryptophan,[24] oxindolylalanine has been postulated by later authors as the single end product of the reaction.[44-48] In actual fact, reaction of the free amino acid

[41] Recently it has been demonstrated that oxidation of tryptophan to oxindolylalanine also occurs on heating the amino acid in the presence of cystine in 6 N HCl [T. Nakai and T. Ohta, *Biochim. Biophys. Acta* **420**, 258 (1976)].

[42] R. J. Simpson, M. R. Neuberger, and T.-Y. Liu, *J. Biol. Chem.* **251**, 1936 (1976).

[43] B. Penke, R. Ferenczi, and K. Kovacs, *Anal. Biochem.* **60**, 45 (1974).

[44] K. Hayashi, T. Imoto, G. Funatsu, and M. Funatsu, *J. Biochem.* (*Tokyo*) **58**, 227 (1965).

[45] A. Holmgren, *J. Biol. Chem.* **248**, 4106 (1973).

[46] W. A. Frazier, R. A. Hogue-Angeletti, R. Sherman, and R. A. Bradshaw, *Biochemistry* **12**, 3281 (1973).

[47] D. J. O'Hern, P. R. Pal, and Y. P. Myer, *Biochemistry* **14**, 382 (1975).

[48] R. L. Heinrikson and K. J. Kramer, *in* "Progress in Bioorganic Chemistry" (E. T. Kaiser, and F. J. Kezdy, eds.), p. 141. Wiley, New York, 1974.

with one equivalent of NBS in 50% acetic acid gives a mixture of oxindolylalanine, dioxindolylalanine, 5-bromooxindolylalanine, and unreacted tryptophan.[15,17]

Tryptophan contains the largest aromatic side chain of all the protein amino acids, and its role is presumably limited to structural interactions such as hydrophobic bonds. Conversion of the indole nucleus into the more hydrophilic oxindole would alter significantly the strength of such bonds and consequently the role of tryptophan in maintaining the integrity of the tertiary structure of a polypeptide chain could be investigated. Other methods of modification seemingly do not fulfill this scope, since the nitroaryl-group(s) introduced using the Koshland reagent[49] or 2-nitrophenylsulfenyl chloride[50] would be expected to increase the hydrophobicity of the indole nucleus and such bulky groups would also significantly alter the structure by steric effects.

Acknowledgment

This work was supported by the Italian Consiglio Nazionale delle Ricerche.

[49] H. R. Horton and D. E. Koshland, Jr., this series Vol. 25, p. 468.
[50] A. Fontana and E. Scoffone, this series Vol. 25, p. 482.

[43] Interconversion of Methionine and Methionine Sulfoxide

By Walter E. Savige[1] and Angelo Fontana

Conversion of methionine to the corresponding sulfoxide residue in proteins has been used to study the influence of the former on the physicochemical and biological properties of proteins.[2-4] The availability of procedures for regenerating methionine from the sulfoxide permits verification of the specificity of the conversion and also allows reversible protection of the thioether group during chemical modification of other amino acid residues. Because of the nucleophilic character of the thioether function, methionine is reactive toward the majority of electrophilic reagents used for chemically modifying amino acid side chains in proteins.[3,4]

[1] This work has been carried out during a stay of W. E. S. at the Institute of Organic Chemistry, University of Padova, Padova, Italy.
[2] N. P. Neumann, this series Vol. 25, p. 393.
[3] G. E. Means and R. E. Feeney, "Chemical Modification of Proteins," p. 162. Holden-Day, San Francisco, California 1971.
[4] R. L. Heinrikson and K. J. Kramer, *in* "Progress in Bioorganic Chemistry" (E. T. Kaiser, and F. J. Kezdy, eds.), p. 141. Wiley, New York, 1974.

Oxidative conversion of methionine to its sulfoxide has been utilized in peptide synthesis as a means of avoiding side reactions that might otherwise occur during chemical manipulations in the synthesis of a polypeptide.[5,6] Furthermore, methionine oxidation is a potentially useful technique in the preparation of overlapping peptides for the purpose of sequence analysis, since methionine sulfoxide peptide bonds are not cleaved in the cyanogen bromide reaction.[7]

Analysis of Methionine Sulfoxide

The accurate determination of the methionine sulfoxide content of a protein has been generally regarded as a problem.[2,8,9] Reduction of the sulfoxide to methionine can occur during hydrolysis with hot 6 N HCl in the absence of air,[10] whereas the presence of oxygen in the hydrolysate can lead to partial oxidation of methionine to sulfoxide. Somewhat inconsistent results have been reported as to the extent of reduction of methionine sulfoxide during hydrolysis of proteins with 6 N HCl.[9,11] The interaction between sulfoxides and haloacids has been studied in detail, and current theories invoke a halosulfonium ion intermediate which can act as a halogenating agent toward a nucleophile with concomitant formation of sulfide.[12-14] The overall reaction is seemingly dependent upon the availability in the reaction mixture of a suitable acceptor of positive halogen such as tryptophan.[15-18]

Because of the relative stability of the sulfoxide function to alkali, basic hydrolysis is usually employed in methods for determination of the

[5] B. Gutte and R. B. Merrifield, *J. Biol. Chem.* **246**, 1922 (1971).

[6] K. Hofmann, W. Haas, M. J. Smithers, and G. Zanetti, *J. Amer. Chem. Soc.* **87**, 631 (1965); see also F. M. Finn and K. Hofmann *in* "The Proteins," 3rd ed. (H. Neurath and R. L. Hill, eds.), Vol. 2, p. 105. Academic Press, New York, 1976.

[7] E. Gross, this series Vol. 11, p. 238.

[8] N. P. Neumann, this series Vol. 11, p. 487.

[9] H. T. Keutmann and J. T. Potts, Jr., *Anal. Biochem.* **29**, 175 (1969) and references therein.

[10] W. J. Ray, Jr., and D. E. Koshland, Jr., *J. Biol. Chem.* **237**, 2493 (1962).

[11] M. S. Doscher and C. H. W. Hirs, *Biochemistry* **6**, 304 (1967).

[12] G. Modena, *Int. J. Sulfur Chem.*, *Part C* **7**, 95 (1972), and references cited therein.

[13] G. Scorrano, *Acc. Chem. Res.* **6**, 132 (1973), and references cited therein.

[14] R. H. Rynbrandt, *Tetrahedron Lett.* 3553 (1971).

[15] W. E. Savige and A Fontana, *J. Chem. Soc., Chem. Commun.* **1976**, 599 (1976).

[16] W. E. Savige and A. Fontana, "Peptides 1976" (A. Loffet, ed.), p. 135. Presse Universitaire, Brussels.

[17] W. E. Savige and A. Fontana, in preparation (1977).

[18] See also A. Fontana and W. E. Savige, this volume [42].

methionine sulfoxide content of a protein. Nevertheless, low recoveries of methionine sulfoxide have sometimes been reported.[2,8]

Indirect methods exploit the fact that methionine sulfoxide residues are resistant to carboxymethylation. After exhaustive alkylation, the protein sample is oxidized using performic acid, which converts methionine sulfoxide to the acid-stable methionine sulfone, which, after acid hydrolysis, is determined by automatic amino acid analysis.[19,20]

Quantitative determination of both methionine and methionine sulfoxide residues in proteins can be achieved by treating the protein with cyanogen bromide, which converts methionine to homoserine, but does not modify methionine sulfoxide.[21] During acid hydrolysis in the presence of mercaptoethanol[9] or dithioerythritol[21] of the cyanogen bromide-treated protein, methionine sulfoxide is reduced back and determined as methionine, while the amount of homoserine (and its lactone) corresponds to that of the unoxidized methionine in the original sample. Alternatively, the amount of methyl thiocyanate, formed upon reaction of cyanogen bromide with methionine residues, can be quantitated by gas chromatography.[22,23]

Both methionine and the sulfoxide are stable to hydrolysis in evacuated sealed tubes with hot 3 N p-toluenesulfonic acid.[15,17,24–26] This is consistent with earlier findings on the stability of sulfoxides to treatment with sulfuric acid.[27,28] Similarly, methanesulfonic acid can be used.[29]

[19] N. P. Neumann, S. Moore, and W. H. Stein, *Biochemistry* 1, 68 (1962).

[20] Methionine sulfoxide is eluted just before aspartic acid and methionine sulfone between aspartic acid and threonine on the amino acid analyzer using the rapid elution system (pH 3.20, 4.10 sodium citrate buffers; Jeol analyzer Model JLC-6AH).

[21] Y. Schechter, Y. Burstein, and A. Patchornik, *Biochemistry* 14, 4497 (1975).

[22] A. S. Inglis and P. Edman, *Anal. Biochem.* 37, 73 (1970).

[23] A. Varadi, A. Patthy, and L. Gràf, *Int. Congr. Biochem. 10th, Abstr. Commun.* p. 682. (1976).

[24] Reagent grade p-toluenesulfonic acid can be used. If additional purification of the acid in needed, it is important that the sample be completely free from HCl contamination. Traces of haloacid will reduce methionine sulfoxide. A chloride test may be performed using a 1% solution of $AgNO_3$ in 50% HNO_3. The hydrolysis for methionine sulfoxide is best carried out without the presence in the hydrolytic mixture of tryptamine hydrochloride[25] or indolyl-3-propionic acid,[26] used as scavengers for tryptophan destruction.

[25] T.-Y. Liu, this series Vol. 25, p. 44.

[26] L. C. Gruen and P. W. Nicholls, *Anal. Biochem.* 47, 348 (1972).

[27] D. Martin, H.-J. Nidas, and A. Weise, *Chem. Ber.* 102, 23 (1969).

[28] D. Martin and H. G. Hauthal, "Dimethyl Sulfoxide," p. 271. Van Nostrand-Reinhold, Princeton, New Jersey, 1976.

[29] R. J. Simpson, M. R. Neuberger, and T.-Y. Liu, *J. Biol. Chem.* 251, 1936 (1976).

Mercaptoethanesulfonic acid, on the other hand, can bring about reduction of methionine sulfoxide to methionine.[30] In conclusion, the use of sulfonic acids as hydrolytic agents should permit the simultaneous estimation of both methionine and its sulfoxide by routine automatic amino acid analysis.

Oxidation of Methionine to the Sulfoxide

Oxidation of methionine residues in peptides to sulfoxide can occur under mild conditions, for example during exposure of dilute peptide solutions to atmospheric oxygen.[31,32] Such oxidation has been often observed during the isolation[33-35] and synthesis[36] of methionine-containing peptides. Native proteins are more resistant to oxidation.

A number of chemical reagents have been used to oxidize methionine in peptides and proteins. Hydrogen peroxide, the most commonly used oxidant, is fairly specific for methionine under acidic conditions. However, under neutral or alkaline conditions, the reagent shows poor selectivity, several other amino acid residues being modified. Dye-sensitized photooxidation also has been used to effect selective modification of methionine in proteins.[37] Compounds containing positive halogen, such as N-bromosuccinimide,[38] 2,4,6-tribromo-4-methylcyclohexadienone,[39] BNPS-skatole,[40-42] N-chlorosuccinimide,[43] Chloramine-T,[43,44] t-butyl hypochlorite,[45] trichloromethanesulfonyl chloride,[46] and others[44] have been used. In some instances, overoxidation to methionine sulfone and other

[30] B. Penke, R. Ferenczi, and K. Kovacs, Anal. Biochem. 60, 45 (1974).
[31] J. I. Harris, this series Vol 11, p. 390.
[32] A. N. Glazer, in "The Proteins," 3rd Ed. (H. Neurath and R. L. Hill, eds.), Vol. 2, p. 2. Academic Press, New York, 1976.
[33] H. B. F. Dixon, Biochim. Biophys. Acta 37, 38 (1960).
[34] T.-B. Lo, J. S. Dixon, and C. H. Li, Biochim. Biophys. Acta 53, 584 (1961).
[35] J. I. Harris and P. Roos, Biochem. J. 71, 434 (1959).
[33] K. Hofmann, F. M. Finn, M. Linetti, J. Montibeller, and G. Zanetti, J. Am. Chem. Soc. 88, 3633 (1966).
[37] E. W. Westhead, this series Vol. 25, p. 401.
[38] B. Witkop, Adv. Protein Chem. 16, 221 (1961).
[39] Y. Burstein and A. Patchornik, Biochemistry 11, 4641 (1972).
[40] Abbreviations used: DMSO, dimethyl sulfoxide; BNPS-skatole, 2-(2-nitrophenyl-sulfenyl)-3-methyl-3-bromoindolenine.
[41] G. S. Omenn, A. Fontana, and C. B. Anfinsen, J. Biol. Chem. 245, 1895 (1970).
[42] A. Fontana, this series Vol. 25, p. 419.
[43] Y. Schechter, Y. Burstein, and A. Patchornik, Biochemistry 14, 4497 (1975).
[44] N. M. Alexander, J. Biol. Chem. 249, 1946 (1974).
[45] C. R. Johnson, and D. McCants, J. Am. Chem. Soc. 87, 1109 (1965).
[46] R. T. Taylor, J. B. Vatz, and R. Lumry, Biochemistry 12, 2933 (1975).

side reactions have been observed. 1-Chlorobenzotriazole,[47] iodobenzene dichloride[48] in aqueous pyridine, bromide complexes[49] of pyridine, quinoline, and in particular 1,4-diazobicyclo[2.2.2]octane in aqueous acetic acid are some other reagents which have been used to effect the selective and high-yield conversion of model thioethers to sulfoxides, and may be useful in protein chemistry.

The transfer of oxygen from a sulfoxide to a sulfide has been reported to be catalyzed by haloacids, whereas little or no catalytic effect has been shown by sulfuric and other acids.[12-14,50,51] The reaction may be written as a formal oxygen transfer thus:

$$R_2'S + R_2'SO \underset{}{\overset{HCl}{\rightleftarrows}} R_2'SO + R_2S$$

If the R's are sufficiently alike, equilibrium can be reached from both sides.[12,52] However, by using an excess of dimethyl sulfoxide (DMSO) the reaction can be shifted toward the formation of methionine sulfoxide and dimethyl sulfide.[15-17,51]

The DMSO/HCl treatment of peptides and proteins can simultaneously oxidize tryptophan, if present, to oxindolylalanine.[15,18] Cysteine is also oxidized to cystine.[15-17,53-56] Other potentially susceptible amino acids, such as tyrosine and histidine, are not affected. The advantages of the method include cheapness and easy availability of reagents, absence of overoxidation to methionine sulfone, and easy isolation of the modified peptide or protein by evaporation under vacuum of the reagents. However, the strongly acidic reaction conditions appear to limit the use of the procedure to methionine-containing peptides for which conformational aspects are not relevant.

[47] R. Harville, and S. F. Reed, Jr., J. Org. Chem. 33, 3976 (1968).
[48] G. Barbieri, M. Cinquini, S. Colonna, and F. Montanari, J. Chem. Soc. C, 659 (1969).
[49] S. Oae, Y. Ohnishi, S. Kozuka, and W. Pagaki, Bull. Chem. Soc. Jpn. 39, 364 (1966).
[50] C. M. Hull and T. W. Bargar, J. Org. Chem. 40, 3152 (1975).
[51] S. H. Lipton and C. E. Bodwell, J. Agric. Food Chem. 24, 26 (1976).
[52] J. C. Paul, S. C. Martin, and E. F. Perozzi, J. Am. Chem. Soc. 93, 6674 (1971).
[53] O. G. Lowe, J. Org. Chem. 40, 2096 (1975).
[54] Cysteine (10 mg) can be quantitatively oxidized to cystine upon incubation in a mixture of DMSO (0.05 ml), 12 N HCl (0.2 ml), and glacial acetic acid (0.2 ml) in 30 min at room temperature. Under much more severe conditions oxidation of cystine to cysteic acid has been observed. Hydrolyses of proteins with 6 N HCl in the presence of DMSO at 110° has been reported[55] to oxidize cysteine and cystine to cysteic acid. Analogously[56], simple thiols and disulfides are oxidized under similar conditions to sulfonic acids.
[55] R. L. Spencer and F. Wold, Anal. Biochem. 32, 185 (1969).
[56] O. G. Lowe, J. Org. Chem. 41, 2061 (1976).

Oxidation of Ribonuclease.[17] In a well-ventilated hood, ribonuclease (7 mg) is dissolved by stirring in a mixture of DMSO (10 μl), 12 N HCl (50 μl), and glacial acetic acid (100 μl). After 15 min at room temperature, the solution is diluted with water, concentrated under vacuum at 25°, and directly applied to a column (1.85 × 35 cm) of Sephadex G-25 equilibrated and eluted with water. The protein is eluted as a single peak, and the corresponding fractions are combined and lyophilized. The yield of oxidized ribonuclease is 5 mg. Amino acid analysis of an acid hydrolyzate with 3 N p-toluenesulfonic acid gives an experimental figure of 4.2 (theory 4) residues of methionine sulfoxide per mole of protein. Methionine sulfone is not detected in the chromatogram. All other amino acids, including cystine, tyrosine, and histidine, are recovered in agreement with theory.

Reduction of Methionine Sulfoxide

Reduction of methionine sulfoxide in peptides and proteins has been usually achieved by treatment in aqueous solution with a thiol such as cysteine, mercaptoacetic acid, mercaptoethanol or dithiothreitol.[2] Incubation at 37° in 0.2 M mercaptoacetic acid has been found to be an effective procedure.[31,32] Mercaptoethanol[9] or dithiothreitol[43] has been included in the usual 6 N HCl solution to effect reduction of methionine sulfoxide during hydrolysis. In view of the reversible redox reaction between sulfides and sulfoxides discussed above,[15] treatment of a methionine sulfoxide-containing peptide with an excess of dimethyl sulfide in the presence of concentrated HCl should be a useful procedure of reduction. In fact, treatment of methionine sulfoxide (5 mg) in 12 N HCl (0.9 ml) with dimethyl sulfide (0.1 ml) for 5 min at 24° gave methionine in 99% yield. Dimethyl sulfide was simultaneously oxidized to DMSO.[51]

It has been shown that treatment of a 1:1 mixture of methionine sulfoxide and tryptophan in concentrated HCl leads to the reduction of the sulfoxide and concomitant oxidation of tryptophan to oxindolylalanine.[16-18] On this basis, it should be possible to reduce methionine sulfoxide with HCl in the presence of a 3-substituted indole as acceptor of the positive halogen released from the halosulfonium ion intermediate.[12] The procedure has been effectively tested with a lysozyme derivative in which both tryptophan and methionine residues were oxidized.[15-18] The oxidized protein (10 mg) was treated with an excess (40 mg; 100 equiv) of indolyl-3-propionic acid in a mixture of glacial acetic acid (200 μl) and 12 N HCl (100 μl) for 10 min at room temperature. The protein was isolated by gel filtration. Amino acid analysis of an acid hydrolysate of the protein sample with 3 N p-toluenesulfonic acid gave 1.9 residues (theory 2) of methionine per mole of protein.[18] The reaction conditions are rather severe and require HCl of high concentrations, but the results obtained with lysozyme[15-18] indicate that peptide bond cleavage does not occur. This technique has an advantage over other available methods in that reduction of methionine sulfoxide residues can be effected without the simultaneous reduction of disulfide bonds.

Acknowledgment

This work was supported by the Italian Consiglio Nazionale delle Ricerche.

[44] Cleavage of the Tryptophanyl Peptide Bond by Dimethyl Sulfoxide–Hydrobromic Acid

By Walter E. Savige[1] and Angelo Fontana

Procedures for effecting specific cleavage of tryptophanyl peptide bonds of proteins have been intensively sought in the last decade.[2-4] The reagents tested include periodate,[5] ozone,[6-8] and in particular brominating agents.[2-4] N-Bromosuccinimide (NBS)[9] was the first reagent proposed for cleavage of tryptophanyl peptide bonds, but various side reactions of this extremely reactive reagent with other amino acid side chains, as well as poor yields in peptide bond fission, hampered practical applications of the reagent.[10-12] 2,4,6-Tribromo-4-methylcyclohexadienone[13] was found to be less reactive than NBS, giving similar yields of cleavage,

[1] This article is based on work carried out during a stay of W. E. S. at the Institute of Organic Chemistry, University of Padova, Padova, Italy.

[2] B. Witkop, Adv. Protein Chem. 16, 221 (1961).

[3] T. F. Spande, B. Witkop, Y. Degani, and A. Patchornik, Adv. Protein Chem. 24, 97 (1970).

[4] A. Fontana and C. Toniolo, Fortschr. Chem. Org. Naturst. 33, 309 (1976).

[5] M. Z. Atassi, Arch. Biochem. Biophys. 120, 56 (1967).

[6] A. Previero, M.-A. Coletti-Previero, and P. Jollès, Biochem. Biophys. Res. Commun. 22, 17 (1966).

[7] A. Previero, M.-A. Coletti-Previero, and P. Jollès, Biochim. Biophys. Acta 124, 400 (1966).

[8] F. M. Veronese, E. Boccù, C. A. Benassi, and E. Scoffone, Z. Naturforsch. Teil B, 24, 294 (1969).

[9] Abbreviations used: NBS, N-bromosuccinimide; DMSO, dimethyl sulfoxide; OIA, oxindolylalanine (2-hydroxytryptophan); DIA, dioxindolylalanine; BNPS-skatole, 2-(2-nitrophenylsulfenyl)-3-methyl-3-bromoindolenine. All amino acids and amino acid derivatives have the L configuration.

[10] A. Patchornik, W. B. Lawson, E. Gross, and B. Witkop, J. Am. Chem. Soc. 82, 5923 (1960).

[11] L. K. Ramachandran and B. Witkop, this series Vol. 11, 283.

[12] R. L. Heinrikson and K. J. Kramer, in "Progress in Bioorganic Chemistry" (E. T. Kaiser, and F. J. Kezdy, eds.), p. 141. Wiley, New York, 1974.

[13] Y. Burstein and A. Patchornik, Biochemistry 11, 4641 (1972).

but extensive modification of tyrosine, histidine, and the sulfur amino acids was observed. Positive halogen reagents such as ICl, iodide oxidized with Chloramine-T, N-iodosuccinimide and active iodine generated with H_2O_2, iodide, and a peroxidase also give moderate yields of cleavage.[14,15]

BNPS-skatole[16-18] was found to be a much more selective agent than those previously used and the cleavage yields of tryptophanyl peptide bonds so far obtained with several peptides and proteins (50–70%) indicate that this reagent is of practical usefulness for fragmentation of proteins for sequence studies, approaching in its utility the cyanogen bromide reaction.

A mixture of dimethyl sulfoxide (DMSO) and concentrated aqueous HBr has been recently used as a brominating agent to effect selective cleavage of tryptophanyl peptide bonds in peptides and proteins.[19,20]

When horse heart cytochrome c was treated with DMSO and concentrated aqueous HCl in glacial acetic acid following a general procedure used to modify tryptophan peptides and proteins[20,21] oxidation of the single tryptophan to an oxindolylalanine residue and of the two methionine residues to their sulfoxides occurred. Simultaneous partial cleavage of the heme group was also observed. When the protein, on the other hand, was treated under analogous conditions by using HBr instead of HCl, selective peptide bond fission occurred at the level of the tryptophan residue in position 59 of the polypeptide chain. Quantitative cleavage of the heme group also occurred. The results, including selectivity and yields of cleavage, were comparable with those already obtained in the fragmentation of the protein using BNPS-skatole.[18]

It was also found that oxindolylalanine-peptides, and in particular the oxindolylalanine derivative of cytochrome c obtained by treatment with DMSO/HCl, could be cleaved by a subsequent treatment with DMSO/HBr, indicating that the oxindole function is an intermediate in the cleavage reaction.[19,20]

[14] N. M. Alexander, *Biochem. Biophys. Res. Commun.* **57**, 614 (1973).

[15] N. M. Alexander, *J. Biol. Chem.* **249**, 1946 (1974).

[16] G. S. Omenn, A. Fontana, and C. B. Anfinsen, *J. Biol. Chem.* **245**, 1895 (1970).

[17] A. Fontana, this series Vol. 25, p. 419.

[18] A Fontana, C. Vita, and C. Toniolo, *FEBS Lett.* **32**, 139 (1973).

[19] A. Fontana, W. E. Savige, C. Vita, and M. Zambonin, *Proc. Int. Congr. Biochem. 10th Hamburg, 25–31 July, Abstr. Commun.* 04-3-403 (1976).

[20] W. E. Savige and A. Fontana, in preparation (1977).

[21] W. E. Savige and A. Fontana, this volume [42].

[22] R. L. Hinman and C. P. Bauman, *J. Org. Chem.* **29**, 1206 (1964).

[23] N. M. Green and B. Witkop, *Trans. N. Y. Acad. Sci. Ser. II*, **26**, 659 (1964).

The procedure recommended to obtain optimum yield of cleavage involves an initial treatment of the protein with DMSO/HCl, followed by later addition of DMSO/HBr. The procedure is specific for tryptophan, similar to the use of BNPS-skatole.[16] Methionine is also oxidized to the sulfoxide and cysteine is converted to cystine. In practice, it has been found convenient to include in the reaction mixture a little phenol as a scavenger for tyrosine modification.

Free tryptophan on treatment with DMSO/HCl/HBr gives a mixture of dioxindolylalanine stereoisomers and the derived spirolactone hydrobromide.[20] A feasible cleavage mechanism, based on the findings so far obtained, is presented in Scheme 1. It is proposed that oxidation of tryptophanyl residues of a peptide or a protein to oxindolylalanine is followed by bromination in the 3-position. Nucleophilic attack by the

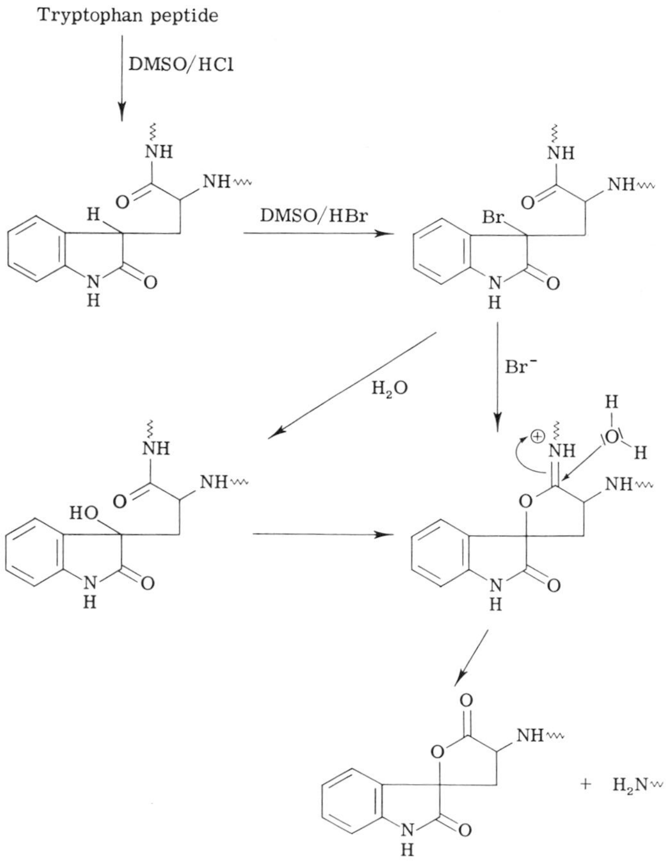

SCHEME 1

carbonyl oxygen of the former tryptophanyl moiety on the γ-carbon bearing the bromine should then bring about the release of bromide and the formation of the iminolactone of dioxindolylalanine. The iminolactone should undergo ready hydrolysis to N-acyldioxindolylalanine lactone, with liberation of the aminoacyl peptide following tryptophan in the peptide chain. Accordingly, a residue of tryptophan which occupied an endo position is converted to dioxindolylalanine lactone (in equilibrium with the open form of the amino acid), which becomes the COOH-terminal amino acid of a peptide fragment.

The hydrolysis of 3-halo-oxindoles in media containing small amounts of water to dioxindoles is well documented in the literature.[22-24] This reaction pathway should apply also for a 3-bromooxindolylalanine peptide, which can be converted not only to the iminolactone, but also to the dioxindolylalanine peptide, without cleavage.

Selective (or preferential) splitting of the peptide bond can be obtained in dioxindolylalanine peptides when heated with dilute acids. The reaction could presumably occur via a carbonium ion intermediate leading to the intermediate iminolactone. The anchimeric assistance of a γ-hydroxyl group in the acid-catalyzed hydrolysis of aliphatic amides is well known.[25-28] A relevant example is the toxic cyclic peptide phalloidin, in which a tertiary γ-hydroxyl group similarly assists easy and selective peptide bond splitting when the peptide is incubated under mildly acidic conditions.[28]

Similar neighboring group effects occur in the chemical cleavage of tryptophan, methionine, and tyrosine peptide bonds.[3] In each case the cleavage of the peptide bond is preceded by the formation of an iminolactone.[29] In addition, it is worthy of mention that the formation of a dioxindolylalanine derivative without cleavage parallels findings with cyanogen bromide cleavage of methionine peptides, in which analogous formation of homoserine can occur without cleavage.[30-32]

Ultraviolet absorption spectra of the peptide fragments can reveal the presence of the COOH-terminal dioxindolylalanine residue or of its lactone.[29] In addition, it is worthy of mention that the formation of a tion maximum near 253 nm (ϵ4070) and a maximum near 287 nm (ϵ1220),

[24] J. C. Powers, J. Org. Chem. 31, 2627 (1966).
[25] T. C. Bruice and F. M. Marquardt, J. Am. Chem. Soc. 84, 365 (1962).
[26] B. Capon, Quart. Rev. 18, 45 (1965).
[27] T. Wieland, C. Lamperstorfer, and C. Birr, Makromol. Chem. 92, 277 (1966).
[28] T. Wieland and W. Schön, Justus Liebigs Ann. Chem. 593, 157 (1955).
[29] G. L. Schmir and B. A. Cunningham, J. Am. Chem. Soc. 87, 5692 (1965).
[30] W. A. Schroeder, J. B. Shelton, and J. R. Shelton, Arch. Biochem. Biophys. 130, 551 (1969).
[31] A. S. Inglis and P. Edman, Anal. Biochem. 37, 73 (1970).
[32] F. H. Carpenter and S. M. Shiigi, Biochemistry 13, 5159 (1971).

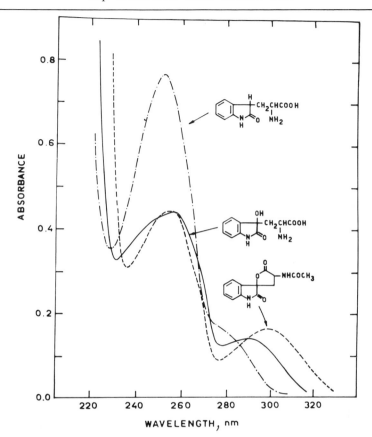

FIG. 1. Absorption spectra of oxindolylalanine (— · —), dioxindolylalanine (——), and *N*-acetyldioxindolylalanine spirolactone (- - -) in 10% acetic acid.

with minima at 230 and 277 nm.[33,34] The corresponding lactone shows similarly a maximum of absorption near 253 nm, but the other maximum occurs at 300 nm. Oxindolylalanine does not show a maximum of absorption in the region 270–300 nm, but a maximum at 250 nm and a shoulder near 280 nm.[35]

Reagents

Dimethyl sulfoxide (DMSO) (Fluka), glacial acetic acid, and 12 *N* HCl are used as supplied by the manufacturers. The aqueous HBr (48%) (Fluka) should be colorless and stored in a dark bottle away

[33] N. M. Green and B. Witkop, *Trans. N. Y. Acad. Sci. Ser. II*, **26**, 659 (1964).
[34] W. E. Savige, *Aust. J. Chem.* **28**, 2275 (1975).
[35] T. F. Spande and B. Witkop, this series Vol. 11, p. 498.

from the light. If necessary it can be redistilled or, to destroy the bromine present, a little phenol can be added to the solution.

Cleavage of Phe-Val-Gln-Trp-Leu

The pentapeptide Phe-Val-Gln-Trp-Leu (5.4 mg; 7 μmol), which corresponds to the 22–26 sequence of glucagon, was dissolved in a preformed mixture of glacial acetic acid (300 μl), 12 N HCl (200 μl), and DMSO (20 μl). After 30 min at room temperature, 48% HBr (60 μl) and additional DMSO (20 μl) were added. The reaction mixture after 30 min was diluted with 0.5 ml of water; the pH was adjusted to $\simeq 2$ with pyridine and diluted to 2 ml. An aliquot of the solution was taken for amino acid analysis. The solution was kept at 60° for 20 hr and then analyzed. The yields of cleavage were estimated from the leucine recovery on the analyzer and were found to be 57% before and 72% after heating at 60°.

The selectivity of cleavage of the tryptophanyl peptide bond was evidenced by the appearance in the chromatogram of the amino acid analyzer of a single peak due to leucine. After heating of the reaction mixture at pH $\simeq 2$, in addition to leucine, peaks for dioxindolylalanine and its lactone were also obtained.[36,37] No traces of other amino acids were evident.

The selective release from the peptide of free dioxindolylalanine (lactone) during mild acid treatment can be best explained by a reaction pathway involving both 5- and 6-membered iminolactone intermediates. These results are related to the mechanistic studies of Carpenter and Shiigi[32] on the cyanogen bromide cleavage of methionine peptides. In particular, analogous selective cleavage both at the amino and carboxyl side has been described by Veronese et al.[38] using mild acid treatment of γ-phenylhomoserine peptides, which are structurally closely related to dioxindolylalanine peptides.

Cleavage of Horse Heart Cytochrome c

Cytochrome c (Sigma, Type III) (17 mg) was dissolved at room temperature, together with 20 mg of phenol, in a preformed mixture of

---

[36] Dioxindolylalanine is eluted on the analyzer (Jeol, Model JLC-6AH) between tyrosine and phenylalanine. The lactone, which is formed under acid conditions is eluted just before the basic amino acids (single-column procedure). The possibility for quantitation of dioxindolylalanine (and its lactone) has not been fully explored, but the method to be used in principle should be similar to that proposed for homoserine in the cyanogen bromide cleavage.[37]

[37] E. Gross and B. Witkop, J. Am. Chem. Soc. 83, 1510 (1961); J. Biol. Chem. 237, 1856 (1962).

[38] E. Margoliash, and A. Schejter, Adv. Protein Chem. 21, 114 (1966).

glacial acetic acid (600 µl), 12 N HCl (300 µl), and DMSO (25 µl). After 30 min, 48% HBr (100 µl) and DMSO (25 µl) were added to the reaction mixture. After an additional 30 min at room temperature, water (2 ml) was added and the solution was extracted several times with ethyl acetate to remove the heme cleaved from the protein and its decomposition products. The aqueous layer was concentrated under vacuum and applied to a column (2 × 144 cm) of Sephadex G-50 SF equilibrated and eluted with 10% formic acid. Three main peaks of peptide material were eluted almost completely separated. The first, preceded by traces of presumably aggregated material (present also in native cytochrome c), consisted of oxidized but uncleaved apocytochrome c, with tryptophan-59 converted to a dioxindolylalanine residue. The second and third peak were fragments 1–59 and 60–104, respectively (Fig. 2).

The yields of cleavage were estimated by spectrophotometric determination of the amount of the COOH-terminal fragment 60–104 obtained, using a molar extinction of 3600 at 275 nm (three tyrosine residues per mole) and found ~60%.

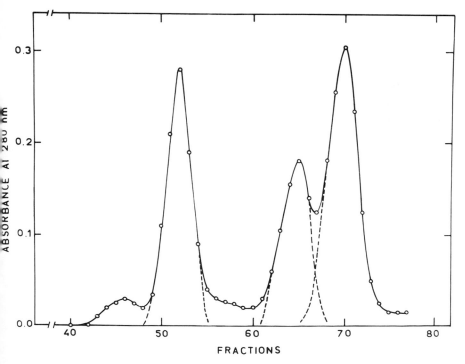

FIG. 2. Gel filtration on Sephadex G-50 SF of the reaction mixture of horse heart cytochrome c treated with hydrochloric and hydrobromic acid in the presence of DMSO. The column (2 × 144 cm) was equilibrated with 10% formic acid and eluted at a flow rate of 10 ml/hr. Fractions of 4 ml were collected and analyzed at 280 nm.

The fractions corresponding to oxidized apocytochrome c (first peak of Fig. 2) were combined, concentrated under vacuum and redissolved in 10% acetic acid (2 ml). The solution was incubated at 60° for 15 hr and then applied to the same Sephadex G-50 SF column as described above. The elution pattern obtained was almost identical to that shown in Fig. 2. An additional 50% of cleavage of the oxidized apocytochrome subjected to the acid treatment at 60° was obtained, accounting for a total cleavage of $\simeq$80%.

The treatment of cytochrome c with DMSO/haloacid results in the complete removal of the heme groups from the protein. Heme is bound to the polypeptide chain through two thioether bridges[38] and the mechanism by which such bonds are cleaved probably involves halogenation by the DMSO/haloacid mixture of the thioether function to a halo-sulfonium intermediate, which can easily then eject the porphyrin ring as a resonance-stabilized carbonium ion. Similar intermediates have been thought to occur in the cleavage of the heme using silver salts,[39] iodine,[40] or 2-nitrophenylsulfenyl chloride.[41,42] During cleavage the two blocked cysteine residues in oxidized apocytochrome c and in fragment 1–59 are apparently converted to a disulfide bridge. No free sulfhydryl groups were detected by the Ellman reagent.[43] Furthermore, the elution position of the peptide fragments from the Sephadex G-50 column (Fig. 2) indicates that the disulfide bonds cannot be intermolecular, and hence must connect cysteine residues 14 and 17.

The identity of the material eluted from the column was determined by amino acid analysis of hydrolysates prepared with 3 N p-toluene-sulfonic acid in evacuated sealed tubes at 110° for 22 hr. The amino acid recovery was in agreement with theory for apocytochrome c and the fragments. The methionine residues in position 65 and 80 of the sequence were fully oxidized to the sulfoxide. Analyses were similar to those already reported for the peptide fragments obtained from cytochrome c using BNPS-skatole.[18]

Figure 3 shows the absorption spectra of oxidized apocytochrome c and the fragments 1–59 and 60–104. Fragment 60–104, containing 3 residues of tyrosine per mole as the ultraviolet absorbing chromophores, shows the typical spectrum of a tyrosine-containing peptide (λ_{max} 275). The spectrum of fragment 1–59 shows a maximum near 250 nm and

[39] K. G. Paul, *Acta Chem. Scand* **5**, 389 (1951).
[40] F. Lederer and J. Tarin, *Eur. J. Biochem.* **20**, 482 (1971).
[41] A. Fontana, *J. Chem. Soc., Chem. Commun.* 976 (1975).
[42] A. Fontana, F. M. Veronese, and E. Boccù, *FEBS Lett.* **32**, 135 (1973).
[43] A. F. S. A. Habeeb, this series Vol. 25, p. 457.

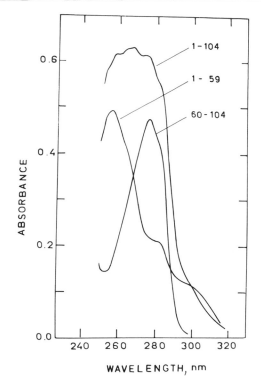

FIG. 3. Absorption spectra of oxidized apocytochrome c (1–104) and fragments 1–59 and 60–104 of cytochrome c in 10% acetic acid.

shoulders near 275-280 nm and 300 nm. The spectrum is consistent with the contribution of the phenolic chromophore of the single tyrosine residue and that of the dioxindolylalanine lactone, which is the COOH-terminal residue of the peptide. The spectrum of apocytochrome 1–104 is interpreted as resulting from the contributions of four tyrosines and one dioxindolylalanine.

Comments

The capability of a sequenator to determine long amino acid sequences (20–60 residues) eliminates the necessity for extensive fragmentation of a protein into a large number of short peptides requiring sequencing by the classical Edman degradation. The protein is best cleaved into a small number of large fragments, which after separation are directly subjected to automated sequence analysis. Only those peptide fragments not sequenced by the sequenator are subsequently analyzed manually.[44]

Among the methods of chemical cleavage of proteins that yield fragments of suitable length for sequenator analysis, fragmentation with cyanogen bromide at methionine residues in the best-known procedure because it is specific and gives high yields of cleavage[45].

Tryptophan is another potential point of attack for attainment of selective fragmentation of proteins. The low content of this amino acid in proteins should permit the achievement of limited fragmentation.

Among the reagents proposed over the years for the fragmentation of tryptophanyl peptide bonds, BNPS-skatole is the best so far available, because of its selectivity and satisfactory yields of cleavage.[17] A disadvantage in the use of BNPS-skatole lies in its high sensitivity to light and moisture. It decomposes on keeping for a few days at room temperature, becoming dark and releasing bromine. In practice, the reagent should be freshly prepared, since recrystallization of an impure sample usually fails to give the desired product, which should be a crystalline yellow compound.[16,17]

The advantages of the DMSO/haloacid procedure here described include cheapness of reagents, ease of performance and also satisfactory yields of cleavage comparable to those obtained with BNPS-skatole. The selectivity of the procedure in modifying, in addition to tryptophan, only cysteine and methionine is another important feature of the reaction. Cysteine is oxidized to cystine and methionine to the sulfoxide.

The cleavage reaction with DMSO/HBr involves formation also of uncleaved dioxindolylalanine-peptide or -protein, which can in turn be cleaved by incubation under mildly acid conditions (10% acetic acid, 60°, 10–15 hr). These are rather mild conditions, but apparently side reactions of cleavage at specially sensitive peptide bonds such as Asp-Pro[46-48] or deamination of asparagine or glutamine residues cannot be ruled out.

In conclusion, the chemical cleavage herein described is a multistep reaction and, although the results so far obtained are satisfactory, it should be possible to improve the overall yields of cleavage by a careful choice of composition of the DMSO/haloacid mixture, time, and

[44] M. A. Hermodson, L. H. Ericsson, K. Titani, H. Neurath, and K. A. Walsh, *Biochemistry* **11**, 4493 (1972).

[45] E. Gross, this series Vol. 11, p. 238.

[46] J. F. Fraser, K. Poulsen, and E. Haber, *Biochemistry* **11**, 4974 (1972).

[47] G. M. Hass, H. Nau, K. Biemann, D. T. Grahn, L. H. Ericsson, and H. Neurath, *Biochemistry* **14**, 1334 (1975).

[48] J. Schultz, this series Vol. 11, p. 255.

temperature of the reaction and, last but not least, by a clear understanding of the detailed mechanism of the reaction.

Acknowledgment

This work was supported by the Italian Consiglio Nazionale delle Ricerche.

[45] Reductive Alkylation of Amino Groups

By GARY E. MEANS

Many simple aldehydes and ketones react rapidly and reversibly with amino groups of proteins, as illustrated in Eq. (1).

$$\text{\textcircled{P}}-NH_2 \;+\; RR'CO \;\rightleftharpoons\; \text{\textcircled{P}}-NH-C\overset{R}{\underset{R'}{<}}OH \;\xrightarrow{-H_2O}\; \text{\textcircled{P}}-N=C\overset{R}{\underset{R'}{<}} \quad (1)$$

Neither the initial adduct or the Schiff base formed upon dehydration is very stable in dilute aqueous solution, but extensive modification of protein amino groups can be obtained by reduction of the Schiff base to a stable secondary amine [Eq. (2)].

$$\text{\textcircled{P}}-N=C\overset{R}{\underset{R'}{<}} \;+\; \tfrac{1}{4}\,NaBH_4 \;\longrightarrow\; \text{\textcircled{P}}-NH-HC\overset{R}{\underset{R'}{<}} \;+\; \tfrac{1}{4}\,NaH_2BO_3 \quad (2)$$

Reductive alkylation of protein amino groups can thus be accomplished with many different aldehydes and ketones under very mild conditions using sodium borohydride as the reductant.[1] Under the conditions described below both α- and ϵ-amino groups are readily modified, but other common protein groups are not affected. Since the modified groups experience only a small change in basicity and, at neutral pH, retain their normal cationic charges, the overall charge and the relative distribution of charged groups in most proteins are not greatly changed by reductive alkylation. Using a large variety of readily available aldehydes and ketones, the size, shape, and hydrophobicity of added substituents can be easily varied.

Reductive Methylation

Using formaldehyde and sodium borohydride, reductive alkylation gives both mono- and dimethyl derivatives of amino groups as illustrated

[1] G. E. Means and R. E. Feeney, *Biochemistry* **7**, 2192 (1968).

in Eq. (3).[1]

$$\textcircled{P}-NH_2 \xrightarrow[\substack{\frac{1}{4} \ NaBH_4}]{H_2CO} \textcircled{P}-NHCH_3 \xrightarrow[\substack{\frac{1}{4} \ NaBH_4}]{H_2CO} \textcircled{P}-N(CH_3)_2 \qquad (3)$$

In moderately alkaline aqueous solutions, formation of the dimethyl derivative rapidly follows formation of the monomethyl derivative, presumably by essentially the same reaction mechanism. Monomethylamino groups are thus normally present in relatively small amounts and, except at very low levels of modification, are a minor product of the reaction. Modification proceeds most readily at approximately pH 9, but it is possible from pH 7 to 10 or above. Below pH 9 modification is limited by the instability of sodium borohydride while higher pH values offer no further advantages and are too severe for many proteins. The reaction requires two equivalents of formaldehyde and 0.5 equiv of sodium borohydride for each amino group, but in practice larger amounts are necessary. Even with very reactive amines, some formaldehyde is reduced directly to methanol and some sodium borohydride is broken down by reaction with solvent, giving rise to small amounts of hydrogen gas.

Both formaldehyde and sodium borohydride alone have been used to modify proteins, but relatively high concentrations of each are normally required. To avoid such reactions the following procedure is designed so that both reagents are used at very low concentrations and so that even then exposure to formaldehyde is very brief.

Procedure.[1] The protein solution is adjusted to approximately pH 9 and cooled in an ice bath. Ammonium ion, or primary and secondary amine buffers should not be present. A drop or two of octanol or another long-chain alcohol can be added to counteract the tendency to foam. Stir slowly and for each milliliter of solution add approximately 0.5 mg of sodium borohydride followed by approximately 2.5 μl of reagent-grade formaldehyde solution (i.e., 37% or 12.1 M formaldehyde) slowly in several aliquots. To avoid high local concentrations and to allow time for the formaldehyde polymerization equilibrium to shift toward more monomer, it may be desirable to use larger volumes of prediluted formaldehyde solution. The reaction is complete a few seconds after each addition. Formaldehyde either reacts with the amino groups and is reduced or is reduced directly to methanol and is not detectable after approximately 4 sec.[2] A slight effervescence is usually observed a few seconds after each addition. A very low concentration of sodium borohydride should remain when the reaction is complete to avoid any possible prob-

[2] F. Galembeck, D. S. Ryan, J. R. Whitaker, and R. E. Feeney, *J. Agric. Food Chem.* **25**, 238 (1976).

lems with excess formaldehyde. The presence of sodium borohydride can be ascertained by a simple colorimetric assay given below.

Using the above procedure for reductive methylation and protein concentrations below 10 mg/ml, usually at least 40%, sometimes 99%, of the amino groups will be modified. Less extensive modification can be easily obtained using smaller amounts of the two reagents. To obtain more extensive modification the given procedure can be repeated several times. Higher concentrations of sodium borohydride and a corresponding increased amount of formaldehyde may also be used, but should usually be avoided as the efficiency of the reaction appears to decrease, since proportional increases in modification are not obtained.[1] With much higher concentrations susceptible disulfide bonds are likely to be reduced or a few labile peptide bonds may be cleaved.[3,4] Sodium borohydride remaining when the modification is complete should be eliminated by brief acidification to pH 5 or below before dialysis or any chromatographic process.

Reductive Alkylation Using Aldehydes or Ketones Other than Formaldehyde

With ketones or aldehydes other than formaldehyde reductive alkylation gives primarily monoalkyl derivatives of the original amino groups according to Eq. (1) and (2). In the reductive alkylation of butylamine using acetaldehyde as the carbonyl component small amounts of the diethyl derivatives are formed only after extensive reaction.[1] Using these generally less reactive carbonyl compounds, which alone do not react irreversibly with proteins, reductive alkylation is best accomplished by reversing the order of addition of reagents as compared to the reaction with formaldehyde. Thus, as in the procedure below, excess carbonyl component is added first, followed by sodium borohydride in a number of small increments sufficient to give the desired extent of modification. In this procedure Schiff base formation repeatedly proceeds to completion, and direct reduction of the carbonyl is minimized.

Procedure.[1] The protein solution, adjusted to approximately pH 9, is cooled in an ice bath and slowly stirred. Acetone is added slowly to avoid high local concentrations to a final concentration of 10% v/v. For each milliliter of solution, add a total of 3–5 mg of sodium borohydride in increments of approximately 0.5 mg at 1-min intervals. After addition

[3] A. Light and N. K. Sinha, *J. Biol. Chem.* **242**, 1358 (1967).
[4] A. M. Crestfield, S. Skupin, S. Moore, and W. H. Stein, *Fed. Proc., Fed. Am. Soc. Exp. Biol.* **19**, 341 (1960).

REDUCTIVE ALKYLATION OF CHICKEN EGG WHITE OVOMUCOID
WITH DIFFERENT CARBONYL COMPOUNDS[a]

Carbonyl compound	Percent amino groups modified
Formaldehyde	100
Butyraldehyde	91
Benzaldehyde	76
Acetone	61
Cyclopentanone	40
Cyclohexanone	72

[a] Unpublished results of K. Fretheim, S. Iwai, and R. E. Feeney using the procedure of Y. Lin, G. E. Means, and R. E. Feeney, *J. Biol. Chem.* **244**, 789 (1969).

of sodium borohydride the protein can be adjusted to another pH and separated from the remaining excess acetone by dialysis or by passage through a column of Sephadex G-25.

As in the procedure for reductive methylation, ammonia and primary amines should not be present during the reaction. As compared to the reductive methylation procedure given above, the reaction with acetone normally gives less extensive modification. Thus in a series of proteins modified by this procedure the extent of modification varied from 73% to 22% of the total amino groups for pancreatic ribonuclease and turkey ovomucoid, respectively.[1]

The preceding procedure can be used for many different aldehydes and ketones with only slight modifications. For those carbonyl compounds with low aqueous solubility corresponding lower concentrations must be used. Reactivities of different carbonyl compounds vary as revealed by different extents of modification using the same conditions. Thus formaldehyde is much more reactive than all others (see the table),[5] and aldehydes are generally more reactive than ketones.

Determination of Sodium Borohydride

In the reductive methylation of proteins, sodium borohydride should be kept in slight excess. The following procedure is convenient for determining its approximate concentration during the modification procedure.

Procedure.[1] A small drop (10–50 μl) of solution to be tested is added to 0.1 ml of ice-cold 0.05% trinitrobenzenesulfonate in 0.2 M NaHCO$_3$

[5] Y. Lin, G. E. Means, and R. E. Feeney, *J. Biol. Chem.* **244**, 789 (1969).

buffer, pH 8.5. The solution is gently mixed and allowed to warm to room temperature. An orange color develops indicating the presence of sodium borohydride-like compounds. The solution can be acidified after the color reaches full intensity ($\sim$15 min) by addition of 0.25 ml of 1 M HCl and quantitated by determining the absorption at its maximum of 480 nm as compared to a standard. The assay is linear up to approximately 0.1 M sodium borohydride.

Properties of Reductively Alkylated Proteins

In contrast to many other procedures for the modification of protein amino groups, reductive alkylation frequently has surprisingly little effect on the physical chemical properties or biological activities of proteins. ϵ-Monoalkylamino groups are generally slightly more basic than the corresponding primary amino groups ($\Delta pK_a \sim +0.3$), and dimethyl amino groups are slightly less basic ($\Delta pK_a \simeq -0.6$).[6] At so-called normal physiological pH, the typical alkylated ϵ-amino group will have essentially a unit positive charge, the same as the unmodified primary amino group. Under these conditions amino groups with low pK_a values, for example most α-amino groups and a few ϵ-amino groups, will have slightly altered net charges. The greatest difference will be at pH values between the original pK_a value and that of the modified group. For most amino groups this maximum difference thus occurs above pH 9. Typically no differences are observed upon electrophoresis under commonly used conditions.

Reductive alkylation not only preserves the positive charges of protein amino groups, but, in contrast to other methods for their modifications, does not change the approximate location of those charges. With small carbonyl compounds like formaldehyde or acetaldehyde, reductive alkylation should have little or no effect on the distribution of charged groups and cause a minimal disturbance of existing electrostatic interactions. Alkyl substitution of hydrogen will, however, necessarily increase both the bulk and hydrophobicity of the amino group and reduce its ability to form hydrogen bonds. Studies using ^{13}C-nuclear magnetic resonance spectroscopy of lysine, N^ε-methyllysine and $N^\varepsilon,N^\varepsilon$-dimethyllysine also appear to indicate a progressive decrease in charge density and an increase in anion affinity of the cationic form with increasing methyl substitution.[7]

[6] J. H. Bradbury and L. R. Brown, *Eur. J. Biochem.* **40**, 565 (1973).
[7] C. S. Baxter and P. Byvoet, *Biochem. Biophys. Res. Commun.* **64**, 514 (1975).

$N^\varepsilon,N^\varepsilon$-Dimethyllysine residues in proteins are not cleaved by trypsin at a significant rate.[1,5] The hydrolysis of N^ε-methyllysine ethyl ester by trypsin is 17 times slower than hydrolysis of the unmethylated compound[8] and N^α-acetyl-$N^\varepsilon,N^\varepsilon$-dimethyllysine methyl ester is completely resistant.[9] Turkey ovomucoid, a lysine-type trypsin inhibitor, loses its trypsin inhibitory activity upon reductive methylation.[1] Reductive methylation of trypsin does not reduce its catalytic activity but reduces its susceptibility to autolysis in the absence of calcium ion.[10] Carboxypeptidase B does not catalyze the hydrolysis of carboxyl-terminal $N^\varepsilon,N^\varepsilon$-dimethyllysine residues.[11]

By using different aldehydes and ketones, reductive alkylation can be used to prepare a wide variety of amino group-modified protein derivatives. Using formaldehyde or acetaldehyde the added substituents are small and generally have the least effect on physical chemical properties. With larger and generally more hydrophobic carbonyl compounds greater changes can be expected. Thus reductive alkylation of bovine pancreatic RNase with acetone gives a derivative with slightly altered optical rotatory parameters and a slightly increased isoelectric point as compared to either unmodified or reductively methylated RNase.[1] Reductive alkylation of chicken egg white lysozyme with cyclohexanone or benzaldehyde gives derivatives with significantly lowered solubility.[12]

In many cases catalytic activities or other biological properties of proteins are unaffected or even enhanced by reductive alkylation. The activity of horse liver alcohol dehydrogenase, for example, increases by more than 10-fold and its stability is enhanced after reductive methylation.[13,14] The modified enzyme retains its subunit structure, but appears to experience a slight change in conformation, is partially desensitized to inhibition by high ethanol concentrations and has increased rates of NAD^+ and NADH dissociation. Reductive alkylation of rabbit muscle glycogen phosphorylase b with a homologous series of aldehydes generally increases its stability with relatively little effect on its physical chemical or catalytic properties.[15,16] The modified enzymes are indistinguishable from unmodified phosphorylase b on the basis of specific

[8] L. Benoiton and J. Deneault, *Biochim. Biophys. Acta* **113**, 613 (1966).
[9] M. Gorecki and Y. Shalitin, *Biochem. Biophys. Res. Commun.* **29**, 189 (1967).
[10] R. H. Rice, G. E. Means, and W. D. Brown, *Biochim. Biophys. Acta* in press (1977).
[11] Y. Lin, G. E. Means, and R. E. Feeney, *Anal. Biochem.* **32**, 436 (1969).
[12] K. Fretheim, S. Iwai, and R. E. Feeney, personal communication.
[13] M. Zoltobrock, J. C. Kim, and B. V. Plapp, *Biochemistry* **13**, 440 (1974).
[14] C. S. Tsai, Y.-H. Tsai, G. Lauzon, and S. T. Chen, *Biochemistry* **13**, 440 (1974).
[15] M. A. Shatsky, H. C. Ho, and J. H.-C. Wang, *Biochim. Biophys. Acta* **303**, 298 (1973).
[16] H. C. Ho, E. Wirch, and J. H.-C. Wang, *Biochim. Biophys. Acta* **317**, 462 (1973).

activity, gel electrophoresis, sedimentation in an ultracentrifuge and, like unmodified enzyme, can be crystallized in the presence of AMP and Mg^{2+} or converted to corresponding derivatives of phosphorylase a by phosphorylase kinase and ATP. *Escherichia coli* ribosomal initiation factor IF-3 is functionally active after reductive methylation,[17] ribosomal proteins L7 and L12 retain their elongation factor G-dependent GTPase activity,[18] and proteins S15 and S17 can be reconstituted into fully functional 30 S particles.[19] Activities of whole ribosomes in poly(U)-directed polyphenylalanine synthesis were found to be unaffected by reductive alkylation with formaldehyde, acetone, benzaldehyde, or 3,4,5-trimethoxybenzaldehyde.[20] Chicken egg white lysozyme retains full catalytic activity, and the aromatic region of its proton magnetic resonance spectrum is unchanged after reductive methylation.[9] Bovine α-chymotrypsin,[1] and trypsin,[10] ovine luteinizing hormone,[21] human haptoglobin,[22] serum transferrin,[23] and wheat germ agglutinin,[24] are all fully functional after reductive methylation. In these cases the absence of an effect on biological activities demonstrates that amino groups are not specifically involved in those activities.

Reductive alkylation of bovine pancreatic ribonuclease reduces its catalytic activity to less than 1%, in keeping with other evidence suggesting a primary role for an amino group in its catalytic mechanism.[1] Extensively methylated RNase binds inorganic phosphate and cytidine 3′-phosphate, but more weakly than unmodified RNase.[25] Its two active site histidine residues are carboxymethylated by bromoacetate anion at only slightly altered rates as compared to unmodified RNase. Proton magnetic resonance studies reveal a slight alteration in the titration behavior of these histidines and in their interactions with phosphate ion.[26] Reductively methylated pancreatic RNase is indistinguishable, however, from unmodified RNase by sedimentation in the ultracentrifuge, optical rotatory dispersion or by gel electrophoresis below pH 8.5.

[17] C. L. Pon, M. Friedman, and C. Gualerzi, *Mol. Gen. Genet.* **116**, 192 (1972).
[18] R. Amos and W. Moller, *Eur. J. Biochem.* **44**, 97 (1974).
[19] W. A. Held, B. Ballou, S. Mizushima, and M. Nomura, *J. Biol. Chem.* **249**, 3103 (1974).
[20] G. Moore and R. R. Crichton, *Biochem. J.* **143**, 6607 (1974).
[21] M. Ascoli and D. Puett, *Biochim. Biophys. Acta* **371**, 203 (1974).
[22] M. T. Chiao and A. Bezkorovainy, *Biochim. Biophys. Acta* **263**, 60 (1972).
[23] R. H. Zschodke, M. T. Chiao, and A. Bezkorovainy, *Eur. J. Biochem.* **27**, 145 (1972).
[24] R. H. Rice and M. E. Etzler, *Biochemistry* **14**, 4093 (1975).
[25] G. E. Means and R. E. Feeney, 156th meeting of the American Chemical Society, Biological Chemistry Division Abstract No. 165.
[26] L. R. Brown and J. H. Bradbury, *Eur. J. Biochem.* **54**, 219 (1975).

Radioactive Labeling

Reductive methylation can be used to prepare radioactively labeled protein derivatives and by comparison to procedures for radioiodination should generally have a milder and more predictable effect on most proteins.[27,28] By using [14]C- or ^{3}H-labeled formaldehyde or NaB^3H_4, reductive methylation can be used to prepare either [14]C- or ^{3}H-labeled proteins and is thus convenient for studies requiring a double label; as both isotopes have relatively long half lives, the labeled proteins can be used over a long period of time. The following procedure was used by Rice and Means[27] to label several proteins and small viruses.

Procedure. The virus or protein (0.1 mg) is adjusted to pH 9 with 0.1 M NaOH or 0.2 M sodium borate buffer, pH 9.0, to a total volume of 0.10 ml. The solution is cooled in an ice bath, and 10 μl of 0.04 M ^{14}C-formaldehyde is added and gently mixed ($\sim$4- to 5-fold excess as compared to the number of amino groups), followed after 30 sec by 4 sequential 2-μl aliquots of sodium borohydride solution (5 mg/ml in water). After an additional minute, another 10 μl of sodium borohydride solution is added to ensure complete reduction of the formaldehyde. The solution is then dialyzed to separate the protein from labeled low molecular weight compounds.

Using [^{14}C]formaldehyde at 10 mCi/mmol, labeled virus or protein can be obtained with a specific activity of 5×10^6 cpm/mg.[1] Under the given conditions, RNA is not labeled. Using high specific activity sodium borotritide (6.4 Ci/mmol), Moore and Crichton[20] have used a similar procedure to prepare fully functional *E. coli* ribosomes labeled to a specific activity greater than 10^7 dpm per A_{260} unit, and Ascoli and Puett[21] also using a similar procedure have prepared a fully active derivative of ovine pituitary luteinizing hormone with a specific activity of $\sim$650 μCi/mg.

Analysis of Reductively Methylated Proteins

ϵ-Mono- and dialkyl derivatives of lysine are stable under conditions normally used for the acid hydrolysis of proteins. In principle, they can be identified and quantitated using an amino acid analyzer, but in practice they elute very close to lysine and in most standard systems usually overlap to some extent with it or with the nearby peaks for ammonia or histidine. The separations vary with different resins and are sensitive to slight differences in temperature, pH, and ionic strength and to the

[27] R. H. Rice and G. E. Means, *J. Biol. Chem.* **246**, 831 (1971).
[28] M. Ottesen and B. Svensson, *C. R. Trav. Lab. Carlsberg* **38**, 445 (1971).

presence of organic solvents. Slight alterations of these conditions can frequently be used to shift peak positions and to obtain the desired separation. Small amounts of added isopropyl alcohol, for example, decrease the elution times of $N^\varepsilon,N^\varepsilon$-dimethyl- and N^ε-isopropyllysine relative to lysine using a Technicon Autoanalyzer and the standard pH 3.8 to pH 5.0 citrate linear buffer-gradient system.[1] Paik and Kim,[29] Delange et al.,[30] and Seeley et al.[31] have devised systems that resolve N^ε-methyl, $N^\varepsilon,N^\varepsilon$-dimethyl and $N^\varepsilon,N^\varepsilon,N^\varepsilon$-trimethyllysine from each other and lysine. Using the ninhydrin assay system of Moore and Stein,[32] color values 88% and 82% of those for lysine have been obtained for N^ε-methyllysine and $N^\varepsilon,N^\varepsilon$-dimethyllysine, respectively.[1] Values 91% and 90% of lysine were obtained in the chromatographic system of Seeley et al.[31] As color intensities appear to differ depending on conditions, exact values should be determined for each new system.

N^ε-Methyllysine and $N^\varepsilon,N^\varepsilon$-dimethyllysine can also be separated from each other and most common amino acids by paper chromatography. Delange et al.[33] have described a system based on that of Stewart[34] with a descending eluant of 88% phenol/m-cresol/0.064 M sodium borate, pH 9.3 (190:165:45), on Whatman 3 MM paper dipped in 0.1 M EDTA, pH 7.0, and dried. After 16–18 hr at room temperature, the amino acids can be revealed by spraying with 0.1% solution of ninhydrin in ethanol/collidine (95:5). $N^\varepsilon,N^\varepsilon$-dimethyllysine gives a blue-violet spot similar to arginine. The reported R_f values are: lysine, 0.18; arginine, 0.30; histidine, 0.43; N^ε-methyllysine, 0.44; and $N^\varepsilon,N^\varepsilon$-dimethyllysine, 0.76.

Paper chromatography can also be used to separate and identify N^α,N^α-dimethyl amino acids generated by hydrolysis from the amino termini of reductively methylated proteins or peptides.[35]

The reaction of protein amino groups with trinitrobenzenesulfonate ion (TNBS) using the procedure of Habeeb[36] or as described in an earlier volume of this series by Fields[37] affords a rapid and convenient means to determine the extent of amino group modification by reductive alkylation. Neither secondary or tertiary amines react with TNBS.[38] A

[29] W. K. Paik and S. Kim, *Biochem. Biophys. Res. Commun.* **27**, 479 (1967).
[30] R. J. Delange, A. N. Glazer, and E. L. Smith, *J. Biol. Chem.* **245**, 3325 (1970).
[31] J. H. Seeley, R. Edattel, and L. Benoiton, *J. Chromatogr.* **44**, 618 (1969).
[32] S. Moore and W. H. Stein, *J. Biol. Chem.* **176**, 367 (1948).
[33] R. J. Delange, A. N. Glazer, and E. L. Smith, *J. Biol. Chem.* **244**, 1385 (1969).
[34] I. Stewart, *J. Chromatogr.* **10**, 404 (1963).
[35] V. M. Ingram, *J. Biol. Chem.* **202**, 193 (1953).
[36] A. F. S. A. Habeeb, *Anal. Biochem.* **14**, 328 (1966).
[37] R. Fields, this series Vol. 25, p. 464.
[38] G. E. Means, W. Congdon, and M. Bender, *Biochemistry* **11**, 3564 (1972).

decrease in TNBS-reactive amino groups as compared to unmodified protein thus reflects the extent of amino group modification, both the mono- and dialkylated derivatives. Sodium borohydride also reacts with TNBS under the same conditions to give an orange product, and in so doing interferes with the determination of amino groups. It can be removed from the reaction solution by brief acidification to approximately pH 5 or below for a few minutes prior to the assay.

Discussion

Reductive alkylation affords a highly specific method to modify the amino groups of proteins under relatively mild conditions. The procedure is thus applicable to most proteins except those containing readily reducible components or prosthetic groups like pyridoxal phosphate and rhodopsin, or those which are not stable at the required alkaline pH values. Using the more acid-stable reductant sodium cyanoborohydride, Friedman et al.[39] have obtained extensive modification of proteins using three nitro-substituted benzaldehydes at approximately pH 5.2.

Using formaldehyde as the carbonyl component, amino groups modified by reductive alkylation undergo minimal changes in character, and many proteins appear to experience relatively minor changes in physical-chemical properties. Loss of biological activity after reductive methylation is thus more clearly attributable directly to the modification and, as compared to other methods for the modification of amino groups, is less likely a result of secondary effects on conformation or disturbances of native electrostatic interactions.

Because many proteins largely retain their native physical chemical and biological properties, reductive alkylation can be used to prepare specifically labeled derivatives closely resembling the unmodified proteins. Thus, amino groups of a protein can be characterized by proton and ^{13}C magnetic resonance studies of N^{ε}-methyl groups in the reductively methylated derivative,[6,7] High-specific activity ^3H- and ^{14}C-labeled proteins can be obtained by reaction with ^3H- or ^{14}C-labeled formaldehyde or NaB^3H_4 and used for physiological, metabolic, and various mechanistic studies.[21,26,27] Labeling viruses, ribosomes, and other complex structural units can be used to facilitate the detection and characterization of minor protein components or to assist in reconstitution studies.[19,26]

[39] M. Friedman, L. D. Williams, and M. S. Masri, *Int. J. Pept. Protein Res.* **6**, 183 (1974).

[46] Active-Site-Directed Reagents of Glycolytic Enzymes[1]

By Fred C. Hartman and I. Lucile Norton

Since its introduction in the early 1960's, affinity labeling has become the most versatile chemical method for the characterization of active sites of enzymes and antigen-combining sites of antibodies.[2-4] As application of affinity labeling requires only a knowledge of the structural requirements for binding of the naturally occurring ligand to its macromolecular binding site, so that reactive structural analogs with an affinity for the binding site can be synthesized, logical extensions of the technique include the selective labeling of receptor sites,[5-9] membrane transport systems,[10-13] and ribosomes.[14-18]

In general, the selectivity of an affinity-labeling reagent is a consequence of its structural similarity to substrate, thereby resulting in the formation of a dissociable enzyme–reagent complex comparable to the

[1] Research from the authors' laboratory was sponsored by the Energy Research and Development Administration under contract with the Union Carbide Corporation.
[2] E. Shaw, *Physiol. Rev.* **50,** 244 (1970).
[3] E. Shaw, *in* "The Enzymes," 3rd ed. (P. D. Boyer, ed.), Vol. 1, p. 91. Academic Press, New York, 1970.
[4] S. J. Singer, *Adv. Protein Chem.* **22,** 1 (1967).
[5] C. E. Guthrow, H. Rasmussen, D. J. Brunswick, and B. S. Cooperman, *Proc. Natl. Acad. Sci. U.S.A.* **70,** 3344 (1973)
[6] A. Karlin and D. Cowburn, *Proc. Natl. Acad. Sci. U.S.A.* **70,** 3636 (1973).
[7] A. Ruoho and J. Kyte, *Proc. Natl. Acad. Sci. U.S.A.* **71,** 2352 (1974).
[8] B. E. Haley, *Biochemistry* **14,** 3852 (1975).
[9] M. E. Wolff, D. Feldman, P. Catsoulacos, J. W. Funder, C. Hancock, Y. Amano, and I. S. Edelman, *Biochemistry* **14,** 1750 (1975).
[10] S. O. Nelson, G. I. Glover, and C. W. Magill, *Arch. Biochem. Biophys.* **168,** 483 (1975).
[11] G. Kaczorowski, H. R. Kaback, nad C. Walsh, *Biochemistry* **14,** 3903 (1975).
[12] J. B. Hays, M. L. Sussman, and T. W. Glass, *J. Biol. Chem.* **250,** 8834 (1975).
[13] M. J. Owen and H. P. Voorheis, *Eur. J. Biochem.* **62,** 619 (1976).
[14] C. R. Cantor, M. Pellegrini, and H. Oen, *in* "Ribosomes" (M. Nomura, A. Tissières, and P. Lengyel, eds.) p. 573. Cold Spring Harbor Laboratory, Cold Spring Harbor, N.Y., 1974.
[15] L. Bispink and H. Matthaei, *FEBS Lett.* **37,** 291 (1973).
[16] I. Schwartz, E. Gordon, and J. Ofengand, *Biochemistry* **14,** 2907 (1975).
[17] J. B. Breitmeyer and H. F. Noller, *J. Mol. Biol.* **101,** 297 (1976).
[18] E. Collatz, E. Küchler, G. Stöffler, and A. P. Czernilofsky, *FEBS Lett.* **63,** 283 (1976).

Michaelis complex or a complex with competitive inhibitor. Complex formation results in a high, localized concentration of reagent at the active site, thus increasing the likelihood of modification of a residue within this region as compared with modification of a like residue elsewhere in the protein molecule. In contrast to ideal, general protein reagents which, due to their chemical nature, are selective for a given functional group, affinity labels are selective for a particular type of binding site and in many instances are reactive toward several functional groups found in proteins. The specificity with respect to enzyme is determined by the substratelike features of the reagent, and the specificity with respect to the kind of residue modified is a consequence of which reactive side chain within the active site is in proper juxtaposition to the leaving group of the reagent. An added dimension of specificity can be obtained by the use of chemically inert compounds that are converted to reactive reagents by the target enzyme.[19-23] This approach clearly has exciting implications in the area of drug design, for it raises the possibility of tailoring a reagent to be selective for a single enzyme within the whole living organism.

For the protein chemist attempting to label active-site residues selectively, and thereby to identify them, absolute specificity for a single enzyme is neither essential nor necessarily the most desirable situation. When one considers the time and effort that might be required for the development of a single new active-site-specific reagent, it is rewarding to find that the reagent can be used effectively to probe the active site of several enzymes. In this regard enzymes involved in the intermediary metabolism of carbohydrates are particularly amenable to active-site characterization with similar affinity-labeling reagents, since the substrates of these enzymes are usually phosphate esters. For most of the enzymes in the glycolytic, hexose monophosphate, and reductive pentosephosphate (Calvin cycle) pathways, electrostatic interactions involving the phosphate anion are major contributors to substrate binding. This is reflected in the observation that inorganic phosphate competitively inhibits many of these enzymes and thus there is a reasonable chance with such enzymes that any small phosphate ester containing a chemically reactive substituent will be of some use as an affinity label.

[19] K. Bloch, *Acc. Chem. Res.* **2**, 193 (1969).

[20] R. R. Rando, *Science* **185**, 320 (1974).

[21] K. Horiike, Y. Nishina, Y. Miyake, and T. Yamano, *J. Biochem.* (*Tokyo*) **78**, 57 (1975).

[22] B. Gärtner, P. Hemmerich, and E. A. Zeller, *Eur. J. Biochem.* **63**, 211 (1976).

[23] A. Schonbrunn, R. H. Abeles, C. T. Walsh, S. Ghisla, H. Ogata, and V. Massey, *Biochemistry* **15**, 1798 (1976).

In this article reagents are described that have been used with varying degrees of success to characterize partially the active sites of glycolytic enzymes. It is hoped that a consideration of these reagents will prompt other investigators to design related, and perhaps more sophisticated, affinity labels so that eventually the protein chemist will have an impressive arsenal of reactive phosphate esters from which to choose.

General Comments

Although not necessarily affinity labels in the strict sense, all the reagents shown in the table have found application or are of potential application in the selective labeling of active sites of glycolytic enzymes. Pyridoxal phosphate has proved to be of general utility in the selective labeling of lysyl residues (reaction of the reagent carbonyl with an ε-amino group to form a Schiff base) in the region of binding sites for phosphate.[24] The Schiff bases are unstable but are reduced to secondary amines by borohydride reduction. Most of the glycolytic enzymes are inactivated by pyridoxal phosphate (see the Table). Since the phenomenon is so general, it appears likely that the phosphate group of pyridoxal phosphate is attracted to the site normally occupied by the substrate phosphate group, and condensation with an amino group occurs subsequent to binding. In the case of pyruvate kinase the dependence of the inactivation rate on the pyridoxal phosphate concentration was determined, and rate-saturation kinetics, indicative of reversible complex formation, were observed.[25]

An especially interesting application of pyridoxal phosphate is its use in the selective labeling of the allosteric site of fructose-1,6-bisphosphatase in the presence of substrate (fructose 1,6-bisphosphate) and selective labeling of the enzyme's active site in the presence of allosteric inhibitor (AMP).[24,26,27]

Two recently described reagents, potassium ferrate(III) and potassium hexacyanoferrate(III), are also of special interest because of their potential general usefulness. The oxidizing agent ferrate, which can be considered a reactive analog of phosphate, inactivates phosphorylase b by the preferential oxidation of tyrosyl residues.[28] The reagent has not yet been tested on glycolytic enzymes. Hexacyanoferrate(III), in combina-

[24] G. Colombo and F. Marcus, *Biochemistry* **13**, 3085 (1974).

[25] G. S. Johnson nad W. C. Deal, *J. Biol. Chem.* **245**, 238 (1970).

[26] T. A. Krulwich, M. Enser, and B. L. Horecker, *Arch. Biochem. Biophys.* **132**, 331 (1969).

[27] G. Colombo, E. Hubert, and F. Marcus, *Biochemistry* **11**, 1798 (1972).

[28] Y. M. Lee and W. F. Benisek, *J. Biol. Chem.* **251**, 1553 (1976).

SITE-SPECIFIC REAGENTS FOR GLYCOLYTIC ENZYMES

Reagent	Structure	Enzyme inactivated	Amino acid residue modified
N-Bromoacetyl-D-galactopyrano-sylamine		Hexokinase	Cys[a]
(R,S)-2',3'-Epoxypropy β-D-glucopyranoside		Hexokinase	Cys[b]
1,2-Anhydro-D-mannitol 6-phosphate		Phosphoglucose isomerase	Glu[c]

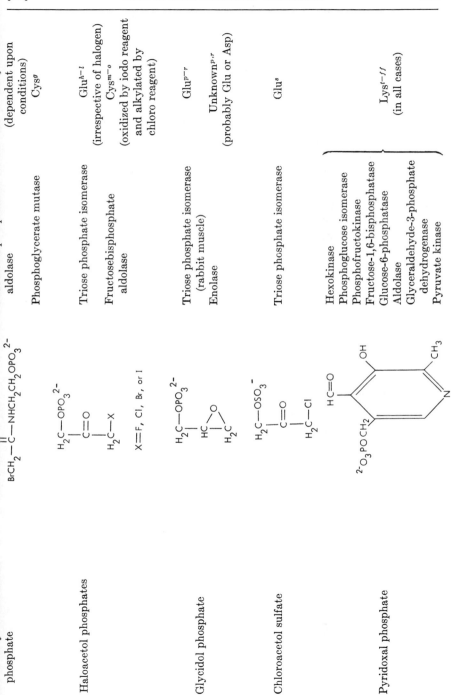

Reagent	Enzyme	Residue modified
phosphate ($BrCH_2\!-\!\overset{\text{O}}{\overset{\|}{C}}\!-\!NHCH_2CH_2OPO_3^{2-}$)	aldolase	Cys^g (dependent upon conditions)
	Phosphoglycerate mutase	
Haloacetol phosphates ($H_2C\!-\!OPO_3^{2-}$, $C\!=\!O$, $H_2C\!-\!X$; $X = F, Cl, Br, \text{ or } I$)	Triose phosphate isomerase	Glu^{h-l} (irrespective of halogen)
	Fructosebisphosphate aldolase	Cys^{m-o} (oxidized by iodo reagent and alkylated by chloro reagent)
Glycidol phosphate ($H_2C\!-\!OPO_3^{2-}$, epoxide)	Triose phosphate isomerase (rabbit muscle)	Glu^{p-r}
	Enolase	$Unknown^{p,r}$ (probably Glu or Asp)
Chloroacetol sulfate ($H_2C\!-\!OSO_3^{-}$, $C\!=\!O$, $H_2C\!-\!Cl$)	Triose phosphate isomerase	Glu^s
Pyridoxal phosphate	Hexokinase, Phosphoglucose isomerase, Phosphofructokinase, Fructose-1,6-bisphosphatase, Glucose-6-phosphatase, Aldolase, Glyceraldehyde-3-phosphate dehydrogenase, Pyruvate kinase	Lys^{t-ff} (in all cases)

SITE-SPECIFIC REAGENTS FOR GLYCOLYTIC ENZYMES (*Continued*)

Reagent	Structure	Enzyme inactivated	Amino acid residue modified
(R,S)-2-Bromo-3-hydroxypropionate 3-phosphate		—	—[gg]
1,3-Bischloroacetylglycerol 2-phosphate		Alkaline phosphatase	Unknown[hh]
Potassium ferrate	K_2FeO_4	Phosphorylase *b*	Tyr[ii]
Potassium hexacyanoferrate(III)	$K_3Fe(CN)_6$	Fructosebisphosphate aldolase Transaldolase Transketolase	Unknown[jj]

1-Guanyl-3,5-dimethylpyrazole
nitrate

Pyruvate kinase

Lys[kk]

[a] S. Otieno, A. K. Bhargava, E. A. Barnard, and A. H. Ramel, *Biochemistry* **14**, 2403 (1975).

[b] E. M. Bessell, *Chem.-Biol. Interact.* **7**, 343 (1973).

[c] E. L. O'Connell and I. A. Rose, *J. Biol. Chem.* **248**, 2225 (1973).

[d] F. C. Hartman, B. Suh, M. H. Welch, and R. Barker, *J. Biol. Chem.* **248**, 8233 (1973).

[e] F. C. Hartman and M. H. Welch, *Biochem. Biophys. Res. Commun.* **57**, 85 (1974).

[f] F. C. Hartman and J. P. Brown, *J. Biol. Chem.* **251**, 3057 (1976).

[g] F. C. Hartman and I. L. Norton, *J. Biol. Chem.* **251**, 4565 (1976).

[h] See this series Vol. 25 [59].

[i] I. L. Norton and F. C. Hartman, *Biochemistry* **11**, 4435 (1972).

[j] S. De La Mare, A. F. W. Coulson, J. R. Knowles, J. D. Priddle, and R. E. Offord, *Biochem. J.* **129**, 321 (1972).

[k] F. C. Hartman and R. W. Gracy, *Biochem. Biophys. Res. Commun.* **52**, 388 (1973).

[l] J. B. Silverman, P. S. Babriarz, K. P. Mahajan, J. Buschek, and T. P. Fondy, *Biochemistry* **14**, 2252 (1975).

[m] F. C. Hartman, *Biochemistry* **9**, 1783 (1970).

[n] Y. Lin, R. D. Kobes, I. L. Norton, and F. C. Hartman, *Biochem. Biophys. Res. Commun.* **45**, 34 (1971).

[o] D. W. Salter, I. L. Norton, and F. C. Hartman, *Biochemistry* **12**, 1 (1973).

[p] I. A. Rose and E. L. O'Connell, *J. Biol. Chem.* **244**, 6548 (1969).

[q] J. C. Miller and S. G. Waley, *Biochem. J.* **123**, 163 (1971).

[r] K. J. Schray, E. L. O'Connell, and I. A. Rose, *J. Biol. Chem.* **248**, 2214 (1973).

[s] F. C. Hartman, G. M. LaMuraglia, Y. Tomozowa, and R. Wolfenden, *Biochemistry* **14**, 5274 (1975).

SITE-SPECIFIC REAGENTS FOR GLYCOLYTIC ENZYMES (*Continued*)

[t] M. A. Grillo, *Enzymologia* **34**, 7 (1968).

[u] K. D. Schnackerz and E. A. Noltmann, *Biochemistry* **10**, 4837 (1971).

[v] K. Uyeda, *Biochemistry* **8**, 2366 (1969).

[w] B. Setlow and T. E. Mansour, *Biochim. Biophys. Acta* **258**, 106 (1972).

[x] G. Colombo and R. G. Kemp, *Biochemistry* **15**, 1774 (1976).

[y] T. A. Krulwich, M. Enser, and B. L. Horecker, *Arch. Biochem. Biophys.* **132**, 331 (1969).

[z] G. Colombo, E. Hubert, and F. Marcus, *Biochemistry* **11**, 1798 (1972).

[aa] G. Colombo and F. Marcus, *Biochemistry* **13**, 3085 (1974).

[bb] V. T. Maddaiah, S. Y. Chen, I. Rezvani, R. Sharma, and P. J. Collipp, *Biochem. Biophys. Res. Commun.* **43**, 114 (1971).

[cc] G. Gold and C. C. Widnell, *J. Biol. Chem.* **251**, 1035 (1976).

[dd] M. Anai, C. Y. Lai, and B. L. Horecker, *Arch. Biochem. Biophys.* **156**, 712 (1973).

[ee] B. G. Forcina, G. Ferri, M. C. Zapponi, and S. Ronchi, *Eir. J. Biochem.* **20**, 535 (1971).

[ff] G. S. Johnson and W. C. Deal, Jr., *J. Biol. Chem.* **245**, 238 (1970).

[gg] P. A. Levene and A. Schormüller, *J. Biol. Chem.* **105**, 547 (1934).

[hh] H. Csopak and G. Fölsch, *Acta Chem. Scand.* **24**, 1025 (1970).

[ii] Y. M. Lee and W. F. Benisek, *J. Biol. Chem.* **251**, 1553 (1976).

[ji] P. Christen, M. Cogoli-Greuter, M. J. Healy, and D. Lubini, *Eur. J. Biochem.* **63**, 223 (1976).

[kk] F. Davidoff, S. Carr, M. Lanner, and J. Leffler, *Biochemistry* **12**, 3017 (1973).

tion with a natural substrate that is converted enzymically to a carbanion, promotes a highly selective modification at the active site of that enzyme.[29] The modifying agent is thought to be a reactive intermediate formed upon oxidation of the enzyme-bound carbanionic form of the substrate. Inactivation is absolutely dependent upon the presence of natural substrate, and the reactive agent formed is enzyme-specific. Thus addition of fructose 1,6-bisphosphate to a mixture of aldolase and transaldolase (or transketolase) in the presence of hexacyanoferrate(III) results in the inactivation of the aldolase only, whereas the addition of fructose 6-phosphate results in the inactivation of the transaldolase (or transketolase) only.[29]

2-Bromo-3-hydroxypropionate 3-phosphate[30] and chloroacetylated derivatives of glycerol 2-phosphate[31] have never been used to probe the active sites of glycolytic enzymes, but might prove useful in future studies.

A lysyl residue in or near the divalent metal-binding sites of pyruvate kinase is subject to affinity labeling by guanidination with 1-guanyl-3,5-dimethylpyrazole nitrate.[32] Although it is difficult to visualize the reagent as an analog of Mn^{2+} (the activating cation), the usual criteria that demonstrate affinity labeling are met.

Synthesis of Selected Reagents

Bromoacetylethanolamine Phosphate (BrAcNHEtOP)[33]

A solution of ethanolamine phosphate (1.41 g, 10 mmol; Sigma Chemical Co.) in 10 ml of water is adjusted to pH 7.0 with 5 N lithium hydroxide. Bromoacetyl bromide (2.2 ml, 25 mmol) is added dropwise, while the pH is kept at 7.0 by the addition of 5 N lithium hydroxide and the temperature is kept at 4°–10° with an ice bath. The product is precipitated by the addition of three volumes of ethanol and is collected by centrifugation. After several washings with ethanol, the residual BrAcNHEtOP (detected with a spray for phosphate esters[34]) appears homogeneous by thin-layer chromatography on cellulose-coated plastic

[29] P. Christen, M. Cogoli-Greuter, M. J. Healy, and D. Lubini, *Eur. J. Biochem.* **63**, 223 (1976).
[30] P. A. Levene and A. Schormüller, *J. Biol. Chem.* **105**, 547 (1934).
[31] H. Csopak and G. Fölsch, *Acta Chem. Scand.* **24**, 1025 (1970).
[32] F. Davidoff, S. Carr, M. Lanner, and J. Leffler, *Biochemistry* **12**, 3017 (1973).
[33] F. C. Hartman, B. Suh, M. H. Welch, and R. Barker, *J. Biol. Chem.* **248,** 8233 (1973).
[34] C. S. Hanes and F. A. Isherwood, *Nature (London)* **164**, 1107 (1949).

sheets (Brinkmann MN-Polygram Cel 300) with R_f of 0.55 in butanol: acetic acid:water (7:2:5 by volume).

[^{14}C]BrAcNHEtOP[33]

N-Hydroxysuccinimide (83 mg, 0.72 mmol) is dissolved, with heating, in 3 ml of ethyl acetate. The solution is cooled to room temperature; upon addition of [1-^{14}C]bromoacetic acid (100 mg, 0.72 mmol; 200 μCi/mmol) (Amersham-Searle Corp.) and dicyclohexylcarbodiimide (150 mg, 0.72 mmol) an immediate exothermic reaction occurs. The reaction mixture is left at room temperature for 2 hr, and the insoluble dicyclohexylurea is then removed by filtration. The filtrate is concentrated to dryness, and the crystalline residue is recrystallized from 5 ml of isopropyl alcohol to give 135 mg (0.57 mmol, 79%) of N-hydroxysuccinimide [^{14}C]bromoacetate,[35] mp 114°–116°.

To an aqueous solution (0.55 ml) of ethanolamine phosphate (50 mg, 0.36 mmol) are added 0.055 ml (0.4 mmol) of triethylamine, 0.75 ml of tetrahydrofuran, and 115 mg (0.48 mmol) of N-hydroxysuccinimide [^{14}C]bromoacetate. Two layers form initially, but after the mixture is stirred for about 1 min, a homogeneous solution results. The reaction mixture is left at room temperature for 1 hr and then cooled to 4° in an ice bath, at which time 300 mg (1 mmol) of barium bromide in 3 ml of methanol is added. The phosphate ester is precipitated by the addition of lutidine (0.15 ml, 1.3 mmol) and ethanol (3 ml). The precipitate is collected by centrifugation, washed 3 times with 3-ml portions of methanol, and dried in a desiccator. After dissolution in 1 ml of water, the material is freed of barium ions by the addition of a slurry of 6 g of Dowex 50 (H+) in methanol (3 ml). The resin is removed by filtration and the filtrate is adjusted to pH 8.0 with 1 N lithium hydroxide. The solvent is removed under a steam of dry nitrogen, and the residue is washed 3 times with 3-ml portions of ethanol and dried to give 70 mg (71%) of the desired compound with a specific radioactivity of 185,000 cpm/μmol. This material is chromatographically indistinguishable from unlabeled BrAcNHEtOP prepared from ethanolamine phosphate and bromoacetyl bromide.

3-Chloroacetol Sulfate[36]

Sulfur trioxide–trimethylamine complex (360 mg, 25.9 mmol; Aldrich Chemical Co.) is added to a solution of chloroacetol (250 mg, 16.2

---

[35] M. Pellegrini, H. Oen, and C. R. Cantor, *Proc. Natl. Acad. Sci. U.S.A.* **69**, 837 (1972).

[36] F. C. Hartman, G. M. LaMuraglia, Y. Tomozowa, and R. Wolfenden, *Biochemistry* **14**, 5274 (1975).

mmol) in 1 ml of dry dimethylformamide. The reaction mixture is stirred overnight, neutralized to pH 8.0 with 1 N NH$_4$OH, and then subjected to chromatography on a 1.2 × 22-cm column of DEAE-cellulose (Whatman DE-52). The column is eluted with a linear gradient composed of 200 ml each of 0.01 M (initial) and 0.2 M (limit) ammonium bicarbonate (pH 8.0). Fractions containing the ammonium salt of chloroacetol sulfate dimethyl ketal, which elutes around 0.02 M ammonium bicarbonate and was detected by spot tests with 2,4-dinitrophenylhydrazine,[37] are pooled and subjected to repeated lyophilization to remove the ammonium bicarbonate completely. The residual ketal (310 mg, 76%) appears homogeneous (R_f = 0.69) by thin-layer chromatography on silica gel (Sil N-HR, Brinkmann Instruments, Inc.) with a solvent composed of butanol:glacial acetic acid:water (7:2:5 by volume). Crystallization of the salt from 5 ml of ethanol gives 220 mg of analytically pure material (mp 114°–116° dec). Dowex 50 (H$^+$) (600 mg) is added to 2.0 ml of 0.22 M chloroacetol sulfate dimethyl ketal ammonium salt, and the resulting mixture is incubated at 40° for 48 hr. During this time an approximate 90% conversion of ketal (R_f = 0.75) to ketone (R_f = 0.62) occurs as shown by thin-layer chromatography (Cell 300, Brinkmann Instruments, Inc.) using the same solvent as above. The exact concentration of chloroacetol sulfate is determined by quantitating its reaction with the sulfhydryl group of glutathione, which is quantitated with Ellman's reagent.[38] An aliquot (0.005 ml) of the stock chloroacetol sulfate solution is added to 1 ml of 2 mM glutathione in 0.1 M Bicine–1 mM EDTA (pH 8.0). At 20 and 60 min, 0.1 ml of the reaction mixture is added to 2.4 ml of 0.4 mM 5,5-dithiobis(2-nitrobenzoic acid) in the same Bicine buffer. The sulfhydryl concentration, determined from the increase in A_{412}, is found to be the same for the 20- and 60-min aliquots. Thus, the reaction between chloroacetol sulfate and glutathione is completed in less than 20 min, and from the observed decrease in sulfhydryl concentration, the chloroacetol sulfate concentration in the stock solution is calculated to be 0.21 M.

N-Bromoacetyl-D-galactopyranosylamine[39]

D-Galactosamine hydrochloride (500 mg) is dissolved in 12 ml of water and 1.2 ml of methanol. Bromoacetic anhydride (800 mg) and

[37] R. A. Gray, *Science* 115, 129 (1952).
[38] G. L. Ellman, *Arch. Biochem. Biophys.* 82, 70 (1959).
[39] S. Otieno, A. K. Bhargava, E. A. Barnard, and A. H. Ramel, *Biochemistry* 14, 2403 (1975).

Dowex 1 (carbonate form; 15 ml wet resin) are then added. The mixture is stirred for 90 min at 0°–5°, then filtered, and the residue is washed with water. The filtrate and washing are passed through a 5-ml column of Dowex 50 (H$^+$), and the column is washed thoroughly with water. The combined eluate and washings are heated just to boiling, and the solution is concentrated at <50° under vacuum by rotary evaporation. The N-bromoacetylgalactosamine formed is crystallized twice from absolute ethanol. After drying, the final yield is 200 mg, mp 104°–110°. The product gives a strongly positive response in the modified Morgan–Elson reaction,[40] and contains no free bromide ion (by silver nitrate reaction). Thin-layer chromatography (acetone:methanol, 10:1 v/v) shows a single component ($R_f = 0.5$).

1,3-Bischloroacetylglycerol 2-Phosphate and 1(3)-Chloroacetylglycerol 2-Phosphate[31]

Sodium glycerol 2-phosphate containing 5.5 mol of water of hydration (3.15 g, 10 mmol) is dissolved in 10 g of molten chloroacetic acid, and chloroacetic anhydride (17.1 g, 100 mmol) is added. The solution is heated for 5 min at 120° and then cooled to room temperature. Dried diethyl ether is added, and the precipitate is collected and washed extensively with ether. The product is dried over potassium hydroxide under vacuum and crystallized from methanol and diethyl ether. The compound (mp 184°) moves as a single spot ($R_f = 0.55$) on thin-layer plates of Kieselgel G with the solvent system 1-butanol:methanol:acetic acid:water (2:1:1:1, v/v/v/v). The monochloroacetylated glycerol phosphate is obtained in the same fashion with the exception that only 2.4 g (15 mmol) of chloroacetic anhydride with 1.6 g (5 mmol) of sodium glycerol 2-phosphate is used. Thin-layer chromatography indicates the presence of some unreacted and some bischloroacetylated glycerol phosphate.

Mixture of 1,2-Anhydro-D-Mannitol 6-Phosphate and 1,2-Anhydro-L-Gulitol 6-Phosphate[41]

The epoxide mixtures are prepared by muscle aldolase-catalyzed condensation of dihydroxyacetone phosphate with D-glycidaldehyde,[42]

[40] J. L. Reissig J. L. Strominger, and L. F. Leloir, *J. Biol. Chem.* **217**, 959 (1955).
[41] E. L. O'Connell and I. A. Rose, *J. Biol. Chem.* **248**, 2225 (1973).
[42] K. J. Schray, E. L. O'Connell, and I. A. Rose, *J. Biol. Chem.* **248**, 2214 (1973).

followed by isolation of the intermediate keto compounds and their reduction by NaBH₄ or NaBT₄ to give the corresponding hexitol products. In a final volume of 5.0 ml, 100 μmol of dihydroxyacetone phosphate, 200 μmol of triethanolamine-HCl (pH 7.5), and 2500 μmol of D-glycidaldehyde are incubated at 25° with 10 units of aldolase that is free of triose phosphate isomerase. After 30 min, less than 3% of the dihydroxyacetone phosphate, as assayed by glycerol phosphate dehydrogenase, remains. Aldolase is inactivated and precipitated by the addition of an equal volume of absolute ethanol. After 10 min at 25°, denatured protein is removed by centrifugation. The supernatant is concentrated to its original volume and then chromatographed on a Sephadex G-10 column (2.5 × 92 cm) equilibrated with 50 mM sodium acetate. The condensation product appears at 1.11 times the column void volume in approximately 20 ml. The condensation product is detected through the formation of dihydroxyacetone phosphate with aldolase and its coupled reduction with glycerol phosphate dehydrogenase and NADH. Attempts to avoid the Sephadex step and purify the condensation product directly by barium precipitation leads to a very impure product.

The condensation product is adjusted to pH 9.0 and maintained at that pH during the stepwise addition of NaBH₄. If the hexitols are to be isotopically labeled, 5 μmol of pyruvate are added as an internal standard for determining tritium specific activity. An appropriate amount of carrier-free NaBT₄ is added, and the loss of condensation product is followed by the coupled assay system described above. The reaction is terminated by dropping the pH to 4.0 with glacial acetic acid to destroy excess NaBH₄. After pH readjustment to 7.5, the volume is reduced to 4.0 ml in a rotary evaporator at 25° or less, and 1 ml of 1 M barium acetate and 16 ml of absolute ethanol are added. A precipitate is slowly formed in an ice bath. The supernatant is used for determination of lactate specific activity. The precipitate is dissolved in 2 ml of water, insoluble material is removed, and the precipitate is reformed by the addition of an equal volume of absolute ethanol. The precipitate, dried by washing with ethanol and ether, weighs 34 mg (75% yield) and contains 75 μmol of organic phosphate. It is stored over desiccant at −70° to prevent decomposition.

DL-Lactate is isolated from Dowex 1 (Cl⁻) by elution with 5 mM HCl and the L component is assayed with lactate dehydrogenase. The tritium specific activity is within 25% of that of the mixed epoxides when the latter are analyzed on the basis of phosphorus content. On this basis the epoxide preparations show a one-to-one correspondence between tritium and phosphorus content and hence should be relatively free of major contaminants containing either tritium or phosphate alone.

2',3'-Epoxypropyl β-D-Glucopyranoside[43]

Allyl β-D-glucopyranoside tetraacetate[44] (6 g) is dissolved in chloroform (20 ml), and 3-chloroperoxybenzoic acid (6 g) in chloroform (70 ml) is added during 30 min. The solution is left at room temperature overnight, during which time a precipitate forms. More chloroform (100 ml) is added, and the chloroform layer is washed with aqueous solutions of sodium sulfite, sodium hydrogen carbonate, sodium thiosulfate, sodium carbonate, and water. The chloroform layer is dried over calcium chloride and evaporated to dryness under vacuum. Crystallization of the residue from ethanol gives 4.8 g of 2',3'-epoxypropyl β-D-glucopyranoside tetraacetate, mp 115°–117°. The compound has an R_f of 0.29 on thin-layer chromatography (Silica Gel F_{254}) with ethyl acetate:light petroleum ether (bp 40°–60°; 3:2, v/v) as solvent and 5% sulfuric acid in ethanol for detection.

The tetraacetate (300 mg) is suspended in dry methanol (5 ml), and 1 M methanolic sodium methoxide (0.2 ml) is added. The suspension is shaken until the solid is dissolved; then it is left for a further 15 min. Solid carbon dioxide is added, and the solution is evaporated to dryness under vacuum. The residue is chromatographed on Whatman 3 MM paper (loading of 2 mg/cm) with the upper layer from butyl alcohol: ethanol:water (4:1:5, v/v/v). The major band of 2,3-epoxypropyl β-D-glucopyranoside is detected on marker strips by using alkaline silver nitrate. Excision, elution, and freeze-drying gives 120 mg of noncrystalline material.

(R,S)-2-Bromo-3-Hydroxypropionate 3-Phosphate[30]

(R,S)-2-Bromo-3-hydroxypropionic acid[45] (17 g) is added, with efficient stirring and cooling, to a mixture of phosphorus pentoxide (4.7 g) and anhydrous phosphoric acid (3.5 g). After 24 hr the reaction mixture is poured into a 500-ml slurry of ice and water. The resulting solution is neutralized with barium hydroxide, and the precipitate is then removed by centrifugation. An equal volume of ethanol is added to the supernatant, and the gelatinous precipitate that forms is collected by centrifugation and extracted with water. The extract is filtered, and the addition of alcohol to the filtrate precipitates the desired product (2.8 g), which is collected by centrifugation, washed with alcohol followed by ether, and dried.

[43] J. E. G. Barnett and A. Ralph, Carbohydr. Res. **17**, 231 (1971).
[44] E. A. Talley, M. D. Vale, and E. Yanovsky, J. Am. Chem. Soc. **67**, 2037 (1945).
[45] C. Neuberg and P. Mayer, Biochem. Z. **3**, 116 (1907).

Affinity Labeling of Selected Enzymes

Phosphoglucose Isomerase

Modification of an Essential Lysyl Residue of the Rabbit Muscle Enzyme with Pyridoxal Phosphate.[46] The enzyme (2–4 mg/ml, isolated according to a published procedure[47]) is incubated for 15 min at 0° in 50 mM Bicene buffer (pH 8.0) with 0.29 mM pyridoxal phosphate (Calbiochem). A 5-μl aliquot is withdrawn to determine the initial enzymic activity (by the method of Noltmann[48]). Prior to reduction with borohydride, a drop of octyl alcohol is added to prevent foaming. A freshly prepared solution of sodium borohydride (0.6 M) in 1 mM sodium hydroxide is added in 10-μl portions, totaling a 50-to 75-fold excess with respect to the pyridoxal phosphate concentration. During the reduction about 80% of the initial activity is lost. If the pyridoxal phosphate is omitted, the isomerase is not inactivated by borohydride. After the reduction, the reaction mixture is passed through a 20 × 50-mm column of Sephadex G-25 (fine) equilibrated with 50 mM Bicene buffer (pH 8.0) at 5°. A flow rate of 2 ml/min is maintained with a peristaltic pump, and the column effluent is monitored at 280 nm. No more than 20 min are required from the first addition of borohydride to complete separation of the reaction products. Based on the absorbancy at 325 nm, the derivatized enzyme contains about 0.8 mol of ε-aminophosphopyridoxyllysine per mole of catalytic subunit.

The lysyl residue that is modified could be involved in protonation of the glycosidic ring oxygen in the preliminary ring-opening step[49,50] or could serve to polarize the carbonyl group in the acyclic form of the substrate.[41]

Modification of an Essential Glutamyl Residue of Rabbit Muscle or Yeast Phosphoglucose Isomerase by a Mixture of 1,2-Anhydro-D-mannitol 6-Phosphate and 1,2-Anhydro-L-gulitol 6-Phosphate.[41] The mixture of epoxides (3.3×10^6 cpm/μmol) to a final concentration of 3 mM is added to a solution of yeast phosphoglucose isomerase (10 mg/ml; Boehringer-Mannheim Corp.) in 20 mM triethanolamine hydrochloride (pH 7.5) at 25°. After 3 hr, only 1–2% of the initial enzymic activity remains,

[46] K. D. Schnackerz and E. A. Noltmann, *Biochemistry* **10**, 4837 (1971). The conditions given are those described in a personal communication and differ slightly from the ones published.

[47] See this series Vol. 9 [98a].

[48] E. A. Noltmann, *J. Biol. Chem.* **234**, 1545 (1964).

[49] J. E. D. Dyson and E. A. Noltmann, *J. Biol. Chem.* **243**, 1401 (1968).

[50] E. A. Noltmann, *in* "The Enzymes," 3rd ed. (P. D. Boyer, ed.), Vol. 6, p. 271. Academic Press, New York, 1972.

and the reaction mixture is dialyzed exhaustively against 10 mM tri-ethanolamine hydrochloride–1 mM EDTA (pH 7.15). Based on radio-activity, the dialyzed enzyme contains 1 mol of reagent per mole of catalytic subunit. From characterization of the reagent moiety released from the protein upon treatment with hydroxylamine, it is shown that 1,2-anhydro-D-mannitol 6-phosphate is the active species and that attack by the glutamyl γ-carboxylate occurs at the C-1 position of the epoxide. The essential glutamate may be the base that abstracts a proton from C-2 of glucose 6-phosphate or C-1 of fructose 6-phosphate, thereby cata-lyzing the conversion of substrate to the intermediate enediol.[41,51] How-ever, such a role for the glutamate has been questioned.[52]

Class I Fructose-1,6-bisphosphate Aldolase

This class of aldolase (the enzyme from rabbit muscle is representa-tive) forms Schiff bases with their ketonic substrates as obligatory inter-mediates.[53,54] The method first used to detect these intermediates, namely reduction of the dihydroxyacetone phosphate Schiff base with boro-hydride,[55] provides a convenient and generally applicable procedure for the highly selective labeling of the essential lysyl residue.[56] The elucida-tion of the complete primary structure of rabbit muscle aldolase places the Schiff-base-forming lysyl residue at position 227.[57] Other residues subject to affinity labeling are Cys-336, His-359, and Lys-146. Cys-336, which is not absolutely required for catalysis but seems to be within the binding domain for fructose 1,6-bisphosphate, is preferentially alkylated by 3-chloroacetol phosphate.[58] His-359 (which may be the acid-base group involved in proton transfer to and from the C-3 position of the enzyme-bound dihydroxyacetone phosphate) and Lys-146 (which is at or near the binding site for the C-1 phosphate of fructose 1,6-bisphosphate) are both subject to preferential alkylation by BrAcNHEtOP.[33,59,60]

[51] I. A. Rose, *Adv. Enzymol. Relat. Areas Mol. Biol.* **43**, 491 (1975).
[52] J. M. Chirgwin and E. A. Noltmann, *J. Biol. Chem.* **250**, 7272 (1975).
[53] B. L. Horecker, O. Tsolas, and C. Y. Lai, *in* "The Enzymes," 3rd ed. (P. D. Boyer, ed.), Vol. 7, p. 213. Academic Press, New York, 1972.
[54] P. Model, L. Ponticorvo, and D. Rittenberg, *Biochemistry* **7**, 1339 (1968).
[55] B. L. Horecker, P. T. Rowley, E. Grazi, T. Cheng, and O. Tchola, *Biochem. Z.* **338**, 36 (1963).
[56] See this series Vol. 11 [77].
[57] C. Y. Lai, N. Nakai, and D. Chang, *Science* **183**, 1204 (1974).
[58] D. W. Salter, I. L. Norton, and F. C. Hartman, *Biochemistry* **12**, 1 (1973).
[59] F. C. Hartman and M. H. Welch, *Biochem. Biophys. Res. Commun.* **57**, 85 (1974).
[60] F. C. Hartman and J. P. Brown, *J. Biol. Chem.* **251**, 3057 (1976).

The alkylation of His-359 by BrAcNHEtOP at pH 6.0 provides evidence of the residue's presence at the active site and participation in catalysis. The reaction is subject to substrate protection and exhibits a rate saturation. The aldolase derivative obtained is devoid of normal fructose bisphosphate cleavage activity but catalyzes the transaldolase reaction in which the proton transfer step is bypassed by the presence in the assay mixture of exogenous aldehyde which condenses nonenzymically with the enzyme-bound carbanion of dihydroxyacetone phosphate and thus allows cleavage of fructose bisphosphate to proceed. Studies on the photoinactivation of aldolase also implicate a histidyl residue in the same proton transfer step.[61]

For reasons that are not fully understood, a change in the pH of the reaction mixture from 6.0 to 8.5 results in an altered specificity of BrAcNHEtOP so that Lys-146 is alkylated preferentially.[60] Since this derivative is devoid of normal cleavage activity, transaldolase activity, and Schiff-base-forming ability, it was suggested that modification of Lys-146 prevents binding of fructose bisphosphate and dihydroxyacetone phosphate. This suggestion seems to be supported by the observation that after modification at Lys-146 the enzyme cannot be modified at His-359 by BrAcNHEtOP (i.e., the enzyme deficient in binding of substrates is also deficient in binding of an affinity label).

Preferential Alkylation of His-359 of Rabbit Muscle Aldolase by BrAcNHEtOP.[33] Aldolase concentration is determined from the absorbance at 280 nm, using an $A_{1 cm}^{1\%}$ value of 9.38[62] and a molecular weight of 160,000.[63] Stock solutions of aldolase (about 25 mg/ml) are prepared by dialysis of the commercial ammonium sulfate suspension (Boehringer-Mannheim Corp.) against 50 mM ammonium acetate–1 mM EDTA (pH 7.0). The enzymic activity of aldolase is measured by the method of Blostein and Rutter[64] with a Beckman Acta V recording spectrophotometer. Each assay solution contains 50 mM glycylglycine (pH 7.5), 1 mM fructose 1,6-bisphosphate, 0.15 mM NADH, 1 mM EDTA, and 28 μg of glycerol phosphate dehydrogenase–triose phosphate isomerase. At 24° the specific activity is 10.1 units/mg. Transaldolase activity[65,66] is measured with the same assay solution containing 25 mM acetaldehyde.

Samples of aldolase (5 mg/ml) in the absence and in the presence of

[61] L. C. Davis, L. W. Brox, R. W. Gracy, G. Ribereau-Gayon, and B. L. Horecker, *Arch. Biochem. Biophys.* 140, 215 (1970).
[62] J. W. Donovan, *Biochemistry* 3, 67 (1964).
[63] K. Kawahara and C. Tanford, *Biochemistry* 5, 1578 (1966).
[64] R. Blostein and W. J. Rutter, *J. Biol. Chem.* 238, 3280 (1963).
[65] I. A. Rose, E. L. O'Connell, and A. H. Mehler, *J. Biol. Chem.* 240, 1758 (1965).
[66] P. D. Spolter, R. C. Adelman, and S. Weinhouse, *J. Biol. Chem.* 240, 1327 (1965).

butanediol bisphosphate (5 mM),[67] a competitive inhibitor of aldolase, are treated with ^{14}C-labeled BrAcNHTtOP (2 mM) at pH 6.0 (50 mM Pipes–1 mM EDTA) for 96 hr. At this time the unprotected sample contains only 5% of the initial fructose 1,6-bisphosphate cleavage activity but retains 71% of its transaldolase activity. The sample protected with butanediol bisphosphate is about 95% active in both assays as compared with a control. The two protein solutions are dialyzed exhaustively against 0.1 M sodium phosphate–1 mM EDTA (pH 7.0) and then assayed for radioactivity. The inactivated aldolase contains about 2.6 mol of reagent per mole of catalytic subunit, whereas the protected enzyme contains 1.5. This difference of about 1.0 mole of reagent per mole of subunit represents the alkylation of His-359 in the inactivated sample as determined by amino acid analyses and characterization of tryptic digests of the two samples. Since the reagent contains an amide linkage, acid hydrolysis of the protein derivatives yields the modified amino acids as carboxymethyl amino acids, which are easily detected and quantitated with the amino acid analyzer.[68]

Preferential Alkylation of Lys-146 of Rabbit Muscle Aldolase by BrAcNHEtOP.[60] The procedures for alkylation of Lys-146 are the same as those described in the preceding paragraph except that the reaction with BrAcNHEtOP is carried out at pH 8.5 (50 mM Bicine–1 mM EDTA) for 48 hr, at which time the protected sample is fully active and the unprotected sample contains only 3–4% of its initial cleavage and transaldolase activities. After dialysis, assays for radioactivity reveal 1.7 and 0.75 mol of reagent per mole of catalytic subunit of the inactivated and protected sample, respectively. Characterization of tryptic peptides shows the difference between the two samples to be alkylation of Lys-146 in the inactivated aldolase.

Phosphoglycerate Mutase (2,3-Bisphosphoglycerate-Dependent)

This enzyme is also subject to affinity labeling by BrAcNHEtOP.[69] The reagent inactivates the mutase by a very rapid and highly selective alkylation of a sulfhydryl group, which, based on the following observations, is at or near the active site. Kinetics of inactivation are pseudo first-order, and the inactivation exhibits a rate saturation. Reversible binding of the reagent to the enzyme is competitive with the binding of 2,3-bisphosphoglycerate. The inactivated enzyme contains a single mole of

[67] F. C. Hartman and R. Barker, *Biochemistry* **4**, 1068 (1965).
[68] See this series Vol. 25 [34a].
[69] F. C. Hartman and I. L. Norton, *J. Biol. Chem.* **251**, 4565 (1976).

reagent per mole of catalytic subunit, and virtually all the incorporated reagent is covalently attached to one sulfhydryl group. Incorporation of BrAcNHEtOP is negligible in the presence of substrate and is reduced 8-fold in the presence of 6 M urea.

Selective Alkylation of an Essential—SH Group in Rabbit Muscle Phosphoglycerate Mutase by BrAcNHEtOP.[69] Concentrations of phosphoglycerate mutase are calculated from the absorbance at 280 nm, using an $A^{1\%}_{1\,cm}$ value of 14.8[70] and a molecular weight of 54,000.[71] Preceding its use in chemical modification studies, the commercial ammonium sulfate suspensions of the mutase (Boehringer-Mannheim Corp.) are dialyzed against 50 mM HEPES–0.1 mM EDTA (pH 7.0). Mutase activity is determined by coupling the reaction to enolase as described by Grisola and Carreras,[72] and the increase in absorbancy is followed with a Beckman Acta V recording spectrophotometer. The only changes from the published procedure are to carry out the assays at 25° instead of 30° and to include 2,3-bisphosphoglycerate (0.33 mM) in the assay mixture, since 3-phosphoglycerate free of the bisphosphate (Sigma Chemical Co.) is used. The specific activity of the commercial mutase is 308 units/mg. Free sulfhydryl groups in phosphoglycerate mutase are quantitated as described by Ellman,[38] but the reaction medium is 8 M guanidinium chloride:0.2 M sodium phosphate:3 mM EDTA:4 mM 5,5′-dithiobis(2-nitrobenzoic acid) (pH 8.0).

Phosphoglycerate mutase (20 mg, 0.74 μmol of subunit) in 10 ml of 50 mM HEPES–0.1 M EDTA (pH 7.0) is treated with 2.0 μmol of ^{14}C-labeled BrAcNHEtOP for 60 min, at which time 5% of the initial enzymic activity remains. The reaction mixture is then dialyzed at 4° against 0.1 M potassium phosphate (pH 7.0), followed by 0.1 M sodium bicarbonate (pH 8.0). On the basis of radioactivity assays, the modified protein contains 0.79 mol of reagent per mole of subunit; on the basis of sulfhydryl titration, the modified protein contains 1.1 fewer —SH groups per subunit than the native enzyme. All the radioactivity in acid hydrolyzates of the inactivated mutase elutes from the amino acid analyzer coincident with Cys(Cm). The number of residues of Cys(Cm) per subunit, as calculated from the ninhydrin profile, is 0.96.

Triose Phosphate Isomerase

The identification, by use of 3-haloacetol phosphates and glycidol phosphate, of a catalytically functional glutamyl γ-carboxylate (Glu-165

[70] R. Czok and T. Bucher, *Adv. Protein Chem.* **15**, 315 (1960).
[71] Z. B. Rose, *Arch. Biochem. Biophys.* **140**, 508 (1970).
[72] See this series Vol. 42 [66].

in the rabbit muscle enzyme)[73] at the active site of triose phosphate isomerase is well documented.[74] In all probability this residue functions as the acid-base group that abstracts a proton from C-3 of dihydroxyacetone phosphate to generate the enediol intermediate and donates a proton to C-2 of the enediol to form D-glyceraldehyde 3-phosphate.[51] By the use of dihydroxyacetone phosphate, stereospecifically tritiated at C-3, it has been shown that most ($\sim$98%) of the tritium exchanges with solvent during the enzyme-catalyzed conversion to glyceraldehyde 3-phosphate and that only 2%, approximately, is found in the product.[75] Thus, the rate of ionization of the essential acid-base group is about 50 times more rapid than the rate at which it transfers the proton to the enediol. With this knowledge and the value of k_{cat} for the overall reaction, one can calculate that the maximal pK_a for the essential group is $\sim$5.[51] If the glutamyl residue, implicated in catalysis by the results of affinity labeling, does effect proton transfer, its pK_a should be no more than 5. Based on the pH-dependence of the inactivation rate of triose phosphate isomerase by glycidol phosphate, Schray et al.[42] found a pK_a of <5, whereas under somewhat different conditions Waley found a pK_a of 6.0.[76] In the latter case, measurement of the pK_a may have been complicated by a variable affinity of the reagent for the enzyme as a result of the change in the ionization state of the reagent over the pH range examined. This potential problem can be circumvented with chloroacetol sulfate, a compound that exists only as a monoanion over the entire pH range at which protein carboxyl groups ionize. Even though this reagent has a very low affinity for triose phosphate isomerase, it selectively esterifies the same glutamyl residue of the enzyme as do glycidol phosphate and haloacetolphosphate.[36] From the pH-dependence of inactivation of the yeast enzyme by chloroacetol sulfate, the pK_a of the essential carboxyl group in the free enzyme is found to be about 3.9,[36] a value consistent with the carboxyl group's postulated role in catalysis.

[73] P. H. Corran and S. G. Waley, *Biochem. J.* **145**, 335 (1975).
[74] See F. C. Hartman, this series Vol. 25 [59].
[75] B. Plaut and J. R. Knowles, *Biochem. J.* **129**, 311 (1972).
[76] S. G. Waley, *Biochem. J.* **126**, 255 (1972).

Section X
Methods of Peptide Synthesis

[47] The Synthesis of Peptides by Homogeneous Solution Procedures

By Panayotis G. Katsoyannis and Gerald P. Schwartz

I. Basic Approaches for Peptide Synthesis

The synthesis of peptides by homogeneous solution procedures is basically similar to the scheme proposed by E. Fischer in the early 1900's. In its most general form this approach involves the following stages.[1]

Stage 1: Synthesis of the "carboxyl component" by blocking the amino group of an amino acid or peptide with a group Y.

$$\underset{\underset{\text{H}_2\text{NCHCOOH}}{|}}{\overset{\text{R}_1}{}} \rightarrow \underset{\underset{\text{Y}\cdot\text{HNCHCOOH}}{|}}{\overset{\text{R}_1}{}}$$

Stage 2: Synthesis of the "amino component" by blocking the carboxyl group of another amino acid or peptide with a group Z.

$$\underset{\underset{\text{H}_2\text{NCHCOOH}}{|}}{\overset{\text{R}_2}{}} \rightarrow \underset{\underset{\text{H}_2\text{NCHCOOZ}}{|}}{\overset{\text{R}_2}{}}$$

Stage 3: Activation of the carboxyl group of the "carboxyl component" (or, very rarely, of the amino group of the "amino component") with a group X and the coupling of the so-formed active intermediate with the "amino component" (or "carboxyl component") to yield a protected peptide.

$$\underset{\underset{\text{Y}\cdot\text{NHCHCOOH}}{|}}{\overset{\text{R}_1}{}} \rightarrow \underset{\underset{\text{Y}\cdot\text{NHCHCOX}}{|}}{\overset{\text{R}_1}{}}$$

$$\underset{\underset{\text{Y}\cdot\text{NHCHCOX}}{|}}{\overset{\text{R}_1}{}} + \underset{\underset{\text{H}_2\text{NCHCOOZ}}{|}}{\overset{\text{R}_2}{}} \rightarrow \underset{\underset{\text{Y}\cdot\text{NHCHCO}}{|}}{\overset{\text{R}_1}{}} \underset{\underset{\text{NHCHCOOZ}}{|}}{\overset{\text{R}_2}{}} + \text{HX}$$

$$\underset{\underset{\text{H}_2\text{NCHCOOZ}}{|}}{\overset{\text{R}_2}{}} \rightarrow \underset{\underset{\text{X}\cdot\text{NHCHCOOZ}}{|}}{\overset{\text{R}_2}{}}$$

$$\underset{\underset{\text{Y}\cdot\text{NHCHCOOH}}{|}}{\overset{\text{R}_1}{}} + \underset{\underset{\text{X}\cdot\text{NHCHCOOZ}}{|}}{\overset{\text{R}_2}{}} \rightarrow \underset{\underset{\text{Y}\cdot\text{NHCHCO}\cdot\text{NHCHCOOZ}}{|}}{\overset{\text{R}_1}{}}\underset{}{\overset{\text{R}_2}{}} + \text{HOX}$$

[1] Useful reviews on peptide synthesis can be found in (a) K. Hofmann and P. G. Katsoyannis, *in* "The Proteins" (H. Neurath, ed.) 2nd ed., Vol. 1, p. 53. Academic Press, New York, 1963; (b) E. Schröder and K. Lübke, "The Peptides," Vol. 1. Academic Press, New York, 1965; (c) H. D. Law, "The Organic Chemistry of Peptides." Wiley (Interscience), New York, 1970; (d) M. Bodanszky and M. A. Ondetti, "Peptide Synthesis." Wiley (Interscience) New York, 1966; (e) P. G. Katsoyannis and J. Z. Ginos, *Annu. Rev. Biochem.* **38,** 710 (1969).

At the present time only a few procedures are available for the activation of the amino group of the "amino component," and peptide synthesis via this route is of no practical significance.

Stage 4: Removal of the blocking groups Y and Z from the protected peptide to form the free peptide.

Synthesis of peptides containing polyfunctional amino acid residues such as cysteine, arginine, lysine, histidine, tyrosine, serine, threonine, glutamic acid, or aspartic acid, requires additional steps. The secondary functional group is often protected during the various synthetic steps and is deprotected in the final step.

II. Elongation of a Peptide Chain

The scheme outlined above describes in essence the synthesis of a dipeptide. In this section methods for the elongation of a peptide chain will be outlined. In general, three approaches are available for this purpose: (1) Stepwise synthesis from the amino terminal of the growing peptide chain; (2) stepwise synthesis from the carboxyl terminal of the growing peptide chain; and (3) fragment condensation, i.e., joining together of small peptide subunits to form a larger peptide.

A. Stepwise Synthesis from the Amino Terminal of the Growing Peptide Chain

In this approach, elongation of the peptide chain is accomplished by the attachment of one amino acid at a time to the amino group of the NH$_2$-terminal amino acid residue of the growing peptide chain. Thus the amino protecting group (Y) of a fully protected peptide (I) is selectively removed and the ensuing amino component (II) is condensed with an activated acylamino acid (III, carboxyl component) to give the respective protected peptide (IV), which in turn is selectively deblocked at the amino end to give a partially protected peptide (V, new amino component), etc.

$$Y \cdot HN\overset{R_2}{\underset{|}{C}}HCO \cdot HN\overset{R_3}{\underset{|}{C}}HCOOZ \xrightarrow[\text{deblocking}]{\text{selective}} H_2N\overset{R_2}{\underset{|}{C}}HCO \cdot HN\overset{R_3}{\underset{|}{C}}HCOOZ \xrightarrow{Y \cdot HN\overset{R_1}{\underset{|}{C}}HCOX \text{ (III)}}$$

(I)   (II)

$$Y \cdot NH\overset{R_1}{\underset{|}{C}}HCO \cdot NH\overset{R_2}{\underset{|}{C}}HCO \cdot NH\overset{R_3}{\underset{|}{C}}HCOOZ \xrightarrow[\text{deblocking}]{\text{selective}}$$

(IV)

$$H_2N\overset{R_1}{\underset{|}{C}}HCO \cdot NH\overset{R_2}{\underset{|}{C}}HCO \cdot NH\overset{R_3}{\underset{|}{C}}HCOOZ \rightarrow, \text{ etc.}$$

(V)

B. Stepwise Synthesis from the Carboxyl Terminal of the Growing Peptide Chain

In this approach elongation of the peptide chain involves the linking of one amino acid at a time to the carboxyl group of the COOH-terminal amino acid residue of the growing peptide chain. The carboxyl protecting group (Z) of a peptide (VI) is selectively removed, and the resulting carboxyl component (VII) is activated by group X. The activated intermediate (VIII) reacts then with an amino component (IX) to yield the protected peptide (X). This in turn is selectively deblocked to give a new carboxyl component (XI), etc.

$$Y \cdot HN\overset{R_1}{\underset{|}{C}}HCO \cdot HN\overset{R_2}{\underset{|}{C}}HCOOZ \xrightarrow[\text{deblocking}]{\text{selective}} Y \cdot HN\overset{R_1}{\underset{|}{C}}HCO \cdot HN\overset{R_2}{\underset{|}{C}}HCOOH \xrightarrow{\text{activation}}$$
$$\text{(VI)} \qquad\qquad\qquad\qquad \text{(VII)}$$

$$Y \cdot HN\overset{R_1}{\underset{|}{C}}HCO \cdot HN\overset{R_2}{\underset{|}{C}}HCO\overset{R_3}{\underset{}{}}X \xrightarrow{H_2N\overset{R_3}{\underset{|}{C}}HCOOZ \ (IX)}$$
$$\text{(VIII)}$$

$$Y \cdot HN\overset{R_1}{\underset{|}{C}}HCO \cdot HN\overset{R_2}{\underset{|}{C}}HCO \cdot HN\overset{R_3}{\underset{|}{C}}HCOOZ \xrightarrow[\text{deblocking}]{\text{selective}}$$
$$X$$

$$Y \cdot NH\overset{R_1}{\underset{|}{C}}HCO \cdot NH\overset{R_2}{\underset{|}{C}}HCO \cdot HN\overset{R_3}{\underset{|}{C}}HCOOH \rightarrow, \text{ etc.}$$
$$\text{(XI)}$$

C. Fragment Condensation

In this approach partially protected peptide fragments serve as the amino and carboxyl components. Condensation of these fragments to produce a larger peptide subunit proceeds via activation of the carboxyl group of the carboxyl component.

III. Problems in Peptide Synthesis and Selection of Methods in Peptide Chain Elongation

The major problems in peptide synthesis by homogeneous solution procedures are the ever-present risk of racemization and the separation of the product from the reactants and by-products formed during the synthesis.

A. Racemization

Racemization may occur during the peptide bond-forming steps, namely, the activation of the carboxyl component

$$
\begin{array}{c}
R \\
| \\
(R'CO \cdot NHCHCOOH)
\end{array}
$$

and its condensation with the amino component, or by direct racemization of certain sensitive amino acid residues under alkaline conditions.

Racemization during the peptide bond-forming steps is thought to occur via the formation of oxazolone[2] intermediates from the activated carboxyl component. The activated carboxyl component (XII) is transformed into the enolate ion (XIII) under basic conditions; in turn, this cyclizes into the oxazolone (XIV), which in the presence of base, tends to racemize very rapidly by proton abstraction at the alpha carbon atom. A reversal of this reaction leads to racemic activated carboxyl component (XII) which, after condensation with an amino component, will afford a racemic product.

(XII)   (XIII)   (XIV)

Alternatively, oxazolone (XIV), itself an activated intermediate, can react with an amino component to produce a racemic product. The degree of racemization via oxazolone depends on the acyl moiety

$$
\begin{array}{c}
O \\
|| \\
(R'-C-)
\end{array}
$$

of the carboxyl component, the basicity of the reaction environment, the nature of the activating group (X), the polarity of the solvent, and the temperature of the reaction. Thus if the acyl moiety of the carboxyl component is an acid amide type (acetyl, benzoyl or aminoacyl amide) formation of oxazolone is favored. On the other hand, urethane type moieties

$$
\begin{array}{c}
O \\
|| \\
(R-O-C-)
\end{array}
$$

---

[2] For a review see P. G. Katsoyannis and J. Z. Ginos, *Annu. Rev. Biochem.* **38,** 710 (1969).

or moieties that do not contain a carbonyl grouping, i.e., trityl [$(C_6H_5)_3C$—] and o-nitrophenylsulfenyl (o-$NO_2C_6H_4S$—) do not form oxazolones. Formation of oxazolone followed by racemization is further facilitated by increased basicity in the reaction environment, the presence of a polar solvent and elevated reaction temperatures. The influence of the activating group (X) on racemization is obscure. Experience has shown that several of the activating groups, under certain conditions, cause no detectable racemization. It must be emphasized that these same activating groups, under different reaction conditions, however, may cause racemization. These aspects will be considered in the experimental section.

Direct racemization of amino acid residues is thought to arise by proton abstraction from the asymmetric alpha carbon atom by base.

$$R'CO \cdot NH\overset{\overset{\displaystyle R_1}{|}}{C}HCOR'' \underset{+H^+}{\overset{-H^+}{\rightleftharpoons}} [R'CO \cdot NH\overset{\overset{\displaystyle R_1}{|}}{C}COR'']^-$$

This type of racemization has been observed for certain amino acids, i.e., tyrosine, phenylalanine, histidine, and tryptophan, and is thought to be due to the inductive effect of the aromatic side chain.[3] Another type of racemization by proton abstraction has been observed for three amino acids, i.e., cysteine, serine, and threonine. Under basic conditions, compound (XV) may undergo a β-elimination reaction to give the dehydro derivative (XVI) which, by the reverse addition reaction, gives the racemized compound (XV).

$$\underset{\text{(XV)}}{R'CO \cdot NH\overset{\overset{\displaystyle R}{|}}{\underset{|}{C}\underset{|}{H}}\,\underset{|}{\overset{CH-XH}{}}COR''} \underset{+HX}{\overset{-HX}{\rightleftharpoons}} \underset{\text{(XVI)}}{R'CO \cdot NH\overset{\overset{\displaystyle R}{|}}{\underset{||}{C}}—COR''}$$

(X = O or S; R = H or CH_3⁻)

Derivatives of these amino acids, with their secondary functional group protected (S- or O-benzyl, etc.), can also be racemized in the same manner. Both types of the aforementioned direct racemization routes are influenced by the basicity of the environment and the polarity of the solvent, and can occur at any time during peptide synthesis when strong alkaline conditions prevail. They are particularly prevalent when the R″ is a carboxyl-activating group.

---

[3] For a review see M. Bodanszky and M. A. Ondetti, "Peptide Synthesis." Wiley (Interscience), New York, 1966.

B. Separation of the Product from the Reaction Mixture

The separation of the product from the reaction mixture is related to the solubility properties of the components of the reaction mixture and to the yield of the desired product.[4] Solubility problems are insignificant in the synthesis of small peptides, i.e., those containing two to six amino acid residues. Indeed, the existing fractionation procedures (recrystallization, solvent extraction, etc.) almost always afford pure products obtained by the stepwise or fragment condensation approaches. In the synthesis of polypeptides, however, solubility problems are often insuperable. Protected polypeptides are usually very insoluble in organic solvents and hence, when such compounds are used as the amino and carboxyl components, the isolation of the product, a large polypeptide itself, is usually difficult and time-consuming and sometimes impossible. The situation is often aggravated by the reduced reactivity of the polypeptide amino and carboxyl components due to a decrease in the effective concentration of the reacting groups, which leads to a low yield of the product.

The stepwise elongation from the carboxyl terminal of the growing peptide chain does not eliminate greatly the aforementioned problems. Its principal disadvantage is the risk of racemization. Indeed, extensive studies have shown that activation of the carboxyl group of acyl peptides may be accompanied by racemization if the COOH-terminal amino acid is other than glycine and proline. Glycine does not have an asymmetric center, and N-acylated proline does not cyclize readily. Obviously then, use of this method for chain elongation would require rigorous tests for racemization after each amino acid addition. There are additional technical disadvantages in this approach that make its use very limited and impractical at this time. It requires a great number of steps for its execution and, as the size of the peptide chain increases, separation of the polypeptide product from the large activated carboxyl component becomes extremely difficult.

Stepwise elongation from the amino terminal of the growing peptide chain minimizes, to a great extent, the aforementioned problems, and is at the present time the method of choice for the construction of moderate-sized polypeptides. One of the advantages of this approach is that the risk of racemization is greatly reduced. Indeed, activation of amino acids protected with the available urethane or nonacyl type N^{α}-protecting groups precludes oxazolone formation, and subsequent coupling with an amino component affords nonracemic products.

Purification of the product obtained by this approach is easily handled, particularly at the early stages of chain elongation. At the

[4] P. G. Katsoyannis, *J. Polym. Sci.* **49**, 51 (1961).

completion of the reaction, the mixture normally consists of a newly formed protected peptide and unreacted amino and carboxyl components. The unreacted carboxyl component (N-protected amino acid) and the unreacted amino component are usually soluble in dilute base and dilute acid, respectively. Since the product, a protected peptide, is insoluble in these solvents, it can be isolated from the starting materials. The situation, however, becomes complicated as the chain length increases. Not only are the reaction yields reduced because of diminished reactivity of the amino component, but also its solubility and that of the product are often similar. To overcome these problems, a large excess of activated carboxyl component is employed to drive the coupling reaction to completion. The activated carboxyl component in large excess may, however, react with the normally sluggish secondary functional groups of amino acid residues found in the amino component. This type of side reaction may be circumvented by protection of the secondary functions with groups that are stable to the subsequent deblocking steps necessary for further chain elongation. An additional distinct disadvantage that limits the application of this procedure to the synthesis of large polypeptides is the great number of deblocking steps required. For example, at least 18 deblocking steps must be performed during the stepwise synthesis of an icosapeptide. Such multiple deblocking, particularly as the length of the peptide chain increases, creates serious problems. The chance of side reactions increases with each step, and the purification of a large number of intermediate peptides constitutes a time-consuming and difficult task. For these reasons the stepwise elongation from the amino terminus is presently used in the synthesis of peptides in the deca- or pentadecapeptide range, although syntheses of peptides containing 23 to 27 amino acid residues have been reported.

The fragment condensation approach, in which the amino and carboxyl components are partially protected peptide fragments, substantially reduces the great number of deblocking steps that the stepwise method imposes on the growing peptide chain and is most often the method of choice for the construction of large polypeptide chains. In essence, the overall scheme followed for the construction of a polypeptide by this approach is as follows: A peptide subunit, consisting of five to ten amino acid residues with the sequence of the COOH terminus of the polypeptide to be synthesized, is constructed by the stepwise approach. This peptide is deblocked at the amino end and serves as the amino component (A) for the subsequent coupling step. The stepwise approach is also employed to construct a peptide fragment which, ideally, contains three to six residues of the amino acid sequence adjacent to that of peptide (A) in the polypeptide to be synthesized. This peptide is deblocked at the

COOH-terminal end and serves as the carboxyl component (B). Components (B) and (A) are condensed by an appropriate procedure to form a larger fragment (B·A), which, following purification, is deblocked at the amino end and serves as the new amino component for the next coupling step. This cycle is repeated until the desired polypeptide is constructed. The success of this approach depends greatly on the correct selection of the individual peptide fragments and the methods used for the subsequent coupling of these fragments. The likelihood of racemization can be eliminated or minimized by selecting, as carboxyl components, fragments whose COOH-terminal residues are amino acids which cannot be racemized or are less sensitive to racemization. Fragments with COOH-terminal glycine or proline are the most obvious choices. The aromatic amino acids, phenylalanine, tyrosine, histidine, and tryptophan, and amino acids subject to β-elimination, should be used with extreme caution. Although detailed studies regarding the sensitivity to racemization of individual amino acids are not available, experience has shown that aliphatic monoamino monocarboxylic amino acids are the least sensitive.[3] Furthermore, it also has been shown that, under properly controlled conditions (to be discussed in the experimental section), several of the existing activating procedures may be employed with little risk of racemization, even when the amino acids that are most prone to racemization are involved.

As previously indicated for the stepwise approach, similar difficulties may be encountered with the fragment condensation procedure, especially as the peptide fragments necessarily become relatively large. The experimental difficulties may again be attributed to two interrelated factors. First, with increasing size of the condensing fragments, their solubility characteristics tend to differ slightly from that of the condensed product, and, second, both the increased chain length and accompanying substantial decrease in their solubilities seriously impede the condensation. Minimization of these difficulties is achieved by the correct choice of fragments and experimental conditions for the condensation. The coupling may be driven, in favorable cases, to near completion by allowing the amino component to react with a large excess of a smaller activated carboxyl component. The final reaction mixture then consists predominantly of the newly formed peptide and unreacted carboxyl component. It then becomes possible, using routine extraction procedures (dilute base, organic solvents) to separate the carboxyl component from the product. To facilitate completion of the coupling reaction it is also advisable to avoid the use of carboxyl components with sterically hindered and hence less reactive amino acids as the COOH-terminal residue, i.e., valine or isoleucine. It must be emphasized, however, that, for the reasons outlined above, the synthesis of very large polypeptides (>50 residues) is, at the present time, an almost insuperable task. For the synthesis of large polypeptides (ca. 25–30 residues), the fragment condensation procedure is the method of choice and has been successfully applied in many instances.

IV. Synthesis[5] of Carboxyl Components (Stage 1)

The synthesis of the carboxyl component constitutes the first stage in the overall scheme for peptide synthesis and, in essence, consists in the protection of the α-amino group of an amino acid (or, very seldom, of a peptide). It should be noted that, if cysteine or lysine are parts of the carboxyl component, their secondary functional groups must always be protected. If other polyfunctional amino acids are parts of the carboxyl component, protection of their secondary functional groups is not always an absolute requirement and depends on the nature of the secondary group and the method of activation of the carboxyl component. The blocking of the secondary functional groups will be discussed in a later section of this chapter.

Protection of the α-Amino Group. An amino protecting group must fulfill certain requirements: its introduction into an amino acid or peptide must proceed smoothly without causing any racemization of the molecule; it must be stable during the various operations required for the synthesis of a peptide; and finally, it must be removed selectively from the product without affecting the rest of the molecule. This last requirement is of crucial importance when the peptide molecule contains protected secondary functions.

In the long history of peptide chemistry, a considerable number of amino protecting groups, which fulfill to different degrees the above-mentioned requirements, have been employed in the synthesis of peptides. However, in this section we will discuss in detail the properties and uses of the benzyloxycarbonyl, *tert*-butyloxycarbonyl, and *o*-nitrophenylsulfenyl groups only. In the authors' experience, these groups are adequate for the synthesis of long polypeptide chains. Certain other amino protecting groups, which in the last two decades have had limited application in the synthesis of biologically important peptides, are cursorily treated.

A. Benzyloxycarbonyl (Carbobenzoxy) Group

The benzyloxycarbonyl group ($C_6H_5CH_2OCO—$), a representative of a urethane-type protecting group (ROCO—), was suggested by Bergmann

---

[5] The syntheses described throughout this chapter will be those that have been carried out reproducibly in the authors' laboratory; often they are modifications of published procedures, which will be cited in the appropriate sections.

and Zervas[6] in 1932 and still remains one of the most important amino protectors in modern peptide chemistry. This group is stable to dilute acids, bases, and hot organic solvents. It can be removed by a variety of procedures. The most elegant of these procedures is catalytic hydrogenation,[6] which distinguishes the benzyloxycarbonyl group (and its modifications) from all other presently available amino group protectors. Hydrogenation proceeds rapidly, usually in a weakly acidic solution, and the by-products of the reaction, carbon dioxide and toluene, are volatile and easily removed from the product. A disadvantage of this method of deblocking is that it is ineffective in the presence of sulfur-containing amino acids which, in general, act as catalyst poisons. An alternative method for cleavage of the benzyloxycarbonyl group, often used with sulfur-containing peptides, involves the use of hydrogen bromide in acetic or trifluoroacetic acids. These latter reagents can also effect cleavage of a variety of other protecting groups. This procedure is frequently used for the stepwise elongation of peptides, provided, of course, that the growing peptide chain does not contain secondary functional groups protected with moieties sensitive to such treatment. Sodium in liquid ammonia or liquid hydrogen fluoride can very effectively cleave N-benzyloxycarbonyl groups; however, both these procedures cleave practically all the existing protecting groups and are presently employed for the total deblocking of the final product.

Benzyloxycarbonylation of amino acids and peptides proceeds readily in high yields and without racemization by allowing the commercially available benzyloxycarbonyl chloride to react with an amino acid or peptide in the presence of base (sodium hydroxide, magnesium oxide, sodium or potassium carbonate, sodium bicarbonate, or triethylamine). In practice, benzyloxycarbonyl chloride is added either directly to the aqueous alkaline solution of the amino acid or peptide (0° to 5°) or as a solution in an organic solvent, such as dioxane. A typical benzyloxycarbonylation proceeds as follows: A solution of an amino acid (0.3 mol) in 76 ml of 4 N sodium hydroxide, in a 500-ml precipitating jar equipped with a mechanical stirrer, is cooled in an ice bath. To the solution is added, under vigorous stirring, alternately benzyloxycarbonyl chloride (46.6 ml) and 4 N sodium hydroxide (80 ml) in five equal portions over a period of 30 min. The pH, checked with litmus paper, should be above 8.0 at all times, adding if necessary an additional amount of 4 N sodium hydroxide. After stirring for an additional 30 min, the solution is extracted twice with 50-ml portions of ether to remove excess reagent, and the cold aqueous layer is acidified to pH 2.0 with concentrated hydro-

[6] M. Bergmann and L. Zervas, Ber. Dtsch. Chem. Ges. 65, 1192 (1932).

chloric acid. The product (which usually separates out as an oil and, for most amino acids, crystallizes on further cooling) is extracted into ethyl acetate (ca. 600 ml). The organic layer is washed with water to remove excess acid, dried over anhydrous magnesium sulfate, and concentrated at 45° under reduced pressure to a small volume. Many N-benzyloxycarbonyl amino acids can be crystallized by addition of petroleum ether to the ethyl acetate solution and can be recrystallized in the same way. A complete list of the physical characteristics of most amino acid derivatives and references for their preparation and purification may be found in a paper by G. A. Fletcher and J. H. Jones.[7] In many cases, the crude residue remaining after complete removal of the ethyl acetate can be converted to N-benzyloxycarbonyl amino acid active ester, thus eliminating the losses that usually accompany crystallization of the protected amino acid.

Polyfunctional amino acids, such as cysteine, lysine, histidine, serine, tyrosine, and arginine, usually require protection of their secondary functions (to be described below) prior to the introduction of the benzyloxycarbonyl group. However, certain of these amino acids, such as serine and arginine, can be selectively acylated without prior protection of their secondary functions. Benzyloxycarbonylation of serine is carried out at pH 9.8 to 10.0 to prevent formation of the N,O-diacylated derivative. Preferably the N-protection is performed in the presence of magnesium oxide.[8] Formation of N^α-protected arginine is accomplished by carrying out the reaction in sodium bicarbonate.[9] Excess benzyloxycarbonyl chloride and highly basic conditions lead to formation of the di- and trisubstituted derivatives of arginine.[10]

B. *tert*-Butyloxycarbonyl Group

This group [$(CH_3)_3C$-OCO—], introduced by McKay and Albertson[11] and by Anderson and McGregor[12] in 1957, has attained at the present time the prominence of the benzyloxycarbonyl group. This urethane-type protecting group has essentially the same advantages as the benzyloxycarbonyl moiety but differs significantly from the latter in a number

[7] G. A. Fletcher and J. H. Jones, *Int. J. Pept. Protein Res.* **4**, 347 (1972).
[8] D. F. DeTar, F. F. Rogers, and H. Bach, *J. Am. Chem. Soc.* **89**, 3039 (1967).
[9] R. A. Boissonnas, S. Guttmann, R. L. Huguenin, P.-A. Jaquenoud, and E. Sandrin, *Helv. Chim. Acta* **41**, 1867 (1958).
[10] L. Zervas, T. T. Otani, M. Winitz, and J. P. Greenstein, *J. Am. Chem. Soc.* **81**, 2878 (1959).
[11] F. C. McKay and N. F. Albertson, *J. Am. Chem. Soc.* **79**, 4686 (1957).
[12] G. W. Anderson and A. C. McGregor, *J. Am. Chem. Soc.* **79**, 6180 (1957).

of its properties. Most *tert*-butyloxycarbonyl amino acid derivatives are crystalline solids readily soluble in most organic solvents. The *N-tert*-butyloxycarbonyl group, as opposed to the *N*-benzyloxycarbonyl moiety, is very stable to catalytic hydrogenation and to treatment with sodium in liquid ammonia. In sharp contrast to the benzyloxycarbonyl group, it is rapidly cleaved under relatively mild acidic conditions. Short exposure, ca. 30–60 min, to trifluoroacetic acid or to solutions of hydrogen chloride in organic solvents or long-term exposure (3 hr) to formic acid suffices to cleave this group. In strongly acid media, such as hydrogen bromide in acetic or trifluoroacetic acid, this group is, of course, cleaved rapidly and quantitatively. The sensitivity of the *N-tert*-butyloxycarbonyl group to acidic media necessitates the cautious handling of peptide derivatives protected with this moiety. Long exposure even to dilute acids during washing procedures, or the presence of even traces of acids during purification or prolonged heating in organic solvents, should be avoided. The introduction of the *tert*-butyloxycarbonyl group proceeds smoothly, under carefully controlled conditions, without racemization and in high yields, by allowing the commercially available *tert*-butoxycarbonyl azide (*tert*-butylazidoformate) to react with an amino acid in aqueous dioxane.[13]

1. A typical *tert*-butyloxycarbonylation of an amino acid or amino acid derivative (but no amino acid ester) is described below.[13] We have devised and describe below an apparatus and procedure that is particularly appropriate when large amounts of N-protected derivatives are required. Since *tert*-butylazidoformate is a very toxic material and the vapors may cause severe headaches, all operations should be conducted in a hood.

The chosen amino acid (0.4 mol) is placed in a 1-liter three-neck round-bottom flask containing dioxane–water (1:1; 140 ml) and to the resulting suspension, stirred with a heavy-duty magnetic stirrer, *tert*-butylazidoformate (55.2 g) is added in one portion. The reaction vessel is then attached to a pH-stat and nitrogen flush line, as shown in Fig. 1. The glass electrode from the pH-stat (Radiometer PHM26c pH meter and TTT116 titrator) is placed in one of the side openings of the reaction flask, and an airtight seal is made by wrapping parafilm around the electrode and flask. A two-hole rubber stopper containing the inlet tube from the reservoir of the pH-stat and the exhaust tube for the nitrogen is attached to the center opening, and the nitrogen flush line is connected to the remaining opening of the reaction flask. Nitrogen gas is passed through water before entering the flask and the exhaust gas is passed through water so that a slow rate of gas flow can be adjusted. It is im-

---

[13] E. Schnabel, *Justus Liebigs Ann. Chem.* **702**, 188 (1967).

tert-BUTYLOXYCARBONYLAMINO ACIDS AND DERIVATIVES. REACTION CONDITIONS AND CRYSTALLIZATION SOLVENTS

Amino acid or derivative	pH of reaction	Crystallization solvent	Melting point (°C)	Reference
L-Valine	9.5	Ether–petroleum ether	77–79	a
L-Tyrosine	9.8	Ethyl acetate–petroleum ether	96–98	a
L-Tryptophan	9.8	Ethyl acetate–petroleum ether	135–137	a
L-Threonine	9.5	Petroleum ether	77–80	a
L-Serine	9.4	Ether–petroleum ether (hydrate)	75–78	a
O-Benzyl-L-tyrosine	10.4	Ethyl acetate–petroleum ether	109–110	a
N^αBenzyloxycarbonyl-L-lysine	9.6	Ethyl acetate–petroleum ether	76–78	a
L-Methionine	9.8	Oil	—	a
L-Phenylalanine	10.2	Ether-petroleum ether	85–87	a
L-Proline	8.6	Ethyl acetate–petroleum ether	135–137	a
S-Benzyl-L-cysteine	9.6	Ether–petroleum ether	63–65	a
S-Diphenylmethyl-L-cysteine	9.8	Petroleum ether	95–98	b
L-Glutamic acid	10.0	Ethyl acetate–petroleum ether	110–112	a
L-Glutamine	10.3	Ethyl acetate	116–118	a
Glycine	10.0	Ethyl acetate–petroleum ether	95–96	a
L-Isoleucine	9.8	Petroleum ether	66–68	a
L-Leucine[c]	9.9	—	78–81	a
N^ϵBenzyloxycarbonyl-L-lysine	10.3	Oil		a
N^ϵ-Tosyl-L-lysine	9.9	Oil		a
L-Aspartic acid	10.2	Ethyl acetate–petroleum ether	115–117	a
L-Asparagine[c]	9.5	—	176	a
O-2,6-Dichlorobenzyl-L-tyrosine	9.8	Ether–petroleum ether	109–111	d
N^ϵ-O-Bromobenzyloxycarbonyl-L-lysine	10.2	Oil	—	d
L-Arginine[e,f]	9.2	—	112–113	g
N^ω-Tosyl-L-arginine	9.8	Ethyl acetate	98–100	a
N^ω-Nitro-L-arginine	9.8	Ethyl acetate	124–125	a
L-Alanine	10.1	Ether–petroleum ether	80–83	a
N^{im}-Benzyl-L-histidine[h]	9.8	Pyridine–ethyl acetate–ether	184–186	a
γ-Benzyl-L-glutamic acid	—	Oil	—	i
β-Benzyl-L-aspartic acid	—	Ether–petroleum ether	99–101	i

[a] E. Schnabel, *Justus Liebigs Ann. Chem.* **702**, 188 (1967).

[b] R. Schwyzer, A. Tun-Kyi, M. Caviezel, and P. Moses, *Helv. Chim. Acta* **53**, 15 (1970).

[c] Crystallizes out upon acidification of the aqueous layer (see text), is filtered, washed, and dried.

[d] D. Yamashiro and C. H. Li, *J. Am. Chem. Soc.* **95**, 1310 (1973).

[e] N^α-Acylation is completed in 3 hr at 45–50°.

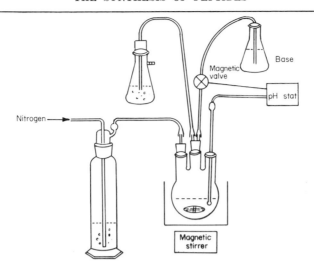

FIG. 1. Apparatus used for the *tert*-butyloxycarbonylation of amino acids.

perative to carry out the reaction in a nitrogen atmosphere to prevent base-uptake due to atmospheric carbon dioxide. The reservoir of the pH-stat is filled with 4 *N* sodium hydroxide. The pH of the reaction is maintained between 8.5 and 10.2, depending on the amino acid used. The table shows the pH ranges used for each amino acid or derivative. The reaction is finished when there is no more base uptake, and this may take more than 24 hr for certain amino acids.

After completion of the reaction, the solution is extracted twice with an equal volume of ether to remove unreacted azide. The extracted aqueous layer is cooled to 4° and acidified to pH 2 by the slow addition of cold 6 *N* hydrochloric acid. Most N-*tert*-butyloxycarbonyl amino acids separate out as oils and are extracted directly into ethyl acetate (1.5 liters). For the derivatives of serine, glycine, proline, threonine, and *N*ω-tosylarginine, however, which have some solubility in water, the aqueous layer must be saturated with sodium chloride in order to quantitatively extract the N-protected product into ethyl acetate. The organic layer is washed three times with water or salt-saturated water (for water-

---

f The pH of the aqueous layer is adjusted to 3, and the product separates out as the monohydrochloride. Addition of sodium chloride facilitates crystallization of the product.

g D. Yamashiro, J. Blake, and C. H. Li, *J. Am. Chem. Soc.* **94**, 2855 (1972).

h The aqueous layer is adjusted to pH 6.0 with solid citric acid and concentrated under reduced pressure until product crystallizes out.

i E. Sandrin and R. A. Boissonnas, *Helv. Chim. Acta* **46**, 1637 (1963).

j D. A. Laufer and E. R. Blout, *J. Am. Chem. Soc.* **89**, 1246 (1967).

soluble derivatives), dried over anhydrous magnesium sulfate, and taken to dryness under reduced pressure. The residue in many cases can be crystallized from ethyl acetate–petroleum ether. Certain amino acid derivatives are oily substances and can be crystallized by conversion to their dicyclohexylamine salts, following essentially the procedure described below for the corresponding o-nitrophenylsulfenyl derivatives. Alternatively, the oily $tert$-butyloxycarbonylated derivatives can be used directly for the preparation of their crystalline active esters (see below).

2. In cases where acylation of amino acid esters are to be performed, i.e., γ-benzyl glutamate, the above-mentioned procedure is unsatisfactory owing to the partial hydrolysis of the ester. An alternative procedure for $tert$-butyloxycarbonylation that gives excellent results with amino acid esters, and which we find convenient for acylation of other amino acid derivatives as well, follows.[14]

To a suspension of the chosen amino acid or amino acid derivative (0.1 mol) in dimethyl sulfoxide (200 ml), triethylamine (28 ml) and $tert$-butylazidoformate (18 ml) are added and the mixture is stirred at room temperature until a solution is obtained. This requires from 20–40 hr of stirring. The solution is diluted with water (750 ml) and extracted three times with equal volumes of ether. The aqueous phase is subsequently worked up as described in the previously mentioned procedure.

A list of the physical properties of the N-$tert$-butyloxycarbonylamino acids and derivatives is given in the table, and references to their original preparation and purification may be found in references 7 and 13. It should be noted that with this particular acylation reagent and with strict control of pH, prior protection of the secondary functions of serine, threonine, tyrosine, and arginine is not required. Histidine, however, cannot be monoacylated by this method. The acylation of histidine is achieved by storing a pyridine solution of histidine methyl ester and $tert$-butylazidoformate at room temperature for several days.[15]

It was reported recently[15a] that $tert$-butylazidoformate is thermally unstable (>80°) and decomposes with apparent detonation. Hence, this reagent now is not readily available commercially in the United States. A nonexplosive replacement for this reagent, 2-$tert$-butyloxycarbonyloxy-imino-2-phenylacetonitrile (BOC-ON),[15b] is now commercially available

[14] J. M. Stewart and J. D. Young, "Solid Phase Peptide Synthesis." Freeman, San Francisco, 1969.

[15] B. O. Handford, T. A. Hylton, K.-T. Wang, and B. Weinstein, $J.$ $Org.$ $Chem.$ **33**, 4251 (1968).

[15a] W. J. Fenlon, Chem. Eng. News **54** [22], 3 (1976); H. C. Koppel, $ibid$ **54** [39] 5 (1976).

[15b] M. Itoh, D. Hagiwara, and T. Kamiya, $Tetrahedron$ $Lett.$ p. 4393 (1975).

from Aldrich Chemical Co. In our experience this new reagent is in many ways superior to the azide and has been used successfully in our laboratory in the last few months for the synthesis of *tert*-butyloxycarbonyl derivatives of several amino acids. Among these are the derivatives of *S*-diphenylmethyl-L-cysteine, *O*-benzyl-L-tyrosine, L-histidine methyl ester, γ-benzyl-glutamic acid, and L-isoleucine. *tert*-Butyloxycarbonylation of amino acids by the use of the new reagent can be carried out in aqueous organic solvents in the presence of base, as described in a publication of Aldrich Chemical Co.[15c] In our laboratory, *tert*-butyloxycarbonylation of amino acids is readily accomplished and in high yields by employing procedure B,2, above. Quantities of reagents used are as follows: amino acid or derivative, 0.1 mol; triethylamine 14 ml; BOC-ON, 0.105 mol; and dimethyl sulfoxide, 100–150 ml.

C. o-Nitrophenylsulfenyl Group

This group (o-NO$_2$C$_6$H$_4$S—) was first employed as an amino protector in peptide synthesis by Zervas and co-workers[16] in 1963, and introduces a new dimension in selective deblocking during peptide synthesis. The removal of the *o*-nitrophenylsulfenyl group, even upon brief exposure to low concentrations of hydrogen chloride in a variety of organic solvents[16] or treatment with nucleophiles[17] (thioacetamide, sodium sulfite, HCN), sharply distinguishes it not only from the *tert*-butyloxycarbonyl moiety but also from all presently available protecting agents. It is especially useful in stepwise elongation when benzyloxycarbonyl, *tert*-butyloxycarbonyl, and a variety of other sensitive agents such as *tert*-butyl esters or ethers (see below) are used as side-chain protectors. Its introduction as an amino protector proceeds smoothly without racemization. In common with the other two N-protectors, the *N-o*-nitrophenylsulfenyl group is stable under basic conditions. The high sensitivity of the *N-o*-nitrophenylsulfenyl group to very mild acidic conditions requires, in the handling of its derivatives, even greater care than that exercised with the *tert*-butyloxycarbonyl group. It is this property of this N-protector that often becomes one of its principal drawbacks. Another possible drawback of this amino protector is encountered in the separation of the by-products formed upon its removal from a peptide chain. These by-products, presumably di-*o*-nitrophenylsulfenyl disulfide and other uncharacterized derivatives, are highly insoluble and often coprecipitate with the peptide product. Indeed, it is not unusual, in spite

[15c] *Aldrichimica Acta* 9[3] (1976).

[16] L. Zervas, D. Borovas, and E. Gazis, *J. Am. Chem. Soc.* **85**, 3660 (1963).

[17] W. Kessler and B. Iselin, *Helv. Chim. Acta* **49**, 1330 (1966).

of repeated washings, to carry such by-products into subsequent steps of chain elongation.

Reaction of commercially available o-nitrophenylsulfenyl chloride with an amino acid or amino acid derivative in aqueous dioxane under alkaline conditions suffices to produce the brightly yellow-colored N-protected derivatives within 15–20 min. The N-protected amino acids prepared in this fashion are isolated from the undesirable by-products formed during the reaction and stored as the stable dicyclohexylammonium salts. Prior protection of secondary groups of serine, threonine, tyrosine, arginine, and histidine is not required for their o-nitrophenylsulfenylation.

A typical N-o-nitrophenylsulfenylation proceeds as follows[16]: The amino acid (0.08 mol) is dissolved in a mixture of 2 N sodium hydroxide (40 ml) and dioxane (100 ml). To this vigorously stirred solution, o-nitrophenylsulfenyl chloride (16.7 g) in ten equal portions and 2 N sodium hydroxide solution (48 ml) are added over a period of 15 min. The sodium hydroxide solution is added dropwise from a dropping funnel. The resulting suspension is diluted with 800 ml of water, filtered, the filtrate cooled to 4° and acidified with 1 N sulfuric acid to pH 2.0. With most amino acids the product separates out as an oil which is extracted into ethyl acetate (800 ml), washed twice with an equal volume of water and dried over anhydrous magnesium sulfate. This solution is concentrated under reduced pressure (35°–40°) to 200 ml and mixed with dicyclohexylamine (16 ml). The N-o-nitrophenylsulfenyl amino acid dicyclohexylamine salt crystallizes out and may be recrystallized from a variety of organic solvents.[16] The N-o-nitrophenylsulfenyl derivatives of asparagine and glutamine crystallize out during the acidification step and can be isolated directly by filtration. After recrystallization from methanol, these two derivatives can be stored at −20°.

The N-o-nitrophenylsulfenylation of γ-benzylglutamic acid and β-benzylaspartic acid proceeds in exactly the same way,[16] the only difference being that these amino acid esters are suspended in water (40 ml) and dioxane (100 ml) and then treated with o-nitrophenylsulfenyl chloride (16.7 g) and 2 N sodium hydroxide (48 ml), as above.

The N^{α}-protection of histidine and arginine proceeds through the following modified version of the general procedure.

N-o-Nitrophenylsulfenyl-L-histidine.[18] To a stirred suspension of L-histidine (9.3 g; 0.06 mol) in a mixture of methanol (75 ml) and water (75 ml) containing triethylamine (16.8 ml), o-nitrophenylsulfenyl chlo-

---

[18] I. Phocas, C. Yovanidis, I. Photaki, and L. Zervas, *J. Chem. Soc. C* p. 1506 (1967).

ride (22.8 g; 0.12 mol) is added over a period of 10 min at 30°. The reaction mixture is subsequently stirred for an additional hour at 40°, cooled to 20°, and the precipitated material, which is a mixture of the desired product and various by-products, is filtered off. This precipitate is suspended in water (60 ml), the pH is adjusted to 9.0 (pH meter) with 1 N ammonium hydroxide, and the undissolved by-product is filtered off. Upon adjustment of the pH of the filtrate to 5.6 with acetic acid, the N-substituted histidine derivative precipitates. After 20 hr at 4°, the precipitated product is filtered, triturated with methanol (30 ml) and dried; yield 5.4 g.

The synthesis of N^α-o-nitrophenylsulfenyl-L-arginine is carried out by the same procedure with the exception that 85% methanol is used as the solvent and the reaction is carried out at 10°.

A listing of the physical characteristics of most amino acid derivatives may be found in reference 7.

D. Other α-Amino Protecting Groups[19]

There are a multitude of other N^α-protecting groups which have been described over the past two decades. With a few exceptions, none of these N-protectors have been used with regularity in polypeptide synthesis. Several substituted benzyloxycarbonyl derivatives, such as the p-nitro, p-chloro, or p-bromo, offer no substantial advantages over the parent group other than the easier crystallizability of their amino acid derivatives and their greater resistance to acidic cleavage. Similarly, the azo-protecting groups, such as the p-(p'-methoxyphenylazo)-benzyloxycarbonyl,[20] which, because of their color, may facilitate purification of certain peptide derivatives, do not offer any distinct advantage over the benzyloxycarbonyl group. A number of urethane-type protecting groups, such as the $tert$-amyloxycarbonyl, p-methoxybenzyloxycarbonyl, etc., have also been described. These derivatives have acid sensitivities similar to those of the $tert$-butyloxycarbonyl moiety but have found no general use in peptide synthesis.

Two moieties that hold some promise in peptide synthesis in solution are the 2-(p-biphenylisopropyloxycarbonyl)-[21] and the α,α'-dimethyl-3,5'-dimethoxybenzyloxycarbonyl[22] functions. The former [$C_6H_5C_6H_4C$-

[19] For a review see (a) E. Schröder and K. Lübke, "The Peptides," Vol. 1. Academic Press, New York, 1965; (b) P. G. Katsoyannis and J. Z. Ginos, *Annu. Rev. Biochem.* **38**, 710 (1969).
[20] R. Schwyzer, P. Sieber, and K. Zatskó, *Helv. Chim. Acta* **41**, 491 (1958).
[21] P. Sieber and B. Iselin, *Helv. Chim. Acta* **52**, 1525 (1969).
[22] C. Birr, W. Lochinger, G. Stahrrke, and P. Lang, *Justus Liebigs Ann. Chem.* **763**, 162 (1972).

$(CH_3)_2$—OCO—] is more sensitive to weak acids and to very low concentrations of strong acids, than the o-nitrophenylsulfenyl group. It is quantitatively cleaved by treatment with 80% aqueous acetic acid[21] (20 hr) or by 0.5% trifluoroacetic acid in methylene chloride[23] (approximately 10 min). The o-nitrophenylsulfenyl group is partially stable under these conditions. The main disadvantages of this reagent are that it is difficult and costly to prepare and the conditions for its removal from large polypeptides have as yet not been adequately defined.[24] The latter group $[(CH_3O)_2C_6H_3C(CH_3)_2$—OCO—] is also fairly sensitive to acid (5% trifluoroacetic acid), though slightly more stable than the former derivative. Of some interest is the observation that it can be removed from its amino acid derivatives photolytically. The triphenylmethyl [trityl: $(C_6H_5)_3C$—] is another group which has been used occasionally in recent years as N^α-protector.[25,26] Its principal advantage resides in its relative instability under acid conditions that permits its selective removal in the presence of the 2-(p-biphenylisopropyloxycarbonyl) moiety.[26a] Because of the difficulties encountered in the tritylation of amino acids and the subsequent activation of such derivatives (due to steric hindrance), the trityl group is often used for N^α-protection of peptide esters.

V. Synthesis of Amino Components (Stage 2)

This stage of the overall scheme for peptide synthesis consists in the protection of the α-carboxyl group of an amino acid (or, rarely, of a peptide).

Protection of the α-Carboxyl Group. A carboxyl protecting group must fulfill all the requirements set forth for amino-protecting functions. This is commonly accomplished by conversion of the carboxyl group to various esters which are alkali- or acid-sensitive. Protection of the α-carboxyl moiety by conversion to an ester is not an absolute requirement and, in its anionic form, may in particular instances facilitate synthesis of peptides. Furthermore, synthesis employing such unprotected fragments not only eliminates the need for deprotection of the COOH-terminal carboxyl

---

[23] S. Wang and R. B. Merrifield, *J. Am. Chem. Soc.* **91**, 6488 (1969).

[24] It is our experience that large polypeptides are not soluble in 80% acetic acid or in methylene chloride. Attempts to remove this group from polypeptides in solutions of dimethylformamide and 0.5% trifluoroacetic acid were not successful.

[25] G. Amiard, R. Heymes, and L. Velluz, *Bull. Soc. Chim. Fr.* p. 191 (1955).

[26] L. Zervas and D. M. Theodoropoulos, *J. Am. Chem. Soc.* **78**, 1359 (1956).

[26a] P. Sieber, B. Kamber, A. Hartmann, A. Jöhl, B. Riniker, and W. Rittel, *Helv. Chim. Acta* **57**, 2617 (1974).

group of the end product, but is especially useful in the synthesis of alkali-sensitive peptides. However, leaving the α-carboxyl group of the amino component unprotected will limit the methods that can be employed for its coupling to the carboxyl component to the azide, mixed anhydride and preformed active esters. Further, the use of an unprotected amino component will make the separation of the end product from the starting carboxyl component (both present as acids) a difficult task.

In general, four types of α-carboxyl protectors have been extensively employed in polypeptide synthesis: methyl or ethyl, benzyl, p-nitrobenzyl, and *tert*-butyl esters.

The methyl or ethyl esters are generally employed for temporary protection of the α-carboxyl group during the stepwise synthesis of a peptide fragment that will eventually be used as the carboxyl component in the fragment condensation approach. Simple saponification or hydrazinolysis produces the free carboxyl peptide and hydrazide, respectively, which can be subsequently activated. A potential drawback of methyl or ethyl esters is the chance of racemization or of damage to the peptide chain in the alkaline medium used for their removal.

The benzyl ester moiety has in general the same properties as the benzyloxycarbonyl group. It is cleaved by hydrogenation,[27,28] sodium in liquid ammonia,[29,30] liquid hydrogen fluoride[31] or hydrogen bromide in acetic acid.[32] The benzyl esters are relatively stable under the acid conditions used to remove the *tert*-butyloxycarbonyl and *o*-nitrophenylsulfenyl groups, and are used to permanently[33] protect the α-carboxyl group, during the course of the synthesis, where *tert*-butyloxycarbonyl or *o*-nitrophenylsulfenyl groups are used as N^α-protectors. This group also, like the methyl or ethyl esters, can be removed by saponification or converted to a hydrazide by hydrazinolysis.

The p-nitrobenzyl ester group was introduced in peptide chemistry to ensure a greater measure of stability to hydrolysis by strong acids.[32] This moiety has all the properties of the benzyl ester group but, in

[27] M. Bergmann, L. Zervas, and L. Salzmann, *Ber. Dtsch. Chem. Ges.* **66**, 1288 (1933).

[28] M. Bergmann and J. S. Fruton, *J. Biol. Chem.* **117**, 189 (1937).

[29] R. H. Sifferd and V. du Vigneaud, *J. Biol. Chem.* **108**, 753 (1935).

[30] P. G. Katsoyannis, D. T. Gish, G. P. Hess, and V. du Vigneaud, *J. Am. Chem. Soc.* **80**, 2558 (1958).

[31] S. Sakakibara, *in* "Chemistry and Biochemistry of Amino Acids, Peptides, and Proteins" (B. Weinstein, ed.), p. 51. Dekker, New York, 1971.

[32] H. Schwarz and K. Arakawa, *J. Am. Chem. Soc.* **81**, 5691 (1959).

[33] The term "permanent" is used to indicate that the group will remain throughout the synthesis and will be removed in a final deprotection step.

contrast to the unsubstituted congener, is stable to hydrogen bromide in acetic acid and to liquid hydrogen fluoride. It is used as a permanent C^α-protector when the benzyloxycarbonyl group is used as the N^α-protector, and consequently must be removed selectively from the growing peptide chain.

The *tert*-butyl ester[34] group has, in general, the same properties as the *tert*-butyloxycarbonyl moiety. It is cleaved by acids, such as trifluoroacetic acid and formic acid, and it is stable to hydrogenation and to treatment with sodium in liquid ammonia. Consequently, it is used as a permanent C^α-protector when the *o*-nitrophenylsulfenyl or other equally acid-labile groups are used as N^α-protectors. Similarly, it is used as a permanent C^α-protector when the benzyloxycarbonyl group is employed as N^α-protector which can be removed by hydrogenation throughout the course of chain elongation. The *tert*-butyl ester moiety is also used as a temporary C^α-protector during the stepwise synthesis of fragments which, upon mild acid treatment, become carboxyl components for fragment condensation. In contrast to all the aforementioned ester groups, this moiety is stable to alkali treatment.

Various other groups, such as the diphenylmethyl, *p*-methoxybenzyl or 2,4,6-trimethylbenzyl ester, have been employed occasionally as C^α-protectors. These derivatives are very similar in their chemistry to the *tert*-butyl group, but offer very little advantage over the latter. An interesting C^α-protector is the *p*-thiomethylphenyl ester,[35] which has the acid stability of the methyl ester group and, on mild oxidation, is transformed into an activated ester.

A. Amino Acid Methyl or Ethyl Esters

Identical procedures are employed for the esterification of all amino acids. Esterification is accomplished by treatment of an alcoholic suspension of the chosen amino acid with hydrogen chloride (Fischer's method) or by addition of the chosen amino acid to a premixed solution of the parent alcohol and thionyl chloride (thionyl chloride method). No special precautions are required for the esterification of the polyfunctional amino acids. Under these conditions acidic amino acids will afford diesters. Selective esterification of the carboxyl groups of the latter amino acids will be discussed in a later section. Certain amino acids are sterically hindered (valine, isoleucine, etc.) or are sparingly soluble in alcoholic hydrogen chloride (lysine). In such cases it is advisable to

---

[34] G. W. Anderson and F. M. Callahan, *J. Am. Chem. Soc.* **82,** 3359 (1960).

[35] B. J. Johnson and P. M. Jacobs, *J. Chem. Soc., Chem Commun.* p. 73 (1968).

carry out the esterification under reflux conditions or to store the alcoholic-hydrogen chloride suspension of such amino acids for at least 24 hr before evaporation of the solvent. If the thionyl chloride method is employed for the esterification of such amino acids, it is advisable to store the alcoholic suspensions more than 2 days before evaporation of the solvent. It is recommended that thin-layer chromatography on silica gel be used to check for the presence of free amino acids. A list of the physical properties of the various amino acid esters and references to their original preparation and purification may be found in reference 7.

1. "Fischer's Method"

The amino acid (0.2 mol) is suspended in 500 ml of absolute methanol (ethanol), and dry hydrogen chloride is passed through the suspension, without cooling, until the amino acid is dissolved. The reaction vessel is a 1-liter round-bottom flask equipped with a reflux condenser and a gas inlet tube and protected from the atmosphere with a calcium chloride drying tube. The reaction vessel is then cooled to room temperature and the solvent is removed under reduced pressure (45°–50°). The residue is redissolved in 500 ml of absolute methanol (ethanol) and hydrogen chloride is passed through the solution for 1 hr. The amino acid ester as the hydrochloride salt may begin to crystallize out at this stage. The mixture is concentrated under reduced pressure to ca. one-half of the original volume, cooled, and mixed with anhydrous ether. The crystalline amino acid ester hydrochloride is collected and recrystallized from methanol (ethanol)-ether.

2. Thionyl Chloride Method[36]

Thionyl chloride (16 ml) is added dropwise, over a period of 30 min and with vigorous stirring, to 100 ml of absolute methanol (ethanol) placed in a round-bottom flask equipped with a calcium chloride drying tube and cooled to −10° (Dry Ice–acetone). To this solution the chosen amino acid (0.2 mol) is added and the mixture is stirred for 1 hr at 4° and stored at room temperature overnight. The reaction mixture is then concentrated to dryness under reduced pressure (45°–50°), 100 ml of methanol (ethanol) is added and the concentration to dryness repeated. This last procedure is repeated twice to secure the removal of excess hydrogen chloride. The crystalline amino acid ester hydrochloride is collected and recrystallized from methanol (ethanol)-ether.

[36] M. Brenner and W. Huber, *Helv. Chim. Acta* **36**, 1109 (1953).

B. Amino Acid Benzyl Esters

In general, such esters are prepared by treating the chosen amino acid with benzyl alcohol in the presence of acid catalysts, such as benzenesulfonic acid[37] or, more commonly, p-toluenesulfonic acid.[38,39] The esterification is carried out in such solvents as benzene, toluene, or carbon tetrachloride, and the water is removed by azeotropic distillation. No special precautions for the esterification of the polyfunctional amino acids are required. Under these conditions the esterification of serine and threonine proceeds in low yields and leads to formation of benzyl ether derivatives. The preparation of such a derivative of threonine is described in a later section.

An amino acid (0.2 mol) and p-toluenesulfonic acid monohydrate (0.205 mol) is placed in a 500-ml round-bottom flask containing benzyl alcohol (100 ml) and benzene or toluene (50 ml). The mixture is heated to reflux and the water produced is removed azeotropically by use of a Dean and Stark trap. When water ceases to distill over, the reaction mixture is cooled to room temperature and anhydrous ether (400 ml) is added. After standing at 4° for several hours, the crystalline amino acid benzyl ester p-toluenesulfonate salt is collected and recrystallized from methanol-ether.

C. Amino Acid p-Nitrobenzyl Esters

These esters are prepared by refluxing the chosen N^α-protected amino acid, such as N-benzyloxycarbonyl amino acid, with p-nitrobenzyl bromide or chloride in ethyl acetate in the presence of a tertiary base, such as triethylamine.[32] The benzyloxycarbonyl group is selectively removed on exposure to hydrogen bromide in acetic acid to give the p-nitrobenzyl ester of the respective amino acid. Polyfunctional amino acids (cysteine, lysine, histidine, etc.), whose secondary function may react with p-nitrobenzyl bromide, must have their secondary group protected prior to esterification. Protection of secondary functions of serine, threonine, or tyrosine, however, is not required prior to esterification.

The chosen N-benzyloxycarbonyl amino acid (0.2 mol) is dissolved in ethyl acetate (250 ml) containing triethylamine (42 ml; 0.3 mol) and p-nitrobenzyl bromide (65 g) is added. The mixture is refluxed for 10 hr and then, after cooling for several hours at 4°, triethylamine hydrobromide is removed by filtration. The filtrate is washed in a separatory

[37] H. K. Miller and H. Waelsch, J. Am. Chem. Soc. **74**, 1092 (1952).

[38] J. D. Cipera and R. V. V. Nicholls, Chem. Ind. (London) p. 16 (1955).

[39] L. Zervas, M. Winitz, and J. P. Greenstein, J. Org. Chem. **22**, 1515 (1957).

funnel with 1 N hydrochloric acid, water, 1 M sodium bicarbonate and water, dried over anhydrous magnesium sulfate and concentrated under reduced pressure to a small volume. Addition of petroleum ether causes the N-benzyloxycarbonyl amino acid p-nitrobenzyl ester to crystallize out. Removal of the N^{α}-protecting group by hydrogen bromide in acetic acid is accomplished as described below.

Esterification of certain benzyloxycarbonyl amino acids, which are not soluble in ethyl acetate, proceeds in a slightly different fashion, as illustrated below in the esterification of the asparagine derivative.

A solution of N-benzyloxycarbonyl-L-asparagine (21.2 g) and p-nitrobenzyl chloride (20.6 g) in dimethylformamide (DMF) (60 ml) containing triethylamine (10.2 ml) is heated at 65° for 4 hr. After the mixture has cooled to room temperature, cold 1 M potassium hydrogen carbonate (500 ml) is added. The precipitated material is collected by filtration, washed with water and cold methanol and dried. The material is triturated with ethyl acetate to induce crystal formation and recrystallized from aqueous methanol.

D. Amino Acid *tert*-Butyl Esters

There are a variety of methods for the introduction of this relatively new, but most important, blocking group. Interaction of amino acids in an organic solvent with isobutylene in the presence of sulfuric acid[40] or with *tert*-butyl acetate in the presence of perchloric acid,[41] leads in many cases to the formation of the *tert*-butyl ester. Owing, however, to the poor solubility of several amino acids in organic solvents, these methods proceed often in poor yields. A superior procedure, which in most cases gives satisfactory yield and is the most commonly used method at the present time, involves N-benzyloxycarbonyl amino acids as the starting material.[34] In essence, this last approach consists in the interaction of an N-benzyloxycarbonyl amino acid in methylene chloride with isobutylene in the presence of a catalytic amount of sulfuric acid. Secondary functional groups, such as hydroxyl, sulfhydryl, under these conditions will form *tert*-butyl derivatives and must be protected prior to esterification. The resulting N-substituted amino acid *tert*-butyl ester is deprotected by catalytic hydrogenation.

A solution or suspension of the chosen N^{α}-benzyloxycarbonyl amino acid (0.1 mol) in methylene chloride (250 ml) is divided into four equal portions and placed into four pressure bottles (capacity ca. 200 ml each).

---

[40] R. Roeske, *Chem. Ind.* (London) p. 1121 (1959).
[41] E. Tashner and C. Wasielewski, *Justus Liebigs Ann. Chem.* **640**, 139 (1961).

To each bottle, concentrated sulfuric acid (0.3 ml) is added, followed, after cooling in a Dry Ice–acetone bath, by liquid isobutylene (25 ml). [Liquid isobutylene is prepared by passing isobutylene gas through a flask equipped with a gas inlet tube and a sodium hydroxide drying tube and immersed in a Dry Ice-acetone bath. For measuring of the liquid isobutylene to be placed in each pressure bottle, we use a graduate cylinder precooled in Dry Ice–acetone.] The bottles are sealed and allowed to stand with occasional shaking (if the amino acid derivative is insoluble) at room temperature for 3–4 days. It is advisable to place the pressure bottles in a covered metal container. After this time period, each pressure bottle is cooled in a Dry Ice–acetone bath before unsealing, and 1 M sodium carbonate (10 ml) is added without mixing. The reaction vessels are then allowed to remain in a hood, unsealed, for 1 hr at room temperature before their contents are mixed. The organic solvent from the combined reaction mixtures is removed under reduced pressure and the remaining viscous liquid is partitioned between 1 M sodium carbonate (50 ml) and ethyl acetate (200 ml). The organic layer is washed with water to neutrality and dried over magnesium sulfate. Following removal of the solvent under reduced pressure, the N^α-benzyloxycarbonyl amino acid *tert*-butyl ester is obtained as an oil. Removal of the N^α-protecting group is accomplished by catalytic hydrogenation in methanol, as described in a later section. The by-products formed during the esterification, presumably polymeric forms of isobutylene, are insoluble in methanol and are removed along with the catalyst after hydrogenation. The free amino acid *tert*-butyl esters are stable liquids and can be distilled without any danger of diketopiperazine or polymer formation.

VI. Protection of Secondary Functional Groups
(Stage 1 and/or Stage 2)

It was emphasized repeatedly in the previous sections that protection of all secondary functional groups is not always an absolute requirement during peptide synthesis. Such protection depends on the reactivity of the functional groups, on the method employed for peptide bond formation, the size of the peptide to be synthesized, and the type of approach employed for chain elongation. A brief elaboration on these points is in order.

The highly reactive sulfhydryl group of cysteine and the ε-amino group of lysine must always be protected. The γ-carboxyl of glutamic acid and the β-carboxyl of aspartic acid can be left unprotected when they are part of the amino component, as discussed above in the section on synthesis of amino components. As part of the carboxyl component, these secondary groups must be protected if any method other than the azide procedure is employed for peptide bond formation.

The strongly basic guanido group of arginine exists in the protonated form under the conditions prevailing during the various stages of peptide synthesis, and conceivably no additional protection is required. Indeed, a number of peptides have been synthesized containing arginine in the protonated form.[42,43] However, the protonated arginine residue, when present in either the amino or carboxyl components, may create, with its highly polar characteristics, considerable problems in the isolation and purification of the synthetic peptide. These difficulties may be further compounded by the possibility that larger water-insoluble peptides which contain protonated arginine residues may act as ion-exchangers. For these reasons it is desirable to protect the guanido function of arginine by groups stable throughout the synthesis and removable in the final deblocking steps.

The presence of histidine in the amino or carboxyl component continues to constitute a serious problem in peptide synthesis. The highly nucleophilic basic imidazole moiety of this amino acid is often involved in a variety of side reactions during the activation of the carboxyl component and its subsequent condensation with the amino component. Acylation of the imidazole nitrogen[44] by the activated carboxyl component, and the subsequent transfer of the acyl moiety to the hydroxyl group of the constituent hydroxyamino acids (esterification by nucleophilic catalysis), are always possibilities. In addition, activation of the carboxyl group of the imidazole-unsubstituted histidine, and its subsequent coupling to the amino component, does not proceed in satisfactory yields by most available methods used for peptide bond formation. Indeed, only the use of the azide method with unsubstituted histidine proceeds in satisfactory yields. Even in this case, however, the presence of the basic imidazole moiety makes the separation of the product from the starting components a difficult task. It is advisable, therefore, to use histidine with the imidazole group protected, in the synthesis of peptides, particularly when smaller peptides are desired.

Tyrosine, serine, and threonine contain a nucleophilic hydroxyl group that is capable of being acylated; hence, all three amino acids present hazards if they are employed in peptide synthesis with their hydroxyl groups free. Indeed, when these amino acids, and especially tyrosine and serine, whose hydroxyl groups are most reactive, are part of the amino

[42] D. T. Gish and V. du Vigneaud, *J. Am. Chem. Soc.* **79**, 3579 (1957).
[43] E. Wünsch, *Z. Naturforsch. Teil B* **22**, 1269 (1967).
[44] D. Yamashiro, J. Blake, and C. H. Li, *J. Am. Chem. Soc.* **94**, 2855 (1972).

component, there is always the possibility that they can be partially acylated by the carboxyl component when activated by most of the available procedures. (It appears that activation by the azide method does not result in O-acylation since a number of polypeptides, with their hydroxyl functions unprotected, have been prepared by this approach with no indication of any by-product formation). Acylation of the hydroxyl functions is most probable in the synthesis of large polypeptides where a large excess of activated carboxyl component is commonly employed. It is, therefore, imperative in such instances that the hydroxyl groups of tyrosine, serine, and even threonine, which has the least reactive group, be protected. The risk of O-acylation is minimized in the synthesis of smaller peptide fragments where equimolecular quantities of amino and carboxyl components are usually employed. In this case the α-amino group of the amino component preferentially reacts with the activated carboxyl component, since it is considerably more reactive than the aforementioned hydroxyl groups.[45] Protection of the hydroxyl group of N^α-substituted tyrosine and serine, and even of threonine, is also recommended when such derivatives are to be used as carboxyl components. Although activation of the unprotected derivatives can be accomplished by several of the available procedures, the interaction of such activated derivatives, particularly those of tyrosine and serine, with amino components does not always proceed in satisfactory yields.

A. Secondary Carboxyl Groups of Glutamic and Aspartic Acids

Protection of these groups can best be accomplished by conversion to the respective benzyl or *tert*-butyl esters.

γ-*Benzyl*-L-*glutamic Acid*[46] or β-*Benzyl*-L-*aspartic Acid*.[47] Into a 3-liter round-bottom flask equipped with a magnetic stirrer and containing 1 liter of cold anhydrous ether, is added, slowly and with vigorous stirring, concentrated sulfuric acid (100 ml). To this solution benzyl alcohol (1 liter) is added, the ether is removed under vacuum, and the chosen amino acid (1 mol) is added. The mixture is stirred at room temperature for 1 day, cooled to 4°, and diluted with 95% ethanol (2 liters) and pyridine (500 ml). After 24 hr at 4°, the crystalline product is collected, washed with cold water, ethanol, and ether. The crude product

---

[45] O-acylation of tyrosine can occur even in such cases if excess base is present [J. Ramachandran and C. H. Li, *J. Org. Chem.* **28**, 173 (1963)]. It can also occur at any time by nucleophilic catalysis if an unprotected histidine is present.

[46] S. Guttmann and R. A. Boissonnas, *Helv. Chim. Acta* **41**, 1852 (1958).

[47] L. Benoiton, *Can. J. Chem.* **40**, 570 (1962).

is recrystallized from boiling water (4 liters) containing pyridine (20 ml). Yield 50–52%; melting point for the glutamic acid ester 174–175° and for the aspartic acid ester 218°–220°.

β-tert-Butyl-L-aspartic Acid and Its N^α-Benzyloxycarbonyl Derivatives.[48] The synthesis of this ester or its N^α-benzyloxycarbonyl derivative proceeds in several steps, as indicated below.

$$C_6H_5CH_2OCONHCHCOOH \overset{\displaystyle CH_2COOH}{\underset{|}{}} + (CH_3CO)_2O \longrightarrow$$

$$\overset{\displaystyle CH_2CO}{\underset{C_6H_5CH_2OCONHCHCO}{|}}\!\!\diagdown\!\!O$$

(XVII)

$$\Big\downarrow p\text{-}NO_2C_6H_4CH_2OH$$

$$C_6H_5CH_2OCONHCHCOOCH_2C_6H_4NO_2 \overset{\displaystyle CH_2COOC(CH_3)_3}{\underset{|}{}} \xleftarrow{(CH_3)_2C=CH_2} C_6H_5CH_2OCONHCHCOOCH_2C_6H_4NO_2 \overset{\displaystyle CH_2COOH}{\underset{|}{}}$$

(XIX)   (XVIII)

$$H_2 \Big\downarrow \diagdown \overset{(C_6H_{11})_2NH}{\underset{OH^-}{}}$$

$$H_2NCHCOOH \overset{\displaystyle CH_2COOC(CH_3)_3}{\underset{|}{}} \diagdown C_6H_5CH_2OCONHCHCOOH \cdot NH(C_6H_{11})_2 \overset{\displaystyle CH_2COOC(CH_3)_3}{\underset{|}{}}$$

(XXI)   (XX)

N-Benzyloxycarbonyl-L-aspartic Acid Anhydride[49] (XVII). Into a 1-liter round-bottom flask, equipped with a calcium chloride drying tube, are placed N^α-benzyloxycarbonyl-L-aspartic acid (150 g), dry tetrahydrofuran (200 ml), and redistilled acetic anhydride (105 ml). The mixture is stirred for 2 hr at 50°; the tetrahydrofuran and acetic anhydride are then removed under reduced pressure using an oil pump (40°). The residual oil is mixed with anhydrous ether (200 ml) and allowed to crystallize overnight at −20°. Petroleum ether (200 ml) is added, the mixture cooled to −20°, and the product is isolated by filtration, washed with cold petroleum ether, and dried under high vacuum over potassium hydroxide (weight 140 g; mp 109–110°).

N-Benzyloxycarbonyl-L-aspartic Acid α-p-Nitrobenzyl Ester[48] (XVIII). A solution of p-nitrobenzyl alcohol (95 g) and dicyclohexylamine (164 ml) in cold ether (450 ml) is added with cooling in a 2-liter flask containing the cyclic anhydride (XVII) (140 g) dissolved in dry tetrahydrofuran (240 ml). The mixture is stirred overnight at room tem-

[48] E. Schröder and E. Klieger, *Justus Liebigs Ann. Chem.* **673**, 208 (1964).
[49] E. Wünsch and A. Zwick, *Hoppe-Seyler's Z. Physiol. Chem.* **328**, 235 (1962).

perature and the product, as the crystalline salt, is collected by filtration and washed with anhydrous ether (weight 220 g). The product is converted to the free acid by suspending the dried salt in a vigorously stirred mixture of 20% citric acid (1 liter) and ethyl acetate (1 liter). After 15 min, the aqueous layer is discarded and the organic layer is washed once with water, dried over magnesium sulfate, and concentrated under reduced pressure to dryness. The residual oil is dissolved in hot 95% ethanol (300 ml), and water is added until the hot solution is slightly turbid. Upon cooling, the product (XVIII) crystallizes out and is collected by filtration (weight 112 g; mp 126–127°).

N-Benzyloxycarbonyl-β-tert-butyl-L-aspartic Acid α-p-Nitrobenzyl Ester[48] (*XIX*). To a suspension of compound (XVIII) (20 g) in methylene chloride (150 ml), placed in a pressure bottle (capacity ca. 350 ml) and cooled in a Dry Ice–acetone bath, is added concentrated sulfuric acid (0.5 ml) and liquid isobutylene (25 ml). The bottle is sealed and shaken for 4 days at room temperature. The reaction vessel is then cooled in a Dry Ice–acetone bath, unsealed and mixed with 1 *M* sodium bicarbonate (20 ml). After warming to room temperature, the methylene chloride and excess isobutylene are removed *in vacuo* and the residue is extracted into ethyl acetate (500 ml). The ethyl acetate solution is washed with 1 *M* sodium bicarbonate and water, dried over magnesium sulfate and concentrated to a small volume. Upon addition of petroleum ether, the product crystallizes out (weight 19 g; mp 93°–94°).

N-Benzyloxycarbonyl-β-tert-butyl-L-aspartic Acid Dicyclohexylamine Salt[48] (*XX*). To a solution of compound (XIX) (19 g) in acetone (150 ml) 1 *N* sodium hydroxide (46 ml) is added. After 1 hr at room temperature, the acetone is removed under reduced pressure; the remaining solution is extracted twice with ether, and the aqueous layer is acidified to pH 2.0 with 20% citric acid. The oil which is formed is extracted into ether (200 ml) and the organic phase is washed once with water and dried over magnesium sulfate. The ether solution is mixed with dicyclohexylamine (10.2 ml), and, after cooling overnight, the crystalline dicyclohexylamine salt (XX) is collected and washed with ether (weight 19 g; mp 124°–126°).

β-tert-Butyl-L-aspartic Acid[48] (*XXI*). Compound (XIX) (10 g) is dissolved in methanol (150 ml) and hydrogenated for 4 hr over 10% palladium–charcoal catalyst (1 g), as described on p. 547. The catalyst is then removed by filtration, and the filtrate is concentrated to a small volume and mixed with ether. After several hours of cooling, the crystalline product is collected by filtration (weight 3.7 g; mp 188°–189°).

γ-*tert-Butyl*-L-*glutamic Acid and Its N$^\alpha$-Benzyloxycarbonyl Derivative*. The overall scheme for the synthesis of these compounds is illustrated below:

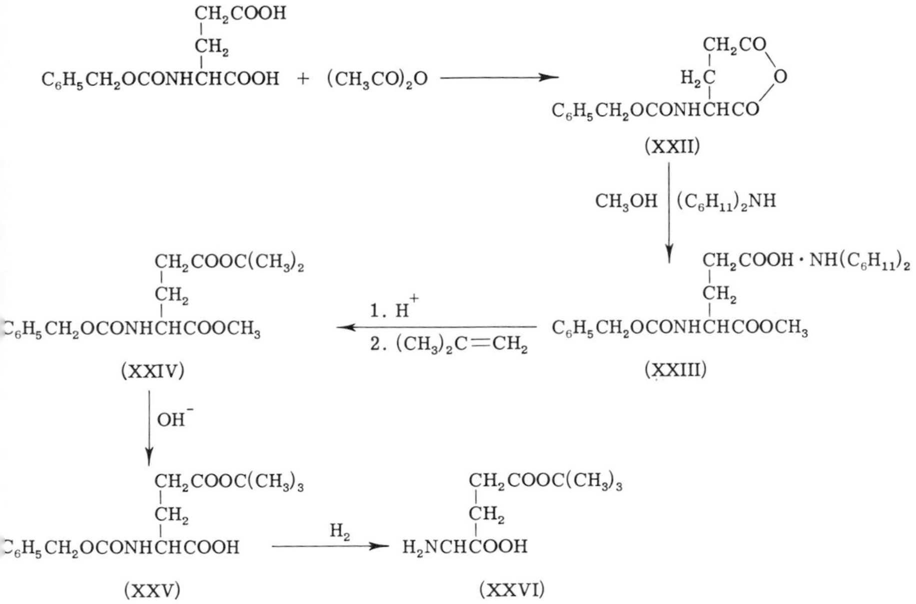

N-*Benzyloxycarbonyl*-L-*glutamic Acid Anhydride*[50] (*XXII*). A suspension of N-benzyloxycarbonyl-L-glutamic acid (56 g) and redistilled acetic anhydride (175 ml), in a 500-ml round-bottom flask equipped with a calcium chloride drying tube, is stored at room temperature with occasional shaking. After 14 hr a solution is obtained and the solvent is removed under high vacuum using an oil pump, at 45°, and to the oily residue, ether (70 ml) is added. It is critical that all the solvent be removed prior to the addition of ether; otherwise the product will not crystallize. The product (XXII) that crystallizes out after standing overnight in the cold, is collected by filtration, washed with ether, and dried under high vacuum over potassium hydroxide (weight 40 g; mp 90°–92°).

N-*Benzyloxycarbonyl*-L-*glutamic Acid-α-methyl Ester, Dicyclohexylamine Salt*[51] (*XXIII*). To a solution of compound (XXII) (40 g) in ether (600 ml), methanol (230 ml) and dicyclohexylamine (36.5 ml) are added. The reaction mixture is stirred overnight at room temperature,

[50] E. Klieger and E. Schröder, *Justus Liebigs Ann. Chem.* **661**, 193 (1963).
[51] E. Klieger, E. Schröder, and H. Gibian, *Justus Liebigs Ann. Chem.* **640**, 157 (1961).

and the crystalline product is collected and recrystallized from methanol (weight 41 g; mp 173°–176°).

N-Benzyloxycarbonyl-γ-tert-butyl-L-glutamic Acid α-Methyl Ester[52] (*XXIV*). Compound (XXIII) (41 g) is suspended in a mixture of water (180 ml) containing concentrated sulfuric acid (19 ml) and ethyl acetate (400 ml). After vigorous stirring for 30 min, the organic layer is separated, washed with water, dried over magnesium sulfate, and concentrated to dryness under reduced pressure. The oily residue is dissolved in methylene chloride (250 ml), placed into four pressure bottles and treated with isobutylene in exactly the same fashion as described above for amino acid *tert*-butyl esters (p. 525). The product (XXIV) is obtained as an oil (29 g).

N-Benzyloxycarbonyl-γ-tert-butyl-L-glutamic Acid[52] (*XXV*). A solution of compound (XXIV) (29 g) in acetone (180 ml) is mixed with 1 N sodium hydroxide (89 ml). After standing at room temperature for 45 min, the solution is neutralized by the addition of 1 N hydrochloric acid (90 ml). The acetone is removed under reduced pressure and the remaining material is extracted into ether. The ether layer is washed with water, dried over magnesium sulfate, and concentrated to dryness to give the product as an oil (26 g).

γ-tert-Butyl-L-glutamic Acid[53] (*XXVI*). Compound (XXV) (26 g) is dissolved in methanol (150 ml) and hydrogenated for 5 hr over 10% palladium–charcoal catalyst (3 g). The catalyst is then removed by filtration, and the filtrate is concentrated to a small volume and mixed with ether. After cooling for several hours, the product crystallizes and is collected by filtration (weight 14 g; mp 186°–187°).

B. The Guanido Group of Arginine

A limited number of protectors for the guanido group have been introduced in peptide chemistry, but only two such groups, namely the nitro and *p*-toluenesulfonyl (tosyl), are presently most commonly used.

N^ω-Nitroarginine[54,55] is readily available commercially and is prepared by treatment of arginine with a mixture of nitric and sulfuric acids. The nitro group is stable under the routine acid and alkaline conditions employed in peptide synthesis. However, recent work indicates that in certain instances, under strongly acidic or basic conditions, the

[52] R. Schwyzer and H. Kappeler, *Helv. Chim. Acta* **44**, 1991 (1961).
[53] E. Schröder and E. Klieger, *Justus Liebigs Ann. Chem.* **673**, 196 (1964).
[54] M. Bergmann, L. Zervas, and H. Rinke, *Hoppe Seyler's Z. Physiol. Chem.* **224**, 40 (1934).
[55] K. Hofmann, W. D. Peckham, and A. Rheiner, *J. Am. Chem. Soc.* **78**, 238 (1956).

nitroguanido group becomes labile and nitroarginyl peptides are converted partially to ornithine-containing peptides.[44] This has been observed during treatment with liquid hydrogen fluoride, hydrogen bromide in acetic acid and during ammonolysis of nitroarginine-containing peptides.[56] Removal of the nitro group can be accomplished readily by liquid hydrogen fluoride.[31] Catalytic hydrogenation has also been used for the removal of the nitro group[54] although, in certain cases, the reduction stops at the level of aminoguanidine. Sodium in liquid ammonia cannot be used effectively for removal of the nitro group, and ornithine is formed as a by-product.[57]

Undoubtedly, the best method to protect the guanido function of arginine is by conversion to N^ω-tosylarginine.[58] N^ω-Tosylation proceeds by reacting N^α-benzyloxycarbonyl arginine in aqueous acetone with tosyl chloride at pH 11.0–11.5, followed by hydrogenolytic removal of the benzyloxycarbonyl group. The N^ω-tosyl group is stable under the acidic and basic conditions used in peptide synthesis and can readily be removed by liquid hydrogen fluoride[31] or sodium in liquid ammonia.

Activation of the N^α-substituted arginine, which has the guanido function protected by either of the above-mentioned moieties, can be accomplished by most of the existing procedures. During such activation, however, there is always the possibility of partial lactam formation with a concomitant decrease in the yield of the product. Lactam formation is more prominent in the N^ω-nitro-substituted derivative, particularly when the "carbodiimide procedure" is the method of activation.

N^α-Benzyloxycarbonyl-N^ω-tosyl-L-arginine.[58] To a suspension of N-benzyloxycarbonyl-L-arginine (25 g) in a mixture of water (100 ml) and acetone (500 ml) at 4°, 4 N sodium hydroxide (40 ml) is added. To this solution, whose pH, as measured by a glass electrode, is 12–13, is added dropwise a solution of tosyl chloride (38 g) in acetone (60 ml) over a period of 30 min with vigorous stirring; the pH is maintained in the range of 11–11.5 by the addition of 4 N sodium hydroxide. The solution is then stirred for another 3 hr (pH 11–11.5), and its pH is adjusted to 8.0 by the addition of 1 N hydrochloric acid. The acetone is removed under reduced pressure, and water (200 ml) is added to the residue. The aqueous solution is extracted three times with ether and the aqueous layer is cooled to 4°, and the pH is adjusted to 3.0 by addition of 6 N hydrochloric acid. The oily product formed is extracted into ethyl acetate (800 ml), and the organic layer is washed twice with 0.1 N hydrochloric acid and with water until neutral, and extracted with

[56] H. Künzi, M. Manneberg, and R. O. Studer, *Helv. Chim. Acta* **57**, 566 (1974).
[57] G. L. Tritsch and D. W. Woolley, *J. Am. Chem. Soc.* **82**, 2787 (1960).
[58] J. Ramachandran and C. H. Li, *J. Org. Chem.* **27**, 4006 (1962).

saturated aqueous sodium bicarbonate (350 ml). The aqueous layer is washed once with ethyl acetate, filtered, and acidified with 6 N hydrochloric acid to pH 3.0. The oily product is extracted into ethyl acetate (800 ml) and the organic layer is washed with water to neutrality, dried over magnesium sulfate and concentrated under reduced pressure to a small volume (ca. 30 ml). Upon cooling, the product crystallizes out. Yield 15–20 g; mp 86°–89°. Crystallization may not occur, particularly if seed crystals are not available. In such case, petroleum ether is added and the precipitated amorphous product is used directly for the subsequent step.

N^{ω}-*Tosyl*-L-*arginine*.[58]N^{α}-Benzyloxycarbonyl-N^{ω}-tosyl-L-arginine (20 g) is dissolved in methanol (300 ml) and hydrogenated for 6 hr over 10% palladium–charcoal catalyst (6 g). The catalyst is removed by filtration, and the filtrate is evaporated to dryness. The residue is suspended in a mixture of water (100 ml) and ethyl acetate (100 ml) and stored at 4° for 20 hr. The crystalline product is collected and recrystallized from 200 ml of hot water (weight 14 g; mp 147°–150°).

C. The N$^{\varepsilon}$-Amino Group of Lysine

The N^{ε}- and N^{α}-amino groups of lysine have similar properties and, hence, all the reagents employed as amino protectors can be used for the masking of either function. The choice, however, of the group to be used as the N$^{\varepsilon}$-protector depends on the agent employed as the N$^{\alpha}$-protector, so that selective deprotection of the latter becomes possible. The most frequently used N$^{\varepsilon}$-protectors are the tosyl, benzyloxycarbonyl and *tert*-butyloxycarbonyl groups. All three moieties can readily be incorporated by the reaction of the copper chelate of lysine with benzyloxycarbonyl chloride, tosyl chloride, and *tert*-butyloxycarbonyl azide, respectively.

The N$^{\varepsilon}$-tosyl group is stable to acidic and basic conditions and can be effectively removed by exposure to sodium in liquid ammonia. Hence, this group, as an N$^{\varepsilon}$-protector, is compatible with all of the N$^{\alpha}$- and C$^{\alpha}$-protectors mentioned throughout this chapter.

The *tert*-butyloxycarbonyl group can be used as N$^{\varepsilon}$-protector if *o*-nitrophenylsulfenyl or other equally labile moieties are used as N$^{\alpha}$-protectors. It is also compatible with the use of the benzyloxycarbonyl group as an N$^{\alpha}$-protector, provided that the latter is removed by catalytic hydrogenation.

At first glance it would appear that the benzyloxycarbonyl group could be ideally suited to serve as an N$^{\varepsilon}$-protector if *o*-nitrophenylsulfenyl, *tert*-butyloxycarbonyl, and other highly acid labile moieties are used as N$^{\alpha}$-protectors. Indeed, this is the case when *o*-nitrophenylsulfenyl and other equally acid labile groups are used. However, certain limitations are imposed if the *tert*-butyloxycarbonyl group is employed as an N^α-protector. It has been shown that exposure to trifluoroacetic acid or treatment with hydrogen chloride in organic solvents, which are the usual methods to remove *tert*-butyl-type protectors, also causes significant cleavage of the N^ε-benzyloxycarbonyl moiety.[59,60] Exposure to formic acid, which also removes *tert*-butyl protectors, although at a lower rate, does not affect the N^ε-benzyloxycarbonyl moiety. An alternative method to protect the N^ε-amino group in the presence of *tert*-butyl-type protectors is by the use of the *o*-bromo-[61] or *o*-chlorobenzyloxycarbonyl[62] moiety as N^ε-protectors. Both of these groups are fairly stable under conditions employed to remove the *tert*-butyl-type groups, but retain all the other properties of the benzyloxycarbonyl moiety.

N^ε-*Benzyloxycarbonyl*-L-*lysine*.[63] A mixture of L-lysine monohydrochloride (60 g) and basic copper carbonate [$2CuCO_3 \cdot Cu(OH)_2$; 100 g] in water (1.5 liters) is refluxed for 1 hr, and the resulting solution of the copper complex is clarified by filtration and cooled to 4° in an ice bath. To this solution is added, under vigorous stirring and over a period of 30 min, benzyloxycarbonyl chloride (65 ml) and enough 4 N sodium hydroxide (ca. 190 ml) to maintain the pH in the range of 8–9. The reaction mixture is stirred for an additional hour, and the precipitated product is collected by filtration and washed on the filter with water, ethanol, and ether.

The copper complex of N^ε-benzyloxycarbonyl-L-lysine is suspended in warm (50°) water (6 liters) containing concentrated hydrochloric acid (45 ml), and a stream of hydrogen sulfide gas is passed through the stirred suspension for 1 hr. The heating is maintained for an additional hour before the copper sulfide is removed by filtration through a Celite filter bed. The clear and colorless filtrate is adjusted to pH 6–7 with concentrated ammonium hydroxide, and the desired product crystallizes out. The crystals are collected by filtration and dried over phosphorus pentoxide in a vacuum desiccator to give 92 g of product (mp 278–280°).

N^ε-*Tosyl*-L-*lysine*.[64] To a solution of the copper complex of L-lysine

[59] G. P. Schwartz and P. G. Katsoyannis, *J. Chem. Soc., Perkin Trans. 1* p. 2890 (1973).

[60] A. Yarn and S. F. Schlossman, *Biochemistry* **7**, 2673 (1968).

[61] D. Yamashiro and C. H. Li, *J. Am. Chem. Soc.* **95**, 1310 (1973).

[62] B. W. Erickson and R. B. Merrifield, *Isr. J. Chem.* **12**, 79 (1974).

[63] A. Neuberger and F. Sanger, *Biochem. J.* (*Tokyo*) **37**, 515 (1943).

[64] R. Roeske, F. H. C. Stewart, R. J. Stedman, and V. du Vigneaud, *J. Am. Chem. Soc.* **78**, 5883 (1956).

prepared as described above [from L-lysine monohydrochloride (24 g), basic copper carbonate (40 g) and water (500 ml)], sodium bicarbonate (42 g) is added, followed by a solution of tosyl chloride (38 g) in acetone (1.5 liters). The reaction mixture is stirred for 10 hr and the precipitated product is collected by filtration and washed with water, acetone, and ether. The dried copper complex is suspended in boiling water (500 ml) and a stream of hydrogen sulfide gas is passed through the suspension for 30 min with vigorous stirring. The boiling is continued for an additional 30 min to remove the excess of hydrogen sulfide. The mixture is then cooled to room temperature, and 6 N hydrochloric acid (15 ml) and charcoal (5 g) are added. After 5 min, the copper sulfide is removed by filtration and the clear filtrate is adjusted to pH 6 with 4 N ammonium hydroxide. After cooling for several hours, the crystalline N^ε-tosyl-L-lysine is collected by filtration, washed with water and ethanol, and dried; yield 22 g; mp 236°–238°.

N^ε-o-Bromobenzyloxycarbonyl-L-lysine.[61] The synthesis of this derivative proceeds through the interaction of o-bromobenzyl-p-nitrophenyl carbonate (o-BrC$_6$H$_4$CH$_2$OCOOC$_6$H$_4$NO$_2$) with the copper chelate of lysine. o-Bromobenzyl-p-nitrophenyl carbonate is prepared by the reaction of o-bromobenzyl alcohol with p-nitrophenyl chloroformate (p-NO$_2$C$_6$H$_4$OCOCl).

To a solution of o-bromobenzyl alcohol (10.8 g) in freshly distilled pyridine (30 ml), p-nitrophenyl chloroformate (9.7 g) is added at 0° with magnetic stirring. After 3 hr at 25°, the reaction mixture is diluted with water (50 ml) and ethyl acetate (250 ml). The organic layer is separated, washed successively with 1 N hydrochloric acid, water, 1 M sodium carbonate, and saturated sodium chloride solution, dried over magnesium sulfate, and concentrated under reduced pressure to dryness. Upon crystallization of the residue from absolute ethanol, 13.5 g of o-bromobenzyl-p-nitrophenyl carbonate is obtained; m.p. 94°.

A suspension of L-lysine monohydrochloride (16.6 g) and basic copper carbonate (10.5 g) in water (250 ml) is heated to boiling for 10 min. The mixture is cooled to room temperature and clarified by filtration through a Celite bed. To this solution, diluted with water (760 ml) and DMF (3 liters), o-bromobenzyl-p-nitrophenyl carbonate (32 g) and sodium bicarbonate (14.2 g) are added. The mixture is stirred for 48 hr at room temperature and the precipitated product is filtered, washed with water (until the filtrate is colorless), 95% ethanol and air-dried. A suspension of this product in water (4 liters) containing the disodium salt of EDTA (40 g), is heated to boiling until a solution is obtained. Upon cooling to 4°, the product crystallizes, is filtered, washed with water and 95% ethanol; yield 22 g; mp 220°–223°.

N^ε-*tert-Butyloxycarbonyl*-L-*lysine.*[65] A mixture of L-lysine monohydrochloride (20 g) and basic copper carbonate (20 g) in water (160 ml) is heated to boiling for 20 min. The hot blue solution is filtered, the precipitate is washed with hot water (80 ml) and the combined filtrates cooled to room temperature. To this solution sodium bicarbonate (20 g) is added, followed by the dropwise addition of a solution of *tert*-butylazidoformate (24 ml) in methanol (300 ml) with vigorous stirring over a period of 1 hr. The reaction mixture is stirred for 30 hr at room temperature and diluted with water (500 ml). The precipitated copper complex of N^ε-*tert*-butyloxycarbonyl-L-lysine is filtered and washed with water (800 ml), methanol (200 ml), and ether; weight 15.7 g; mp 240°–244°.

The finely powdered copper complex is suspended in water (300 ml), and 2 N ammonium hydroxide (55 ml) is added. Through this suspension, under vigorous magnetic stirring, hydrogen sulfide gas is passed for 3 hr. The mixture is then cooled in an ice bath and diluted with 2 N acetic acid (80 ml). The excess hydrogen sulfide is removed by passing a stream of air through the reaction mixture. The precipitated copper sulfide is filtered off and washed with water; the combined filtrates (ca. 800 ml) are concentrated under reduced pressure (25°–34°) to approximately 100 ml. After cooling, the precipitated crystalline N^ε-*tert*-butyloxycarbonyl-L-lysine is filtered, washed with cold water (30 ml) and dried; weight 11 g; mp 255°.

D. The Sulfhydryl Group of Cysteine

Protection of the sulfhydryl group of cysteine is imperative in the synthesis of cysteine-containing peptides. The classical method of protection of this moiety involves its conversion to the S-benzyl thioether.[66] The benzylation proceeds smoothly by reaction of cysteine with benzyl chloride in dilute alcohol in the presence of sodium hydroxide.[67] The S-benzyl group is stable under acid and alkaline conditions commonly used in the synthesis of peptides and can be removed effectively only on exposure to sodium in liquid ammonia.[29] Upon treatment with liquid hydrogen fluoride at room temperature, only partial removal of this group is effected.[31] Hence, the S-benzyl group is compatible with all N^α- and C^α-protecting groups mentioned in this chapter. This great stability of

[65] R. Schwyzer and W. Rittel, *Helv. Chim. Acta* **44**, 159 (1961).

[66] J. L. Wood and V. du Vigneaud, *J. Biol. Chem.* **130**, 109 (1939).

[67] M. Frankel, D. Gertner, H. Jacobson, and A. Zilkha, *J. Chem. Soc.* (*London*) **1960**, 1390 (1960).

the S-benzyl group is perhaps its only drawback, since its final removal can be effected only by sodium in liquid ammonia, a procedure that is not always the method of choice for final deprotection of long polypeptide chains.

Conversion of cysteine to the S-diphenylmethyl derivative offers another way for protection of the secondary function of this amino acid.[68] The S-diphenylmethyl group is stable in alkali, and it is also stable under acidic conditions usually employed for the removal of the acid labile N-protectors, *tert*-butyloxycarbonyl, o-nitrophenylsulfenyl, namely formic acid, trifluoroacetic acid, etc., and it is removed by sodium in liquid ammonia and liquid hydrogen fluoride.[31] However, the behavior of the S-diphenylmethyl group in acid media requires amplification. It appears that in acid media such as trifluoroacetic acid, S-diphenylmethylcysteine is in equilibrium with the diphenylmethyl cation and free cysteine, and the equilibrium is in favor of the former.[68] The presence of cation scavengers such as phenol or anisole in excess, shifts the equilibrium toward free cysteine and, hence, S-diphenylmethylcysteine can be cleaved in trifluoroacetic acid in the presence of phenol.[68] It is therefore important that in S-diphenylmethylcysteine-containing peptides, removal of the N^α-protectors be effected in the absence of cation scavengers. The S-diphenylmethyl group is only partially stable to solutions of hydrogen bromide in acetic acid and, hence, use of the benzyloxycarbonyl moiety as an N^α-protector must be excluded for chain elongation.

The S-trityl group has also been used as an S-protector.[68] It is more acid-labile than the S-diphenylmethyl moiety and could be best used in combination with o-nitrophenylsulfenyl and similar acid-labile N^α-protectors. Its removal can be readily effected upon treatment with silver or mercury cations.[69] Alternatively, this group can be removed by iodine in certain organic solvents[70] or by exposure to thiocyanogen.[71] Deprotection by the latter two methods results in the formation of cystine derivatives. The combination of trityl and o-nitrophenylsulfenyl moieties as S- and N^α-protectors, respectively, poses one problem. Cleavage of the N-o-nitrophenylsulfenyl group by hydrogen chloride in organic solvents leads to the formation of yellow impurities, presumably due to partial removal of the S-trityl moiety.[72,72a] Similar impurities were observed in

[68] I. Photaki, J. Taylor-Papadimitriou, C. Sakarellos, P. Mazarakis, and L. Zervas, *J. Chem. Soc.* (*C*) p. 2683 (1970).

[69] L. Zervas and I. Photaki, *J. Am. Chem. Soc.* **84**, 3887 (1962).

[70] B. Kamber and W. Rittel, *Helv. Chim. Acta* **51**, 2061 (1968).

[71] R. G. Hiskey and W. P. Tucker, *J. Am. Chem. Soc.* **84**, 4794 (1962).

[72] R. G. Hiskey, E. T. Wolters, G. Ülkii, and V. R. Rao, *J. Org. Chem.* **37**, 2478 (1972).

[72a] It was found recently [A. Fontana, *J. Chem. Soc., Chem. Commun.* p. 976

our own laboratory in the cleavage of the N-o-nitrophenylsulfenyl group from S-diphenylmethylcysteine-containing peptides. It is our experience the formation of such yellow by-products can be avoided in this case by effecting the removal of o-nitrophenylsulfenyl groups with nucleophiles (i.e., thioacetamide) under mild acidic conditions.

The p-methylbenzyl group has been recently introduced as an S-protector.[73] It appears that this group has similar properties to the S-benzyl moiety, with the additional advantage that it can be cleaved with liquid hydrogen fluoride.

S-Benzyl-L-cysteine.[67] To a stirred solution of L-cysteine hydrochloride (46.8 g) in a mixture of 2 N sodium hydroxide (300 ml) and ethanol (360 ml), benzyl chloride (42 ml) is added. After stirring for 30 min at room temperature, the pH is adjusted to 6.0 with concentrated hydrochloric acid, and the suspension is cooled overnight at 4°. The crude product is collected, washed successively with cold water, ethanol and ether, and recrystallized from boiling water (weight 43 g; mp 214°–215°).

S-Diphenylmethyl-L-cysteine.[68] Into a 500-ml round-bottom flask equipped with a calcium chloride drying tube is added L-cysteine hydrochloride (32 g), diphenylmethanol (37 g), and trifluoroacetic acid (120 ml). After standing for 20 min at room temperature, the trifluoroacetic acid is removed under reduced pressure and the residue is mixed with cold ether (200 ml) and enough saturated sodium acetate solution to make the pH 5.0. The suspension is cooled for several hours and the precipitated S-diphenylmethylcysteine is collected by filtration and washed with cold water and ether (weight 59 g; mp 200°–201°). Product of this purity can be used for further synthetic work. Recrystallization of the crude product can be accomplished as follows: The product is dissolved with warming in 95% ethanol (900 ml) containing 5 N hydrochloric acid (50 ml) and, after cooling to room temperature, pyridine (40 ml) is added and the mixture is cooled overnight. The crystals are collected and washed with a small amount of cold 95% ethanol and ether.

E. The Imidazole Group of Histidine

A number of groups have been suggested for the protection of the imidazole moiety (N^{im}) of histidine; among these the benzyl and tosyl groups are the most commonly used at the present time.

(1975)] that o-nitrophenylsulfenyl chloride is an effective reagent for the displacement of the S-trityl or S-diphenylmethyl protecting groups.

[73] B. W. Erickson and R. B. Merrifield, *J. Am. Chem. Soc.* **95**, 3750 (1973).

The introduction of the benzyl group at the imidazole moiety of histidine takes place by interaction of benzyl chloride with histidine in liquid ammonia in the presence of sodium.[74] The N^{im}-benzyl group is stable under basic and acidic conditions and can be removed only by reduction with sodium in liquid ammonia or at a slower rate by catalytic hydrogenation. For this reason, this group is compatible with all the above-mentioned protecting groups. It should be pointed out that basicity of the imidazole moiety is not suppressed totally by benzylation and it can form salts in strongly acid media. Recent work indicates that activation of N^{im}-benzyl histidine derivatives is often accompanied by racemization.[75] It appears, however, that racemization is suppressed if activation of the carboxyl group is via conversion to the 1-hydroxybenzotriazole ester.[75]

Protection of the imidazole moiety of histidine with a tosyl group is another useful method available at the present time.[76] In contrast to the N^{im}-benzyl, the N^{im}-tosyl derivatives of histidine are more soluble in organic solvents, are less prone to racemization during activation, and their imidazole moiety is less basic. N^{im}-Tosylation is accomplished by reaction of N^{α}-protected histidine with tosyl chloride in aqueous acetone buffered with sodium carbonate.[76] The N^{im}-tosyl group is stable to hydrogenolysis and to trifluoroacetic acid, and is readily cleaved upon exposure to liquid hydrogen fluoride and sodium in liquid ammonia. Hence, for chain elongation, it is compatible with the *tert*-butyloxycarbonyl group and other highly acid labile N^{α}-protectors, and the benzyloxycarbonyl group if hydrogenolysis is the method of deprotection. However, the N^{im}-tosyl group is unstable to halogen acids in organic solvents,[76] to alkali, and to other nucleophiles such as pyridine hydrochloride[77] or 1-hydroxybenzotriazole.[78] Obviously, then, activation of carboxyl components containing N^{im}-tosyl histidine by the use of 1-hydroxybenzotriazole should be avoided. In spite of the advantages of the N^{im}-tosyl group, its high instability to a variety of agents makes its widespread use as a protecting group questionable. In our own laboratory, for example, in the stepwise synthesis of an N^{im}-tosyl histidine-containing peptide, we have detected partial detosylation even though we have employed routine manipulations common in the synthesis of peptides.[79]

[74] V. du Vigneaud and O. K. Behrens, *J. Biol. Chem.* **117**, 27 (1937).
[75] J. L. M. Syrier and H. C. Beyerman, *Recl. Trav. Chim. Pays-Bas* **93**, 117 (1974).
[76] T. Fugii and S. Sakakibara, *Bull. Chem. Soc. Jpn.* **43**, 3954 (1970).
[77] H. C. Beyerman, J. Hirt, P. Kranenburg, J. L. M. Syrier, and A. von Zon, *Recl. Trav. Chim. Pays-Bas* **93**, 256 (1974).
[78] T. Fugii and S. Sakakibara, *Bull. Chem. Soc. Jpn.* **47**, 3146 (1974).
[79] G. P. Schwartz and P. G. Katsoyannis, *J. Chem. Soc. Perkin Trans. 1* p. 2894 (1973).

N^{im}-*Benzyl*-L-*histidine*.[74] In a three-neck round-bottom flask fitted with a drying tube containing sodium hydroxide, a separatory funnel, and a magnetic stirrer, liquid ammonia (300 ml) is collected. Collection of the liquid ammonia is accomplished by passing ammonia gas through the reaction vessel, which is cooled in a Dry Ice–acetone bath. L-Histidine monohydrochloride (20 g) is then placed in the reaction flask, and small pieces of sodium are added until a permanent blue color persists (ca. 9 g of sodium). The blue color (excess of sodium) is discharged by the addition of a small amount of histidine monohydrochloride, and, to the resulting mixture, benzyl chloride (12 ml) is added dropwise with vigorous stirring. The mixture is stirred for an additional 30 min, the Dry Ice–acetone bath is removed and the ammonia is allowed to evaporate off. The residue is dried under vacuum for 15 min and dissolved in cold water (100 ml). This solution is extracted twice with ether, filtered, and its pH is adjusted to 8.0–8.5 by the addition of dilute sulfuric acid. The precipitated product is filtered, washed with water, and recrystallized from 78% ethanol; weight 15 g; mp 248°–249°.

N^{α}-*tert-Butyloxycarbonyl*-N^{im}-*tosyl*-L-*histidine*.[78,79] To a solution of N-*tert*-butyloxycarbonyl-L-histidine methyl ester[15] (13.5 g) in dioxane–acetone (1:1; 100 ml), 1 N sodium hydroxide (50 ml) is added. After 1 hr the solution is diluted with 2 M potassium carbonate (30 ml) and cooled to 0°. To this solution is then added during 30 min, with vigorous stirring, a solution of tosyl chloride (11.5 g) in acetone (30 ml). After stirring for an additional hour the mixture is diluted with cold water (100 ml) and extracted four times with ether. The aqueous layer is then acidified to pH 3.5 with solid citric acid and the oil formed is extracted into ethyl acetate. The organic layer is washed with cold water, dried (magnesium sulfate), and concentrated under reduced pressure. The residue is solidified upon trituration with petroleum ether and crystallized from ethyl acetate–petroleum ether; weight 15.5 g; mp 133°–134°.

F. The Hydroxyl Group of Tyrosine

Several blocking groups for the phenolic moiety of tyrosine have been introduced in peptide synthesis. Among these the O-benzyloxycarbonyl,[80] O-tosyl,[81] O-benzyl,[82] 2,6-dichlorobenzyl,[73] and O-*tert*-butyl[83] may be mentioned. Of all of the aforementioned protecting groups, the O-benzyl has been most extensively employed in peptide synthesis. Benzylation of

[80] B. G. Overell and V. Petrow, *J. Chem. Soc. London* p. 232 (1955).
[81] P. G. Katsoyannis, *J. Amer. Chem. Soc.* **79**, 109 (1957).
[82] E. Wünsch, G. Fries, and A. Swick, *Chem. Ber.* **91**, 542 (1958).
[83] E. Wünsch and J. Jentsch, *Chem. Ber.* **97**, 2490 (1964).

the phenolic group of tyrosine is readily accomplished by the interaction of the copper complex of tyrosine with benzyl chloride in the presence of sodium hydroxide. The O-benzyl group of tyrosine is stable under routine manipulations employed in peptide synthesis, i.e., exposure to dilute acid and alkali. It can be removed by catalytic hydrogenation, by reduction with sodium in liquid ammonia, or by exposure to hydrogen bromide in acetic acid or liquid hydrogen fluoride. It is also partially cleaved on exposure to trifluoroacetic[73] acid, but not by formic acid.[84] It would appear that the O-benzyl moiety, with the exception of the benzyloxycarbonyl group, is compatible with all the N$^\alpha$-protectors mentioned in this chapter. However, besides its instability in trifluoroacetic acid, there is an additional drawback that limits the use of the benzyl group as a universal phenolic function protector. During acidolytic cleavage of the O-benzyl moiety from tyrosine-containing peptides, there is the risk of alkylation of the aromatic ring of tyrosine by the benzyl cation which is formed.[73] The risk of such a side reaction can be diminished, but not eliminated, by carrying out the reaction in the presence of anisole or other cation scavengers. To ascertain the absence of such a side product, it is important to carry out accurate determination of tyrosine in the synthetic polypeptide. Since acid hydrolysis of peptides usually leads to extensive destruction of tyrosine, it is advisable to analyze for tyrosine content after enzymic hydrolysis, using, for example, leucine aminopeptidase or aminopeptidase M.[79] Recent studies indicate that the O-2,6-dichlorobenzyl group, which is similar to the O-benzyl function, has more desirable properties than the latter moiety. It is stable in trifluoroacetic acid and during its acidolytic cleavage in the presence of cation scavengers is less prone than the benzyl moiety to alkylate the aromatic ring of tyrosine.[73]

A number of peptides have been synthesized in recent years using the *tert*-butyl group as the protector of the phenolic group of tyrosine. The O-*tert*-butyl ether group has similar properties to the O-*tert*-butyl ester moiety and it thus has the same compatibility with N$^\alpha$-protectors as the latter. The O-*tert*-butylation of tyrosine and of the other hydroxy amino acids (serine and threonine) consists in the interaction of the respective N$^\alpha$-benzyloxycarbonyl amino acid p-nitrobenzyl ester with isobutylene, as described previously in the synthesis of the *tert*-butyl esters,[83] followed by hydrogenolytic removal of the benzyloxycarbonyl and p-nitrobenzyl moieties.

*O-Benzyl-*L-*tyrosine.*[82] A solution of L-tyrosine (109 g) in 2 N sodium hydroxide (300 ml) is mixed with a solution of copper sulfate hydrate

---

[84] G. P. Schwartz and P. G. Katsoyannis, unpublished data.

($CuSo_4 \cdot 5H_2O$; 75 g) in water (180 ml). The mixture is heated at 50° in a water bath for 15 min, and the precipitated copper complex of tyrosine is collected by filtration and washed with water until the filtrate is colorless. The copper complex is then dissolved in a mixture of methanol (2.2 liters) and 2 N sodium hydroxide (300 ml), and benzyl bromide (75 ml) is added. The mixture is shaken overnight at room temperature and the precipitated product is filtered, washed with a methanol–water mixture (3.5:1; 700 ml) and ether. The blue solid is transferred into a beaker and triturated twice with warm (37°) 1 N hydrochloric acid (ca. 1 liter) to break the copper complex. The white product which is formed is filtered, washed with water, and suspended in water (ca. 400 ml). The pH of the mixture is adjusted to 7.0 with 1 N ammonium hydroxide (ca. 100 ml), stirred for a few minutes and the white product is filtered out, washed with water, 95% methanol and ether; weight 90 g; mp 244–247°).

O-2,6-Dichlorobenzyl-L-tyrosine.[61] A solution of tyrosine (30 g) in 2 N sodium hydroxide (168 ml) is mixed with a solution of copper sulfate hydrate ($CuSO_4 \cdot 5H_2O$; 20.4 g) in water (84 ml). The mixture is heated to 60° in a water bath for 10 min, cooled to room temperature, and diluted with methanol (720 ml). To this solution is added, with vigorous stirring, 2,6-dichlorobenzyl bromide (45 g). After stirring (in the hood) for 20 hr, the precipitated product is filtered and washed successively with 25% aqueous methanol (500 ml), methanol (500 ml), and acetone (500 ml). This solid is dissolved, by heating, in 50% aqueous ethanol (5 liters) containing EDTA disodium salt (50 g) and the hot solution is decanted from the insoluble by-products. Upon cooling the solution overnight, the *O*-2,6-dichlorobenzyl-L-tyrosine crystallizes out, is filtered and washed with water and ethanol; weight 18 g; mp 212°–216°.

G. The Hydroxyl Groups of Serine and Threonine

Masking of the hydroxyl groups of serine and threonine can be accomplished by acetylation, benzylation, or *tert*-butylation. Of these three methods, however, only the latter two are commonly employed in peptide synthesis. O-Benzylation of serine is accomplished by a multistep chemical synthesis that cannot be easily adapted for routine use in a peptide synthesis laboratory. Alternatively, it can be prepared by the interaction of N^{α}-*tert*-butyloxycarbonyl serine with benzyl bromide in liquid ammonia in the presence of sodium.[85] O-Benzyl-L-serine is now commercially available.

O-Benzyl-L-threonine is prepared by saponification of the corres-

[85] V. J. Hruby and K. W. Ehler, *J. Org. Chem.* **35**, 1690 (1970).

ponding benzyl ester.[86] The latter compound is produced by interaction of threonine with benzyl alcohol in the presence of p-toluenesulfonic acid.

The O-benzyl ether bond is stable under alkaline conditions and can be split by the same procedures used for the removal of the benzyloxycarbonyl group.

The O-*tert*-butylation of serine and threonine is carried out as described in the case of tyrosine.

O-Benzyl-L-threonine Benzyl Ester Hemioxalate.[86] A mixture of L-threonine (35.7 g), toluene (750 ml), benzyl alcohol (300 ml), and p-toluenesulfonic acid monohydrate (74.1 g) is refluxed with a Dean–Stark trap for 18 hr. After cooling the reaction mixture to room temperature, ethyl acetate (500 ml) is added, followed by 0.5 M sodium carbonate sufficient to raise the pH of the aqueous phase to 9.0. The organic layer is separated and washed once with water. The combined aqueous phases are reextracted with ethyl acetate (350 ml), and the combined organic phases are dried over magnesium sulfate and mixed with a solution of oxalic acid dihydrate (36 g) in methanol (180 ml). The mixture is cooled for 4 hr, then the crude hemioxalate salt is collected by filtration, washed with cold ethanol, and recrystallized from ethanol; weight 23 g; mp 169–170°.

O-Benzyl-L-threonine.[86] O-Benzyl-L-threonine benzyl ester hemioxalate (23 g) is partitioned between ethyl acetate (500 ml) and 1 M potassium carbonate (500 ml). The organic layer which contains the free base is separated, washed with water, dried over magnesium sulfate, and evaporated under reduced pressure (25°–32°). The remaining oily residue is dissolved in methanol (300 ml), and to this solution 1 N sodium hydroxide (71 ml) is added over a period of 10 min. After 2 hr at room temperature, the reaction mixture is concentrated under reduced pressure (25°–32°) to one-fourth of its original volume, its pH is adjusted to 6.6 with 1 N acetic acid, and it is stored at 4° overnight. The precipitated product is filtered off and washed with cold water, n-propanol and ether; weight 10 g; mp 201°–202°.

VII. Methods of Removal of Protecting Groups during Chain Elongation

In this section we describe experimental procedures for removal of protecting groups from the amino functions of peptides. It should be pointed out that the same procedures are applicable if the protecting group in question, or similar-type protectors, are employed for masking

---

[86] T. Mizoguchi, G. Levin, D. W. Woolley, and J. M. Stewart, *J. Org. Chem.* **33**, 903 (1968).

other functions in the peptide. In the latter case, only the isolation procedure of the deprotected peptide may differ.

A. Removal of *tert*-Butyl-Type Protectors (*N-tert*-Butyloxycarbonyl, *tert*-Butyl Esters, *tert*-Butyl Ethers)

1. The Trifluoroacetic Acid Method[52,87]

A peptide containing a *tert*-butyl-type N-protector (0.01 mol) is mixed with anhydrous trifluoroacetic acid (30 ml) in a round-bottom flask equipped with a calcium chloride drying tube. After standing for 30–60 min at room temperature, the solution is poured into cold anhydrous ether (500 ml) and the mixture is stored in the cold for 1–2 hr. The precipitated trifluoroacetic acid salt of the deblocked peptide is collected by filtration,[88,89] washed with ether, and dried under reduced pressure over potassium hydroxide. The trifluoroacetic acid method should not be applied if the peptide in question contains N^ε-benzyloxycarbonyl lysine and/or O-benzyltyrosine residues. Removal of the *tert*-butyl-type protectors in these cases should be performed by the formic acid method.

2. The Formic Acid Method.[59,79,90]

The protected peptide is dissolved in 98% formic acid (150 ml) in a round-bottom flask equipped with calcium chloride drying tube. After 3 hr at room temperature, the solvent is removed under reduced pressure at 25°. The residue is crystallized by trituration with ether,[88,89] filtered, washed with ether, and dried under vacuum over potassium hydroxide. If the formic acid salt of the peptide is subsequently used as the amino component for chain elongation, it is imperative that it be converted to the free base prior to its interaction with the activated carboxyl compo-

[87] R. Schwyzer, W. Rittel, H. Kappeler, and B. Iselin, *Angew. Chem.* **72**, 915 (1960).
[88] Often the salts of small peptides are oils or do not precipitate by the addition of ether. In these cases the deprotection solvent is removed under reduced pressure, and the residue is partitioned between cold ethyl acetate and 1 M sodium carbonate (with formic acid salts, 1 N ammonium hydroxide is recommended). The organic layer is washed quickly with cold water until neutral, dried over magnesium sulfate for a few minutes with cooling, and concentrated to dryness (25°) to give the peptide as the free base.
[89] If the deprotected peptide also contains a free carboxyl group, it can be isolated as its dipolar ion as follows: After removal of the deblocking solvent, the residue is mixed with a cold saturated solution of sodium acetate until the pH of the mixture is approximately 6.0. After cooling for a few hours, the insoluble peptide is collected by filtration and dried.
[90] B. Halpern and D. E. Nitecki, *Tetrahedron Lett.* p. 3031 (1967).

nent. We have found that interaction, in the presence of triethylamine, of a formic acid salt of a deblocked peptide with an activated carboxyl component leads occasionally to the formylation of the deblocked peptide.[59] The free base of the deblocked peptide is obtained as follows: If the free base is soluble in ethyl acetate the procedure in footnote 88 is followed. Otherwise,[59] a cold solution of the formic acid salt of the deblocked peptide (0.02 mol) in DMF (70 ml) is mixed with 1 N ammonium hydroxide (30 ml) and immediately poured into cold water (1 liter) saturated with sodium chloride. The pH of the mixture should be greater than 9.0. The precipitated free base of the peptide derivative is collected, washed with water until neutral, and dried over phosphorus pentoxide under high vacuum.

B. Removal of the N-Benzyloxycarbonyl, Benzyl Ether, and Benzyl Ester Groups

1. Hydrogen Bromide in Acetic Acid Method[91,92]

The reagent used in this method, hydrogen bromide-saturated acetic acid, is prepared as follows: Distilled glacial acetic acid (800 ml) is placed in a 1-liter bottle with ground-glass fitted with a gas inlet tube and a calcium chloride drying tube. A slow stream of hydrogen bromide gas, which passes first through a gas washing bottle containing tetralin, is bubbled into the acetic acid at room temperature. After 1 hr the reagent bottle is cooled in an ice bath and the passage of the hydrogen bromide gas is continued until saturation. This stock reagent (ca. 4 N hydrogen bromide) can be kept in the refrigerator for several months if it is tightly stoppered.

In a round-bottom flask fitted with a calcium chloride drying tube, a protected peptide (0.01 mol) is placed and acetic acid (20 ml) and acetic acid saturated with hydrogen bromide (20 ml) are added. After standing for 1 hr the mixture is poured into 1 liter of cold anhydrous ether. The suspension is cooled for several hours before the deblocked peptide is collected by filtration, washed with ether, and dried under reduced pressure over sodium hydroxide.[88,89]

Exposure of serine- or, occasionally, threonine-containing peptides to hydrogen bromide in acetic acid invariably causes acetylation of the hydroxyl function.[46] This can be avoided if hydrogen bromide in trifluoroacetic acid is the cleaving reagent.[46] This alternative method of cleavage

[91] G. W. Anderson, J. Blodinger, and A. D. Welcher, *J. Am. Chem. Soc.* **74**, 5309 (1952).

[92] D. Ben-Ishai and A. Berger, *J. Org. Chem.* **17**, 1564 (1952).

proceeds as follows: The peptide (0.01 mol) is dissolved in trifluoroacetic acid (25 ml) containing water (1 ml) in a round-bottom flask equipped with calcium chloride drying tube and a gas inlet tube. The flask is cooled to 4°, and hydrogen bromide gas (which is passed first through tetralin) is passed through the solution for 1 hr before the mixture is diluted with cold ether (600 ml). The precipitate is collected by filtration, washed with cold ether, and dried under vacuum over solid sodium hydroxide.

The hydrogen bromide cleavage is a harsh treatment that may cause a variety of side reactions and decomposition of the product. Methionine residues, if present, may be alkylated and form S-benzylsulfonium derivatives.[92a] Tryptophan is highly sensitive to strong acid conditions and under such treatment may decompose. Tyrosine residues may be brominated if free bromine is present or, similarly to methionine, may be alkylated in the aromatic ring. Damage to methionine and tyrosine can be minimized if cation scavengers, such as anisole, phenol, or methyl ethyl sulfide, are added to the reaction mixture.[17,46,92b] Tryptophan destruction may be minimized if reducing agents, such as β-mercaptoethanol, are added to the reaction medium.[92c]

2. Catalytic Hydrogenation Method[6]

This is probably the most elegant method for removal of the aforementioned groups. It proceeds smoothly at room temperature and at atmospheric pressure without any damage, even to the more sensitive groups present. In spite of the variation that exists from batch to batch, 10% palladium on charcoal or palladium black catalysts are adequate to perform routine hydrogenolysis of peptides. Both catalysts are available commercially. A hydrogenation apparatus can be readily assembled from the glassware usually found in the laboratory. It consists of a round-bottom or conical flask equipped for magnetic stirring and fitted with a gas inlet and exhaust tubes. All operations are performed in a hood.

The protected peptide (0.01 mol) is dissolved in methanol or ethanol (150 ml), depending on whether the protected peptide is a methyl or ethyl ester, containing one equivalent of hydrochloric acid, or acetic acid if secondary functions are protected with *tert*-butyl groups. Hydrogenolysis of protected peptides containing a *tert*-butyl ester group as the C^α-protector can be performed in the absence of any acid. Protected pep-

[92a] S. Guttmann and R. A. Boissonnas, *Helv. Chim. Acta* **42**, 1257 (1959).
[92b] E. Wünsch, *Collect. Czech. Chem. Commun.* **24**, 60 (1959).
[92c] J. Blake and C. H. Li, *J. Am. Chem. Soc.* **90**, 5882 (1968).

tides that are not soluble in alcohol are dissolved in DMF. To the solution, flushed with nitrogen gas for 5 min, the catalyst (ca. 2 g) is then added. The use of palladium black is recommended in hydrogenations performed in DMF solutions. It is our experience that palladium on charcoal forms colloidal dispersions in DMF and the separation of the catalyst at the end of the reaction becomes a difficult task. After the addition of the catalyst, the reaction vessel is flushed again with nitrogen for another 5 min and then connected to the hydrogen line. Hydrogenation proceeds with a moderate stream of hydrogen gas and good stirring. The exhaust gases are allowed to escape through a flask containing saturated barium hydroxide solution. Formation of a barium carbonate precipitate indicates the course of the reaction; the end of the reaction is signified by the absence of formation of barium carbonate. Hydrogenation of N-benzyloxycarbonyl derivatives is usually completed within 2 hr. For the hydrogenation of O-benzyl esters or O-benzyl ethers, where no carbon dioxide is produced, passage of the hydrogen gas is allowed to proceed for 4–5 hr. After completion of the reaction, the vessel is flushed again with nitrogen for 10 min, and the catalyst is removed from the reaction mixture by filtration. The filtrate is concentrated to dryness under reduced pressure and the acid salt of the deprotected peptide is solidified by trituration with ether. If the hydrogenation was carried out in the absence of acid (*tert*-butyl esters), evaporation of the filtrate yields the partially deblocked peptide as the free base.

C. Removal of the N-o-Nitrophenylsulfenyl Group

The facile and selective removal of the o-nitrophenylsulfenyl group can be accomplished upon treatment of the peptide with ca. 2–3 equivalents of hydrogen chloride in organic solvents, such as acetone, ethyl ether, or DMF.[16] If the peptide contains tryptophan or S-diphenylmethyl or S-trityl cysteine, the by-product of the deprotection reaction, o-nitrophenylsulfenyl chloride reacts with the aforementioned amino acid residues to yield o-nitrophenylsulfenyltryptophan[93,94] (indole moiety substitution) and yellow impurities with the cysteine derivatives.[92a] These side reactions are avoided by deprotection with thioacetamide in alcohol and in the presence of a weak acid. Deprotection of larger peptides, which are insoluble in methanol, is effected in a mixture of DMF–methanol.

---

[93] A. Fontana, F. Marchiori, R. Rocchi, and P. Pajetta, *Gazz. Chim. Ital.* 96, 1301 (1966).

[94] A review of the use of the o-nitrophenylsulfenyl group may be found in an article by A. Fontana and E. Scoffone *in* "Mechanisms of Reactions of Sulfur Compounds," Vol. 4, p. 15. Intra Science Res. Found. 1969.

1. The Hydrogen Chloride-Acetone Method[95]

An N-o-nitrophenylsulfenyl peptide (0.01 mol) is dissolved in acetone (120 ml) and 6 N hydrochloric acid (7 ml) is added. After standing for 5 min, the solvent is removed under reduced pressure (30°), and cold ether (700 ml) is added to the remaining material. The precipitated peptide hydrochloride salt is collected by filtration, washed with ether, and dried. In some cases a yellow, insoluble by-product, presumably o-nitrophenyl disulfide, contaminates the product. This disulfide can be removed by dissolving the peptide hydrochloride in a small amount of 2-propanol followed by filtering off the disulfide. The peptide salt is recovered by reprecipitation with ether.

2. The Thioacetamide Method[17]

The N-o-nitrophenylsulfenyl peptide (0.01 mol) is dissolved in a mixture of DMF (36 ml) and methanol (18 ml), and to this solution thioacetamide (1.6 g) and acetic acid (4.4 ml) are added. After stirring for 30 min at room temperature, the acetate salt is precipitated by the addition of ether (600 ml).

VIII. Stepwise Synthesis of Peptides

As we mentioned above in the section on problems in peptide synthesis, the stepwise synthesis from the amino terminal of the growing peptide chain is the preferred method for synthesis of peptides containing about 15 amino acid residues. A great number of peptides have been synthesized by this method, and it is used very often to prepare the peptide fragments employed in the fragment condensation approach. It is best to keep all secondary functions protected, not only to minimize side reactions, but also to simplify the isolation and purification of the product from the reactants. Synthesis with minimal side-group protection is also possible, but the manipulations for successful isolation of the product become more difficult, especially if the acidic (glutamic and aspartic) and basic (arginine and histidine) amino acids remain unprotected. In addition, certain coupling procedures cannot be used and the presence of excess carboxyl component cannot be tolerated in many cases. Consequently, the chain length that can be achieved with minimal side-group protection, is limited.

Several procedures are available for the activation of N^α-protected

[95] G. C. Stelakatos, A. Paganou, and L. Zervas, *J. Chem. Soc.* (*C*) p. 1191 (1966).

amino acids and the coupling of the activated derivatives to amino components in the stepwise approach. Of these, the carbodiimide, the mixed anhydrides, the isoxazolium salt and the most commonly used, "activated ester" methods will be described below. No matter, however, which one of the aforementioned activating methods is used, the same general procedure for the condensation of the activated carboxyl component to the amino component, and the isolation of the product, is employed.

A. General Procedure for the Condensation of an N^α-Protected Amino Acid to the Growing Peptide Chain and Isolation of the Product

The amino component, as its ammonium salt,[96] or the free base, is dissolved in a suitable solvent. DMF is commonly employed since most N^α-deprotected peptides, particularly of larger size, are sparingly soluble in other solvents. However, solvents such as tetrahydrofuran, methylene chloride, acetonitrile are often used with small peptides, as indicated in the specific examples cited below. One equivalent of a tertiary base, usually triethylamine,[97] is added followed by one equivalent of the desired N^α-protected amino acid[98] and the activating agent. In many cases the N^α-protected amino acid is activated before mixing with the amino component. The volume of solvent is adjusted so that both the carboxyl and amino components are at a concentration of at least 0.1 M at the beginning of the reaction. If required, a few drops of triethylamine are added to ensure that the reaction mixture is slightly alkaline; this is readily checked by holding a wet pH-paper over the mixture. Reaction times are usually 12–24 hr.

At the end of the reaction period, the mixture is poured into ethyl acetate (if the product is soluble in this solvent), and the organic layer is washed with dilute aqueous base, i.e., 1 M sodium bicarbonate or 1 N ammonium hydroxide, to remove unreacted carboxyl component. The unreacted amino component is removed by washing the organic layer with

---

[96] The amino component is an amino acid ester in the first coupling step and, thereafter, the growing peptide chain which has been N^α-deprotected by one of the above-mentioned procedures.

[97] If the free base of the amino component is used, addition of triethylamine is omitted.

[98] N-o-Nitrophenylsulfenyl amino acids are usually stored as the dicyclohexylamine salts. If such compounds are to be used as the carboxyl component, they must be first converted to the free acids by the following procedure [L. Zervas, D. Borovas and E. Gazis, *J. Am. Chem. Soc.* **85**, 3660 (1963)]: The chosen salt is partitioned between ethyl acetate and 1.1 equivalents of 0.2 N sulfuric acid. The organic layer is separated, washed with water until neutral, dried over magnesium sulfate, and concentrated to dryness (30°).

dilute aqueous acid, i.e., 1 N hydrochloric acid, 0.2 N sulfuric acid, or 0.5 N acetic acid. The acetic acid washing is recommended when acid-labile protectors of secondary functions are present. The organic layer is dried over magnesium sulfate, and the solvent is removed under vacuum. The residue is purified by crystallization or precipitation from organic solvents.

If the product is insoluble in ethyl acetate, the reaction mixture is diluted with aqueous base (1 M sodium bicarbonate or 1 N ammonium hydroxide) and the precipitated product is washed on the filter with dilute base, acid, and water, dried and purified by crystallization or re-precipitation from organic solvents. The purity of the product is ascertained by thin-layer chromatography of the N^α-deprotected material and by elemental analysis. Yields, in general, vary from 60 to 90%.

As the chain length increases, peptides lose much of their solubility in the common organic solvents, i.e., alcohol, acetonitrile, ethyl acetate, and their N^α-deprotected derivatives are insoluble in dilute acids. Thus washing with acid will not remove unreacted amino component. Chain elongation of larger polypeptides requires the use of an excess carboxyl component to drive the coupling reaction to completion. In this case, the concentration of the carboxyl component is greater than 0.1 M throughout the reaction period. The course of the reaction is monitored by performing a ninhydrin test on the mixture to ascertain whether unreacted amino component is present.[99] A negative ninhydrin test signifies completion of the reaction. The mixture is then poured into a solvent in which the carboxyl component is soluble, and the precipitated product is collected and reprecipitated or triturated with the same solvent. Large protected peptides are in general apolar and of such limited solubility that simple chromatographic methods of purification and/or purity check are not feasible. Large peptides, in general, are highly solvated, and removal of the bound solvent requires vigorous drying at elevated temperature, a procedure that usually results in alteration of the peptide material. Elemental analysis is, therefore, of little value in ascertaining the purity of such compounds. It appears then that amino acid analysis after acid or enzymic hydrolysis (with leucine aminopeptidase or aminopeptidase M) and average recovery of the constituent amino acids, is the only cri-

---

[99] A small quantity of the reaction mixture is mixed with a suitable solvent to effect precipitation of the unreacted amino component and of the product. This precipitate is redissolved in DMF, and a few drops are spotted on a piece of filter paper, sprayed with ninhydrin, and heated in an oven at 110°. A negative color test indicates the absence of free amino component within the limits of the test. It must be realized that, as the length of the amino component increases, the ninhydrin test for its detection becomes less sensitive.

terion available of purity of large polypeptides. A pertinent discussion of this problem may be found in two articles by Hofmann et al.[99a,99b]

As indicated previously, there are limitations in the use of this general procedure of peptide bond formation and isolation of the product if partial side-group protection is employed. Unprotected C^α-carboxyl or secondary carboxyl groups of aspartic and glutamic acids precludes the purification step involving washing and dilute base, and unprotected secondary functions of arginine and histidine do not permit washing with dilute acid. However, since peptide products with limited side-group protection are polar compounds, their purification may be feasible by application of various chromatographic procedures.

B. Peptide Bond-Forming Procedures in the Stepwise Synthesis

Amino acids protected by the aforementioned protectors are not prone to racemization on activation. The activation methods described below can be used without any special precautions other than that the reaction medium *must* be neutral or very slightly basic.

1. Carbodiimide Method[100]

Principle. N,N'-Dicyclohexylcarbodiimide, which is commercially available, is the reagent most frequently employed in this method. The overall reaction can be expressed by the following equation:

$$
\begin{array}{cc}
R_1 & R_2 \\
| & | \\
Y \cdot HNCHCOOH + H_2NCHCOOZ + RN{=}C{=}NR \rightarrow
\end{array}
$$

$$
\begin{array}{ccc}
R_1 & R_2 & O \\
| & | & \| \\
Y \cdot NHCHCO \cdot NHCHCOOZ + RNHCNHR
\end{array}
$$

It is thought that the reaction may proceed by the addition of the acyl amino acid to the carbodiimide with formation of an O-acylisourea derivative (XXVII)

$$
\begin{array}{c}
R_1 \\
| \\
Y \cdot NHCHCOO \\
| \\
RN{=}C{-}NHR \\
\text{(XXVII)}
\end{array}
$$

[99a] J. Beacham, G. Dupuis, F. M. Finn, H. T. Storey, C. Yanaihara, N. Yanaihara, and K. Hofmann, *J. Am. Chem. Soc.* **93**, 5526 (1971).

[99b] R. Cambel, G. Dupuis, K. Kawasaki, H. Romovacek, N. Yanaihara, and K. Hofmann, *J. Am. Chem. Soc.* **94**, 2091 (1972).

[100] J. C. Sheehan and G. P. Hess, *J. Am. Chem. Soc.* **77**, 1067 (1955).

The latter, an activated intermediate, then reacts with the amino component to yield a peptide and the symmetrically substituted dicyclohexylurea. Or, alternatively, the activated O-acylisourea derivative (XXVII) may react with a second molecule of an acyl amino acid to form the active symmetrical anhydride (XXVIII)

$$\overset{\overset{\displaystyle R_1}{\displaystyle |}}{(Y \cdot NHCHCO)_2O}$$
(XXVIII)

and dicyclohexylurea. The anhydride in turn acylates the amino component to form the peptide. The principal side reaction of carbodiimides is the rearrangement of the activated intermediate (XXVII) to give the N-acylurea derivative (XXIX)

$$\overset{\overset{\displaystyle R_1}{\displaystyle |}}{Y \cdot NHCHCO} \overset{\displaystyle O}{\underset{\displaystyle RN-C-NHR}{\overset{\displaystyle \|}{|}}}$$
(XXIX)

The latter is fairly stable, does not react with the amino component, and is often obtained as a side product in the carbodiimide procedure.

 Comments. The dicyclohexylurea, which is always formed during activation by dicyclohexylcarbodiimide, is highly insoluble in most organic solvents except alcohols, and its separation from the peptide product requires additional purification steps. The use of water-soluble carbodiimides containing polar substituents offers an advantage in such cases.[101] Formation of the N-acylurea side product (XXIX) is favored under alkaline conditions and elevated temperatures. Its formation can be minimized by carrying out the reaction at a low temperature (ca. 4°) in solvents such as methylene chloride or acetonitrile. It is obvious that the carbodiimide method cannot be applied with the secondary carboxyl groups of acidic amino acids unprotected. For the same reason, acetic acid or trifluoroacetic acid salts of the amino component cannot be used, and only the reisolated free base is employed. If equimolecular amounts of carboxyl and amino components are used in the carbodiimide procedure, protection of the various secondary functions, other than carboxyl group, is not imperative. Activation of N^α-protected glycine or serine by this method is not recommended since with both of these amino acids formation of the N-acylurea by-product is favored. Activation of N^α-protected glutamine and asparagine by carbodiimide should be avoided, as extensive dehydration of the amide groups of both amino

[101] J. C. Sheehan and J. J. Hlavka, *J. Org. Chem.* **21**, 439 (1956).

acids occurs with the concomitant formation of the corresponding cyano derivative.[102,103]

Example: N-Benzyloxycarbonyl-L-threonyl-L-alanine Methyl Ester.[104] A solution of *N*-benzyloxycarbonyl-L-threonine (5 g, 0.02 mol) and L-alanine methyl ester hydrochloride (2.8 g, 0.02 mol) in methylene chloride or acetonitrile (100 ml) is cooled to 4°, and triethylamine (2.8 ml) and dicyclohexylcarbodiimide (4.08 g) are added. After stirring for 1 hr at 4°, and at room temperature for 18 hr, the mixture is cooled for several hours and the precipitated dicyclohexylurea by-product is filtered off. The filtrate is taken to dryness under reduced pressure, and the residue is dissolved in ethyl acetate (500 ml). This solution is then washed successively with 1 *N* hydrochloric acid, water, and 1 *M* sodium bicarbonate, dried over anhydrous magnesium sulfate, and concentrated to one-half of its original volume. Addition of petroleum ether (300 ml) causes the precipitation of the crystalline product, which is filtered and dried; weight 4.8 g (70%); mp 130°–131°.

2. Mixed Anhydride Method[105-107]

Principle. The interaction of N^α-protected amino acids with acid chlorides in the presence of one equivalent of a tertiary base yields mixed anhydrides:

$$\underset{\underset{R_1}{|}\underset{O}{\|}}{Y\cdot HNCHCOH} + \underset{\underset{O}{\|}}{ClC\cdot R'} \xrightarrow[\text{solvent}]{\textit{tert}\text{-base}} \underset{\underset{R_1}{|}\underset{O}{\|}\underset{O}{\|}}{Y\cdot HNCHCOC\cdot R'}$$

"Mixed anhydride"

The mixed anhydrides react with amino components in a reaction that may proceed in two directions; peptide bond forming or the acylation of the amino component:

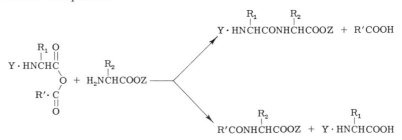

[102] D. T. Gish, P. G. Katsoyannis, G. P. Hess, and R. J. Stedmann, *J. Am. Chem. Soc.* **78**, 5954 (1956).

[103] C. Ressler, *J. Am. Chem. Soc.* **78**, 5956 (1956).

[104] P. Cruickshank and J. C. Sheehan, *J. Am. Chem. Soc.* **86**, 2070 (1964).

[105] T. Wieland and H. Bernhard, *Justus Liebigs Ann. Chem.* **572**, 190 (1951).

[106] R. A. Boissonnas, *Helv. Chim. Acta* **34**, 874 (1951).

[107] J. R. Vaughan, Jr., and R. L. Osato, *J. Am. Chem. Soc.* **74**, 676 (1952).

Mixed anhydrides of N^α-protected amino acids and certain half-esters of carbonic acid, such as ethyl, butyl, and isobutyl carbonate or with sterically hindered organic acids, such as isovaleric acid, upon interaction with amino components, predominantly follow the peptide bond forming reaction. Mixed anhydrides of N^α-protected amino acids with the mono-esters of carbonic acids are most widely used since they have the advantage that, on interaction with amino components they yield, in addition to the desired peptide, only carbon dioxide and the parent alcohol:

$$\begin{array}{c} R_1 \\ | \\ Y\cdot HNCHCO \\ \diagdown \\ O \ + \ H_2NCHCOOZ \ \longrightarrow \ Y\cdot HNCHCONHCHCOOZ \ + \ R'OH \ + \ CO_2 \\ \diagup \\ R'OC \\ || \\ O \end{array}$$

Comments. For the formation of the mixed anhydride, tetrahydrofuran is the solvent of choice. DMF should be avoided since the alkyl chloroformate may react with this solvent. Triethylamine is most commonly used as the tertiary base. Formation of the aforementioned anhydrides is a very fast reaction and is usually carried out at $-10°$ to $-15°$. Interaction of the mixed anhydride with the amino component is usually performed at temperatures ranging from $0°$ to $25°$, and the solvent does not influence the course of the reaction. Mixed anhydrides are highly reactive intermediates and, when present in excess, may acylate unprotected hydroxyl groups of the amino component. It must be emphasized that the above-mentioned mixed anhydrides, upon interaction with amino components, yield predominantly, but not exclusively, peptides. A small amount of acylated amino component may also be formed, and the amount of this by-product depends greatly on the amino acid sequence of the amino component. In our hands, for example, interaction of the mixed anhydride of N^α-butyloxycarbonyl-L-proline and isobutyl carbonate with an amino component containing proline as the NH_2 terminus formed almost exclusively isobutyloxycarbonylated amino component.

Example: N^α-*Benzyloxycarbonyl*-L-*glutaminyl*-S-*benzyl*-L-*cysteine Methyl Ester.*[30]* S-Benzyl-L-cysteine methyl ester hydrochloride (10.4 g; 0.04 mol) is suspended in tetrahydrofuran, and triethylamine (6.15 ml) is added. The mixture is stirred and cooled and, after a few minutes, the triethylamine hydrochloride is filtered off and washed with tetrahydrofuran. Evaporation of the combined filtrates under reduced pressure yields the free base as an oil, which is redissolved in tetrahydrofuran (100 ml). In a 500-ml three-neck round-bottom flask equipped with a magnetic stirrer and fitted with a calcium chloride drying tube and a dropping funnel (the whole apparatus is dried for 1 hr in an oven at 110° prior to its use), benzyloxycarbonyl-L-glutamine (11.2 g; 0.04 mol) is dissolved in anhydrous tetrahydrofuran (80 ml), the solution is cooled to −15°, and triethylamine (5.6 ml) is added. Isobutyl chloroformate (5.2 ml) dissolved in tetrahydrofuran (15 ml) is added dropwise over a period of 8 min to the stirred mixture. The reaction is continued for an additional 3 min, and the tetrahydrofuran solution (0°) of the free ester, prepared as described above, is added. After 3 hr at room temperature, the crystalline precipitate is collected by filtration and washed with two 50-ml portions of tetrahydrofuran and then with ether. The crystalline product is then washed successively with 1 N hydrochloric acid, water, 1 M sodium bicarbonate, water, and dried to give 17.2 g (89%) of the protected dipeptide ester; mp 199°–202°. On recrystallization from 50% aqueous acetic acid, a mp of 203°–204° is obtained.

3. Isoxazolium Salts Method[108,109]

Principle. Interaction of N^α-protected amino acids with 3-unsubstituted isoxazolium salts such as 2-ethyl-5-phenyloxazolium 3′-sulfonate (XXX; Woodward's reagent K) in the presence of one equivalent of a tertiary base, affords enol esters (XXXI). The latter derivatives are highly active, and on interaction with amino components readily form peptides:

Comments. The Woodward's reagent K is commercially available. The most commonly used solvents for the formation of the active enol esters (XXXI) are acetonitrile or nitromethane. This reagent can be used to activate N^α-protected amino acids which have unprotected hydroxyl functions. Similarly, it can activate N^α-glutamine and asparagine without

[108] R. B. Woodward, R. A. Olofson, and H. Mayer, *J. Am. Chem. Soc.* **83**, 1010 (1961).

[109] R. B. Woodward, R. A. Olofson, and H. Mayer, *Tetrahedron*, Suppl. 8, Part I, p. 321 (1967).

the risk of side reactions (nitrile formation). One major side reaction in the synthesis of peptides by the use of this reagent is the rearrangement of the active intermediate enol ester (XXXI) to give the unreactive imide (XXXII).

(XXXI)   (XXXII)

This side reaction is favored by polar solvents such as DMF and hence the use of this reagent in the synthesis of large peptides is limited.

Example. N^α-*Benzyloxycarbonyl-N^ε-tosyl-L-arginylglycyl-L-phenylalanyl-L-phenylalanyl-L-tyrosyl-L-threonyl-L-prolyl-N^ε-tosyl-L-lysyl-L-threonine Methyl Ester.*[110] N^α-Benzyloxycarbonyl-N^ε-tosyl-L-arginine (10.4 g; 0.022 mol) is dissolved in a mixture of DMF (60 ml) and acetonitrile (300 ml); after cooling to 4°, triethylamine is added followed by Woodward's reagent K (5.95 g). The mixture is stirred in the cold until all the reagent has dissolved (approximately 1 hr) and the hydrochloride salt of glycyl-L-phenylalanyl-L-phenylalanyl-L-tyrosyl-L-threonyl-L-prolyl-N^ε-tosyl-L-lysyl-L-threonine methyl ester dissolved in a mixture of DMF (110 ml), and acetonitrile (110 ml) containing triethylamine (3.3 ml) is added. [The peptide salt is prepared by hydrogenation of the N^α-benzyloxycarbonyl octapeptide (27 g) for 4 hr over 10% palladium-charcoal catalyst (16 g) in a solution of methanol (400 ml) containing 2 N hydrochloric acid (12 ml). The catalyst is filtered off and the solvent is removed under reduced pressure. The residue is redissolved in methanol and taken to dryness before use.] The mixture is stirred for 24 hr, then cold 0.5 M sodium bicarbonate (500 ml) is added and the precipitated product is collected by filtration, washed with water, 1 N hydrochloric acid, and water. The product is reprecipitated from a methanolic solution (160 ml) by the addition of ether; weight 26.6 g (80%), mp 156°–161°).

4. The Active Esters Method

Principle. Esters of carboxylic acids whose alcohol moiety bears substituents (electronegative) that increase the acidity of their hydroxyl

[110] P. G. Katsoyannis, J. Ginos, and M. Tilak, *J. Am. Chem. Soc.* **93**, 5866 (1971).

functions, are known as "activated" or "active esters."[111-113] These esters readily react with amino components to form acylated peptide derivatives. In peptide synthesis the most widely used are the p-nitrophenyl,[114] 2,4,5-trichlorophenyl,[115] and N-hydroxysuccinimidyl[116]

$$\begin{array}{c} H_2C-CO \\ | \quad\quad\, \diagdown \\ \quad\quad\; NO- \\ | \quad\quad\, \diagup \\ H_2C-CO \end{array}$$

esters of N^α-protected amino acids. Interaction of these esters of N^α-protected amino acids with amino components leads to the formation of peptides. This approach is the most commonly used method for chain elongation at the present time.

Comments. In practice, a solution of N^α-protected amino acid interacts with the corresponding "alcohol" in the presence of one equivalent of N,N'-dicyclohexylcarbodiimide. Ethyl acetate is often used in the preparation of the "phenol" esters and acetonitrile is the solvent of choice in the synthesis of the N-hydroxysuccinimide esters. Removal of the dicyclohexylurea by-product, followed by evaporation of the solvent, affords the respective N^α-protected amino acid active ester which is recrystallized, usually from 95% ethyl alcohol or isopropanol. The active esters of most amino acids are stable compounds and can be stored in a desiccator (0°) for long periods of time. Active esters of N^α-protected amino acids react almost exclusively with the amino group of the amino component. In large excess, however, and particularly under basic conditions, they may react, although to a much lesser extent than the mixed anhydrides, with unprotected hydroxyl or phenolic functions of the amino component.

It is not imperative that active esters of N^α-protected amino acids be isolated prior to the condensation reaction with the amino component. Very often, particularly in the case of N-hydroxysuccinimide esters, the latter compounds are formed *in situ* and then react directly in the same reaction vessel with the amino component (prior activation procedure). Alternatively, the formation of the active esters takes place in the presence of the amino component with which they eventually interact.[117]

[111] M. Bodanszky, *Nature (London)* **175**, 685 (1955).
[112] R. Schwyzer, B. Iselin, and M. Feurer, *Helv. Chim. Acta* **38**, 69 (1955).
[113] Th. Wieland, W. Schafer, and E. Bokelmann, *Justus Liebigs Ann. Chem.* **573**, 99 (1951).
[114] M. Bodanszky and V. du Vigneaud, *J. Am. Chem. Soc.* **81**, 5688 (1959).
[115] J. Pless and R. A. Boissonnas, *Helv. Chim. Acta* **46**, 1609 (1963).
[116] G. W. Anderson, J. E. Zimmermann, and F. M. Callahan, *J. Am. Chem. Soc.* **86**, 1839 (1964).
[117] F. Weygand, D. Hoffman, and E. Wünsch, *Z. Naturforsch. Teil B* **21**, 426 (1966).

In the latter case, a solution of the N^α-protected amino acid, the amino component and the N-hydroxysuccinimide is treated with the activating agent, the N,N'-dicyclohexylcarbodiimide. The last two procedures minimize mechanical losses incurred in the isolation of the activated esters. In addition, the latter procedure, in particular, minimizes potential side reactions which may occur with the N-hydroxysuccinimide esters of certain N^α-protected amino acids. For example, activated N^α,N^ω-disubstituted arginine may form an internal lactam. Since active ester formation involves activation by carbodiimide, all the restrictions regarding the carbodiimide method, which were discussed previously, are relevant in the active ester method as well. An additional technical problem inherent in this method is the removal of unreacted active ester upon completion of the reaction. It often requires extensive washings of the reaction mixture with dilute ammonia or repeated extractions with organic solvents. Whenever such purification procedures fail to yield a pure product, it is advisable to use, at the start of the reaction, a slight excess of the amino component.

Example 1: Synthesis of N^α-Protected Amino Acid p-Nitrophenyl or 2,4,5-Trichlorophenyl Esters. To a solution of the chosen N^α-protected amino acid (0.06 mol) in ethyl acetate (300 ml) the desired "phenol" (0.061 mol) is added, followed, after cooling the mixture to 4°, by N,N'-dicyclohexylcarbodiimide (12 g, ca. 0.06 mol). After 20 hr stirring (4°) the precipitated dicyclohexylurea is filtered off and washed with ethyl acetate (50 ml). Evaporation of the combined filtrates under reduced pressures affords the crude active ester which usually is crystallized upon mixing with 95% ethanol (150 ml) or preferably isopropanol. Recrystallization is carried out from warm 95% ethanol or isopropanol. A list of the physical properties of the active esters of most protected amino acids may be found in Fletcher and Jones.[7]

The active esters of N^α-protected glutamine and asparagine are prepared by a slight modification of the above procedure.[114,118] The reaction is carried out in DMF for 4–5 hr at 4°. Subsequently the dicyclohexylurea is filtered off and the filtrate is poured into cold water. The precipitated product is collected by filtration, washed with water, dried, and crystallized from 95% ethanol. Since both these amino acid derivatives may be contaminated with the respective nitrile by-product, their physical properties must be carefully determined. Further recrystallization of these active esters may be necessary.

Example 2: Synthesis of N^α-Protected Amino Acid N-Hydroxysuccinimide Esters. To a solution of the chosen N^α-protected amino acid

---

[118] M. Bodanszky, G. S. Denning, and V. du Vigneaud, *Biochem. Prep.* **10**, 122 (1963).

(0.05 mol) in acetonitrile (150 ml), N-hydroxysuccinimide (0.05 mol; commercially available) is added, followed, after cooling the mixture to 4°, by N,N'-dicyclohexylcarbodiimide (10 g; ca. 0.05 mol). After 20 hr stirring at 4°, the dicyclohexylurea is filtered off and the filtrate is evaporated under reduced pressure. The residue is crystallized by trituration with isopropanol. For the synthesis of the respective derivatives of the N^α-protected glutamine and asparagine, the modification discussed previously in the synthesis of the corresponding "phenol" esters is applied.

PEPTIDE BOND FORMATION USING ISOLATED ACTIVE ESTERS. (i) *Via p-nitrophenyl esters: N-Benzyloxycarbonyl-O-benzyl-L-tyrosyl-S-benzyl-L-cysteinyl-L-asparagine p-nitrobenzyl ester.*[119] N-Benzyloxycarbonyl-S-benzyl-L-cysteinyl-L-asparagine p-nitrobenzyl ester (1.8 g) is suspended in acetic acid (5 ml) and 4 N hydrogen bromide in acetic acid (5 ml) is added. After standing for 1 hr at room temperature, the resulting solution is poured into cold ether (500 ml), and the precipitated hydrobromide salt is collected by filtration, washed with ether, and dried over potassium hydroxide. The dipeptide salt is dissolved in cold DMF (10 ml) and to this solution triethylamine (0.6 ml) is added followed by N-benzyloxycarbonyl-O-benzyl-L-tyrosine p-nitrophenyl ester (1.5 g). After stirring for 24 hr at room temperature, the mixture is poured into cold 1 N ammonium hydroxide. The precipitated product is filtered off and washed with 0.5 N ammonium hydroxide until the filtrate is colorless. The colorless product is then washed with 1 N hydrochloric acid and water and recrystallized by addition of water to an acetic acid solution of the tripeptide; weight 1.5 g (62%); mp 209°–211°.

(ii) *Via 2,4,5-Trichlorophenyl esters: N-tert-Butyloxycarbonyl-N^{im}-tosyl-L-histidyl-L-leucyl-L-valyl-γ-benzyl-L-glutamyl-L-alanine tert-butyl ester.*[79] A solution of o-nitrophenylsulfenyl-L-leucyl-L-valyl-γ-benzyl-L-glutamyl-L-alanine *tert*-butyl ester (5.4 g) in acetone is mixed with 6 N hydrochloric acid (2.8 ml). After standing for 5 min the solvent is removed under reduced pressure and the residue is triturated with cold ether. The precipitated tetrapeptide hydrochloride is collected by filtration and purified by suspending the material in isopropanol (30 ml) and diluting the mixture with ether (300 ml). The partially deprotected peptide, after drying, is dissolved in DMF (50 ml) containing triethylamine (1 ml) and N-*tert*-butyloxycarbonyl-N^{im}-tosyl-L-histidine 2,4,5-trichlorophenyl ester (4 g) is added. After 20 hr the mixture is poured into ethyl acetate (500 ml) and this solution is washed once with 0.2 N ammonium

[119] P. G. Katsoyannis, A. M. Tometsko, and C. Zalut, *J. Am. Chem. Soc.* **88**, 5618 (1966).

hydroxide, water until neutral, 0.2 N sulfuric acid, and water until neutral, dried over magnesium sulfate, and concentrated to dryness under vacuum. The residue is dissolved in ether (40 ml), and the product is precipitated by the addition of petroleum ether; weight 6.2 g (87%); mp 180–192° (indefinite).

(iii) *Via N-hydroxysuccinimide esters: N-tert-Butyloxycarbonyl-L-valyl-S-diphenylmethyl-L-cysteinylglycyl methyl ester.*[59] *N-tert*-Butyl-oxycarbonyl-S-diphenylmethyl-L-cysteinylglycine methyl ester (22 g) is dissolved in 98% formic acid (200 ml). After standing for 3 hr at room temperature, the solvent is removed under reduced pressure and the residue is dissolved in ethyl acetate (400 ml) and cold water (200 ml) is added. The pH of the aqueous layer is adjusted to pH 9.0 by the addition of cold 2 N ammonium hydroxide. The organic layer is separated and washed with water until neutral, dried over magnesium sulfate, and concentrated to dryness under vacuum. The residue (free base of the dipeptide ester) is dissolved in DMF (100 ml) and N-*tert*-butyloxycarbonyl-L-valine N-hydroxysuccinimide ester (16 g) is added. After 48 hr the mixture is poured into cold water (1 liter) containing 1 N ammonium hydroxide (20 ml). The precipitate is filtered off and washed with 1 N ammonium hydroxide (150 ml), water, 2% acetic acid, and water. The wet residue is crystallized from hot 95% ethanol; weight 23 g (83%); mp 159–161°.

PEPTIDE BOND FORMATION WITHOUT ISOLATION OF THE ACTIVE ESTER (*in situ* ACTIVATION PROCEDURE). (i) *Via prior activation of the carboxyl component: N-tert-Butyloxycarbonyl-O-benzyl-L-seryl-N^{im}-tosyl-L-histidyl-L-leucyl-L-valyl-γ-benzyl-L-glutamyl-L-alanine.*[79] *N-tert*-Butyloxycarbonyl-N^{im}-tosyl-L-histidyl-L-leucyl-L-valyl-γ-benzyl-L-glutamyl-L-alanine *tert*-butyl ester (3 g) is mixed with trifluoroacetic acid (20 ml) and after standing for 1 hr at room temperature the solution is poured into cold ether (300 ml). The precipitated amino and carboxyl-deprotected pentapeptide, as the trifluoroacetic salt, is collected by filtration and washed with cold ether. This material is suspended in isopropanol (30 ml), and ether (300 ml) is added. After cooling for 1 hr, the purified peptide is filtered off and dried over potassium hydroxide. The dried peptide salt is dissolved in DMF (20 ml) containing triethylamine (0.5 ml), and this solution is added to *N-tert*-butyloxycarbonyl-O-benzyl-L-serine which is activated as follows. To a solution of *N-tert*-butyloxy-carbonyl-O-benzyl-L-serine (950 mg) and N-hydroxysuccinimide (348 mg) in cold acetonitrile (20 ml) is added dicyclohexylcarbodiimide (600 mg). After stirring for 3 hr at 4°, the dicyclohexylurea by-product is filtered off and the filtrate is concentrated to dryness. To the residue is added the solution of the deprotected pentapeptide and after stirring for

24 hr at room temperature, the mixture is poured into cold ether (300 ml). After cooling for a few hours, the partially protected hexapeptide is isolated by filtration and reprecipitated once from a mixture of acetone–ether and once from a mixture of DMF–water; weight 2.1 g (61%); mp 130–150° (indefinite).

(ii) *Via activation of the carboxyl component in the presence of the amino component: N-tert-Butyloxycarbonyl-L-threonyl-L-prolyl-Nε-benzyloxycarbonyl-L-lysyl-L-alanine benzyl ester.*[120] *N-tert-*Butyloxycarbonyl-L-prolyl-N^ε-benzyloxycarbonyl-L-lysyl-L-alanine benzyl ester (25 g) is dissolved in 98% formic acid (200 ml). After standing for 3 hr, the solvent is removed under reduced pressure (25°) and the free base is isolated by partitioning between ethyl acetate and aqueous ammonia, as described in reference 88. The organic layer containing the free base is concentrated to a small volume and ether (200 ml) is added. After cooling for 1 hr, the solid tripeptide, as the free amine, is collected, dried, and dissolved in acetonitrile (500 ml). To this solution *N-tert-*butyloxycarbonyl-L-threonine (8.8 g) and *N*-hydroxysuccinimide (4.6 g) are added, followed, after cooling to 4°, by dicyclohexylcarbodiimide (8.3 g). The mixture is stirred for 1 hr in the cold and overnight at room temperature. The reaction mixture is then cooled for a few hours, the precipitated dicyclohexylurea by-product is removed by filtration, and the filtrate is evaporated to dryness. The residue is dissolved in ethyl acetate and this solution is washed successively with 0.2 *N* ammonium hydroxide, water, 0.1 *N* hydrochloric acid and water, and dried over magnesium sulfate. Upon evaporation of the solvent to dryness under reduced pressure the product is obtained as a heavy oil; weight 30 g.

IX. Synthesis of Peptides by Fragment Condensation

As stated above in the section on problems in peptide synthesis, the synthesis of large polypeptides is best accomplished by the fragment condensation approach. Peptide fragments are prepared by the stepwise approach. The fragments that will serve as carboxyl components usually have their C^α-carboxyl functions protected as methyl, ethyl, or benzyl esters. Upon exposure to hydrazine or sodium hydroxide these substances are converted to the corresponding hydrazides and free acid derivatives, respectively. Occasionally carboxyl components are synthesized having the C^α-carboxyl group unprotected or in the form of a *tert*-butyl ester. The latter derivative is converted to the free acid by

[120] Prepared in our laboratory and not yet published.

mild acidolysis. The fragment employed as carboxyl component should ideally contain three to six amino acid residues and, if possible, should have a nonracemizable amino acid, i.e., proline or glycine, at the COOH-terminal end. If this is not possible, the COOH-terminal residue should be one of the aliphatic amino acids (i.e., alanine or leucine) since these are the least sensitive to racemization during the activation and coupling steps.

Fragment condensation of peptide subunits with all their secondary groups protected eliminates to a great extent side reactions. Fully protected polypeptide products, however, are highly apolar substances which defy purification by the usual techniques. Indeed, such derivatives are generally soluble only in DMF, hexamethylphosphoramide, or dimethyl sulfoxide, and are practically insoluble in all the other commonly employed solvents, i.e., ethyl acetate, acetone, alcohol, etc. The carboxyl component is used in excess to drive the reaction to completion (as monitored by the ninhydrin assay).[99] The final reaction mixture will then consist of the newly formed polypeptide fragment and the unreacted carboxyl component. In many cases the carboxyl component in the size range specified above is soluble in the common organic solvents and thus can be separated from the larger, less-soluble polypeptide product.

Limited side-group protection of fragments generally affords a more polar product. In most cases, however, these partially protected polypeptides also have very limited solubilities, restricting the application of the usual purification techniques. It is, therefore, necessary, even with limited secondary group protection, to use excess carboxyl component to ensure reaction of all the amino component. In these cases the activation of the carboxyl component is usually performed via the azide method, since azides react exclusively with amino functions.

All the activation methods discussed in the stepwise procedure have been used for the activation of the carboxyl component in the fragment condensation approach. At the present time, however, the "azide method" and the "active ester method," via N-hydroxysuccinimide or 1-hydroxybenzotriazole, are the most commonly employed for fragment activation. Experience has shown that fragment condensation by these procedures under the experimental conditions described below, proceeds usually with no detectable racemization. The synthesis of the vast majority of biologically important peptides which have been accomplished thus far by the stepwise or fragment condensation approaches utilizes triethylamine for the generation of the free base of the amino component. In most of these syntheses no detectable racemization has been observed. In certain cases, however, racemization has been observed, even in syntheses via the azide method,[121] and this may be attributed to the use of triethylamine. Substitution of this base by N-methyl- or N-ethylmorpholine or N,N-diisopropylethylamine has led to satisfactory results.[122]

A. Azide Method[123]

Principle. Interaction of a protected amino acid or peptide ester with hydrazine hydrate (usually a commercially available 99% solution) results in the formation of the respective hydrazide. Upon exposure to nitrous acid the hydrazide is converted to the corresponding azide. The nitrous acid used for this reaction is generated by the action of acid, usually hydrochloric acid, on sodium nitrite or acid-labile organic esters of nitrous acid. The most commonly used esters are the commercially available *tert*-butyl nitrite or isoamyl nitrite.

$$
\underset{\text{Y·HN}\overset{|}{\text{C}}\text{HCONH}\overset{|}{\text{C}}\text{HCOOZ}}{\overset{R_1\quad\;\; R_2}{}} \xrightarrow{\;H_2N\cdot NH_2\;} \underset{\text{Y·HN}\overset{|}{\text{C}}\text{HCONH}\overset{|}{\text{C}}\text{HCONHNH}_2}{\overset{R_1\quad\;\; R_2}{}} \xrightarrow{\;HNO_2\;}
$$

$$
\underset{\text{Y·HN}\overset{|}{\text{C}}\text{HCONH}\overset{|}{\text{C}}\text{HCON}_3}{\overset{R_1\quad\;\; R_2}{}}
$$

The amino acid or peptide azides are highly reactive compounds and produce peptides in good yields upon interaction with amino components.

$$
\underset{\text{Y·HN}\overset{|}{\text{C}}\text{HCONH}\overset{|}{\text{C}}\text{HCON}_3}{\overset{R_1\quad\;\; R_2}{}} + \underset{\text{H}_2\text{N}\overset{|}{\text{C}}\text{HCOOZ}}{\overset{R_3}{}} \rightarrow
$$

$$
\underset{\text{Y·HN}\overset{|}{\text{C}}\text{HCONH}\overset{|}{\text{C}}\text{HCONH}\overset{|}{\text{C}}\text{HCOOZ}}{\overset{R_1\quad\;\; R_2\quad\;\; R_3}{}} + HN_3
$$

As with almost all the peptide bond-forming reactions, the azide method has its own limitations. Two of the most prominent side reactions of this method need to be mentioned. The most troublesome is the Curtius rearrangement[124] of the azide resulting in the formation of the respective α-isocyanate derivative. The latter compound may then react with the amino component to form a "urea linkage," not a peptide bond.

$$
\underset{\text{Y·HN}\overset{|}{\text{C}}\text{HCONH}\overset{|}{\text{C}}\text{HCON}_3}{\overset{R_1\quad\;\; R_2}{}} \xrightarrow{\;-N_2\;} \underset{\text{Y·HN}\overset{|}{\text{C}}\text{HCONH}\overset{|}{\text{C}}\text{H·N}=\text{C}=\text{O}}{\overset{R_1\quad\;\; R_2}{}} \xrightarrow{\;H_2N\overset{R_3}{\overset{|}{\text{C}}}\text{HCOOZ}\;}
$$

$$
\underset{\text{Y·HN}\overset{|}{\text{C}}\text{HCONH}\overset{|}{\text{C}}\text{HNH·}\overset{||}{\text{C}}\text{·NH}\overset{|}{\text{C}}\text{HCOOZ}}{\overset{R_1\quad\;\; R_2\quad\;\; O\quad\; R_3}{}}
$$

[121] P. Sieber, B. Riniker, M. Brugger, B. Kamber, and W. Rittel, *Helv. Chim. Acta* **53**, 2135 (1970).

[122] P. Sieber and B. Riniker *in* "Peptides 1971" (H. Nesvadba, ed.), p. 49. North-Holland Publ., Amsterdam, 1973.

[123] T. Curtius, *Ber. Dtsch. Chem. Ges.* **35**, 3226 (1902).

[124] P. A. S. Smith, *Org. Reactions* **3**, 337 (1946).

The second side reaction takes place during the formation of the azide and results in the conversion of the peptide hydrazide to the corresponding peptide amide.[125]

The risk of Curtius rearrangement is always present during the azide coupling, but can be minimized by working at low temperatures. The second side reaction, amide formation, is extensively suppressed by forming the azide with esters of nitrous acid in pure organic solvents (less than 5% water) at low temperatures (below −10°). This modification of the azide method is known as Rudinger's Method.[126]

Comments. The azide method is the favored procedure for the fragment condensation approach and usually proceeds, as compared to all the other methods, with the least racemization.[127,128] It is the method of choice for coupling of peptide fragments with free functional side groups. The main drawback of this method is the formation of by-products (urea derivatives) which very often are extremely difficult to remove. The Rudinger modification, which involves *in situ* formation of the azide, without need for its isolation, is clearly the method of choice. The original method[123] of azide formation, by interaction of peptide hydrazides in aqueous acid media with sodium nitrite, is still in use when the peptide azide can be isolated as a solid. In these cases the isocyanate content of the solid azide can be ascertained by infrared spectroscopy.[52,129] Azides exhibit a characteristic absorption band at 4.75 μm, their principal decomposition products, the isocyanates, show a sharp band at 4.50 μm.

1. Formation of Hyrazides

Peptides Soluble in Alcohol. A solution of a protected peptide, methyl, ethyl, or benzyl ester (0.03 mol) in absolute alcohol (100 ml) is mixed with hydrazine hydrate (4 ml) and refluxed for 1 hr. The mixture is left to stand overnight and the crystalline peptide hydrazide is collected by filtration and washed with ethanol and ether. In some cases ether is added to effect precipitation of the hydrazide. If the peptide is sensitive to heating in the basic medium, a larger excess (8–10 ml) of hydrazine hydrate is used, and the reaction is carried out at room temperature. The yield ranges from 70 to 90%.

[125] V. Prelog and P. Wieland, *Helv. Chim. Acta* **29,** 1128 (1946).
[126] J. Honzl and J. Rudinger, *Coll. Czech. Chem. Commun.* **26,** 2333 (1961).
[127] M. B. North and G. T. Young, *Chem. Ind. (London)*, p. 1597 (1955).
[128] N. A. Smart, G. T. Young, and N. W. Williams, *J. Chem. Soc. London* **1960,** 3902 (1960).
[129] P. G. Katsoyannis, A. M. Tometsko, C. Zalut, and K. Fukuda, *J. Am. Chem. Soc.* **88,** 5625 (1966).

Peptides Insoluble in Alcohol. A solution of a protected peptide methyl, ethyl, or benzyl ester (0.005 mol) in DMF (150 ml) is mixed with hydrazine hydrate (8 ml). After stirring for 24–48 hr at room temperature, the mixture is poured into ice water (400 ml), and the pH of the suspension is adjusted to 6.0 with acetic acid. The precipitate is collected by filtration and washed with water. After drying, the hydrazide is reprecipitated from dimethyl sulfoxide–water; the yield ranges from 70 to 80%.

2. Formation of Azide and Coupling with the Amino Component

Azides Soluble in Water-Immiscible Solvents: N-Benzyloxycarbonylglycyl-L-seryl-L-leucyl-L-isoleucine Methyl Ester.[120] N-Benzyloxycarbonylglycyl-L-serine hydrazide (6.2 g; 0.02 mol) is suspended in water (100 ml) containing concentrated hydrochloric acid (2 ml), and the peptide is dissolved by the addition of acetic acid (30 ml). To this solution, cooled to —5° in a salt–ice bath, is added, with vigorous stirring, a solution of sodium nitrite (140 mg) in cold water (5 ml). After 10 min, the mixture is extracted twice with 25-ml portions of ethyl acetate. The organic layer is washed twice with cold water, once with 1 *M* sodium bicarbonate, and finally with water. The azide solution is dried for a few minutes (0°) over magnesium sulfate and filtered into a flask containing a solution of L-leucyl-L-isoleucine methyl ester hydrochloride (0.02 mol) in DMF (50 ml) containing triethylamine (2.8 ml). After stirring in the cold for 48 hr, the mixture is diluted with ethyl acetate (500 ml) and the solution is washed successively with 1 *N* hydrochloric acid, water, saturated sodium bicarbonate, water and dried over magnesium sulfate. Upon concentration to 50–75 ml under reduced pressure and addition of petroleum ether, the product precipitates; weight 7.5 g; mp 131°–133°.

Azides Insoluble in Water-Immiscible Solvents: N-Benzyloxycarbonyl-L-glutaminyl-S-benzyl-L-cysteinyl-S-benzyl-L-cysteinyl-L-threonyl-L-seryl-L-isoleucyl-S-benzyl-L-cysteinyl-L-seryl-L-leucyl-L-tyrosyl-L-glutaminyl-L-leucyl-L-glutamyl-L-asparaginyl-L-tyrosyl-S-benzyl-L-cysteinyl-L-asparagine p-Nitrobenzyl Ester.[130] N-Benzyloxycarbonyl-L-isoleucyl-S-benzyl-L-cysteinyl-L-seryl-L-leucyl-L-tyrosyl-L-glutaminyl-L-leucyl-L-glutamyl-L-asparaginyl-L-tyrosyl-S-benzyl-L-cysteinyl-L-asparagine p-nitrobenzyl ester is deblocked by the hydrogen bromide in trifluoroacetic acid method (page 546). Specifically, the protected dodecapeptide (1 g) is dissolved in trifluoroacetic acid (25 ml) containing water

---

[130] P. G. Katsoyannis, A. M. Tometsko, and C. Zalut, *J. Am. Chem. Soc.* **89,** 4505 (1967).

(0.6 ml), and hydrogen bromide gas is passed through the solution for 1 hr at 4°. The N^α-deprotected dodecapeptide is precipitated by the addition of cold ether (300 ml) and isolated by centrifugation. The precipitate is washed in the centrifuge several times with ether and dried over potassium hydroxide under reduced pressure. This product is dissolved in DMF (25 ml), triethylamine (0.4 ml) added (to make it slightly basic), followed by the solid pentapeptide azide which is prepared as follows. To a solution of N-benzyloxycarbonyl-L-glutaminyl-S-benzyl-L-cysteinyl-S-benzyl-L-cysteinyl-L-threonyl-L-serine hydrazide (0.6 g) in DMF (35 ml) containing 2 N hydrochloric acid (3 ml) and cooled to −15° (Dry Ice-acetone bath), a solution of sodium nitrite (48 mg) in water (0.5 ml) is added. The mixture is stirred at −15° for 5 min and then diluted with a cold half-saturated sodium chloride solution (150 ml). The precipitated azide is isolated (4°) by filtration and washed on the filter successively with cold 1 N sodium bicarbonate and water, and dried for 1 hr over phosphorus pentoxide under reduced pressure at 4°. (An infrared spectrum of the dried azide reveals the extent of decomposition to the isocyanate.) The dried azide is added to the solution of the deblocked dodecapeptide and the mixture is stirred at 4° for 3 days and then poured into methanol (250 ml) containing 1 N hydrochloric acid (0.5 ml). The precipitated product is collected by filtration, washed with methanol, 50% aqueous methanol, and water, and reprecipitated from a solution in DMF by the addition of water; weight 0.96 g (70%); mp 274°–275°.

Rudinger's Modification: N^α-tert-Butyloxycarbonyl-L-seryl-L-leucyl-L-tyrosyl-L-glutaminyl-L-leucyl-γ-benzyl-L-glutamyl-L-asparaginyl-L-tyrosyl-S-benzyl-L-cysteinyl-L-asparagine p-Nitrobenzyl Ester.[131] N^α-tert-Butyloxycarbonyl-γ-benzyl-L-glutamyl-L-asparaginyl-L-tyrosyl-S-benzyl-L-cysteinyl-L-asparagine p-nitrobenzyl ester (3.14 g) is dissolved in trifluoroacetic acid (11 ml); after the mixture has stood for 1 hr at room temperature, it is diluted with cold ether (500 ml). The precipitated N^α-deblocked pentapeptide is isolated by filtration, washed with cold ether, and dried over potassium hydroxide under vacuum. This trifluoroacetate salt is dissolved in DMF (10 ml), the solution cooled to 4°, then triethylamine (0.64 ml) is added, followed by the solution of the pentapeptide azide prepared as follows. To a solution of N^α-tert-butyloxycarbonyl-L-seryl-L-leucyl-L-tyrosyl-L-glutaminyl-L-leucine hydrazide (2.2 g) in DMF (10 ml), cooled to −10°, a solution of 1 N hydrochloric acid in DMF (6.6 ml) is added, followed by isoamyl nitrite (0.4 ml). After 5 min at −10° the mixture is cooled to −30°, neutralized with triethyl-

[131] P. G. Katsoyannis, Y. Okada, and C. Zalut, *Biochemistry* 12, 2516 (1973).

amine (0.4 ml), and added to the solution of deprotected pentapeptide ester prepared as described above. After 2 days at 4°, the mixture is concentrated under reduced pressure (35°) to dryness and the residue is mixed with water. The solid residue is collected and reprecipitated from DMF–methanol; weight 7.4 g (74%); mp 222°–225°.

B. Active Ester Method

Principle. The protected peptide fragment which is to serve as the carboxyl component must have only the C^α-carboxyl group unprotected. If it is in the form of an ester (methyl, ethyl, or benzyl) it is first saponified. Activation of the carboxyl component is accomplished, in general, according to the *in situ* active ester procedures discussed previously (pp. 561). For this purpose N-hydroxysuccinimide[117] or, more commonly, 1-hydroxybenzotriazole[132] esters are formed. In practice the peptide fragment interacts with one equivalent of commercially available N-hydroxysuccinimide or 1-hydroxybenzotriazole

in the presence of N,N'-dicyclohexylcarbodiimide. As was stated previously, activation by conversion to the aforementioned esters, at low temperatures (4°) and even in the presence of polar solvents, proceeds usually with no detectable racemization. This is of critical importance since large polypeptide fragments are soluble only in polar solvents such as DMF. In order to minimize racemization, activation of the carboxyl component is performed in the presence of the amino component. However, prior activation of the carboxyl component may be carried out if its C^α terminus is glycine or proline.

1-Hydroxybenzotriazole esters are, in general, recommended for fragment activation. It appears that, in contrast to the N-hydroxysuccinimide esters, the degree of racemization of these esters is independent of solvent.[133] Furthermore, whereas N-hydroxysuccinimide interacts in a side reaction with N,N'-dicyclohexylcarbodiimide to give rise eventually to β-alanine derivatives of the amino component,[134] the 1-hydroxybenzotriazole does not react with this reagent. Formation of β-alanine derivatives of the amino component is observed during activation of carboxyl

[132] W. König and R. Geiger, *Chem. Ber.* **103**, 788 (1970).
[133] D. S. Kemp, H. Trangle, and K. Trangle, *Tetrahedron Lett.* p. 2695 (1974).
[134] F. Weygand, W. Steglich, and N. Chytil, *Z. Naturforsch. Teil B* **23**, 1391 (1968).

components which possess a sterically hindered or otherwise sluggishly reactive α-carboxyl group. A potential drawback of the 1-hydroxybenzotriazole esters was discussed previously (see p. 539).

1. Saponification of Peptide Esters

To a stirred solution of the chosen protected peptide ester (methyl, ethyl, or benzyl; 0.01 mol) in a mixture of dioxane–acetone or alcohol–acetone (1:1; 100 ml) cooled to 4°, 1 N sodium hydroxide (0.011 mol) is added over a period of 30 min. The reaction mixture is stirred at room temperature for an additional 45 min, diluted with cold water (150 ml) and acidified with 1 N hydrochloric acid (0.012 mol). The precipitated partially protected peptide is filtered, washed with water, and reprecipitated from appropriate solvents. The yield ranges from 75 to 85%. In cases where the protected peptide is not soluble in the aforementioned solvents, DMF is the recommended solvent. If the saponified peptide does not precipitate after acidification, the reaction mixture is concentrated under reduced pressure to remove the organic solvents and the residue is extracted with ethyl acetate. The organic layer is separated, washed with water, and dried over magnesium sulfate. Upon concentration of the solution to a small volume and addition of petroleum ether, the product precipitates.

2. In Situ Activation of the Carboxyl Component and Coupling with the Amino Component

Prior activation of the carboxyl component: N^α-*tert-Butyloxycarbonyl* - L - *leucyl-O-benzyl*-L-*tyrosyl*-L-*leucyl*-L-*valyl-S-diphenylmethyl*-L-*cysteinylglycyl* -γ- *benzyl*-L-*glutamyl-N^ω-nitro*-L-*arginylglycyl*-L-*phenylalanyl* - L - *phenylalanyl* - O - *benzyl*-L-*tyrosyl*-L-*threonyl*-L-*prolyl-N^ε-benzyloxycarbonyl*-L-*lysyl-O-benzyl*-L-*threonine Benzyl Ester.*[59] *N-tert*-Butyloxycarbonyl - γ - benzyl - L - glutamyl - N^ω - nitro - L - arginylglycyl - L - phenylalanyl - L - phenylalanyl - O - benzyl - L - tyrosyl -L-threonyl-L-prolyl-N^ε-benzyloxycarbonyl-L-lysyl-O-benzyl-L-threonine benzyl ester (6.0 g) is dissolved in 98% formic acid (150 ml) and, after standing for 3 hr at room temperature, the solvent is removed under reduced pressure (25°) and the residue is solidified by trituration with cold ether. This solid is dissolved in cold DMF (60 ml), and 1 N ammonium hydroxide (15 ml) is added. The mixture is immediately poured into cold water saturated with sodium chloride, and the pH of the mixture is adjusted to 9.5 with 1 N ammonium hydroxide. The precipitated decapeptide ester, as the free base, is isolated by filtration, washed with cold water until the filtrate is neutral, and dried over phosphorus pentoxide under reduced pressure. This solid (amino derivative) is then added to a solution in DMF of the protected hexapeptide *N-tert*-butyloxycarbonyl-L-leucyl-*O*-benzyl-L-tyrosyl-L-leucyl-L-valyl-*S*-diphenylmethyl-L-cysteinylglycine (carboxyl component) which is activated as follows. To a cold solution of the hexapeptide (5.8 g) in DMF (35 ml), 1-hydroxybenzotriazole (1 g) is added, followed by *N,N'*-dicyclohexylcarbodiimide (1.2 g). After stirring for 1 hr at 4° and 1 hr at room temperature, the free base of the decapeptide ester is added. The mixture is stirred for 24 hr at 4° and then poured into a saturated solution of sodium chloride (500 ml) containing 1 *N* ammonium hydroxide (20 ml). The precipitated hexadeca-peptide is isolated by filtration and washed with water and ether. The dried product is reprecipitated from DMF–isopropanol and from DMF–water to give the purified hexadecapeptide; weight 7.1 g (80%); mp >260°.

Activation of the Carboxyl Component in the Presence of the Amino Component: N-*tert-Butyloxycarbonyl-O-benzyl-*L*-seryl-*N^{im}-*tosyl-*L*-histidyl-*L*-leucyl-*L*-valyl-*γ*-benzyl-*L*-glutamyl-*L*-alanyl-*L*-leucyl-O-benzyl-*L*-tyrosyl-*L*-leucyl-*L*-valyl-S-diphenylmethyl-*L*-cysteinylglycyl-*γ*-benzyl-*L*-glutamyl-*N^{ω}-nitro-*L*-arginylglycyl-*L*-phenylalanyl-*L*-phenylalanyl-O-benzyl-*L*-tyrosyl-*L*-threonyl-*L*-prolyl-*N^{ε}-benzyloxycarbonyl-*L*-lysyl-O-benzyl-*L*-threonine Benzyl Ester.*[79] The protected hexadeca-peptide *N-tert*-butyloxycarbonyl-L-leucyl-*O*-benzyl-L-tyrosyl-L-leucyl-L-valyl-*S*-diphenylmethyl-L-cysteinylglycyl-*γ*-benzyl-L-glutamyl-*N^{ω}-nitro-L-arginylygycyl-L-phenylalanyl-L-phenylalanyl-*O*-benzyl-L-tyrosyl-L-threonyl-L-prolyl-*N^{ε}-benzyloxycarbonyl-L-lysyl-*O*-benzyl-L-threonine benzyl ester (800 mg) is deblocked at the amino terminus with 98% formic acid (30 ml) following the same procedure described in the previous preparation. The resulting N^{α}-deblocked derivative (amino component) is dissolved in hexamethylphosphoramide (5 ml) and DMF (5 ml) and to this solution, cooled to 4°, the hexapeptide derivative (carboxyl component) *N-tert*-butyloxycarbonyl-*O*-benzyl-L-seryl-*N^{im}-tosyl-L-histidyl-L-leucyl-L-valyl-*γ*-benzyl-L-glutamyl-L-alanine (1 g) is added. The mixture is stirred for a few minutes and to the resulting solution *N*-hydroxysuccinimide (116 mg) and *N,N'*-dicyclohexylcarbodiimide (200 mg) are added. After 24 hr stirring at 4°, the reaction mixture is diluted with hexamethylphosphoramide (20 ml), stirred for an additional 5 hr and poured into cold saturated aqueous sodium chloride (400 ml) containing 2 *M* sodium carbonate (20 ml). The precipitated protected docosapeptide is collected by filtration, washed successively with water, acetone, and ether, and reprecipitated from DMF–isopropanol; weight 1 g (93%); mp >260°.

X. Final Deprotection of Synthetic Peptides

The final product in peptide synthesis is usually a polypeptide with the N^α- and C^α-terminal functional groups and several or all of its secondary moieties protected with a variety of blocking agents. All these protectors have been selected so that they can resist a great number of deblocking steps required in the synthesis. Most of these deblocking steps involve removal of N^α-protectors for the preparation of the amino components in the stepwise or fragment condensation.

A great number of polypeptides of biological interest contain sulfur-amino acid residues. As a result, in several stages of the multistep synthetic scheme, it becomes necessary to utilize as N^α-protectors moieties that can be removed readily by acidolytic reactions. This necessitates the use, as protectors for the secondary functions, of only those groups that resist acidolytic cleavage. For example, the employment of the benzyloxycarbonyl moiety as the N^α-protector during the various synthetic steps necessitates the use of the acid-stable tosyl and benzyl groups for permanent protection of the secondary functions of lysine (or arginine) and histidine (or cysteine) respectively, and the p-nitrobenzyl moiety for the permanent protection of the carboxyl functions. Final deprotection of a polypeptide bearing the above-mentioned permanent protectors can be accomplished only with sodium in liquid ammonia.[29]

On the other hand, the use of the *tert*-butyloxycarbonyl or the o-nitrophenylsulfenyl groups as N^α-protectors allows the use of a different variety of permanent protectors for the secondary functions (N^ω-tosyl for arginine, N^ε-o-bromobenzyloxycarbonyl for lysine, S-diphenylmethyl for cysteine and O-benzyl for serine, threonine, or tyrosine). Final deprotection of the synthetic product in this case will require exposure to liquid hydrogen fluoride.[31] It is, therefore, apparent that the final deprotection step of polypeptides with sulfur-containing residues requires treatment with sodium in liquid ammonia or with hydrogen fluoride. There are examples, however, where extensive use of acid-labile protectors for the various functions allows the final deprotection to be carried out under mild acidic conditions.[26a,43]

The situation differs, however, when the final product is a peptide which does not contain sulfur-amino acids and contains only a small number of secondary functional groups. A judicious combination of a limited number of protectors (i.e., benzyloxycarbonyl or o-nitrophenylsulfenyl for the α-amino group, N^ε-*tert*-butyloxycarbonyl for lysine, *tert*-butyl for the carboxyl functions, benzyl for hydroxyl groups, etc.) will permit the use of mild acidolysis and/or catalytic hydrogenation for final deprotection.

A. Deprotection by Sodium in Liquid Ammonia

The classical method, sodium in liquid ammonia, for removing a number of protecting groups (benzyloxycarbonyl, benzyl, tosyl, and their derivatives) was introduced by Sifferd and du Vigneaud almost 40 years ago.[29] In spite of several serious drawbacks, it remains an indispensable tool in the synthesis of large polypeptides. In practice, a protected peptide is dissolved in anhydrous liquid ammonia, and to this solution metallic sodium is added until a light blue color, indicative of excess of sodium, persists throughout the solution. The blue color is discharged by adding an ammonium salt (chloride, iodide, acetate) or acetic acid, and the ammonia is allowed to evaporate spontaneously. Isolation of the deblocked peptide from the large amounts of salts present is accomplished by appropriate desalting techniques (i.e., dialysis, molecular-sieve or ion-exchange chromatography, etc.). Sodium in liquid ammonia treatment of a polypeptide is often accompanied by a variety of side reactions. A major side reaction of this procedure is the reductive cleavage of acyl proline bonds present in the peptide chain.[135-137] Threonine–proline and serine–proline bonds are especially labile. In the synthesis of the B chain of insulin, for example, sodium in liquid ammonia treatment resulted in the near-quantitative cleavage of the threonyl-proline bond (B^{27}–B^{28}) of that chain.[138] This side reaction can be effectively inhibited by carrying out the reduction in the presence of large excess of sodamide.[138] It is thought that the sodamide, as a strong base, reacts with proton donors present or formed during the sodium in liquid ammonia reaction, and thus prevents the reductive cleavage of the threonyl–proline bond.

A second drawback of the sodium in liquid ammonia treatment is the slow or often incomplete removal of tosyl groups.[139] This problem can be overcome by the addition of "acids" in the ammonia system.[140]

[135] K. Hofmann and H. Yajima, *J. Am. Chem. Soc.* **83**, 2289 (1961).
[136] S. Guttmann *in* "Peptides: Proceedings of the Fifth European Peptide Symposium" (G. T. Young, ed.), p. 41. Pergamon, Oxford, 1963.
[137] W. F. Benisek, M. A. Raftery, and R. D. Cole, *Biochemistry* **6**, 3780 (1967).
[138] P. G. Katsoyannis, C. Zalut, A. Tometsko, M. Tilak, S. Johnson and A. C. Trakatellis, *J. Am. Chem. Soc.* **93**, 5871 (1971).
[139] J. Meienhofer and C. H. Li, *J. Am. Chem. Soc.* **84**, 2434 (1962).
[140] J. Rudinger, *in* "The Chemistry of Polypeptides; Essays in Honor of L. Zervas" (P. G. Katsoyannis, ed.), p. 87. Plenum, New York, 1973.

Ammonium chloride, ethyl alcohol, acetamide, or urea have been used for this purpose.

A number of other side reactions have also been observed during the sodium in liquid ammonia treatment. Among these, conversion of arginine to ornithine,[136] cysteine to alanine,[141] cleavage of serine, threonine,[142] and tryptophan,[143] and partial racemization of the peptide product,[144] may be mentioned. The incidence of some of these side reactions can be reduced by controlling the course of addition of sodium to the liquid ammonia solution of the peptide. The first attempt in this direction involved the use of a sodium-containing narrow glass tube.[145] By dipping the tube below the surface of the solution, the metal is gradually dissolved out of the end of the glass tube until the blue color appears throughout. More recently,[146] a better control of the reaction conditions and avoidance of high local concentration of sodium were achieved by the use of a special apparatus shown in Fig. 2. This apparatus permits the dropwise addition of a solution of sodium in liquid ammonia into the reaction mixture and functions as a titrimeter as well.

Example: Human Insulin B Chain S-Sulfonate.[146] The fragment condensation approach was applied for the synthesis of the protected triacontapeptide embodying the amino acid sequence of the B chain of human insulin. The following protecting groups are present: N^{α}-benzyloxycarbonyl, S-benzyl, N^{im}-benzyl (histidine), N^{ω}-tosyl (arginine), and N^{ε}-tosyl (lysine). The thoroughly dried, crude, protected triacontapeptide (300 mg) is dissolved in anhydrous liquid ammonia (300 ml) in a 500-ml round-bottom flask fitted for magnetic stirring, and to this solution solid sodamide (ca. 120 mg) is added. The reaction vessel is removed from the cooling bath (Dry Ice–acetone) and the subsequent reaction is carried out at the boiling point of the solution. The special apparatus shown in Fig. 1 is attached and reduction is carried out by the dropwise addition of a liquid ammonia solution of sodium into the reaction system. The faint blue color, which is obtained eventually and persists for 30 sec, is then discharged by the addition of 2–3 drops of glacial acetic acid. The solution is allowed to evaporate at atmospheric pressure to about

---

[141] P. G. Katsoyannis, K. Fukuda, A. Tometsko, K. Suzuki, and M. Tilak, *J. Am. Chem. Soc.* **86**, 930 (1964).

[142] M. Brenner, *Coll. Czech. Chem. Commun.* **24**, 141 (Special Issue 1959).

[143] S. Bajusz and K. Medzihradszky, *in* "Peptides: Proceedings of the Fifth European Peptide Symposium" (G. T. Young, ed.), p. 49. Pergamon, Oxford, 1963.

[144] D. B. Hope and J. F. Humphries, *J. Chem. Soc. London* p. 869 (1964).

[145] V. du Vigneaud, C. Ressler, J. M. Swan, C. W. Roberts, and P. G. Katsoyannis, *J. Am. Chem. Soc.* **76**, 3115 (1954).

[146] P. G. Katsoyannis, J. Ginos, C. Zalut, M. Tilak, S. Johnson, and A. C. Trakatellis, *J. Am. Chem. Soc.* **93**, 5877 (1971).

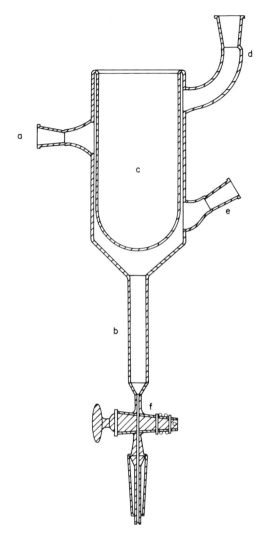

FIG. 2. Glass apparatus for sodium in liquid ammonia reaction. Liquid ammonia is introduced through a and is collected in the graduated tube b after condensation with the aid of Dry Ice–acetone contained in c. The system is vented through a sodium hydroxide-containing drying tube at d. When the desired amount of ammonia is collected, a piece of sodium metal is introduced through e, which is kept stoppered. The addition of the liquid ammonia solution of the sodium from b, into a suitable glass flask containing the liquid ammonia solution of the compound to be reduced, is controlled by manipulation of stopcock f. From P. G. Katsoyannis, J. Ginos, C. Zalut, M. Tilak, S. Johnson, and A. C. Trakatellis, *J. Am. Chem. Soc.* **93**, 5877 (1971). Reprinted with permission. Copyright by the American Chemical Society.

10 ml and the reaction flask is immediately evacuated (using a water pump) and the contents swirled until frozen. The remaining mixture is subsequently lyophilized, using the water pump. The residue is dissolved in 8 M guanidine hydrochloride (40 ml) and to this solution, adjusted to pH 8.9 (with acetic acid or dilute ammonium hydroxide, depending on the pH of the solution), sodium sulfite (3 g) and freshly prepared sodium tetrathionate (1.5 g) are added. After stirring for 3 hr (during which the sulfhydryl groups of the peptide chain are converted to the S-sulfonated form: $-SH \rightarrow -S \cdot SO_3^-$) the mixture is placed in an 18/32 Visking dialysis tubing and dialyzed against four changes of distilled water (4 liters each) at 4° for 24 hr. Upon lyophilization of the dialyzate, the crude human B chain S-sulfonate is obtained as a white fluffy material; weight 130 mg. Highly purified B chain S-sulfonate is obtained by chromatography of the crude product on a CM-cellulose column (4 × 60 cm) equilibrated and eluted with an 8 M urea–acetate buffer, pH 4.0.

B. Deprotection by Liquid Hydrogen Fluoride

The recently introduced liquid hydrogen fluoride procedure[31] for removing a variety of blocking groups has been used successfully for final deprotection of several synthetic polypeptides. In practice, a solution of the chosen protected peptide in anhydrous liquid hydrogen fluoride, containing anisole as the scavenger of the cations which are formed during the cleavage, is stored for 1 hr, preferably at 0°. In isolated cases (see example below) it becomes necessary to raise the temperature of the reaction to 5–10° to ensure solubilization of the protected peptide. The solvent is subsequently removed by vacuum distillation (0°) and the residue is dried over potassium hydroxide under high vacuum (0°). The crude material is washed with an organic solvent (i.e., ethyl acetate) and subjected to appropriate purification procedures. Use of hydrogen fluoride, which is a highly toxic and hazardous chemical, requires the utilization of special equipment. A convenient apparatus for this purpose is constructed of Kel-F and is commercially available (Peninsula Laboratories, 1105 Laurel Street, San Carlos, 94070).

As with the sodium in liquid ammonia technique, however, a number of side reactions are associated with this new procedure. Among these, intra- or intermolecular esterification between hydroxyl and carboxyl groups,[31] rearrangement of α-aspartyl to β-aspartyl peptides,[147] Friedel-

---

[147] S. S. Wang, C. C. Yang, I. D. Kulesha, M. Sonenberg, and R. B. Merrifield, *Int. J. Pept. Protein Res.* **6**, 103 (1974).

Crafts acylation of the anisole by carboxyl groups,[148] and conversion of glutamyl residues to pyrrolidone derivatives,[148] may be mentioned. With the exception of the α-aspartyl to β-aspartyl rearrangement, it appears that these side reactions can be minimized by carrying out the deprotection at low temperature (0°) and for shorter time periods (<1 hr).

Example: Human Insulin B-Chain S-Sulfonate.[79] The protected triacontapeptide with the amino acid sequence of the human B chain bears the following protecting groups: N^{α}-*tert*-butyloxycarbonyl, S-diphenylmethyl, O-benzyl, N^{im}-tosyl, N^{ω}-nitro (arginine) and N^{ε}-benzyloxycarbonyl (lysine). The thoroughly dry (high vacuum over phosphorus pentoxide), crude protected triacontapeptide (200 mg) is placed in the reaction vessel of the special apparatus along with anisole (1 ml). To this vessel liquid hydrogen fluoride (10 ml), which has been redistilled over cobalt trifluoride, is added. The mixture is stirred at 10° for 1 hr and the hydrogen fluoride is removed at 0° with the aid of a vacuum pump. The latter is protected with a liquid nitrogen trap and a trap containing solid potassium hydroxide. The residue is dried under high vacuum (over potassium hydroxide) for 24 hr and then washed thoroughly with ethyl acetate. The sulfitolysis of this product and its isolation in highly purified form is accomplished as described in the previous example; weight 120 mg.

XI. Purification and Assessment of Chemical and Stereochemical Homogeneity of the Final Product

The final deprotection of the synthetic polypeptide is followed by purification and the establishment of its homogeneity. There are no special rules to follow for carrying out this task. Low-molecular-weight peptides are often obtained as crystalline solids and can be purified by recrystallization from various organic solvents. For the larger peptides the procedures which are employed in the purification of naturally occurring polypeptides and proteins are usually applied. Partition and ion-exchange chromatography, a variety of electrophoretic techniques and countercurrent distribution are some of the techniques which have been applied for the purification of a great number of synthetic polypeptides. In general, success in this endeavor depends greatly on the nature of the synthetic peptide, the physical facilities available and the skill of the investigators.

[148] R. S. Feinberg and R. B. Merrifield, *in* "Peptides: Chemistry, Structure and Biology" (R. Walter and J. Meienhofer, eds.), p. 455. Ann Arbor Sci. Publ., Ann Arbor, Michigan, 1975.

Procedures similar to those employed for the assessment of homogeneity of naturally occurring polypeptides and proteins are also utilized for synthetic peptides. Amino acid analyses after acid hydrolysis and calculation of the ratios and the percentage recovery of the constituent amino acids provide evidence, although not absolute proof, of homogeneity. Single spots on thin-layer chromatography in various solvent systems or in high voltage thin-layer electrophoresis at different pH values are important criteria of homogeneity of peptides. Similarly, discrete elution peaks in the chromatographic pattern obtained on ion-exchange chromatography under various experimental conditions (i.e., urea buffers, use of gradients, etc.) provide strong evidence for the homogeneity of the synthetic product. In general, any of the vast number of analytical techniques available at the present time for assessing homogeneity of proteins may be used also for synthetic peptides.

The classical approach for establishing the stereochemical homogeneity of a synthetic peptide is based on its complete digestability by proteolytic enzymes. The intermediate peptide fragments, utilized for the fragment condensation synthesis, and the final product are digested, usually with leucine aminopeptidase[149] or aminopeptidase M,[150] and the enzymic digests are subjected to amino acid analysis. Calculation of the molar ratios and the average recovery of the constituent amino acid residues and comparison with the theoretically expected values provide a satisfactory measure of the optical purity of the synthetic product. It must be recognized, however, that the limitations of the analytical techniques employed do not permit the detection of a small degree of racemization.

Utilization of L- or D-amino acid oxidases provides another enzymic technique for assessing the optical purity of peptides.[151,152] This approach, however, has a number of limitations and has not been used extensively for this purpose.

A combination of two newer techniques can now detect, to better than 1%, racemization of most amino acid residues in a synthetic peptide.[153] The first technique[154] involves the reaction of an optically active amino acid N-carboxyanhydride with the amino acid residues of an acid hydrol-

[149] R. L. Hill and E. L. Smith, J. Biol. Chem. 228, 577 (1957).
[150] G. Pfleiderer, P. G. Celliers, M. Stanulovic, E. D. Wachsmuth, H. Determann, and G. Braunitzer, Biochem. Z. 340, 552 (1964).
[151] T.-Y. Liu and E. C. Gotschlich, J. Biol. Chem. 238, 1928 (1963).
[152] S. Ishii and B. Witkop, J. Am. Chem. Soc. 85, 1832 (1963).
[153] J. M. Manning, A. Marglin, and S. Moore, in "Progress in Peptide Research" (S. Lande, ed.), p. 173. Gorden & Breach, New York, 1972.
[154] J. M. Manning and S. Moore, J. Biol. Chem. 243, 5591 (1968).

ysate of the synthetic peptide. Resolution of the resulting diastereomeric dipeptides on an amino acid analyzer permits the detection of 0.01% of optical impurity. The second technique[155] consists in the use of tritiated hydrochloric acid for the hydrolysis of the synthetic peptide and the calculation of the incorporation of the label into each amino acid. This latter technique permits one to establish the degree of racemization during the acid hydrolysis of the peptide. Analysis of the data from both these techniques establishes the optical purity of the synthetic product.

XII. Concluding Remarks

The purpose of this chapter is not to give a thorough review of the field of peptide chemistry. It is recognized that a considerable number of elegant investigations have not been given their rightful due. This brief account is intended only as an overall guide for peptide synthesis and the choice of material was governed primarily by the authors' own experience in this field.

[155] J. M. Manning, *J. Am. Chem. Soc.* **92**, 7449 (1970).

---

[48] Solid-Phase Peptide Synthesis

By MARILYNN S. DOSCHER

Introduction

The synthesis of peptides by the successive addition of the appropriately protected amino acid residues to a completely insoluble growing point was first proposed by Merrifield.[1] The idea was compelling and, perhaps to some, irritating in its simplicity. By the device of having the growing peptide chain anchored to a matrix, i.e., solid phase, which has no significant solubility in the various solutions required to perform the synthesis, it becomes possible to effect quantitative removal of each successive amino acid derivative and any auxiliary reagents that are required simply by repeatedly extracting the insoluble peptidyl solid-phase compound with an appropriate solvent.

Several general strategies are possible in the application of this idea to peptide synthesis. The chain may be built up residue by residue, i.e.,

[1] R. B. Merrifield, *J. Am. Chem.* **85**, 2149 (1963).

stepwise, proceeding from the COOH-terminal[2] to the NH_2-terminal residue or stepwise in the opposite direction, i.e., from the NH_2-terminal to the COOH-terminal residue. Alternatively, fragments composed of two or more residues may be successively added, again with the possibilities of progression from or to the COOH-terminal residue. Up to the present time the great majority of peptides prepared by solid-phase techniques have been synthesized in a stepwise fashion with progression from the COOH-terminal to the NH_2-terminal residue, and this article is confined to a discussion of those methods that utilize this strategy. The reader will find a discussion of methods and results for the other general strategies in a recent review by Erickson and Merrifield.[3]

A schematic presentation of the minimum number of reactions involved in a stepwise tripeptide synthesis is given in Fig. 1. The first reaction (step 1) results in attachment of the COOH-terminal residue to the solid support. The loading of a chloromethylated solid support by reaction with the cesium salt of the Boc derivative of the COOH-terminal amino acid[4] has been used to illustrate step 1; the exact reaction employed will depend upon the nature of the solid support chosen for the synthesis (see Choice of Solid Support). The nature of the group X with which the α-NH_2 groups of the amino acids are temporarily protected will also be dictated to some extent by the choice of the solid support (see discussion of α-NH_2 protection under The Synthetic Cycle). Removal of group X, i.e., deprotection (step 2), uncovers the α-NH_2 group of the COOH-terminal residue for subsequent peptide bond formation with the carboxyl group of the penultimate residue, i.e., for the coupling step (step 4). Essentially all the α-NH_2 protecting groups currently used in solid-phase syntheses are removed by acidolysis, and it is necessary in this case to deprotonate the α-NH_2 group in a neutralization step (step 3) before proceeding to the coupling step. Activation of the carboxyl group of the penultimate residue by Y may be carried out *in situ*, directly before the coupling reaction in a separate vessel, or by formation of a derivative which is sufficiently stable to be isolated and stored (see peptide bond formation under The Synthetic Cycle). Steps 2, 3, and 4 are

---

[2] Unless otherwise designated, amino acid names refer to the L-stereoisomer. Abbreviations follow the recommendations of the IUPAC–IUB Commission on Biochemical Nomenclature [*J. Biol. Chem.* **247**, 977 (1972)]. Other abbreviations are Bpoc, 2-(p-biphenyl)-2-propyloxycarbonyl; Ppoc, 2-phenyl-2-propyloxycarbonyl; DCC, dicyclohexylcarbodiimide; CHA, cyclohexylamine; DCHA, dicyclohexylamine; DMF, dimethylformamide; HOAc, acetic acid; TFA, trifluoroacetic acid.

[3] B. W. Erickson and R. B. Merrifield, *in* "The Proteins," 3rd ed. (H. Neurath and R. L. Hill, eds.), Vol. 2, p. 255. Academic Press, New York, 1976.

[4] B. F. Gisin, *Helv. Chim. Acta* **56**, 1476 (1973).

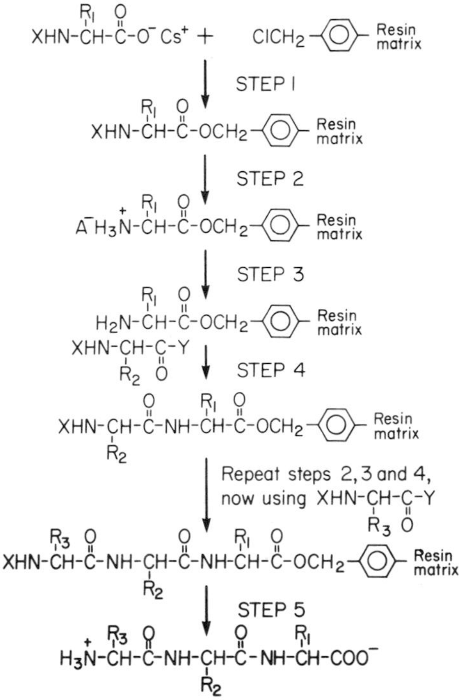

Fig. 1. The minimum number of reactions involved in the stepwise synthesis of a tripeptide by solid-phase methods.

then carried out for each successive residue in the peptide until the addition of the NH$_2$-terminal residue is completed. At this point the completed peptide is cleaved from the solid support (step 5). The specific reaction used again depends upon the choice of the support; the structure of the desired peptide as well as the nature of the side-chain blocking groups will also be factors in determining the nature of the cleavage reaction (see Cleavage from the Solid Support and Removal of Side Chain Blocking Groups). Many of the structures which are used to block the side chain functional groups of the trifunctional amino acids are removed by the various reactions that cleave the finished peptide from the solid support. Several are not, however (see Cleavage from the Solid Support and Removal of Side Chain Blocking Groups as well as Appendix B for specific examples); these blocking groups may be removed prior to or directly after step 5 depending on the particular case.

The success of a stepwise solid-phase peptide synthesis is critically dependent upon the attainment of very high yields at steps 2, 3, and 4. If such yields are not attained, the peptidyl solid phase will soon contain significant amounts of a large number of side products, many of which have a close structural similarity to the desired peptide. These side products are formed from the permanent cessation at some point of further reaction by a given chain, resulting in the formation of a termination peptide, or the temporary failure of a given chain to react, resulting in the formation of a deletion peptide. The termination and deletion peptides which may arise from incomplete reaction during the synthesis of a pentapeptide are shown in Fig. 2. The difficulty encountered in removing such side products from the desired peptide will depend upon the nature of the residues in that peptide, but it may be noted that most deletion peptides will be more difficult to remove than most termination peptides. It may be noted also that the larger the peptide, the more serious the problem of removing all such side products is likely to become.

An appreciation of the necessity for the attainment of very high yields at each step in the synthetic cycle has led to the development of a number of procedures that permit semiquantitative and quantitative measurement of reaction yield in the deprotection and coupling steps (see Monitoring Procedures). The more frequently encountered problem is a low yield in a coupling step. Often the repetition of the coupling step (after a repetition of the neutralization step) raises the yield to an acceptable value. Treatment of the refractory peptidyl solid phase with a solvent

I. TERMINATION PEPTIDES

$$H_3\overset{+}{N}-CH(R_2)CO-NHCH(R_I)COO^-$$

$$H_3\overset{+}{N}-CH(R_3)CO-NHCH(R_2)CO-NHCH(R_I)COO^-$$

$$H_3\overset{+}{N}-CH(R_4)CO-NHCH(R_3)CO-NHCH(R_2)CO-NHCH(R_I)COO^-$$

2. DELETION PEPTIDES

$$H_3\overset{+}{N}-CH(R_5)CO-NHCH(R_4)CO-NHCH(R_3)CO-NHCH(R_I)COO^-$$

$$H_3\overset{+}{N}-CH(R_5)CO-NHCH(R_4)CO-NHCH(R_2)CO-NHCH(R_I)COO^-$$

$$H_3\overset{+}{N}-CH(R_5)CO-NHCH(R_3)CO-NHCH(R_2)CO-NHCH(R_I)COO^-$$

Fig. 2. The termination and deletion peptides that may arise from incomplete reaction during the synthesis of a pentapeptide. More minor side products that would arise from combinations of two or more incomplete reactions are not included.

such as *tert*-butanol or isopropanol, which causes marked conformational change, i.e., shrinkage, of the solid support followed by repetition of the neutralization and coupling steps has improved yields.[5] If the peptidyl solid phase ceases to react as judged by the plateauing of yield after repeated reaction, most, if not all, of the unreacted α-NH$_2$ groups may be permanently blocked by reaction with an acylating or alkylating agent (see termination of peptide chains under Monitoring Procedures). This treatment reduces the number of potential deletion sequences by converting many to termination sequences, generally a less undesirable type of side product, as noted above. Low yields in deprotection steps have been encountered, but much less often than with coupling steps. This lower frequency may reflect in part, however, a lower frequency in the application of monitoring procedures to deprotection steps. Simple repetition of the deprotection step as well as repetition in conjunction with shrinkage and swelling of the peptidyl solid phase by solvent variation have been utilized to remedy low deprotection yields. Interestingly, monitoring of the completeness of the neutralization step has been reported in one instance only.[5]

The formation of side products may also occur through a number of reactions involving unwanted interaction of the various amino acid derivatives and auxiliary reagents with one or another of the side chain structures in the growing peptide, with elements of the peptide backbone, or with the α-NH$_2$ group. Discussions of such reactions appear under A Side Reaction Independent of Sequence, Side Reactions Partly Dependent on Sequence, and Difficult Sequences and in connection with the description of commonly used derivatives of the individual amino acids given in Appendix B.

Racemization is a potential problem in peptide synthesis as essentially all syntheses of interest involve the use of pure enantiomers of the amino acids and the aim is to preserve the optical integrity in the product. Synthetic procedures are now available which *generally* avoid racemization, but it is still standard practice to test the final product for racemized residues.

The small number and apparent simplicity of the manipulations required to perform a solid-phase peptide synthesis may tempt the uninitiated to proceed without obtaining a background in the chemistry of the common amino acids and the methods of peptide synthesis in solution. It cannot be overemphasized how serious an error this is likely to be. Many textbooks of organic chemistry now contain a substantial discus-

[5] W. S. Hancock, D. J. Prescott, P. R. Vagelos, and G. R. Marshall, *J. Org. Chem.* **38**, 774 (1973).

sion of the chemistry of the common amino acids. The classic, three-volume contribution by Greenstein and Winitz[6] may be consulted for greater detail on specific points. Two textbooks on peptide synthesis, which are of moderate length, are available,[7,8] as is a more comprehensive compendium of methods and classical syntheses;[9] however, the last of these sources is rapidly becoming outdated. Comprehensive reviews of both peptide synthesis in solution[10] and peptide synthesis on a solid phase[3] have recently appeared. Other reviews, which cover aspects of peptide synthesis both in solution and on solid phases, have appeared in the intervening period.[11-14] The reader may also wish to consult an older review of solid-phase peptide synthesis by Merrifield, as it was encyclopedic at the time of its publication.[15] A unique contribution to the methodology of solid-phase peptide synthesis by Stewart and Young[16] should be consulted; several of the newer procedures are absent from this volume, but it contains numerous preparative and analytical techniques that will not soon be outdated. Marshall and Merrifield have prepared a list of the 328 peptides that had been synthesized on a solid support as of 1969.[17] An extensive compendium of techniques for the synthesis of peptides, both in solution and on a solid support, has recently been published in German.[18]

Many syntheses are cited throughout the text in connection with the discussion of specific points. In addition to these examples, the reader

[6] J. P. Greenstein and M. Winitz, "Chemistry of the Amino Acids," 3 vol. Wiley, New York, 1961.
[7] M. Bodanszky, Y. S. Klausner, and M. A. Ondetti, "Peptide Synthesis," 2nd ed. Wiley (Interscience), New York, 1976.
[8] H. D. Law, "The Organic Chemistry of Peptides." Wiley (Interscience), New York, 1970.
[9] E. Schröder and K. Lübke, "The Peptides," 2 vol. Academic Press, New York, 1965, 1966.
[10] F. M. Finn and K. Hofmann, in "The Proteins," 3rd ed. (H. Neurath and R. L. Hill, eds.), Vol. 2, p. 105. Academic Press, New York, 1976.
[11] A. Marglin and R. B. Merrifield, Annu. Rev. Biochem. 39, 841 (1970).
[12] E. Wünsch, Angew. Chem., Int. Ed. 10, 786 (1971).
[13] J. Meienhofer, in "Hormonal Proteins and Peptides" (C. H. Li, ed.), Vol. 2, p. 45. Academic Press, New York, 1973.
[14] M. Fridkin and A. Patchornik, Annu. Rev. Biochem. 43, 419 (1974).
[15] R. B. Merrifield, Adv. Enzymol. 32, 221 (1969).
[16] J. M. Stewart and J. D. Young, "Solid Phase Peptide Synthesis." Freeman, San Francisco, 1969.
[17] G. R. Marshall and R. B. Merrifield, in "Handbook of Biochemistry and Molecular Biology," 3rd ed. (G. D. Fasman, ed.), Proteins, Vol. 1, p. 374. CRC Press, Cleveland, Ohio, 1976.
[18] E. Wünsch, ed., "Houben-Weyl:Methoden der organischen Chemie," Vol. 15, Part 1 and Part 2. "Synthese von Peptiden." Thieme, Stuttgart, 1974.

may wish to study syntheses of certain smaller linear peptides[19-21] as well as two cyclic peptides.[22,23] Examples of syntheses of somewhat larger molecules are those of monkey and human β-melanotropins,[24] containing 18 and 22 residues, respectively, and the fragment containing residues 37–55 of ovine prolactin.[25] Finally, the achievements and problems connected with some of the most ambitious and successful syntheses so far undertaken may be appreciated by study of the papers reporting the syntheses of human adrenocorticotropin (39 residues)[26] and bovine pancreatic ribonuclease A (124 residues).[27]

Choice of Solid Support

The choice of solid support will depend principally upon the structure of the desired peptide, but other considerations such as commercial availability or the requirements for the stability of the peptide–solid phase bond may also be relevant.

Chloromethylated Polystyrene

For the synthesis of simple peptides, i.e., those devoid finally of any labile substituents such as side-chain blocking groups or an α-NH$_2$ protecting group, the most frequently used solid support has been chloromethylated polystyrene (I). In certain cases this resin is also suitable for the synthesis of peptides having a COOH-terminal amide (see discussion below under Resins for peptides with a COOH-terminal amide). This resin is commercially available under the designation of Merrifield resin from Lab Systems, Inc. (San Mateo, CA); Vega-Fox Biochemicals (Tucson, AZ); Pierce Chemical Co. (Rockford, IL); and Calbiochem (La Jolla, CA). Lab Systems markets both 1% and 2% cross-linked preparations, and each is available at several levels of chloromethylation. Early samples of 1% cross-linked preparations were too fragile to be useful,[1] but the

[19] D. Yamashiro and C. H. Li, *Int. J. Pept. Protein Res.* **4**, 181 (1972).
[20] R. Carraway and S. E. Leeman, *J. Biol. Chem.* **250**, 1912 (1975).
[21] J. Burton, K. Poulsen, and E. Haber, *Biochemistry* **14**, 3892 (1975).
[22] B. F. Gisin, R. B. Merrifield, and D. C. Tosteson, *J. Am. Chem. Soc.* **91**, 2691 (1969).
[23] B. F. Gisin and R. B. Merrifield, *J. Am. Chem. Soc.* **94**, 6165 (1972).
[24] K.-T. Wang, J. Blake, and C. H. Li, *Int. J. Pept. Protein Res.* **5**, 33 (1973).
[25] R. L. Noble, D. Yamashiro, and C. H. Li, *J. Am. Chem. Soc.* **98**, 2324 (1976).
[26] D. Yamashiro and C. H. Li, *J. Am. Chem. Soc.* **95**, 1310 (1973).
[27] B. Gutte and R. B. Merrifield, *J. Biol. Chem.* **246**, 1922 (1971).

properties of subsequent batches were more satisfactory. The 1% cross-linked preparations are preferable to the 2% cross-linked preparations, as the sites of synthesis within the resin bead should be more available for reaction. If desired, the nominal degree of chloromethylation in a commercial sample can be readily checked by displacing the chlorine with a tertiary amine and measuring the resulting chloride ion.[28-30]

Chloromethylation of copolystyrene divinylbenzene may be carried out in the laboratory if a preparation of desired substitution, cross-linking, or mesh size is not commercially available.[23,28] Unfortunately, most current procedures utilize chloromethyl methyl ether, and it is now appreciated that this compound (and even more so the 1,3-dichloromethyl methyl ether which commonly contaminates chloromethyl methyl ether preparations) is carcinogenic. One should be equipped to meet the OSHA handling requirement for chloromethyl methyl ether[31] before working with this compound. Recently, Sparrow has suggested replacement of chloromethyl methyl ether by chloromethyl ethyl ether, although the relative carcinogenicity of this latter compound is evidently unknown.[32]

Hydroxymethylated Polystyrene

This resin (II) may be used for the synthesis of the same types of peptides as those prepared with the chloromethylated resin. Although it is not commercially available, it has the advantage over resin (I) that any functional groups remaining as a result of incomplete loading of the COOH-terminal amino acid can be rendered inactive by acylation. The procedures for loading the COOH-terminal amino acid differ for the two resins, but subsequent manipulations are the same, as the benzyl ester linkage formed by the various loading procedures is identical. (See, however, the discussion of the consequences of incomplete loading under Attachment of the COOH-Terminal Amino Acid.)

The resin may be prepared from chloromethylated polystyrene in two steps via the acetoxymethyl resin. The conversion of the acetoxymethyl resin to the hydroxymethyl resin has been carried out with alcoholic sodium hydroxide,[33] diethylamine,[23] hydrazine,[34] or LiAlH$_4$ reduction.[34]

[28] R. S. Feinberg and R. B. Merrifield, *Tetrahedron* **30**, 3209 (1974).
[29] L. C. Dorman, *Tetrahedron Lett.* 2319 (1969).
[30] J. M. Stewart and J. D. Young, "Solid Phase Peptide Synthesis," p. 27. Freeman, San Francisco, 1969.
[31] *Fed. Regist.* **39**, 3768 (1974).
[32] J. T. Sparrow, *Tetrahedron Lett.* p. 4637 (1975).
[33] M. Bodanszky and J. T. Sheehan, *Chem. Ind.* **1966**, 1597 (1966).
[34] S. S. Wang, *J. Org. Chem.* **40**, 1235 (1975).

$$\text{ClCH}_2-\langle\bigcirc\rangle-\text{resin} \xrightarrow{\text{KOAc}} \text{CH}_3\text{COOCH}_2-\langle\bigcirc\rangle-\text{resin}$$

(I)

$$\text{HOCH}_2-\langle\bigcirc\rangle-\text{resin}$$

(II)

The marginal stability of the benzyl ester linkage to the repeated acidolysis required to remove some types of α-NH$_2$ protection, in particular the Boc group, was appreciated from the start[1] and has since been amply documented.[27] The lability of the benzyl ester linkage not only lowers the yield of the desired peptide, but also introduces the possibility that shorter peptides representing NH$_2$-terminal portions of the complete sequence will be synthesized. This type of side product can result if hydroxymethyl groups, which can be formed on the resin upon acidolysis of the anchoring linkage, subsequently react with an incoming amino acid during a coupling step.[23,35] In response to this problem a resin has now been prepared in which the benzyl group carries in the para position the electron-withdrawing acetamidomethyl moiety (III).[36] The benzyl ester linkage between this resin and Leu-Ala-Gly-Val is one hundred times more stable toward 50% TFA-CH$_2$Cl$_2$ than is the linkage between the simple

$$\text{HOCH}_2-\langle\bigcirc\rangle-\text{CH}_2\text{CONHCH}_2-\langle\bigcirc\rangle-\text{resin}$$

4-(Hydroxymethyl)-
phenylacetamidomethyl-resin
(Pam-resin)

(III)

resins, (I) and (II), and this peptide.[36] At present the Pam-resin appears to be a more suitable polystyrene support for the synthesis of peptides of the type previously synthesized using the chloromethylated or hydroxymethylated resin. It is not commercially available, but may be prepared by direct amidomethylation of copolystyrene–1% divinylbenzene (Bio-Rad Laboratories) in a four-step reaction sequence.[36a] For shorter peptides or for syntheses which utilize α-NH$_2$ protection not requiring 50% TFA-

[35] M. C. Khosla, R. R. Smeby, and F. M. Bumpus, *J. Am. Chem. Soc.* **94**, 4721 (1972).
[36] A. R. Mitchell, B. W. Erickson, M. N. Ryabtsev, R. S. Hodges, and R. B. Merrifield, *J. Am. Chem. Soc.* **98**, 7357 (1976).
[36a] A. R. Mitchell, S. B. H. Kent, B. W. Erickson, and R. B. Merrifield, *Tetrahedron Lett.* p. 3795 (1976).

CH$_2$Cl$_2$ for removal, the investigator may find the investment in the synthesis of the Pam-resin is not worthwhile.

Resins for Peptides with a COOH-Terminal Amide

The chloromethylated and hydroxymethylated resins described above may be used for the synthesis of peptides with a COOH-terminal amide group. In these cases the completed peptide is removed from the resin by ammonolysis rather than by acidolysis (see Cleavage from the Solid Support and Removal of Side-Chain Blocking Groups). Among the successful syntheses utilizing this approach are those of oxytocin[37] and of inhibitors of luteinizing hormone-releasing hormone.[38] If the peptide contains aspartyl or glutamyl residues and the β- and γ-COOH groups of these residues are blocked with the most commonly used blocking group, i.e., the benzyl moiety, the ammonolysis will also convert these residues to asparagine and glutamine, respectively. To avoid this problem, syntheses have been devised where the blocking groups for these residues have been removed prior to the ammonolysis.[39]

Another approach involves the use of a benzhydrylamine resin (IV).[40] Acidolytic removal of the completed peptide from this resin by anhydrous

Benzhydrylamine-resin

(IV)

HF results in the formation of the COOH-terminal amide structure. The resin may be purchased from Lab Systems, Inc. (San Mateo, CA) or Beckman, Spinco Division, (Palo Alto, CA). The original synthesis[40] proceeded by ammonolysis of the corresponding bromo- or chlorobenzhydryl resin prepared previously by Southard *et al.*[41] from copolystyrene–2% divinylbenzene. Two-step syntheses from copolystyrene–1% divinylben-

[37] M. Manning, *J. Am. Chem. Soc.* **90**, 1348 (1968).
[38] Y. P. Wan, J. Humphries, G. Fisher, K. Folkers, and C. Y. Bowers, *J. Med. Chem.* **19**, 199 (1976).
[39] C. T. Wang, I. D. Kulesha, P. L. Stefko, and S. S. Wang, *Int. J. Pept. Protein Res.* **6**, 59 (1974).
[40] P. G. Pietta and G. R. Marshall, *J. Chem. Soc., Chem. Commun.* **1970, 650** (1970).
[41] G. L. Southard, G. S. Brooke, and J. M. Pettee, *Tetrahedron Lett.* p. 3505 (1969).

zene have been reported more recently.[42,43,43a] Orlowski *et al.*[43a] have noted the formation of a large percentage of secondary amine when the synthesis was via oxime reduction.[43] Recent representative syntheses utilizing the benzhydrylamine resin are those of luteinizing hormone-releasing hormone,[42] COOH-terminal segments of secretin,[44] and an analog of substance P.[45]

Resins for the Preparation of Protected Peptide Fragments

It has been appreciated for some time that protected peptide fragments could be synthesized on a solid support and that such fragments could then be coupled using solution procedures or could be incorporated into a second solid-phase synthesis. One approach to this goal has involved the use of polystyrene resins in which the peptidyl–resin bond is sufficiently labile to permit its rupture under conditions mild enough to permit the retention of side-chain blocking groups and α-NH_2 protection.[34,46,47]

A 1,1-dimethylpropyloxycarbonylhydrazide resin (V), first prepared by Wang and Merrifield,[46] permits the synthesis of a protected peptide hydrazide, which can be subsequently incorporated into a larger peptide by an azide condensation.[48] A new synthesis of (V) starting from chloromethylated resin (I) and avoiding the use of HF has now been pub-

$$NH_2NHCOOC(CH_3)_2CH_2CH_2 \!-\!\!\left\langle\!\bigcirc\!\right\rangle\!-\! resin$$

1, 1-Dimethylpropyloxycarbonyl
hydrazide-resin

(V)

lished.[47] None of the resins designed specifically for fragment preparation is presently available commercially. Wang and Merrifield[49] have synthesized a protected decapeptide hydrazide using (V) while Wolters *et al.*[47]

[42] M. W. Monahan and J. Rivier, *Biochem. Biophys. Res. Commun.* **48**, 1100 (1972).

[43] P. G. Pietta, P. F. Cavallo, K. Takahashi, and G. R. Marshall, *J. Org. Chem.* **39**, 44 (1974).

[43a] R. C. Orlowski, R. Walter, and D. Winkler, *J. Org. Chem.* **41**, 3701 (1976).

[44] B. Hemmasi and E. Bayer, *Hoppe-Seyler's Z. Physiol. Chem.* **355**, 481 (1974).

[45] G. H. Fisher, K. Folkers, B. Pernow, and C. Y. Bowers, *J. Med. Chem.* **19**, 325 (1976).

[46] S. S. Wang and R. B. Merrifield, *J. Am. Chem. Soc.* **91**, 6488 (1969).

[47] E. T. M. Wolters, G. I. Tesser, and R. J. F. Nivard, *J. Org. Chem.* **39**, 3388 (1974).

[48] P. G. Katsoyannis and G. P. Schwartz, this volume [47].

[49] S. S. Wang and R. B. Merrifield, *Int. J. Pept. Protein Res.* **4**, 309 (1972).

have used it to prepare an α-NH$_2$-protected hexapeptide hydrazide comprising the A$_{14-19}$ fragment of insulin.

A second resin (VI), closely related structurally to (V) and also useful for the synthesis of protected hydrazides, has recently been introduced by Wang.[34]

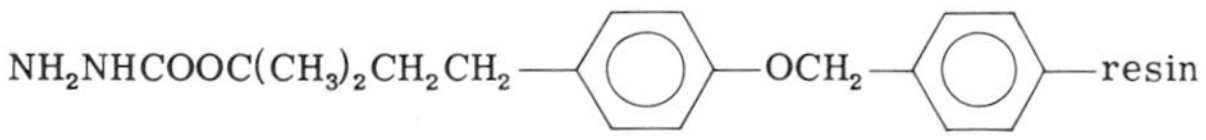

$$NH_2NHCOOC(CH_3)_2CH_2CH_2 —⟨◯⟩— OCH_2 —⟨◯⟩— resin$$

3-(p-Benzyloxyphenyl)-1,1-dimethylpropyl-
carbonylhydrazide-resin

(VI)

A second resin (VI), closely related structurally to (V) and also useful known to form too labile a peptide-resin bond to be useful for many syntheses.[51,52]

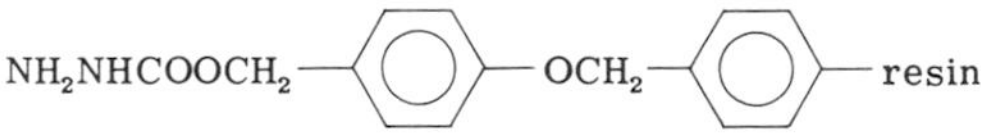

$$NH_2NHCOOCH_2 —⟨◯⟩— OCH_2 —⟨◯⟩— resin$$

p-(Benzyloxyphenyl)methyloxycarbonyl
hydrazide-resin

(VII)

A hydroxymethylated resin (VIII) which forms a peptide–resin bond more labile than that formed by the simple hydroxymethylated resin (II) has been prepared by Wang.[50]

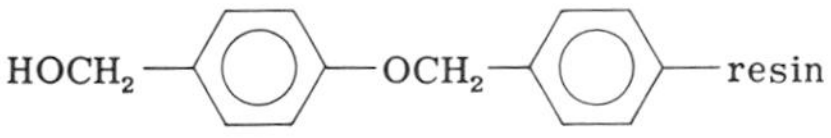

$$HOCH_2 —⟨◯⟩— OCH_2 —⟨◯⟩— resin$$

p-(Benzyloxyphenyl)methanol-resin

(VIII)

Use of this resin permits the synthesis of protected peptides with free COOH-terminal carboxyl groups; such peptides then can be incorporated into larger fragments using established techniques for peptide α-COOH group activation without racemization.[48] The resin has been used for the synthesis of seven protected peptides which span the ovine pituitary growth hormone sequence 96–135.[53]

The recent introduction of resins which permit photolytic cleavage of protected peptides[53a,53b] represents a very promising departure as the

[50] S. S. Wang, *J. Am. Chem. Soc.* **95**, 1328 (1973).
[51] S. S. Wang, private communication, 1975.
[52] M. S. Doscher, unpublished observation, 1975.
[53] S. S. Wang and I. D. Kulesha, *J. Org. Chem.* **40**, 1227 (1975).
[53a] D. H. Rich and S. K. Gurwara, *J. Am. Chem. Soc.* **97**, 1575 (1975).
[53b] S. S. Wang, *J. Org. Chem.* **41**, 3258 (1976).

well-tested Boc group may be used for α-amino protection (see The Synthetic Cycle) in these syntheses.

While the use of special polystyrene resins for the solid-phase synthesis of protected peptides has received most attention so far, other approaches have been tried with some success. Ohno and Anfinsen[54] were able to prepare protected peptide hydrazides by hydrazinolysis of protected peptides prepared on chloromethylated resin (I). Another route involves the transesterification by dimethylaminoethanol of a protected peptide prepared on chloromethylated resin (I).[55,56] Ready hydrolysis of the resulting dimethylaminoethyl ester at room temperature in aqueous DMF provides the protected peptide with a free α-COOH group. Both the hydrazinolysis and transesterification procedures require, of course, the absence of other susceptible groups, such as side-chain benzyl esters.

Attachment of the COOH-Terminal Amino Acid

The reaction used for the attachment of the COOH-terminal amino acid depends upon the type of resin chosen for use. The consequences of incomplete substitution of resin functional groups by the COOH-terminal amino acid as well as procedures for dealing with this situation will depend also upon resin type.

Chloromethylated Polystyrene. Esterification of this resin derivative is best effected by reaction with the cesium salt of the suitably protected COOH-terminal amino acid.[4] This procedure avoids the formation of quaternary ammonium sites on the resin. Such sites can exchange anions and thereby provide inaccurate values with certain monitoring procedures (see discussion under Monitoring Procedures). Gisin notes also that trifluoroacetate ions from a deprotection step might bind at such sites and then be released during a subsequent coupling step with resultant trifluoroacetylation of α-NH$_2$ groups.[4] The use of cesium salts has allowed essentially complete substitution of chloromethyl groups in some cases; subsequent formation of quaternary ammonium sites by reaction of remaining chloromethyl groups with the tertiary amine used in the neutralization steps is thereby avoided. Roeske and Gesellchen[56a] have recently reported virtually complete esterification of chloromethylated polystyrene by a variety of bulky residues in the presence of stoichiometric amounts of the crown ether, 18-crown-6.

Substitution of chloromethylated polystyrene has also been carried out

[54] M. Ohno and C. B. Anfinsen, *J. Am. Chem. Soc.* **89**, 5994 (1967).
[55] M. A. Barton, R. U. Lemieux, and J. Y. Savoie, *J. Am. Chem. Soc.* **95**, 4501 (1973).
[56] J. Y. Savoie and M. A. Barton, *Can. J. Chem.* **52**, 2832 (1974).
[56a] R. W. Roeske and P. D. Gesellchen, *Tetrahedron Lett.* p. 3369 (1976).

with Boc amino acids and potassium *t*-butoxide.[57] This procedure is rapid and avoids the quaternization reaction, but complete substitution of chloromethyl groups has not yet been reported.

Hydroxymethylated Polystyrene Derivatives. Esterification of those resin derivatives in which the functional group is hydroxymethyl [(II), (III), and (VIII)] may be achieved by an activation of the protected COOH-terminal amino acid mediated by carbonyldiimidazole.[23,33] Complete substitution of hydroxymethyl groups is not generally achieved by this procedure. As the remaining hydroxymethyl groups will almost certainly react to some extent in subsequent coupling steps with the resultant formation of NH_2-terminal segments of the desired peptide[35] they must be blocked by acetylation[23,36] or benzoylation[50] after attachment of the COOH-terminal residue.

Polystyrene Derivatives Containing a Primary Amino Group. The COOH-terminal amino acid may be attached to the resins designed for the synthesis of COOH-terminal amides, (IV), and protected hydrazides, (V) and (VI), by the use of DCC.[44-46] If the acylation of amino groups is not complete, those remaining should be acetylated[40,45] or otherwise terminated (see Monitoring Procedures).

Analysis of the Load of the COOH-Terminal Residue on the Resin. Amino acid analysis of a weighed sample of the protected aminoacyl-resin will provide the degree of substitution achieved by the above procedures. Methods of hydrolysis for aminoacyl- and peptidyl-resins are discussed under Monitoring Procedures. With resin derivatives containing a primary amino group the possibility exists of retention by ion exchange of molecules of protected amino acids[58]; this phenomenon will result in an erroneously high value for the COOH-terminal residue load. Such molecules can be removed by deprotecting the substituted resin prior to the removal of the analytical sample. Such deprotection forecloses the possibility of a second round of loading, but if a second round is contemplated, deprotection can be confined to the analytical sample.

Alternatively, the degree of loading may be estimated by subjecting a weighed sample of the deprotected aminoacyl-resin to analysis by the picric acid[59] or imidazolium picrate methods[60] (see Monitoring Procedures). These procedures determine the number of amino groups per weight of resin and, therefore, will give an accurate indication of the load of COOH-terminal amino acid, assuming all free amino groups are

[57] M. W. Monahan and C. Gilon, *Biopolymers* **12**, 2513 (1973).
[58] K. Esko and S. Karlsson, *Acta Chem. Scand.* **24**, 1415 (1970).
[59] B. F. Gisin, *Anal. Chim. Acta* **58**, 248 (1972).
[60] R. S. Hodges and R. B. Merrifield, *Anal. Biochem.* **65**, 241 (1975).

indeed α-NH$_2$ groups and are not derived from incomplete substitution of a resin originally containing primary amino groups, viz., resins (IV), (V), and (VI). Application of these monitoring procedures for this purpose also requires the assumption that deprotection of the COOH-terminal residue has been quantitative.

The Synthetic Cycle

After the solid support has been selected and the COOH-terminal amino acid attached to it, the desired weight of protected aminoacyl-resin is placed in a reaction vessel (see Equipment) and the cycle of reactions required to effect the addition of each successive residue in the peptide is begun. The precise sequence and type of washes and additions to be used will depend principally upon the nature of the α-NH$_2$ protecting group to be removed in the deprotection step and the nature of the reaction used to achieve peptide bond formation in the coupling step. A schedule which is suitable for the most frequently applied combination of α-NH$_2$ protection and coupling reaction is given in Appendix A.

α-*Amino Protecting Groups.* The overwhelming majority of solid-phase syntheses have used the Boc group (IX) for α-NH$_2$ protection. First introduced for use in peptide synthesis in solution[61,62] it proved to

$$(CH_3)_3-C-O-CO$$

tert-Butyloxycarbonyl group
(Boc group)

(IX)

be useful early in the development of solid-phase procedures[63] and continues as the dominant choice. Boc derivatives of all the common amino acids are commercially available from a number of sources: Bachem, Inc. Marina Del Rey, CA); Beckman Instruments, Inc., Bioproducts Dept. (Palo Alto, CA); Peninsula Laboratories, Inc. (San Carlos, CA); Research Plus Laboratories, Inc. (Denville, NJ); Vega-Fox Biochemicals (Tucson, AZ). Although these derivatives are quite stable they should be checked before use for chemical purity by thin-layer chromatography and melting-point determination and for optical purity by rotation measurements or by application of the Manning-Moore procedure to a deprotected sample.[64,65] The presence of any underivatized amino acid in

[61] F. C. McKay and N. F. Albertson, *J. Am. Chem. Soc.* **79**, 4686 (1957).
[62] G. W. Anderson and A. C. McGregor, *J. Am. Chem. Soc.* **79**, 6180 (1957).
[63] R. B. Merrifield, *Biochemistry* **3**, 1385 (1964).
[64] J. M. Manning, this series Vol. 25, p. 9 (1972).
[65] J. M. Manning and S. Moore, *J. Biol. Chem.* **243**, 5591 (1968).

such a preparation may result in the introduction into the growing peptide of more than one equivalent of the desired residue. Occasionally, commercial preparations which are the DCHA salt of a Boc derivative have been mislabeled as the free acid.[66,67] Such errors will be revealed by a melting-point determination, but not necessarily by chromatographic analysis (where acidic solvents may convert the samples of the salt to the free acid). These salts will not react readily with DCC, so the unwitting use of them will result in the synthesis of a deletion peptide.

The Boc group is generally removed by exposing the protected aminoacyl- or peptidyl-resin to a 50% solution of TFA in CH_2Cl_2 for a period of 20 min.[68] When it became evident that these conditions caused detectable acidolysis of the peptide-resin ester linkage with resins I and II as well as loss of certain side-chain blocking groups (see individual amino acids in Appendix B), attempts were made to achieve the deprotection with 20% TFA in CH_2Cl_2. It now appears that these latter conditions are sometimes too mild to obtain the necessary quantitative reaction.[69] Fortunately, in the interim period, a resin which forms an ester linkage more resistant to acidolysis has been devised (resin III)[36] and more stable side chain blocking groups have been found (see Appendix B). Consequently, the demonstrated necessity to use an acid as concentrated as 50% TFA to remove the Boc group need not any longer result in a number of undesirable side reactions.

With the resins designed for the synthesis of protected peptide fragments (resins V, VI, and VIII) the acidity required to remove the Boc group is adequate to cleave the peptidyl-resin bond as well. Consequently, a more labile α-NH_2 protecting group is required when these resins are used. The Bpoc group (X) has most frequently been used in these circumstances. It may be removed by a 10-min treatment with

2-(p-Biphenyl)-2-propyloxycarbonyl
group (Bpoc group)

(X)

---

[66] J. M. Stewart, private communication, 1972.

[67] M. S. Doscher, unpublished observations, 1972.

[68] Alkylation of susceptible groups by the *tert*-butyl carbonium ion generated in this reaction can be prevented by including a carbonium ion scavenger, such as anisole in the deprotection solution (F. Weygand and W. Steglich, *Z. Naturforsch. Teil B* **14**, 472 (1959).

[69] R. E. Reid, *J. Org. Chem.* **41**, 1027 (1976).

0.5% TFA in CH_2Cl_2,[34,70] conditions mild enough to maintain the integrity of the peptidyl-resin linkage. Certain drawbacks attend the use of this protecting group, however. Derivatives are not commercially available (although Bachem, Inc., will prepare them on special order) so they must be synthesized.[49,70-72] Moreover, the lability of the Bpoc group makes storage of the derivatives as free acids risky, if not quite impractical.[73] Consequently, it has been recommended that they be stored as CHA or DCHA salts and converted into the free acid prior to use.[71,72] This operation adds significantly to the labor required for a given synthesis. It is possible to remove the Bpoc group with pyridine hydrochloride.[74] This finding suggests that imidazolium picrate, used to monitor the completeness of deprotection and coupling reactions (see Monitoring Procedures), might cause removal of Bpoc protection; actual detection of this reaction has not been reported, however.

The Ppoc group (XI) is another possibility for α-NH_2 protection in circumstances where greater lability than that exhibited by the Boc group is required or desired. It may be removed by treatment for 30 min

$$\langle O \rangle - C(CH_3)_2O - CO -$$

2-Phenyl-2-propyloxycarbonyl
group (Ppoc group)

(XI)

with 1-2% TFA in CH_2Cl_2.[75] As the group is approximately five times more stable than the Bpoc[76], storage of derivatives as the free acids at low temperatures may be feasible[75]. No derivatives are commercially available (Bachem will prepare them on special order), but the properties of derivatives of many of the common amino acids have been published.[75,77]

While protecting groups such as Bpoc and Ppoc must be used with resins (V), (VI), and (VIII), they may sometimes be used to advantage with other resins. Even with resin (III), in which stability of the pep-

[70] S. S. Wang and R. B. Merrifield, *Int. J. Protein Res.* **1**, 235 (1969).

[71] P. Sieber and B. Iselin, *Helv. Chim. Acta* **51**, 622 (1968).

[72] R. S. Feinberg and R. B. Merrifield, *Tetrahedron* **28**, 5865 (1972).

[73] Experiences vary, however. Wang found no deterioration in a preparation of the free acid of Bpoc-tryptophan which had been stored first at 4° for 1 year and then subsequently at −15° for 5½ years (S. S. Wang, private communication, 1976).

[74] H. Klostermeyer and E. Schwertner, *Z. Naturforsch. Teil B* **28**, 334 (1973).

[75] B. E. B. Sandberg and U. Ragnarsson, *Int. J. Pept. Protein Res.* **6**, 111 (1974).

[76] P. Sieber and B. Iselin, *Helv. Chim. Acta* **51**, 614 (1968).

[77] B. E. B. Sandberg and U. Ragnarsson, *Int. J. Pept. Protein Res.* **7**, 503 (1975).

tidyl-resin linkage is not a problem, their use increases the range of lability which can be tolerated in the side chain blocking groups; with resins I and II the acidolytic loss of peptide chains from the resins is reduced, as well.

Side Chain Blocking Groups. A majority of the common amino acids contain a side chain functional group that has the potential for involvement in one or another of the reactions utilized for the synthesis of a peptide. Consequently, these groups must be blocked during the synthesis by some moiety which can then be removed upon completion of the synthesis. Which side chains require blocking depends mainly upon the nature of the coupling and cleavage reactions that will be utilized. The relationships between the types of reactions used in a synthesis and the necessity for blocking is discussed for the individual amino acids in Appendix B.

Peptide Bond Formation. With acidolytic deprotection procedures, which are used almost exclusively for deprotection in solid-phase peptide synthesis, it is necessary to neutralize the protonated α-NH_2 amino group that results from this type of reaction before proceeding with the coupling reaction. The agent used for neutralization was, until quite recently, a 10% solution of triethylamine in CH_2Cl_2,[78] DMF,[1] or $CHCl_3$.[79] When it was realized that alkylation of this amine by unsubstituted chloromethyl groups on resin (I) could occur and result in the formation of quaternary ammonium sites, the use of diisopropylethylamine (5% solution in methylene chloride) was begun,[70] as it seemed that steric hindrance would result in a lower rate of alkylation of this base. It is an expensive compound relative to triethylamine, however, so the investigator may wish to use triethylamine, particularly if resin preparations devoid of chloromethylated groups are used. It is imperative that all traces of base used in the neutralization step be removed before initiation of the coupling reaction, as it has been well documented that salts of tertiary amines catalyze racemization.[7,9]

As with the use of chloromethylated polystyrene for the solid support and the Boc group for α-NH_2 protection, the use of DCC has so far been the dominant method in effecting peptide bond formation in solid-phase syntheses. The standard procedure has involved the introduction of an amount of protected amino acid equivalent to three or four times the moles of α-NH_2 groups present (either based on the original load of COOH-terminal residues or on subsequent monitoring measurements after deprotection steps) into the reaction vessel containing the neutralized,

[78] P. Kusch, *Kolloid-Z., Z. Polym.* **208**, 138 (1966).
[79] V. A. Najjar and R. B. Merrifield, *Biochemistry* **5**, 3765 (1966).

washed peptidyl-resin, the protected amino acid having been dissolved in half to three-quarters of the total eventual volume of solvent. After allowing the derivative to diffuse into the resin beads for 7–10 min, where protonation of α-NH$_2$ groups and consequent retention by ion exchange of an equivalent of protected amino acid presumably occurs,[58] an amount of DCC equivalent to the moles of protected amino acid used is introduced into the reaction vessel as a solution in the remainder of the solvent.

The first activated compound formed by reaction of equivalent amounts of protected amino acid and DCC is the O-acyl isourea (XII).[9,80] The lifetime of this type of derivative is thought to be rela-

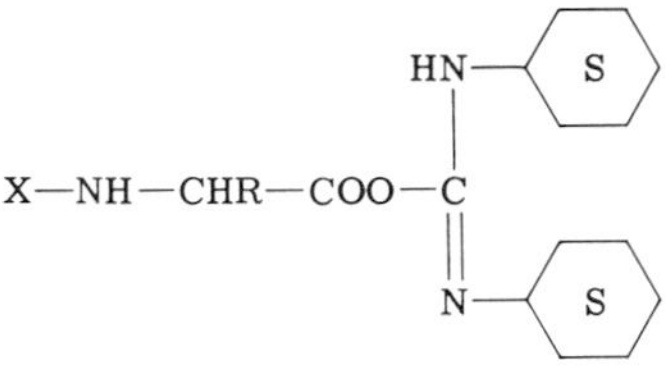

X—NH—CHR—COO—C

(XII)

tively short, with rearrangement to the inactive N-acyl urea (XIII) occurring if the isourea is not consumed by an alternative reaction, e.g.,

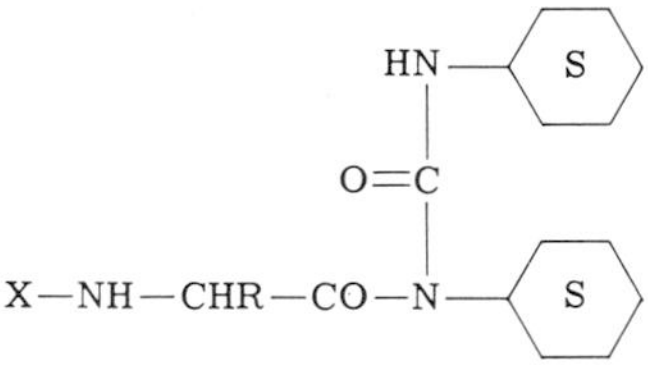

X—NH—CHR—CO—N

(XIII)

peptide bond formation by nucleophilic attack of the α-NH$_2$ group of a peptide or amino acid. This view has led to the practice of limiting re-action time in the coupling step to 2 hr or even less. Recently, however, Dorman[81] has demonstrated that a substantial percentage of the incoming amino acid remains in an activated state for many hours under conditions commonly used for the coupling reaction. This finding would seem to suggest that longer reaction times could be advantageous, but Dorman cites the study of Merrifield and co-workers,[82] which reported substantial incorporation of diglycyl units into peptidyl moieties if the protected symmetrical anhydride of this amino acid was allowed to preincubate at room temperature, particularly in the presence of a tertiary base (tri-

[80] H. G. Khorana, *Chem. Rev.* **53**, 145 (1953).

[81] L. C. Dorman, *Biochem. Biophys. Res. Commun.* **60**, 318 (1974).

[82] R. B. Merrifield, A. R. Mitchell, and J. E. Clarke, *J. Org. Chem.* **39**, 660 (1974).

ethylamine); he then goes on to note that these conditions may correspond closely to those obtaining in standard solid-phase coupling reactions as there is evidence that equimolar amounts of Z-Gly-OH and DCC react to form the symmetrical anhydride and one-half equivalent of dicyclohexylurea, with one-half equivalent of DCC unchanged, the last conceivably able to function in a manner analogous to triethylamine.[83] In conclusion, Dorman advises continuance of the standard practice of reaction times of 2 hr. Nevertheless, before the possibility of longer reaction times is permanently abandoned, more information is needed about the nature and rate of formation of side products when equimolar amounts of amino acid and DCC react, as well as the consequences of using preformed symmetrical anhydrides[84] of amino acids other than glycine and allowing them to react for periods longer than 2 hr.

Peptide bond formation by the reaction between the p-nitrophenyl ester of a protected amino acid and the α-NH$_2$ group has been used in solid-phase syntheses, particularly in connection with the introduction of asparagine and glutamine. These two types of residue are dehydrated to the corresponding ω-nitriles if direct activation with DCC is attempted.[85] The best response to this problem until quite recently has involved the use of the p-nitrophenyl ester derivatives.[33] The reaction rates are slow; consequently a 5 M or greater excess of derivative is used, and reactions are allowed to proceed for as long as 24 hr. Acceleration of rates of reaction of Boc asparagine and Boc glutamine p-nitrophenyl esters by the addition of 1-hydroxybenzotriazole has been reported very recently, however.[86]

The use of p-nitrophenyl esters for the introduction of residues other than asparagine and glutamine has been contraindicated for some time now[1,87,88] owing to the problem of low reactivity, and, recently, several alternative methods of introducing asparagine and glutamine as well have been used. The availability of derivatives in which the amide nitrogen has been substituted (see Appendix B) and thereby rendered resistant to dehydration permits the use of DCC for the coupling step reaction. This method would appear to be ideal, as the blocking groups may be removed with TFA or HF (see Cleavage from the Solid Support and Removal of Side Chain Blocking Groups); however, incomplete reaction has been

[83] D. F. DeTar, R. Silverstein, and F. F. Rogers, Jr., *J. Am. Chem. Soc.* **88**, 1024 (1966).

[84] T. Wieland, C. Birr, and F. Flor, *Angew. Chem., Int. Ed.* **10**, 336 (1971).

[85] C. Ressler and H. Ratzkin, *J. Org. Chem.* **26**, 3356 (1961).

[86] D. A. Upson and V. J. Hruby, *J. Org. Chem.* **41**, 1353 (1976).

[87] S. Karlsson, G. Lindeberg, and U. Ragnarsson, *Acta Chem. Scand.* **24**, 337 (1970).

[88] U. Ragnarsson, G. Lindeberg, and S. Karlsson, *Acta Chem. Scand.* **24**, 3079 (1970).

observed, probably owing to steric hindrance by the blocking groups, all of which are bulky.[89] Hemmasi and Bayer[44] have coupled unblocked glutamine as the symmetrical anhydride on the basis of a cogent argument that use of a 2:1 mole ratio of amino acid to DCC for formation of the anhydride precludes significant dehydration. Their assertion that preformation of the anhydride with concomitant consumption of all the DCC eliminates introduction of any nitrile that may have been formed is incorrect, however, as the nitrile may be present as the anhydride or as half of a mixed anhydride. Nevertheless, ion-exchange chromatography of a leucine aminopeptidase digest of the crude peptides revealed no detectable 2-amino-4-cyanobutyric acid. Wang *et al.*[39] have used *N*-ethoxycarbonyl-2-ethoxy-1,2-dihydroquinoline (EEDQ) to introduce both glutamine and asparagine without side-chain blocking, while reports from two laboratories[90,91] indicate that a combination of DCC and 1-hydroxybenzotriazole can also be used successfully to introduce both types of residue without side-chain blocking. With both EEDQ and DCC-1-hydroxybenzotriazole the absence of nitrile was determined by measurement of infrared (IR) spectra; this technique probably would not detect nitrile if it were present at a level of 5% or less.

Monitoring Procedures

It is essential to have methods with which to measure the completeness of each coupling step and each deprotection step in a solid-phase synthesis, as otherwise even a complete failure at some point may not be evident until workup and characterization of the product are performed. Syntheses in which certain steps did not go to completion have been well documented.[5,92-94]

Amino Acid Analysis. A moderately precise estimate of the extent of a coupling reaction may be obtained from an amino acid analysis of a hydrolysate of a sample of the peptidyl-resin.[95,96] The peptidyl-poly-

[89] R. B. Merrifield, private communication, 1976.
[90] V. J. Hruby, F. Muscio, C. M. Groginsky, P. M. Gitu, D. Saba, and W. Y. Chan, *J. Med. Chem.* 16, 624 (1973).
[91] E. T. M. Wolters, G. I. Tesser, and R. J. F. Nivard, *J. Org. Chem.* 39, 3388 (1974).
[92] E. Bayer, H. Eckstein, K. Hägele, W. A. König, W. Brüning, H. Hagenmaier, and W. Parr, *J. Am. Chem. Soc.* 92, 1735 (1970).
[93] H. Hagenmaier, *Tetrahedron Lett.* p. 283 (1970).
[94] C. Birr and R. Frank, *FEBS Lett.* 55, 68 (1975).
[95] Obviously the precision obtainable will depend, in part, upon the total number of residues of the type in question that are present in the peptide.
[96] The sample should be taken after the next deprotection step or should itself be

styrene beads are contracted in the aqueous 6 N HCl routinely used for the hydrolysis of peptides and proteins, with the result that this reagent is not suitable for the hydrolysis of peptidyl-resins. Several mixtures have been devised, however, in which the beads are expanded and a satisfactory rate of hydrolysis does occur. Merrifield and co-workers[27,70] have successfully utilized a mixture of 12 N HCl, acetic acid, and phenol. Scotchler et al.[97] originally proposed the use of a 6 N HCl-propionic acid solution for the analysis of the extent of loading of the first amino acid on the resin, but the mixture has also been used with peptidyl-resins.[98]

In addition to indicating mole ratios of the amino acids present in a peptide, the analysis of a peptidyl-resin sample of known weight provides the amount of peptide per unit weight of peptidyl-resin. As the weight ratio of peptide to resin constantly increases during the synthesis a useful comparison of peptide yields can be made only if the data are expressed as amount of peptide per gram of *unsubstituted* resin.[79] This may be done according to the formula $S = a/(1 - ae)$, where S = substitution in microequivalents per gram, a = concentration from the analysis of the substituent in microequivalents per gram, and e = equivalent weight of the substituent bound to the benzyl resin in grams per microequivalent.[23]

Reaction with Ninhydrin or Fluorescamine. A semiquantitative and rapid estimate of the extent of a coupling reaction can be gained by subjected a sample of the peptidyl-resin to a photometric reaction involving residual free amino groups. Analyses based upon reaction with ninhydrin[99] or fluorescamine[100] are among those that have been developed.

A procedure in which peptide chains are first cleaved from the resin by saponification with sodium methoxide and then allowed to react with fluoroescamine has the potential of providing a more quantitative estimate,[101] but does require that the peptide obtained by the saponification procedure be soluble in aqueous ethanol solutions.

Reaction with Picric Acid or Imidazolium Picrate. Gisin has developed

---

deprotected as otherwise a high value may be obtained owing to ion-exchange retention of noncovalently bonded molecules of protected amino acid from the coupling step [K. Esko and S. Karlsson, *Acta Chem. Scand.* **24**, 1415 (1970)]. Of course, deprotection of the main batch precludes the use of a second coupling cycle.

[97] J. Scotchler, R. Lozier, and A. B. Robinson, *J. Org. Chem.* **35**, 3151 (1970).

[98] In the author's laboratory the recovery of serine is low with this procedure, and satisfactory recovery of tryosine is obtained only if a crystal of phenol is added to the solution.

[99] E. Kaiser, R. L. Colescott, C. D. Bossinger, and P. I. Cook, *Anal. Biochem.* **34**, 595 (1970).

[100] A. M. Felix and M. H. Jimenez, *Anal. Biochem.* **52**, 377 (1973).

[101] A. M. Tometsko and E. Vogelstein, *Anal. Biochem.* **64**, 438 (1975).

a monitoring procedure based on the temporary retention of picrate anions by residual free amino groups which occurs when peptidyl-resin is treated with a CH_2Cl_2 solution of picric acid (the amino groups having been first deprotonated by prior treatment of the peptidyl-resin with a tertiary amine such as diisopropylethylamine).[59] Subsequent quantitative displacement of the picrate with a CH_2Cl_2 solution of a tertiary amine, such as diisopropylethylamine, and measurement of the absorbance of the resulting diisopropylethylamine picrate solution provides a quantitative estimate of unreacted amino groups. The procedure can be used to monitor both coupling and deprotection steps. It is theoretically completely nondestructive of the peptidyl-resin preparation and therefore can be used with the main batch; this fact greatly increases the potential sensitivity of the method, a critical consideration if one wishes to monitor precisely coupling reactions that have yields of 99% or better.[60] For the measurement of the extent of a deprotection step, one may wish to use a small sample to avoid the possibility of insufficient amounts of reagents for the relatively enormous number of amino groups in the main batch as well as to avoid the necessity for extreme dilutions of the eluted diisopropylethylamine solution.

The success of the method depends upon the absence of other groups that retain picrate and thereby provide spuriously high values for free amino groups. The major problem in this regard occurs if the peptide contains one or more histidine residues. Even if the pK value of the imidazole nitrogen system has been depressed by substitution with a dinitrophenyl or tosyl moiety, the residue is titrated with picric acid.[60] Hodges and Merrifield[60] therefore replaced the picric acid with imidazolium picrate and picrate retention by dinitrophenyl or tosylhistidyl residues was greatly reduced, although not completely eliminated.

Quaternary ammonium groups, introduced to a significant degree by older methods of loading the COOH-terminal amino acid on the chloromethylated resin[102] and formed continuously, but at a much lower rate, during the course of a synthesis using chloromethylated resin by the reaction of unsubstituted chloromethylated groups with the tertiary amine used in the neutralization step[103] will interfere and may be the basis for the rising background observed with lengthy syntheses.[60]

As the picrate method is nondestructive, it has been possible to use it in an automatic feedback control system.[60]

Termination of Peptide Chains with Low Reactivity in a Coupling Step. If application of a monitoring procedure reveals that a coupling re-

---

[102] Discussed by Gisin.[4]

[103] P. Fankhauser and M. Brenner, *Helv. Chim. Acta* **56**, 2516 (1973).

action is incomplete, repeated exposure to fresh solutions of the protected, incoming amino acid and coupling reagent may result in adequate reaction eventually, or a combination of this repetition with intervening solvent washes which shrink and swell the resin[5] may suffice. Instances do occur, however, when amino groups remain which are reactive to the monitoring reagent, but are unreactive in the coupling step.[93] To minimize the synthesis of deletion peptides which would occur if these groups should become reactive again at a later stage of the synthesis, various reagents have been used to effect a more certain termination of the chains.

Of the reagents which have been used, N-acetylimidazole[104] or fluorescamine[105] has resulted in the most complete termination of such chains. Unfortunately, the routine use of fluorescamine in a moderately large-scale synthesis would be an expensive practice.

One can incorporate a termination reaction routinely into a synthetic cycle, and the practice has some merit in view of the possibility that unreacted amino groups may remain at some stage in a synthesis and be unreactive toward the monitoring reagent as well. This practice would not be a panacea, however, as such groups may also be unreactive toward the terminating reagent.[5,93] Moreover, the use of an additional reagent always introduces the possibility of additional side reactions.

Synthesis of Leucylalanylglycylvaline. An excellent way to check procedures, reagents, or solvents is to perform a practice synthesis of the tetrapeptide, leucylalanylglycylvaline.[106] Dorman *et al.* have devised a system whereby each of the eight possible nonracemized products of this synthesis, i.e., valine, the tetrapeptide and six shorter peptide fragments, may be determined quantitatively using an automatic amino acid analyzer. This practice is quite efficient if large batches of reagent or solvent are tested.

This discussion of monitoring procedures has been highly selective; the reader desiring a more comprehensive review may refer to Erickson and Merrifield.[3]

Cleavage from the Solid Support and Removal of Side Chain Blocking Groups

Many of the reagents used to effect cleavage of the completed peptide from the solid support also cause removal (deblocking) of many side

[104] L. D. Markley and L. C. Dorman, *Tetrahedron Lett.* p. 1787 (1970).
[105] A. M. Felix, M. H. Jimenez, R. Vergona, and M. R. Cohen, *Int. J. Pept. Protein Res.* **7**, 11 (1975).
[106] L. C. Dorman, L. D. Markley, and D. A. Mapes, *Anal. Biochem.* **39**, 492 (1971).

chain blocking groups. This overlap of reactivities has the advantage of efficiency. If the combination of solid support and blocking groups required for a particular peptide does not permit simultaneous removal, then one or more additional steps must be added to the synthesis. This requirement is not always completely disadvantageous. The added intermediate may provide the opportunity for application of a type of purification partly dependent on the presence of the blocking group; in addition, it may provide another structure of pertinence to the question under investigation.

Anhydrous Hydrogen Fluoride. The introduction of the use of HF for the deblocking of peptides[107,108] has had profound effects upon the subsequent evolution of solid-phase peptide synthesis as it has encouraged the utilization of blocking groups and peptidyl-solid phase linkages having ever greater stability to acidolysis. Aside from the difficulties and danger involved in handling this compound (see Equipment), HF seemed for a time to be the ideal cleavage-deblock reagent, one which permitted the use of blocking groups and a peptidyl-solid phase linkage impervious to the conditions of the deprotection and coupling steps required for the synthesis and then cleanly removed upon completion of the peptide chain. The pursuit of reagents which would provide cleavage of specific peptide bonds in proteins had already revealed that serine- and threonine-containing peptides undergo an N- to O-acyl shift in HF,[109] but it appeared that this reaction might not be significant under conditions used for the cleavage and deblocking of peptides,[110] or, if it were, that it could be reversed by treatment with dilute base.[109]

Not surprisingly, increasing use of HF revealed the presence of additional side reactions. A side reaction or reactions involving glutamyl residues and dependent in part upon the presence in the cleavage mixture of the carbonium ion scavenger anisole was reported.[111] Subsequent work has elucidated the nature of these reactions.[112,113] It does appear that these reactions can be minimized or eliminated by carrying out the cleavage-deblock at 0° for short periods of time, e.g., 30 min.[112] Such strictures do now limit, however, the stability of side chain blocking

[107] S. Sakakibara and Y. Shimonishi, *Bull. Chem. Soc. Jpn.* **38**, 1412 (1965).
[108] S. Sakakibara, Y. Shimonishi, Y. Kishida, M. Okada, and H. Sugihara, *Bull. Chem. Soc. Jpn.* **40**, 2164 (1967).
[109] S. Sakakibara, K. H. Shin, and G. P. Hess, *J. Am. Chem. Soc.* **84**, 4921 (1962).
[110] J. Lenard and A. B. Robinson, *J. Am. Chem. Soc.* **89**, 181 (1967).
[111] S. Sano and S. Kawanishi, *Biochem. Biophys. Res. Commun.* **51**, 46 (1973).
[112] R. S. Feinberg and R. B. Merrifield, *J. Am. Chem. Soc.* **97**, 3485 (1975).
[113] S. Sano and S. Kawanishi, *J. Am. Chem. Soc.* **97**, 3480 (1975).

groups and peptidyl-solid phase linkages which can be used with peptides containing glutamyl residues.

The cyclization and rearrangement of certain aspartyl-containing sequences, already known to occur during cleavage and deblock with HBr-TFA, has now been detected with HF as well.[113,115] See Difficult Sequences for further discussion of this particular problem.

The generation of carbonium ions which occurs upon deblocking of certain groups, e.g., benzyl or benzyloxycarbonyl, raises the possibility of unwanted alkylation of various structures in the peptide. Such alkylation can be minimized by the inclusion of a carbonium ion scavenger, such as anisole,[116] but in the case of O-benzyltyrosyl residues alkylation at the 3-position of the phenyl ring by benzyl cations is so strongly favored that scavengers are only partially effective[117]; the use of O-blocking groups that are sterically hindered has alleviated this problem, however (see Appendix B).

The stability of unblocked tryptophan residues toward HF (in the presence of anisole scavenger) appears to be acceptable[108,118] in spite of the apparent lability of this structure in anhydrous TFA or HBr-TFA.[118]

HBr-TFA and HBr-HOAc. Cleavage of peptide chains from resins (I), (II), and (V)–(VIII) may be carried out using solutions of HBr in TFA or HOAc; many side chain blocking groups are also removed by these reagents (see Appendix B). The HBr may be bubbled through a suspension of the peptidyl-resin in neat TFA or in 50% TFA in CH_2Cl_2,[27] or the peptidyl-resin may be treated with 30% HBr in acetic acid[23] or a solution of 1 volume of 32% HBr in acetic acid and 1 volume of TFA.[119]

If the peptide contains serine or threonine residues, TFA should be used rather than HOAc, as the latter acid may acetylate the hydroxyl groups of these residues.[120,121]

If the peptide contains tyrosine or tryptophan, the HBr should be scrubbed free of bromine by passage through a solution of resorcinol or anisole in TFA.[122] Even so, unblocked tryptophan residues have been

[114] S. S. Wang, C. C. Yang, I. D. Kulesha, M. Sonenberg, and R. B. Merrifield, *Int. J. Pept. Protein Res.* 6, 103 (1974).

[115] C. C. Yang and R. B. Merrifield, *J. Org. Chem.* 41, 1032 (1976).

[116] F. Weygand and W. Steglich, *Z. Naturforsch. Teil B* 14, 472 (1959).

[117] B. W. Erickson and R. B. Merrifield, *J. Am. Chem. Soc.* 95, 3750 (1973).

[118] G. R. Marshall, *in* "Pharmacology of Hormonal Polypeptides and Proteins" (N. Back, L. Martini, and R. Paoletti, eds.), p. 48. Plenum, New York, 1968.

[119] B. F. Gisin, cited in A. R. Mitchell, B. W. Erickson, M. N. Ryabtsev, R. S. Hodges, and R. B. Merrifield, *J. Am. Chem. Soc.* 98, 7357 (1976).

[120] S. Guttmann and R. A. Boissonnas, *Helv. Chim. Acta* 42, 1257 (1959).

[121] E. D. Nicolaides and H. A. DeWald, *J. Org. Chem.* 28, 1926 (1963).

[122] A. Marglin and R. B. Merrifield, *J. Am. Chem. Soc.* 88, 5051 (1966).

reported to undergo degradation in HBr-TFA (15% loss in 30 min with Boc-tryptophan) in spite of extensive scrubbing of the HBr by passage through a solution of indole in TFA.[118]

As with HF, the cyclization and rearrangement of certain sequences containing an aspartyl residue occurs in HBr-TFA or HBr-HOAc and the synthetic strategies discussed under Difficult Sequences should be applied if one of these sequences is present.

A carbonium ion scavenger such as anisole or methionine should also be present when HBr-TFA or HBr-HOAc is used.

50% TFA in CH₂Cl₂. Cleavage of peptide chains from resins (V)– (VIII) occurs readily with a 50% solution of TFA in CH_2Cl_2.[34,46,50] Most of the standard side chain blocking groups are stable to the length of exposure required for cleavage these resins, but a side chain Boc group will be removed simultaneously, as will the Boc or Bpoc group which has been used for the α-NH₂ protection of the NH₂-terminal residue. If removal of any such groups is anticipated, a carbonium ion scavenger should be included in the cleavage solution.

Ammonia. As noted under Choice of Solid Support, treatment with ammonia of peptides which have been synthesized using resins (I) and (II)—and presumably (III) and (VIII) as well—can result in cleavage with the formation of the amide of the COOH-terminal residue, although a case has been reported in which ammonolysis in methanol resulted in intermediate formation of the COOH-terminal methyl ester.[33] The reader is referred to the syntheses cited under Choice of Solid Support for a description of reaction conditions.

Hydrazine. Hydrazine may be used in the same fashion as ammonia to provide in this case the COOH-terminal hydrazide.[54]

Alcohols. Considerable effort has been expended to obtain cleavage of peptides from the resin by transesterification, with varying degrees of success. The reader is referred to Erickson and Merrifield[3] for a discussion of these experiments.

Side Chain Blocking Groups Removed by Special Conditions. As noted, not all the commonly used side chain blocking groups are removed by acidolysis (and most are not removed by aminolysis or alcoholysis). Descriptions of the conditions under which these groups are removed are given in Appendix B.

Equipment

Reaction Vessels. Syntheses in which all operations are to be carried out manually may be performed with very simple apparatus. Fundamentally, the requirement is for a vessel fitted with a fritted disk to permit removal of solvents from the solid phase by filtration and a shaker to agitate the vessel and keep the solid phase circulating in the various solutions used in the synthesis. A suitable all-glass vessel and shaker, used in the initial solid-phase synthesis, has been illustrated in various publications.[15,16,123] Screw-capped Pyrex culture tubes have been modified to obtain reaction vessels with 5–30-ml capacities.[124] An instrument designed for manual operation in which solutions are drawn into the reaction vessel by suction and can thereby be protected from atmospheric moisture by drying tubes has been described by Samuels and Holybee.[125]

A number of systems that incorporate varying degrees of automation are in use in various laboratories. Fourteen such types of apparatus have been tabulated recently,[3] and several of these models are commercially available·(Beckman Instruments, Spinco Division, Palo Alto, CA; Vega Engineering, Tucson, AZ).

Cleavage Vessels. Cleavage of the peptide from the solid phase may be carried out directly in the reaction vessel if the reagent is a solution.[126] Suitable vessels for cleavage by gaseous HBr-TFA have been illustrated.[16]

A more elaborate apparatus is required for cleavage reactions utilizing HF owing to the properties of this reagent, primarily its toxicity and its capacity to dissolve borosilicate glass. Robinson has described a suitable apparatus made from Teflon and Kel-F,[127] which is commercially available (Peninsula Laboratories, Inc., PO Box 205, San Carlos, CA). Although the polyfluoroethylenes are the only plastics known to be completely stable to HF, polyethylene and polypropylene are not much affected at room temperature.[128] A relatively inexpensive apparatus incorporating components of polyethylene and polypropylene has been described[129]; experience in the author's laboratory with a model closely based on this design indicates that it is quite adequate for the volumes

---

[123] R. B. Merrifield and M. A. Corigliano, *Biochem. Prep.* **12**, 98 (1968).

[124] B. F. Gisin and R. B. Merrifield, *J. Am. Chem. Soc.* **94**, 3102 (1972).

[125] R. B. Samuels and L. D. Holybee, *in* "Solid Phase Peptide Synthesis" (J. M. Stewart and J. D. Young), p. 68. Freeman, San Francisco, 1969.

[126] Note, however, the possibility of pressure buildup with such solutions as 32% HBr in HOAc:TFA, 1:1 (v/v) [B. F. Gisin, cited in A. R. Mitchell, B. W. Erickson, M. N. Ryabtsev, R. S. Hodges, and R. B. Merrifield, *J. Am. Chem. Soc.* **98**, 7357 (1976)] or methanol saturated with NH₃.

[127] A. B. Robinson, *in* "Solid Phase Peptide Synthesis" (J. M. Stewart and J. D. Young), p. 41. Freeman, San Francisco, 1969.

[128] S. Sakakibara, *in* "Chemistry and Biochemistry of Amino Acids, Peptides and Proteins" (B. Weinstein, ed.), Vol. 1 p. 51. Dekker, New York, 1971.

[129] L. M. Pourchot and J. H. Johnson, *Org. Prep. Proced.* **1** (2), 121 (1969).

of HF (10–20 ml) commonly required. A very useful discussion of both the chemistry and handling of HF is available.[128]

A Side Reaction Independent of Sequence

Using ^{14}C-labeled HOAc Brunfeldt and co-workers demonstrated that carryover of acid from the deprotection step (done in their studies with 1 N HCl in HOAc) to the next coupling step results in termination of peptide chains by acetylation.[130,131] Spontaneous termination to the extent of 1% of the chains per cycle has also been observed when TFA is used for deprotection.[132] Termination by trifluoroacetylation has been observed with automated equipment.[133] Gisin has suggested that quaternary ammonium groups, if present on a resin, could retain trifluoroacetate anions from a deprotection step and then release them during the next coupling step with consequent termination.[4]

Side Reactions Partly Dependent on Sequence

Variation of reaction rates in both coupling and deprotection steps as a function of both chain length and type of amino acid reacting has been documented.[5,93] Monitoring and chain termination techniques can detect and reduce the deletion and termination peptides resulting from such variation in rates, but as discussed under Monitoring Procedures, elimination of such products is not thereby guaranteed.

Alkylation of proline residues by unsubstituted chloromethyl groups on the resin has been reported.[134] Although the study was confined to a single sequence, it seems likely that the extent of such a reaction would be dependent in part upon sequence.

Difficult Sequences

The ready cyclization of the aspartyl residue in an aspartylglycyl sequence with consequent formation of both the imide and β-aspartyl

[130] K. Brunfeldt, T. Christensen, and P. Villemoes, *FEBS Lett.* **22**, 238 (1972).

[131] K. Brunfeldt, T. Christensen, and P. Roepstorff, *FEBS Lett.* **25**, 184 (1972).

[132] M. E. Bush, S. S. Alkan, D. E. Nitecki, and J. W. Goodman, *J. Exp. Med.* **136**, 1478 (1972).

[133] D. Yamashiro and C. H. Li, *Proc. Natl. Acad. Sci. U.S.A.* **71**, 4945 (1974).

[134] O. Schou, D. Bucher, and E. Nebelin, *Hoppe-Seyler's Z. Physiol. Chem.* **357**, 103 (1976).

structures has been reported a number of times (see discussion and references in Yang and Merrifield[115]). It was shown to occur upon cleavage with HF of the nonapeptide comprising residues 123–131 of bovine pituitary growth hormone (Asp[127]–Gly[128])[114]; moreover, the mixture of α- and β-aspartyl peptides resulting from hydrolysis of the cyclic imide structure could not be separated by the usual chromatographic methods.

Fairly satisfactory solutions to this problem have emerged from the realization that cyclization occurs perhaps ten times more slowly if the β-COOH is unsubstituted rather than blocked at the time the peptide is exposed to the cyclizing conditions.[135] Wang et al.[114] were able to realize this situation by incorporating the aspartyl residue as the β-tert-butyl ester; deblocking by relatively mild acid provided the free β-COOH grouping prior to cleavage with HF. Protection of α-NH$_2$ groups was achieved with the Bpoc grouping. In a more recent study, blocking of the β-COOH with a phenacyl moiety, which is removed by 1 M sodium thiophenoxide in DMF, but unaffected by 50% TFA-CH$_2$Cl$_2$, permitted use of the more common Boc group for α-NH$_2$ protection with retention of the capacity to deblock β-COOH groups prior to HF treatment.[115] While cyclization of the aspartylglycyl structure has been most commonly observed, other aspartyl-containing sequences such as aspartylseryl and aspartylalanyl also cyclize to a significant extent in HF.[115]

The asparaginylglycyl sequence can form the β-aspartylglycyl structure[135] although this conversion has not yet been reported to occur during a solid-phase synthesis.

Significant cyclization of the prolylprolyl moiety in prolyprolyl-resin preparations has been reported from three laboratories.[35,124,130] Catalysis of the reaction by the carboxyl group of the third amino acid in the peptide occurs during the coupling of this residue to the dipeptide; this was demonstrated to be the case with D-valylprolyl-resin, a structure that cyclizes at approximately the same rate as prolylprolyl-resin.[124] By reversing the order of addition of amino acid and DCC in the coupling step the yield of cyclized product was reduced from 70% to 7–9%. Cyclization of several other dipeptidyl-resins composed of one or two bulky residues was observed, but the rate of formation was at least an order of magnitude lower than in the cases of prolylprolyl-resin and D-valylprolyl-resin.[124]

One may wish to quibble with the implication that NH$_2$-terminal

[135] E. E. Haley, B. J. Corcoran, F. E. Dorer, and D. L. Buchanan, *Biochemistry* **5**, 3229 (1966).

glutamine is a sequence, but one cannot quibble with the fact that it undergoes fairly ready cyclization to the pyroglutamyl moiety under mildly acid or basic conditions.[136] The ease with which this reaction can occur may be appreciated from the case of glutaminylhistidylprolyl amide where 60% conversion to the pyroglutamyl structure occurred after three days at 25° in 0.1% HOAc-*n*-butanol-pyridine (11:5:3).[137] Admittedly this may be an extraordinary example, as the pyroglutamyl structure is thyrotropin-releasing hormone, but many examples of cyclization have been documented[136] and it is best to avoid extensive manipulations, e.g., chromatographic purifications, with peptides having an NH$_2$-terminal glutamine.

Purification and Characterization of the Product

The investigator will want to subject the crude product obtained from the cleavage-deblocking reaction(s) to a number of the techniques that have been developed for the purification of peptides and proteins, regardless of the prognosis which has emerged from the application during the synthesis of various monitoring procedures. These techniques include crystallization, gel filtration, ion-exchange chromatography, partition chromatography, countercurrent distribution, and affinity chromatography. Articles on almost all these techniques have appeared in various volumes of this series (in particular, in Vol. 11, Vol. 25, and the present volume). The investigator may also refer to the purification schemes described for the syntheses cited in the Introduction.

While the purification will itself provide partial characterization of the product, the application of additional analytical techniques will almost certainly be warranted. These techniques include amino acid analysis, thin-layer chromatography in several solvents for smaller peptides, polyacrylamide gel electrophoresis, analysis for optical integrity, end-group analysis, chromatographic behavior after enzymic or chemical degradation, and quantitation of biological activity, if appropriate. Again, articles on essentially all these techniques have appeared in various volumes of this series. A recent report describes a modification of the standard polyacrylamide gel electrophoresis technique which permits analysis of peptides containing as few as three residues.[138]

---

[136] B. Blombäck, this series Vol. 11, p. 398 (1967).
[137] K. Folkers, J. K. Chang, B. L. Currie, C. Y. Bowers, A. Weil, and A. V. Schally, *Biochem. Biophys. Res. Commun.* **39**, 110 (1970).
[138] M. S. Rosemblatt, M. N. Margolies, L. E. Cannon, and E. Haber, *Anal. Biochem.* **65**, 321 (1975).

Appendix A

SEQUENCE OF OPERATIONS FOR A SYNTHETIC CYCLE IN WHICH α-NH$_2$
PROTECTION IS THE BOC GROUP AND BOND FORMATION IS
EFFECTED WITH DCC^a

Operation	Solvent	Number of times performed	Dwell times (min)
Deprotection	50% TFA-CH$_2$Cl$_2$	1	5
Deprotection	50% TFA-CH$_2$Cl$_2$	1	25
Wash	CH$_2$Cl$_2$	3	1
Neutralization	5% DIEA-CH$_2$Cl$_2$^{b}	2	1
Neutralization	5% DIEA-CH$_2$Cl$_2$	1	5
Wash	CH$_2$Cl$_2$	5	1
Wash	CH$_3$CHOHCH$_3$^{c}	3	1
Wash	CH$_2$Cl$_2$	5	1
Coupling	Boc-amino acid	1	5
Coupling	DCC	1	120
Wash	CH$_2$Cl$_2$	5	1
Wash	DMF	3	1
Wash	CH$_2$Cl$_2$	3	1

^{a} This sequence in many respects reflects one used by Feinberg and Merrifield [F. S. Feinberg and R. B. Merrifield, *J. Am. Chem. Soc.* **97**, 3485 (1975)], but it also differs in some steps.
^{b} Triethylamine may be used, particularly if chloromethyl groups are absent.
^{c} The isopropanol washes shrink the resin and possibly thereby increase the number of reaction sites [W. S. Hancock, D. J. Prescott, P. R. Vagelos, and G. R. Marshall, *J. Org. Chem.* **38**, 774 (1973)].

Appendix B

Derivatives of Individual Amino Acids Commonly Used in Solid-Phase Peptide Synthesis

This discussion is limited to those derivatives that are commonly used in solid-phase peptide synthesis. The vast majority of syntheses reported so far has utilized, interestingly enough, only very few of the α-NH$_2$ protecting and side chain blocking groups available to the peptide chemist, so this limitation is not as severe as it might at first appear to be. The reader is referred to Erickson and Merrifield[3] for a more encompassing discussion of derivatives used in solid-phase synthesis. Several recent compilations of derivatives used in both solid-phase and solution syntheses are available.[139-142] The investigator should refer to these

[139] G. R. Pettit, "Synthetic Peptides," Vols. 1 and 2, Van Nostrand-Reinhold, Princeton, New Jersey. 1970, 1971.

compilations for references to the synthesis of derivatives that are commercially available, as these references are not given below.

Proline, Glycine, Alanine, Valine, Isoleucine, Leucine, and Phenylalanine. The side chains of these seven amino acids are not reactive under any of the conditions obtaining in peptide synthesis and therefore never require side chain blocking.

Lysine. The benzyloxycarbonyl group (Z) [see (XIV)], incorrectly but commonly referred to as the carbobenzoxy group, was used almost exclusively for blocking the ϵ-NH$_2$ group of lysine until a few years ago.

$$\langle O \rangle - CH_2OCONH(CH_2)_4 - CH(NH_3^+) - COO^-$$

N^ϵ - Benzyloxycarbonyllysine

(XIV)

It is readily removed by the usual HF, HBr-TFA, or HBr-HOAc cleavage conditions and, in fact, is removed at a noticeable rate by acidolytic deprotection conditions such as 1 N HCl in HOAc[143] or 50% TFA–CH$_2$Cl$_2$.[144] This lability to the conditions required for removal of the Boc α-NH$_2$ protecting group, i.e., premature deblocking, results in the synthesis of branched side products. To avoid this type of reaction, ϵ-NH$_2$ blocking groups with greater resistance to acidolysis have been introduced; they include the 2-bromobenzyloxycarbonyl[145] and 2-chlorobenzyloxycarbonyl[144] groups. Although a number of additional structures have been used or proposed (see discussion in Erickson and Merrifield[3]), it appears that the 2-halogenated benzyloxycarbonyl structures may exhibit the best compromise between stability to Boc deprotection conditions and adequate reactivity to the milder HF deblocking conditions now indicated for peptides containing glutamic acid.[112]

For the synthesis of a relatively short peptide or one that contains the lysyl residues near the NH$_2$-terminal end, the use of the classical benzyloxycarbonyl groups may not result in a serious accumulation of side products; the methodical determination of pertinent rate constants[144] permits the investigator to calculate the likely magnitude of the problem for a particular structure. The advantage of continuing to use derivatives containing this blocking groups is their wide commercial availability at

[140] G. R. Pettit, "Synthetic Peptides," Vol. 3. Academic Press, New York, 1975.

[141] G. A. Fletcher and J. H. Jones, *Int. J. Pept. Protein Res.* **4**, 347 (1972).

[142] G. A. Fletcher and J. H. Jones, *Int. J. Pept. Protein Res.* **7**, (1975).

[143] A. Yaron and S. F. Schlossman, *Biochemistry* **7**, 2673 (1968).

[144] B. W. Erickson and R. B. Merrifield, *J. Am. Chem. Soc.* **95**, 3757 (1973).

[145] D. Yamashiro, R. L. Noble, and C. H. Li, *in* "Chemistry and Biology of Peptides. Proceedings of the Third American Peptide Symposium" (J. Meienhofer, ed.), p. 197. Ann Arbor Sci. Publ., Ann Arbor Michigan, 1972.

reasonable cost. Certain of the newer derivatives are now also commercially available (Bachem, Inc., Marina Del Rey, CA; Beckman Instruments, Inc., Bioproducts Dept., Palo Alto, CA).

Histidine. The most satisfactory blocking groups for the imidazole ring of histidine are the dinitrophenyl (DNP) (XV) and the *p*-toluenesulfonyl (tosyl) (XVI).

CH$_2$—CH(NH$_3^+$)COO$^-$

N$^\tau$ N$^\pi$

—NO$_2$

O$_2$N

N^{im}-DNP-histidine

(XV)

CH$_2$—CH(NH$_3^+$)COO$^-$

N$^\tau$ N$^\pi$

H$_3$C—⟨O⟩—SO$_2$

N^{im}-Tosylhistidine

(XVI)

The DNP moiety is not removed by acid and, in fact, a 95% recovery of N^{im}-DNP-histidine from 24 hr in 6 N HCl at 110° is observed.[146] A separate deblocking step must therefore be included when this group is used. Satisfactory deblocking is obtained by thiolysis[147] with thiophenol[148,149] or by ammonolysis.[148] A direct demonstration that the DNP moiety is bonded to the N^τ rather than the N^π has recently appeared.[150]

The tosyl grouping may be removed by HF, HBr-TFA, HBr-HOAC or by ammonolysis. The acid lability of this group seems to be comparable to that of the N^ε-benzyloxycarbonyl structure, as there are indications of significant deblocking in 50% TFA–CH$_2$Cl$_2$.[60] Both N^α-Boc-N^{im}-DNP-histidine and N^α-Boc-N^{im}-tosylhistidine are commercially available.

Arginine. Satisfactory blocking of the δ-guanidino group of arginine is obtained with the tosyl moiety (XVII). This blocking group may be removed by HF,[151] contrary to an earlier report[108]. The N^α-Boc-N^G-tosyl

H$_3$C—⟨O⟩—SO$_2$—NH—C(=NH)—NH(CH$_2$)$_3$CH(NH$_3^+$)COO$^-$

N^G-Tosylarginine

(XVII)

[146] P. Henkart, *J. Biol. Chem.* **246**, 2711 (1971).
[147] S. Shaltiel and M. Fridkin, *Biochemistry* **9**, 5122 (1970).
[148] J. M. Stewart, M. Knight, A. C. M. Paiva, and T. Paiva, *in* "Progress in Peptide Research" (S. Lande, ed.), Vol. 2, p. 59. Gordon & Breach, New York, 1972.
[149] M. C. Lin, B. Gutte, D. G. Caldi, S. Moore, and R. B. Merrifield, *J. Biol. Chem.* **247**, 4768 (1972).
[150] J. R. Bell and J. H. Jones, *J. Chem. Soc., Perkin Trans. 1* **1974**, 2336 (1974).
[151] R. H. Mazur and G. Plume, *Experientia* **24**, 661 (1968).

derivative has limited solubility in CH_2Cl_2. One may use a mixed solvent such as $DMF–CH_2Cl_2$ (1:3)[60] or the more soluble N^α-t-amyloxycarbonyl derivative, which is available from Beckman Instruments, Inc. (Bioproducts Dept., Palo Alto, CA).

Aspartic Acid. The standard blocking group for the β-COOH group of aspartic acid is the benzyl moiety (see XVIII). It is removed by

$$\langle\bigcirc\rangle\!\!-\!CH_2-OOC-CH_2-CH(NH_3^+)-COO^-$$

β-Benzylaspartic acid

(XVIII)

HF^{108}, HBr–TFA,[152] or HBr–HOAc.[153] Deblocking in 50% $TFA–CH_2Cl_2$ has been detected, but the rate is sufficiently low to make the reaction a tolerable one for most syntheses.[117] The N^α-Boc-β-benzyl derivative is commercially available.

Formation of the β-amide, viz., asparagine, will occur upon ammonolysis of a peptidyl-resin containing β-benzylaspartyl residues (see Choice of Solid Support). Use of the *tert*-butyl group to block the β-COOH[154] allows formation of the COOH-terminal amide without concomitant asparagine formation as this ester does not undergo ammonolysis; its great acid lability requires the use of very labile α-NH$_2$ protection such as the Bpoc or Ppoc groups, however. The β-*tert*-butyl ester is commercially available (Bachem, Inc., Marina Del Rey, CA).

The use of the *tert*-butyl and β-phenacyl groups to prevent rearrangement of particular aspartyl-containing sequences is discussed under Difficult Sequences.

Asparagine. If asparagine is to be incorporated by the direct action of DCC (see discussion under Bond Formation), protection of the β-carboxamido structure from dehydration to the nitrile is necessary.[85] Suitable blocking groups are the bis-(4-methoxyphenyl)methyl (Mbh)[155] (XIX) and the 2,4-dimethoxybenzyl (Dmb)[156] (XX). Both these groups

$$\left(CH_3O\!-\!\langle\bigcirc\rangle\right)_{\!2}\!\!CH-NH-CO-CH_2-CH(NH_3^+)-COO^-$$

N^γ-Bis(4-methoxyphenyl)methylasparagine

(XIX)

[152] G. R. Marshall and R. B. Merrifield, *Biochemistry* **4**, 2394 (1965).
[153] D. Ben-Ishai and A. Berger, *J. Org. Chem.* **17**, 1564 (1952).
[154] R. Schwyzer and H. Dietrich, *Helv. Chim. Acta* **44**, 2003 (1961).
[155] W. König and R. Geiger, *Chem. Ber.* **103**, 2041 (1970).
[156] P. G. Pietta, P. F. Cavallo, and G. R. Marshall, *J. Org. Chem.* **36**, 3966 (1971).

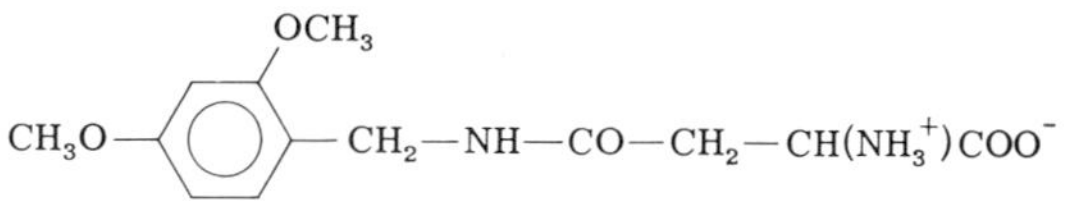

OCH$_3$

CH$_3$O—⟨◯⟩—CH$_2$—NH—CO—CH$_2$—CH(NH$_3^+$)COO$^-$

N^γ-2, 4-Dimethoxybenzylasparagine

(XX)

are removed at a significant rate by 50% TFA–CH$_2$Cl$_2$,[155,157] but this lability does not detract seriously from their usefulness as dehydration of the unprotected β-carboxamido group does not occur to a measurable extent after incorporation of the asparaginyl residue into the peptide chain.[158] Complete removal of these groups may be effected by HF under the usual cleavage conditions.

N^α-Boc-N^γ-Mbh-asparagine has a limited solubility in CH$_2$Cl$_2$; it has been used in DMF-CH$_2$Cl$_2$, 1:3.[60]

Threonine and Serine. Protection of the hydroxyl groups of these amino acids is routinely obtained by formation of the O-benzyl ether [(XXI), (XXII)].

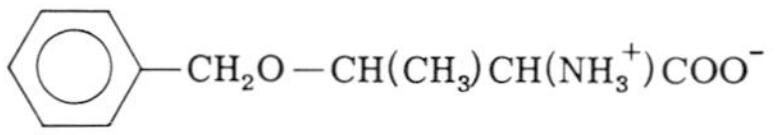

⟨◯⟩—CH$_2$O—CH(CH$_3$)CH(NH$_3^+$)COO$^-$

O-Benzylthreonine

(XXI)

⟨◯⟩—CH$_2$O—CH$_2$—CH(NH$_3^+$)COO$^-$

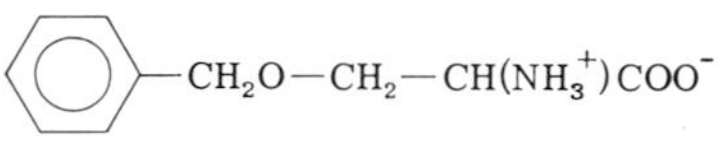

O-Benzylserine

(XXII)

The benzyl moiety is smoothly removed by HF, HBr-TFA, or HBr-HOAc. Detectable but very slight deblocking occurs in 50% TFA–CH$_2$Cl$_2$.[117] An improved synthesis of O-benzylthreonine has recently been published[114]; both N^α-Boc-O-benzyl derivatives are commercially available, however.

Glutamic Acid. The γ-COOH of this amino acid is generally blocked by formation of the benzyl ester (XXIII).

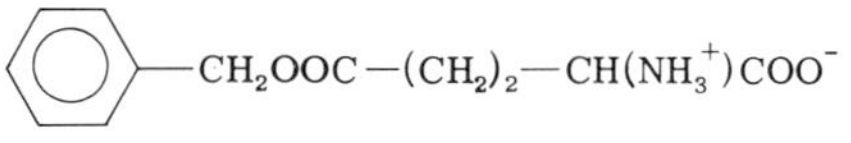

⟨◯⟩—CH$_2$OOC—(CH$_2$)$_2$—CH(NH$_3^+$)COO$^-$

γ-Benzylglutamic acid

(XXIII)

The group exhibits an acceptable stability toward 50% TFA–CH$_2$Cl$_2$,[117] but is readily removed by HF, HBr-TFA, or HBr-HOAc. The N^α-Boc-γ-benzyl derivative is commercially available.

[157] P. G. Pietta, P. A. Biondi, and O. Brenna, *J. Org. Chem.* **41**, 703 (1976).
[158] M. Bodanszky and V. du Vigneaud, *J. Am. Chem. Soc.* **81**, 5688 (1959).

As with the β-benzyl derivative of aspartic acid, removal of a peptide from a resin by ammonolysis can convert the γ-benzyl ester to the corresponding amide, viz., glutamine. Use of the *tert*-butyl ester[159] will avoid this problem, but owing to the lability of this ester in acid, α-NH$_2$ protection must then be by a group such as Bpoc or Ppoc.

Glutamine. As with asparagine, incorporation of glutamine by the mediation of DCC requires blocking of the carboxamido group to prevent dehydration to the nitrile.[85] The bis(4-methoxyphenyl)methyl (Mbh)[155] [see (XIX)] and 2,4-dimethoxybenzyl (Dmb)[157] [see (XX)] groups are suitable for glutamine as well as asparagine.

N^α-Boc-N^δ-Mbh-glutamine has a limited solubility in CH$_2$Cl$_2$; it has been used in DMF-CH$_2$Cl$_2$, 1:3.[60]

Cysteine. The benzyl group has been the standard blocking group for cysteine since its introduction in 1930[160] (XXIV). Unfortunately, the rate of deblocking by HF is rather slow.[108] On the basis of their experi-

$$\text{C}_6\text{H}_5-\text{CH}_2-\text{S}-\text{CH}_2-\text{CH}(\text{NH}_3^+)\text{COO}^-$$

S-Benzylcysteine

(XXIV)

ence with the use of this group during the synthesis of ribonuclease A, Gutte and Merrifield[27] suggested its possible replacement by the 4-methoxybenzyl moiety.[161] However, this latter group proved to be rather labile to standard deprotection conditions, i.e., 50% TFA-CH$_2$Cl$_2$[145,162] and consequently, the 3,4-dimethylbenzyl and 4-methylbenzyl[145] groups were proposed as having promising intermediate reactivities. The details of the synthesis of *S*-(4-methylbenzyl)cysteine[117] and N^α-Boc-*S*-(3,4-dimethylbenzyl)cysteine[163] have been published; the latter derivative is available from Bachem, Inc.

Methionine. Successful syntheses have been carried out with no blocking group on the sulfur of methionine. Alkylation by the carbonium ions generated during deprotection and cleavage-deblocking steps appears to be minimized or avoided by the addition of scavengers such as methyl

[159] R. Schwyzer and H. Kappeler, *Helv. Chim. Acta* **44**, 1991 (1961).
[160] V. du Vigneaud, L. F. Audrieth, and H. S. Loring, *J. Am. Chem. Soc.* **52**, 4500 (1930).
[161] S. Akabori, S. Sakakibara, Y. Shimonishi, and Y. Nobuhara, *Bull. Chem. Soc. Jpn.* **37**, 433 (1964).
[162] B. W. Erickson and R. B. Merrifield, *in* "Chemistry and Biology of Peptides. Proceedings of the Third American Peptide Symposium" (J. Meienhofer, ed.), p. 191. Ann Arbor Sci. Publ., Ann Arbor, Michigan, 1972.
[163] D. Yamashiro, R. L. Noble, and C. H. Li, *J. Org. Chem.* **38**, 3561 (1973).

ethyl sulfide,[164] although a recent study indicates that formation of the metastable *tert*-butyl sulfonium derivative may occur, even in the presence of a scavenger.[25]

The sulfoxide derivatives have been used to avoid alkylation reactions; at the end of the synthesis, methonine is regenerated by reduction with mercaptoethanol.[27,165] The sulfoxide derivatives are available from Research Plus Laboratories, Inc. (Denville, NJ); the synthesis of N^α-Boc-methionine-*d*-sulfoxide as well as N^α-Boc-methionine-*dl*-sulfoxide has been reported.[166]

Tyrosine. Blocking of the phenolic hydroxyl of tyrosine can be done with the benzyl moiety [see (XXV)], but the chemistry of this grouping

$$\langle\bigcirc\rangle\!-\!CH_2O\!-\!\langle\bigcirc\rangle\!-\!CH_2CH(NH_3^+)COO^-$$

O-Benzyltyrosine

(XXV)

has two serious flaws. It is labile to a significant extent in 50% TFA-CH_2Cl_2,[117] and when it is removed by acidolysis, intramolecular alkylation at the 3-position of the ring occurs to the extent of 13–20% (anhydrous HF), even in the presence of massive amounts of anisole to scavenge carbonium ions.[117]

The problem of acid lability has been solved by the introduction of benzyl derivatives carrying electron-withdrawing groups such as 3-bromobenzyl[19] and 2,6-dichlorobenzyl.[26] Both groups may be smoothly removed in HF, but 3-alkylation is still detectable, e.g., 5% with the 2,6-dichlorobenzyl group.[117] N^α-Boc-*O*-2,6-dichlorobenzyltyrosine is available from Bachem, Inc. (Marina Del Rey, CA).

Tryptophan. Tryptophan has been used without side chain protection in a number of successful syntheses,[70,167] but the potential for oxidative decomposition of this residue under acid conditions, i.e., during acidolytic deprotection and cleavage reactions, requires that special precautions be taken. As noted under Cleavage from the Solid Support and Removal of Side Chain Blocking Groups, HBr must be thoroughly scrubbed to remove bromine and even so some decomposition may occur during cleavage with this reagent.[118] Acceptable recovery from HF cleavage is

[164] R. B. Merrifield, *J. Org. Chem.* **29**, 3100 (1964).

[165] B. Iselin, *Helv. Chim. Acta* **44**, 61 (1961).

[166] K. Hofmann, W. Haas, M. J. Smithers, and G. Zanetti, *J. Am. Chem. Soc.* **87**, 631 (1965).

[167] M. W. Draper, R. B. Merrifield, and M. A. Rizack, *J. Med. Chem.* **16**, 1326 (1973).

observed.[108,118] In this connection, a 50% recovery of lysozyme, which contains six tryptophan residues, from 1-hr exposure to anhydrous HF at 0° has been reported recently.[168] Oxidative decomposition during deprotection with 50% TFA in CH_2Cl_2 has been minimized by the addition of dithiothreitol[167,169] or mercaptoethanol.[170] Even when the use of the Bpoc group for α-NH_2 protection has permitted the use of 1.5% TFA in CH_2Cl_2 for deprotection, mercaptoethanol has been added.[70]

Blocking of the indole nitrogen (N^i) with a formyl group has been utilized to stabilize the residue against acidic deprotection and cleavage conditions.[171-173] Deblocking is reported to occur upon treatment of the peptide with 1 M NH_4HCO_3, pH 9.1 (24 hr at 24°) or with liquid NH_3 (2 hr at −60°),[173] although deformylation in NH_4HCO_3 at pH 9 of the pentakontapeptide corresponding to positions 42–91 of ovine β-lipotropin has been reported to occur too slowly to be useful, necessitating the use of liquid NH_3.[133]

Tryptophan is destroyed by the conditions conventionally used to hydrolyze free peptides for amino acid analysis, viz., 6 N HCl at 110° for 22 hr, and by the acidic mixtures (see Monitoring Procedures) used to hydrolyze aminoacyl- and peptidyl-resins. Hydrolysis of free peptides with 6 N HCl containing 2–4% thioglycolic acid,[174] with 4 N methanesulfonic acid,[175] or with 4.2 N NaOH[176] will provide accurate tryptophan values.

α-NH_2 Protection. With the exception of a few amino acids, the side chain is blocked first and the α-NH_2 group is then modified with the desired protecting group. Detailed general procedures for the preparation of Boc derivatives have been published[16,177,178] although, as noted under The Synthetic Cycle, many of the commonly used intermediates carrying the Boc protecting group are commercially available. The preparation of many of the Bpoc derivatives of the amino acids and blocked inter-

[168] S. Aimoto and Y. Shimonishi, Bull. Chem. Soc. Jpn. 48, 3293 (1975).

[169] C. H. Li and D. Yamashiro, J. Am. Chem. Soc. 92, 7608 (1970).

[170] J. Blake and C. H. Li, J. Med. Chem. 17, 233 (1974).

[171] M. Ohno, S. Tsukamoto, S. Makisumi, and N. Izumiya, Bull. Chem. Soc. Jpn. 45, 2852 (1972).

[172] M. Ohno, S. Tsukamoto, S. Sato, and N. Izumiya, Bull. Chem. Soc. Jpn. 46, 3280 (1973).

[173] D. Yamashiro and C. H. Li, J. Org. Chem. 38, 2594 (1973).

[174] H. Matsubara and R. M. Sasaki, Biochem. Biophys. Res. Commun. 35, 175 (1969).

[175] R. J. Simpson, M. R. Neuberger, and T.-Y. Liu, J. Biol. Chem. 251, 1936 (1976).

[176] T. E. Hugli and S. Moore, J. Biol. Chem. 247, 2828 (1972).

[177] U. Ragnarsson, S. M. Karlsson, B. E. Sandberg, and L. E. Larsson, Org. Synth. 53, 25 (1973).

[178] E. Schnabel, Justus Liebigs Ann. Chem. 702, 188 (1967).

mediates described above has been described.[49,70-72] A smaller percentage of the corresponding Ppoc derivatives has been described.[75,77]

Acknowledgments

I am greatly indebted to R. B. Merrifield, A. R. Mitchell, and S. S. Wang, who read a draft of this manuscript and provided much helpful advice concerning its contents.

The cost of manuscript preparation was met in part with funds from NIH Grant No. RR 05384.

Author Index

Numbers in parentheses are reference numbers and indicate that an author's work is referred to although his name is not cited in the text.

Subject Index

A

H

A
B
C
D
E
F
G
H
I
J